# Engineering Tools for Environmental Risk Management –2

# Engineering Tools for Environmental Risk Management –2

## Environmental Toxicology

*Editors*

# Katalin Gruiz
*Department of Applied Biotechnology and Food Science, Budapest University of Technology and Economics, Budapest, Hungary*

# Tamás Meggyes
*Berlin, Germany*

# Éva Fenyvesi
*Cyclolab, Budapest, Hungary*

**CRC Press**
Taylor & Francis Group
Boca Raton London New York

CRC Press is an imprint of the
Taylor & Francis Group, an **informa** business

A BALKEMA BOOK

First published 2015 by CRC Press

Published 2019 by
CRC Press/Balkema
P.O. Box 447, 2300 AK Leiden, The Netherlands
e-mail: Pub.NL@taylorandfrancis.com
www.crcpress.com – www.taylorandfrancis.com

First issued in paperback 2020

© 2015 Taylor & Francis Group, London, UK
CRC Press/Balkema is an imprint of the Taylor & Francis Group, an informa business

No claim to original U.S. Government works

ISBN 13: 978-0-367-57589-2 (pbk)
ISBN 13: 978-1-138-00155-8 (hbk)

Typeset by MPS Limited, Chennai, India

*British Library Cataloging in Publication Data*

A catalogue record for this book is available from the British Library

*Library of Congress Cataloging-in-Publication Data*

Environmental toxicology (CRC Press)
    Environmental toxicology / editors Katalin Gruiz, Department of Applied Biotechnology and Food Science, Budapest University of Technology and Economics, Budapest, Hungary, Tamás Meggyes, Berlin, Germany, Éva Fenyvesi, Cyclolab, Budapest, Hungary.
        pages cm. – (Engineering tools for environmental risk management ; 2)
    Includes bibliographical references and index.
    ISBN 978-1-138-00155-8 (hardback : alk. paper) – ISBN 978-1-315-77877-8 (ebook)
1. Environmental toxicology. 2. Pollutants–Toxicity testing. I. Gruiz, Katalin. II. Meggyes, T. (Tamás) III. Fenyvesi, Éva. IV. Title.
    RA1226.E56855 2015
    615.9′02–dc23
                                                                    2014048002

# Table of contents

# 3   Human toxicology                                                             125

K. GRUIZ

# Preface

This is the second volume of the five-volume book series "Engineering Tools for Environmental Risk Management". The book series deals with the following topics:

1 Environmental deterioration and pollution, management of environmental problems
2 Environmental toxicology – a tool for managing chemical substances and contaminated environment
3 Assessment and monitoring tools, risk assessment
4 Risk reduction measures and technologies
5 Case studies for demonstration of the application of engineering tools

The authors aim to describe interactions and options in risk management by

- providing a broad scientific overview of the environment, its human uses and the associated local, regional and global environmental problems;
- interpreting the holistic approach used in solving environmental protection issues;
- striking a balance between nature's needs and engineering capabilities;
- understanding interactions between regulation, management and engineering;
- obtaining information about novel technologies and innovative scientific and engineering tools;
- providing a broader perspective for engineers and explaining engineering solutions to environmental managers, owners and other decision makers

*This second volume* provides an overview on environmental toxicology, the main concepts, methods and applications, focusing on environmental knowledge and its conscious and structured application in environmental engineering, management and decision making.

The main topics of this second volume include

- the legal and managerial context of environmental toxicology;
- the chemical and biological as well as *in silico* models for toxicology;
- fate and behavior of chemical substances in the environment and their influence on the actual effects;
- human, aquatic and terrestrial toxicology – test organisms and test methods;
- ecotoxicology and ecological assessment;

- alternative test methods to reduce the use of animals in research and testing;
- new trends in environmental analytics;
- bioaccessibility and bioavailability of chemical substances;
- interactive and dynamic testing of environmental solid phases;
- developments of soil ecotoxicology;
- microcosm models and experiments;
- evaluation and interpretation of environmental fate and effect data;
- statistics of environmental toxicology.

# Abbreviations

| | | |
|---|---|---|
| 2-D GC | – | two-dimensional gas chromatography |
| 3R | – | replacement, reduction and refinement in animal testing |
| 4CP | – | 4-chlorophenol, a chlorinated pesticide/biocide |
| AA-EQS | – | annual average environmental quality criterion |
| ADE | – | absorption, distribution and elimination |
| ADI | – | acceptable daily intake |
| ADME | – | absorption, distribution, metabolism, and elimination |
| AF | – | assessment factor |
| AMD | – | acid mine drainage |
| ANCOVA | – | analysis of covariance |
| ANOVA | – | analysis of variance |
| APHA | – | American Public Health Association |
| ARD | – | acid rock drainage |
| ATO | – | aquatic test organism |
| ATP | – | adenosine triphosphate |
| AS | – | allometric scaling factor |
| ASTM | – | American Society for Testing and Materials |
| ASTM | – | International the new name of ASTM |
| AVS | – | acid-volatile sulfide |
| AWWA | – | American Water Works Association |
| BAF | – | bioaccumulation factor |
| BAFK | – | kinetic bioaccumulation factor |
| BALB/c 3T3 | – | albino mouse fibroblast cell line for cytotoxicity testing; the standard fibroblast cell line |
| BAT | – | best available technology |
| BCC | – | bioaccumulative chemicals of concern |
| BCF | – | bioconcentration factor |
| BCOP | – | bovine corneal opacity and permeability test |
| BCR | – | Commission of the European Communities Bureau of Reference |
| BCR | – | extraction sequential extraction method for soil/sediment |
| Bhas42 | – | immortalized rodent cells line |
| BIOWIN | – | biodegradation probability program for estimating aerobic and anaerobic biodegradability of organic chemicals |
| BLOBs | – | binary large objects |
| BMD | – | benchmark dose |

$BMD_{10}$ – benchmark dose, associated with a 10% response in animal tests

BMDL – lower limit of BMD

$BMDL_{10}$ – lower limit of the benchmark dose expected to produce a 10% reduction in the response

BMF – biomagnification factor

BMR – benchmark response

BOD – biological oxygen demand

BPA – bisphenol-A

BSAF – biota-sediment accumulation factor

BTEX – benzene, toluene, xylenes, alkyl benzenes

CA – correspondence analysis

Caco-2 – human epithelial colorectal adenocarcinoma cells

CAS – Chemical Abstracts Service

Cb – background concentration

CB – closed-bottle test

CBR – critical body residue (theory)

cDNA – DNA sequences synthetized *in vitro* from messenger RNA

CDT – cyclodextrin technology (cyclodextrin-enhanced bioremediation of contaminated soil)

CE – capillary electrophoresis

CEC – cation exchange capacity

CECs – contaminants of emerging concern

CEN – European Committee for Standardization

CFU – colony-forming units

C3H – a mouse strain used as an animal model for research

cl – confidence limits

CLH – European counterpart of Globally Harmonized System of Classification and Labeling of Chemicals (GHS)

CLP – European regulation on classification, labeling and packaging of substances and mixtures

CMR – chemical substances with carcinogenic, mutagenic and reprotoxic effects

CR – cancer risk

CSA – chemical safety assessment

CSIRO – Commonwealth Scientific and Industrial Research Organization, Australia

CTA – cell transformation assay

$C_x$ – the concentration causing x% reduction in the response

CZE – capillary zone electrophoresis

DBALM – DataBase service on ALternative Methods to animal experimentation

DBNPA – 2,2-dibromo-3-nitril-propionamide, a biocide

DBP – dibutyl phthalate, an industrial chemical

D – dose

DBNPA – 2,2-dibromo-3-nitrilopropionamide, a quick-kill biocide

DDE – dichlorodiphenyldichloroethane, a toxic breakdown product of DDT

| | | |
|---|---|---|
| DDT | – | dichlorodiphenyltrichloroethane, a persistent CMR pesticide |
| DEB | – | dynamic energy budget |
| DEH | – | dehydrogenase enzyme activity |
| DGT | – | diffusive gradients in thin films |
| DNA | – | deoxyribonucleic acid |
| DNA-SIP | – | stable isotope probing, a PCR method utilizing $^{13}C$ isotopes |
| DNAPL | – | dense non-aqueous phase liquides |
| DMEL | – | derived minimal effect level |
| DNEL | – | derived noeffect level |
| DOC | – | dissolved organic carbon |
| DOM | – | dissolved organic matter |
| DSD | – | Dangerous Substance Directive in Europe |
| $DT_{50}$ | – | half-life of a chemical substance |
| DTA | – | direct toxicity testing |
| EAMD | – | Ecological Assessment Methods Database, USA |
| EBTC | – | Evidence-Based Toxicology Collaboration |
| EC | – | effective concentration |
| EC | – | European Commission |
| ECHA | – | European Chemicals Agency |
| ECVAM | – | European Centre for the Validation of Alternative Methods |
| $EC_x$ | – | effective concentrations that cause x% decrease in the response |
| $EC_{20}$ | – | effective concentrations that cause 20% decrease in the response |
| $EC_{50}$ | – | effective concentrations that cause 50% decrease in the response |
| $ED_x$ | – | effective dose that causes x% decrease in the response |
| $ED_{20}$ | – | effective dose that causes 20% decrease in the response |
| $ED_{50}$ | – | effective dose that causes 50% decrease in the response |
| ED | – | effective dose |
| EDCs | – | endocrine disrupting chemicals |
| EDSP | – | endocrine disruptor screening program |
| EDTA | – | ethylenediaminetetraacetic acid, a chelating agent |
| ED-XRF | – | energy dispersive X-ray fluorescence |
| EFDB | – | environmental fate database |
| EFSA | – | European Food Safety Authority |
| EPH | – | extractable petroleum hydrocarbons |
| EPH GC | – | extractable petroleum hydrocarbon content measured by GC-FID |
| EQC | – | environmental quality criteria |
| EQS | – | Environmental Quality Standards |
| ERA | – | environmental risk assessment |
| ERAPharm | – | environmental risk assessment of pharmaceuticals |
| $E_rC_{20}$ | – | an effective concentration causing 20% reduction in the growth rate, calculated from the slope of the growth curve |
| $E_rC_{50}$ | – | an effective concentration causing 50% reduction in the growth rate, calculated from the slope of the growth curve |
| ERDC | – | Engineer Research and Development Center, US Army |
| EROD | – | ethoxyresorufin-O-deethylase, a biomarker for chemical exposure |
| ESAC | – | ECVAM's Scientific Advisory Committee |
| ESIS | – | European Chemical Substances Information System |
| EsD | – | effective sample dose |
| $EsD_{20}$ | – | effective sample dose causing 20% decrease in the measured response |

| | | |
|---|---|---|
| $EsD_{50}$ | – | effective sample dose causing 50% decrease in the measured response |
| $EsM_x$ | – | effective sample mass, causing x% decrease in the measured response |
| $EsV_x$ | – | effective sample volume causing x% decrease in the measured response |
| ETV | – | electrothermal vaporization |
| EU | – | European Union |
| EURL ECVAM | – | EU Reference Laboratory for Alternatives to Animal Testing |
| FA | – | fly ash |
| FAME | – | factorial extrapolation method |
| FDA | – | Food and Drug Administration (US) |
| FETAX | – | frog embryo teratogenesis assay on *Xenopus* |
| FGETS | – | food and gill exchange of toxic substances |
| FID | – | flame ionization detection |
| FIFRA | – | Federal Insecticide, Fungicide, and Rodenticide Act (US) |
| FQAI | – | floristic quality assessment index |
| FRAME | – | Fund for the Replacement of Animals in Medical Experiments in the UK |
| GC | – | gas chromatography |
| GC-FID | – | gas chromatography with flame ionization detector |
| GC-MS | – | gas chromatography with mass spectrometry detector |
| GEE | – | generalized estimating equations |
| GFE | – | gastric fluid extraction methods |
| GFP | – | green fluorescent protein |
| GHS | – | Globally Harmonized System of Classification and Labeling of Chemicals, UN |
| GIS | – | geographic information system |
| GJIC | – | gap junction intercellular communication |
| GLM | – | generalized linear models for logistic regression |
| GPL | – | general public license |
| GUI | – | graphical user interface |
| H | – | dimensionless Henry's law constant: the ratio of the concentration of a chemical substance in water to the concentration in the equilibrium gas phase ($c_{aq}/c_{gas}$) |
| Henry | – | Henry's law constant with the dimension of L × Pa/mol: the ratio of the partial pressure of gas above the solution (in Pa) to the concentration (molarity) of gas in solution (in mol/L) |
| H% | – | inhibition rate of the luminescent light emission |
| H | – | null hypothesis in hypothesis testing |
| $H_A$ | – | alternative hypothesis in hypothesis testing |
| HAC | – | hierarchical agglomerative clustering |
| HDPE | – | high density polyethylene |
| HEH GC | – | hydrocarbons extracted by aqueous hydroxypropyl beta-cyclodextrin solution and measured by GC-FID |
| HEP | – | habitat evaluation procedure |
| HepG2 | – | liver cell line (hepatoma) for testing cell growth and cytotoxicity |
| HET-CAM | – | hen's egg test – chorioallantoic membrane assay |
| HHPN | – | hydraulic high pressure nebulizers |

| | | |
|---|---|---|
| HL-60 | – | human acute promyelocytic leukemia cell line |
| HPLC-MS | – | high-performance liquid chromatograph–mass spectrometer |
| HQ | – | human hazard quotient |
| HSDB | – | hazardous substances database |
| HTS | – | high throughput screening |
| HxCDD | – | 1,2,3,7,8,9-hexachlorodibenzo-p-dioxin, persistent organic toxicant |
| I% | – | inhibition rate: difference in response to the effect of a toxicant compared to the control, given in the % of the control |
| IBI | – | index of biological integrity |
| ICCVAM | – | Interagency Coordinating Committee on the Validation of Alternative Methods |
| ICE | – | isolated chicken eye assay |
| ICP-AES | – | inductively coupled plasma atomic emission spectroscopy |
| ICP-OES | – | inductively coupled plasma optical emission spectrometry |
| ICP-MS | – | inductively coupled plasma with mass spectrometry |
| IHCP | – | Institute for Health and Consumer Protection, JRC, EC |
| INT | – | 2-(p-iodophenyl)-3-(p-nitrophenyl)-5-phenyl tetrazolium chloride, an indicator dye for microbial respiration bioassays |
| IOBC | – | International Organization for Biological Control |
| IPCS | – | International Programme on Chemical Safety |
| IPPC | – | Integrated Pollution Prevention and Control, EU Directive |
| IR | – | infrared light |
| IRE | – | isolated rabbit eye assay |
| IRIS | – | Integrated Risk Information System US EPA |
| ISCO | – | *in-situ* chemical oxidation |
| ISO | – | International Organization for Standardization |
| IT | – | information technology |
| IVG | – | *in vitro* gastrointestinal digestion model |
| JaCVAM | – | Japanese Center for the Validation of Alternative Methods |
| JRC | – | Joint Research Centre, EC |
| $K_d$ | – | equilibrium partitioning of charged molecules / contaminants between water and solid phases |
| $K_{liquid\text{-}gas}$ | – | equilibrium partitioning of the chemical substance between liquid and gas |
| $K_{oc}$ | – | equilibrium partitioning of the contaminant between the soil's organic and water content |
| $K_{ow}$ | – | octanol-water partition coefficient, the ratio of the concentration of a chemical in octanol to the concentration in water ($c_o/c_w$) at equilibrium |
| $K_p$ | – | equilibrium partitioning of neutral contaminants between solid and water phases |
| LAD | – | least sum of absolute deviations |
| $LC_{20}$ | – | lethal concentrations that cause 20% mortality of the test organisms |
| $LC_{50}$ | – | lethal concentrations that cause 50% mortality of the test organisms |
| $LD_{20}$ | – | lethal doses that cause 20% mortality of test organisms |
| $LD_{50}$ | – | lethal doses that cause 50% mortality of test organisms |
| LDPE | – | low-density polyethylene |
| LE | – | solution of ammonium lactate and acetic acid applied first by Lakanen & Erviö for soil extraction to imitate plants' uptake |

| | | |
|---|---|---|
| LLC-PK1 | – | kidney proximal tubule cell line for testing cell damage |
| LNAPL | – | light non-aqueous phase liquids |
| LOEAsD | – | lowest effective sample dose calculated from the area under the growth curve |
| LOEC | – | lowest observed effect concentration |
| LOEL | – | lowest observed effect level |
| LOErsD | – | growth ratebased lowest effect sample dose calculated from the slope of the growth curve |
| LOEsD | – | lowest tested sample dose showing an effect |
| LOEsV | – | lowest tested sample volume or mass showing an effect |
| MAC | – | maximum allowable concentration |
| MAE | – | microwave assisted extraction |
| MANCOVA | – | multivariate analysis of covariance |
| MANOVA | – | multivariate analysis of variance |
| MATC | – | maximum allowable toxicant concentration |
| MAWI | – | multi-scale assessment of watershed integrity, ERDC, US Army |
| MCA | – | multiple correspondence analysis |
| MDCK | – | (Madin-Darby) canine kidney epithelial cells |
| MDGC | – | multidimensional gas chromatography |
| METI | – | Japanese Ministry of Economy, Trade and Industry, before 2001: MITI |
| MFC | – | mixed-flask culture mesocosm |
| MI | – | maturity index |
| MINISSA | – | Michigan-lsrael-Nijmegen lntegrated Smallest Space Analysis |
| MITI | – | Japanese Ministry of International Trade and Industry, from 2001: METI |
| ML | – | maximum likelihood |
| MML | – | marginal maximum likelihood |
| MNA | – | mean number of class attributes |
| MNT | – | mammalian cell micronucleus test |
| MoA | – | mode of action |
| MoE | – | margin of exposure |
| MPA | – | maximum permissible addition |
| MPC | – | maximum permissible concentration |
| MPN | – | most probable number, a statistical method |
| mRNA | – | messenger RNA |
| MS | – | mass spectrometry |
| MTD | – | maximum tolerated dose |
| MW | – | microwave |
| NADPH | – | the reduced form of nicotinamide adenine dinucleotide phosphate |
| NCBI | – | National Center for Biotechnology Information, US |
| NEC | – | no-effect concentration the concentration of a substance that will not adversely affect the species or the community exposed to it |
| NHK | – | normal human keratinocyte cells |
| NICEATM NTP | – | Interagency Center for the Evaluation of Alternative Toxicological Methods, National Toxicology Program, National Institutes of Health, US |

| | | |
|---|---|---|
| NIH | – | National institute of Health, US |
| NIST | – | National Institute for Standardization and Technology US |
| NIST/SEMATECH | – | Working Group Elaborating an E-Handbook for Engineering Statistics |
| NGO | – | non-governmental organization |
| NOAEC | – | no observed adverse effect concentration |
| NOAEL | – | no observed adverse effect level |
| NOEC | – | no-observed effect concentration |
| NOEL | – | no observed effect level |
| NOEAsD | – | no observed effect sample dose, based on the area under the growth curve |
| NOEsV | – | highest tested sample volume with no observed effect on the response |
| NOEsD | – | highest tested sample dose with no observed effect on the response |
| NPDS | – | National Pollutant Discharge Elimination System, US |
| NRU | – | neutral red uptake test |
| NSAID | – | non-steroidal anti-inflammatory drug |
| NTP | – | National Toxicology Program, US |
| OECD | – | Organization for Economic Cooperation and Development |
| PAHs | – | polycyclic aromatic hydrocarbons, frequent environmental CMR pollutants |
| PBASE | – | sequential extraction of metals modeling the potential bioavailability |
| PBET | – | physiologically based extraction test for predicting metals bioavailability |
| PBPKs | – | physiologically based pharmacokinetic models |
| PBTs | – | persistent, bioaccumulative and toxic chemicals |
| PCA | – | principal components analysis |
| PCBs | – | polychlorinated biphenyls |
| PCP | – | pentachlorophenol |
| PCR | – | polymerase chain reaction, a technique to amplify a piece of DNA |
| PEC | – | predicted environmental concentration |
| PICT | – | pollution-induced community tolerance |
| PID | – | photoionization detector |
| PLE | – | pressurized liquid extraction |
| PLFA | – | phospholipid fatty acid analysis |
| PNEC | – | predicted no-effect concentration |
| POM | – | polyoxymethylene membrane |
| POPs | – | persistent organic pollutants |
| PPCPs | – | pharmaceuticals and personal care products |
| PPY | – | proteose peptone yeast extract |
| PSM | – | primary biodegradation model |
| psRNAs | – | putative sRNAs |
| PVC | – | polyvinylchloride |
| QSAR | – | quantitative structure–activity relationship |

QSPR       – quantitative structure–permeability relationship
(Q)SAR     – both methods of structure–activity relationship and
             quantitative structure–activity relationship
RAMEB      – randomly methylated beta-cyclodextrin
RBALP      – relative bioaccessibility leaching procedure
RBA        – relative bioavailability
RBP        – rapid bioassessment procedure
RC         – primary biodegradability
RCR        – risk characterization ratio
RDA        – redundancy analysis
REACH      – Registration, Evaluation, Authorization and Restriction of Chemicals,
             EU regulation
REH GC     – hydrocarbons extracted by aqueous RAMEB solution and measured
             by GC-FID
RF         – radiofrequency
RfD        – human oral reference dose
RHE        – reconstructed human epidermis
RIVM       – Dutch National Institute for Public Health and the Environment
RNA        – ribonucleic acid
ROC        – rat keratinocyte organotypic culture
RQ         – risk quotient
RS         – remote sensing
SA         – solubilizing agent
SAM        – standardized aquatic microcosm
SAR        – structure–activity relationship
SAS        – Statistical Analysis System a software package
SB         – soil biota
SBET       – simplified bioaccessibility extraction test
SCGE       – single cell gel electrophoresis
sD         – sample dose
s.e.       – standard error
SEM        – solvent extractable mass
SETAC      – Society of Environmental Toxicology and Chemistry
SFE        – supercritical fluid extraction
SHE        – Syrian hamster embryo cell line
SHIME      – simulator of the human intestinal microbial ecosystem
SIFT       – mouse skin integrity function test
SIN        – substrateinduced nitrification
SIR        – substrate-induced respiration technique
SIRC       – rabbit corneal cells
SIT        – skin irritation test
SME        – multi-step sequential extraction method
SOM        – self-organizing map
SPE        – solid–liquid extraction
SPMD       – semi-permeable membrane device
SPSS       – Statistical Package from IBM
SQT        – sediment quality triad
sRNAs      – small RNAs
SS         – sum of squares method for regression analysis

| | | |
|---|---|---|
| SSC | – | soil screening concentration |
| SSD | – | species sensitivity distribution |
| SSE | – | selective sequential extraction |
| STT | – | soil testing triad |
| STTA | – | stably transfected transactivation assay |
| $T_{25}$ | – | carcinogenicity potency estimate the chronic dose rate which will give tumors to 25% of the animals |
| TAM | – | thermo activity monitor |
| TCDD | – | 2,3,7,8-tetrachlorodibenzo-$p$-dioxin, persistent organic toxicant |
| TCE | – | trichloroethylene, a toxic solvent |
| $TD_{50}$ | – | 50% toxic dose |
| TDI | – | tolerable daily intake |
| TECAM | – | triolein-embedded cellulose acetate membrane |
| TEF | – | toxicant equivalent factor |
| Tenax TA beads | – | porous polymer resin based on 2,6-diphenylene oxide |
| $TEQ_{4CP}$ | – | toxicity-equivalent value expressed in mg 4-chlorophenol/kg soil |
| $TEQ_{Cu}$ | – | toxicity-equivalent value expressed in mg Cu/kg soil |
| TER | – | toxicity/exposure ratio |
| TG | – | test guideline |
| TGD | – | technical guidance document |
| THQ | – | target hazard quotient |
| TIE | – | toxicity identification evaluation for sediments |
| TIM | – | TNO intestinal model |
| TMC | – | terrestrial microcosm chamber |
| TME | – | terrestrial model ecosystem |
| TO | – | transformer oil |
| ToA | – | Treaty of Amsterdam |
| TOFMS | – | time-of-flight mass spectrometer |
| TPF | – | triphenylformazan |
| TPH | – | total petroleum hydrocarbons |
| TTC | – | 2,3,5-triphenyl-tetrazolium-chloride, an indicator dye, used for microbial respiration bioassays |
| TTO | – | terrestrial test organism(s) |
| UAE | – | ultrasonic assisted extraction |
| UBA | – | Umweltbundesamt, the Federal Environment Agency in Germany |
| UBM | – | unified bioaccessibility method |
| UCLA | – | University of California, Los Angeles |
| UDS | – | unscheduled DNA synthesis test |
| UMAM | – | uniform mitigation assessment method |
| UNECE | – | United Nations Economic Commission for Europe |
| US EPA | – | Environmental Protection Agency of the United States |
| USM | – | ultimate biodegradation model |
| UV-VIS | – | ultraviolet and visible light |
| vPvBs | – | very persistent and very bioaccumulative chemical substances |
| VARMINT | – | variables for assessing reasonable mitigation in new transportation |
| VMPs | – | veterinary medicinal products |

WD-XRF – wavelength dispersive X-ray fluorescence
WHAP – wildlife habitat appraisal procedure
WET – whole effluent toxicity, direct toxicity testing of effluents
WFD – Water Framework Directive, EU Directive
WoE – weight of evidence
WPCF – Water Pollution Control Federation
WVA – wetland value assessment methodology
XAD – crosslinked polystyrene copolymer resin adsorbent
XAFS – X-ray absorption fine structure
XANES – X-ray absorption near edge structure
XRF – X-ray fluorescence
YES – yeast estrogen screen assay

# About the editors

**Katalin Gruiz** graduated in chemical engineering at Budapest University of Technology and Economics in 1975, received her doctorate in bioengineering and her Ph.D. in environmental engineering. Her main fields of activities are: teaching, consulting, research and development of engineering tools for risk-based environmental management, development and use of innovative technologies such as special environmental toxicity assays, integrated monitoring methods, biological and ecological remediation technologies for soils and waters, both for regulatory and engineering purposes. Prof. Gruiz has published 35 papers, 25 book chapters, more than hundred conference papers, edited 6 books and a special journal edition. She has coordinated a number of Hungarian research projects and participated in European ones. Gruiz is a member of the REACH Risk Assessment Committee of the European Chemicals Agency. She is a full time associate professor at Budapest University of Technology and Economics and heads the research group of Environmental Microbiology and Biotechnology.

**Tamás Meggyes** is a research co-ordinator specialising in research and book projects in environmental engineering. His work focuses on fluid mechanics, hydraulic transport of solids, jet devices, landfill engineering, groundwater remediation, tailings facilities and risk-based environmental management. He contributed to and organised several international conferences and national and European integrated research projects in Hungary, Germany, United Kingdom and USA. Tamás Meggyes was Europe editor of the Land Contamination and Reclamation journal in the UK and a reviewer of several environmental journals. He was invited by the EU as an expert evaluator to assess research applications and by Samarco Mining Company, Brazil, as a tailings management expert. In 2007, he was named Visiting Professor of Built Environment Sustainability at the University of Wolverhampton, UK. He has published 130 papers including eleven books and holds a doctor's title in fluid mechanics and a Ph.D. degree in landfill engineering from Miskolc University, Hungary.

**Éva Fenyvesi** is a senior scientist, a founding member of CycloLab Cyclodextrin Research and Development Ltd. She graduated as a chemist and received her PhD in chemical technology at Eötvös University of Natural Sciences, Budapest. She is experienced in the preparation and application of cyclodextrin polymers, in environmental application of cyclodextrins and in gas chromatography. She participated in several national and international research projects, in the development of various environmental technologies applying cyclodextrins. She is author or co-author of over 50 scientific papers, 3 chapters in monographs, over 50 conference presentations and 14 patents. She is an editor of the Cyclodextrin News, the monthly periodical on cyclodextrins.

# Chapter 1

# Environmental toxicology – A general overview

*K. Gruiz*

*Department of Applied Biotechnology and Food Science,*
*Budapest University of Technology and Economics, Budapest, Hungary*

## ABSTRACT

The core issues of this book are chemical substances and their adverse effects on the environment and humans. Every aspect of environmental management is linked to the topic of chemical pollution. Millions of people die and millions of hectares of land are degraded every year because of chemical pollution of the environment, thus dealing with pollution is one of the most important global challenges.

Environmental toxicology – capable of directly measuring the adverse effects of chemicals on individual organisms or complex communities – is the scientific tool of sustainable environmental management and is a rapidly developing field in science and engineering. Now that it has been found that chemical models exhibit numerous short-comings in the description of the complex system of the environment, environmental toxicology has moved to the forefront of attention.

This chapter summarizes the basics of environmental toxicology, the science of measuring and using data on adverse effects from the point of view of environmental management such as legislation, monitoring, risk assessment and risk reduction.

The test organisms and test methods used for assessing the adverse effects of chemical substances or contaminated land on the environment and humans will be discussed. Humans are considered as part of the environment, being in direct contact with the atmosphere, the water and soils as well as in indirect contact through the food chain. The topics will range from the basic theory to engineering practice, including legislative toxicology, standardization and site-specific tools.

Traditional methods will be discussed extensively, but special emphasis will be placed on innovative procedures, for example, the alternatives to animal testing, rapid and cost-saving bioassays and molecular methods such as *in vitro* and *in silico* models and the tiered approach to ensure time and cost efficiency, to save animal lives and avoid overextending the use of laboratory equipment and materials.

## I INTRODUCTION, BASIC DEFINITIONS

Millions of chemical substances are produced and used world-wide and different scales of precaution and prevention are applied to them, depending on region, country or substance type. According to the CAS Registry Number – an identifier for chemical

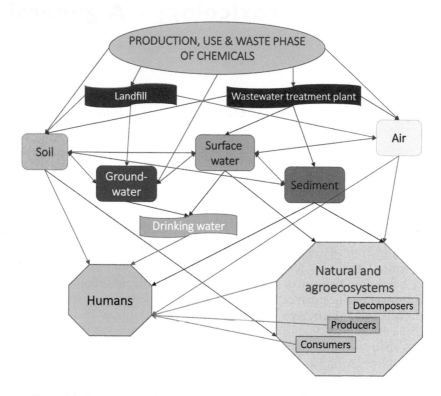

*Figure 1.1* Distribution of man-made chemical substances in the environment.

substances assigned by the Chemical Abstracts Service (CAS, 2013) – 309 000 inventoried/regulated substances and 65 million commercially available organic and inorganic chemicals were listed in the registry by May 2014 (CAS, 2014). In 2011, the number of inventoried substances had been 282 000 and 48 million commercial substances were available.

The last century showed that all chemical substances that are produced and used are emitted into the environment, and, depending on their fate and behavior, they reach the ecosystem and affect humans sooner or later. Some others such as pesticides are produced especially for killing insects and are spread directly into the environment.

Figure 1.1 shows the sources and transport pathways of chemical substances produced and used for industrial, agricultural and household purposes.

Many of the chemical substances produced and used in large volumes are so-called xenobiotics which differ from natural molecules and have structures alien to living organisms. Other chemical substances are of natural origin, but they occur in the wrong place and in an abnormal concentration.

The cause of the risk due to chemicals in the environment is their hazardousness; including their behavior in the environment and their adverse effects on the ecosystem and humans.

The adverse effects of hazardous chemical substances can be observed and measured in the environment or tested in the laboratory. The probable damage to the natural or built environment including workplaces and to users of the environment can be forecast based on the type and extent of the adverse effects and the chemicals' fate and behavior. The probability of an adverse effect multiplied with the extent of future damage is called risk, in this case environmental risk.

Adverse effects of contaminated environmental compartments, i.e. air, water and soil can be measured directly in the environment, *in situ* or on environmental samples, *ex situ*. For example, one can put well-controlled test organisms into the surface water (mussels in a cage) or deliver water samples to the lab and test them on the same mussels or on fish or daphnids. Besides direct toxicity testing of environmental samples, another possibility is the prediction of the adverse effects of a chemical substance (e.g. on aquatic ecosystems), either based on the concentration (measured by chemical analytical methods) or the toxicity (measured by standardized toxicity testing methods) of the pure chemical. The difference between the two is that the hazardous chemical substance in the environment always occurs together with other substances, whether they are contaminants or the natural constituents of the environmental medium. The interactions between the contaminant and other physical, chemical and biological agents influence its adverse effect on a large scale. The environmental relevance of chemical models based on concentration or biological models based on the toxicity of the pure chemical is generally poor. On the other hand, their reproducibility and precision are much better than that of environmental assessments. Some other models such as microcosms and mesocosms give priority to environmental reality at the expense of reproducibility and precision. The concept and design of the environmental assessment should determine the right set of models and test methods to fulfill the requirements of the assessment goal and tier.

The results of the testing of adverse effects can be used to calculate the risk of the contaminated environment, more precisely the risk which the chemical in the environment poses to the users (receptors) of the contaminated environment. Environmental toxicology equally applies mathematical models, chemical analyses and a wide range of biomarkers in biochemical, physiological, organism-, population- and community-level tests.

Environmental risk assessment of chemicals integrates chemical and toxicological methodologies for the identification of the presence, fate and adverse effects of chemical substances in the environment (Figure 1.2).

## 1.1  Toxicology and its role

*Toxicology* involves all aspects of the adverse effects of chemicals on living systems. Adverse effects are those effects which are damaging to either the survival or normal function of living organisms or their communities.

*Environmental toxicology* is the science and engineering of the adverse effects – mainly of chemicals and other man-made agents – in the environment and through the environment. Both the ecosystem and humans may be the targeted receptors. Environmental toxicology also includes studying the chemical substances – which may be potential and real contaminants – polluting air, water, soil and food, their impacts upon the structure and function of ecological systems, including people as well as the use of these results for decision making and environmental management.

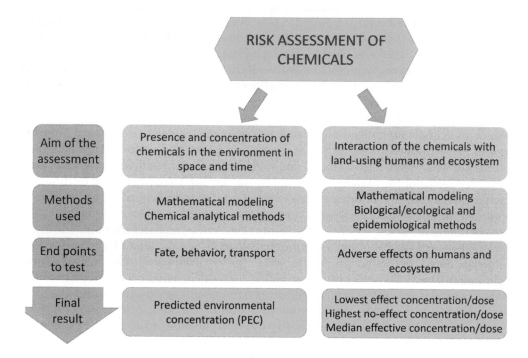

Figure 1.2  The role of chemical and biological methods in environmental risk assessment.

*The results of environmental toxicology* are mainly used to predict hazard and risk due to certain chemical substances or the contaminated environment at local, regional and global scales. The key function of environmental toxicology is to support decision making in environmental management and policy by setting risk-based priorities, establishing environmental quality criteria, designing monitoring systems, selecting risk reduction measures, establishing specific land use target values and performing risk-based management activities. Environmental toxicity results are suitable for direct decision making: decisions are based only on the type and scale of adverse effects. The use of fate and hazard data and information will be discussed and demonstrated in Chapter 2.

*Chemical substances* are materials with a specified chemical composition. Based on this specification, they can be pure chemicals, mixtures of chemicals, or products with a known chemical composition. The production and use of chemical substances are likely to create the main sources of risk in the civilized world. The hazard of chemical substances derives from their structure, their intrinsic physico-chemical, biological and environmental fate properties, but foremost their potential to adversely affect living organisms, the members of the ecosystem and humans. The actual impact is based on their interaction with the other party – the properties of the environment and the living organisms. When preparing a forecast one can be faced with a known or as yet unknown environment and receptors. When the target environment is unknown (e.g. the chemical product is not on the market yet), only the chemicals' potential behavior and effect data are given: a hazard assessment can be done. One can calculate the environmental risk when the target environment and its users, their sensitivity and resistance are known and can be taken into account. The influence of the exposed party on the appearance of the effect can also be estimated. The proper management of chemical substances demands scientific and engineering tools, which are only available to a limited extent. Managers and politicians have to understand that the scientific and engineering basis of their activities and decisions is in constant development and must be further developed and refined. The long-term effects, non-dose-related effects, accumulation, changes and interactions in the environment or in the bodies of organisms are not yet fully understood and quantitatively characterized. In addition to all of these aspects, data acquisition, evaluation and interpretation must be standardized so as to be able to utilize the available measured data for decision making and management of the environment; otherwise, they are useless. Understanding and using the mathematical and statistical evaluation tools, dealing with uncertainties, validation of the results and verification of the methods and technologies are essential for environmental management, legislation and political decision making. Managing the dynamic system of the environment in combination with the dynamic system of knowledge – necessary for good decision making – cannot be reduced to a mechanical choice between two alternatives.

*Xenobiotics* are substances foreign to a biological system. They are artificial substances which had not existed in nature before being synthesized by humans. The term stems from Greek and means foreigner, stranger. Xenobiotic substances can mimic natural molecules and in this way, they may partly or fully substitute biotic molecules in the metabolic pathways. They can be degraded, modified or utilized by microorganisms or higher organisms. This false metabolism sometimes leads to the production of more hazardous metabolites and secondary effects such as the destruction of the endocrine or immune systems.

*Ecological system* is defined as a complex system ranging from the molecular level through individual organisms and communities to the whole ecosystem. An ecosystem comprises the organisms and their habitat; it means all of the organisms living in a particular area and the non-living, physical components of the environment with which the organisms interact, i.e. air, water, soil, sediment and other environmental elements such as sunlight, wind, precipitation, temperature, elevation and so on.

*Aquatic and terrestrial ecosystems* differ to a great extent, and a more or less clear line of division can be drawn between these two habitats. The size and form of ecosystems is subjective, it depends on our interest; it may be a microecosystem within a drop of water, a flock of sewage sludge or the rhizosphere of a plant, the local ecosystem of a contaminated site, smaller or larger regions such as the Mátra Hill in Hungary, the watershed of the Danube in Europe, or the entire earth (global ecosystem). When the deterioration of an ecosystem is measured, the habitat, the components of the ecosystem and the ecosystem services such as regulating (climate, nutrient and water balance), provisioning (water, food) and cultural (science, spiritual, ceremonial, recreation, aesthetic) services are assessed and evaluated by the environmental manager.

*Environment* itself is a complex entity: the environmental compartments – in spite of the differences in the methods needed for monitoring and managing their risks – should be handled in an integrated way because environmental compartments are in close connection, interaction and interrelationship with each other. The state of the earth's ecosystem and its capacity can only be estimated if the global environmental/ecological trends are understood. These trends and the risky changes necessitate monitoring, predictions and forecast. Scientific knowledge and engineering tools are available to support correct short-term environmental risk management and the connecting decision making. A typical problem is that short-term socio-economic reasoning overrides environmental interest. The lack of knowledge about long-term global trends results in locally justifiable decisions and environmental management measures that often fail to harmonize with the needs of regions or the entire earth. Hydroelectric power plants can solve the energy problems of a small country but may cause huge problems for the downstream environment of another country. Opening a mine provides social advantage for the local community but causes environmental damage to the whole of the watershed. If we were positive about global warming as a natural trend, we would not spend huge financial resources on preventing it. Moving 'dirty' industrial and mining activities from industrialized countries to third-world countries with less stringent environmental criteria is extremely harmful from a global perspective. The 'result' is that the global state of the environment will suffer.

The **IPPC Directive** of the EU (Integrated Pollution Prevention and Control, 2008/1/EC) (IPPC, 2008) reflects the scientific and managerial experience concerning the fact that individual environmental compartments cannot be managed on their own: preventing only water from being polluted at the expense of sediments and soil, treating waste water without taking into account the treatment or utilization of sewage sludge, or cleaning the local environment without dealing with its neighborhood do not qualify as sustainable management of the environment.

**Environmental toxicology** needs an integrated approach and requires a multidisciplinary scientific basis to a variety of specialist fields as listed here:

- Analytical chemistry
- Biology
- Biochemistry
- Biometrics
- Chemistry, chemical engineering
- DNA and molecular techniques
- Ecology
- Evolutionary biology
- Limnology
- Marine biology and oceanography
- Mathematical and computer modeling
- Meteorology
- Microbiology
- Molecular biology and genetics
- Pharmacokinetics
- Physiology
- Population biology
- Risk assessment
- Risk management
- Statistics
- Toxicology
- Toxicokinetics.

The application of toxicology is widespread. Apart from the management of the environment, many other areas of human activity manage risk based on toxicological results, such as human healthcare, drug development and occupational health and safety as well as legislation.

*Occupational toxicology* – also called industrial toxicology – deals with (potential) toxic effects at workplaces on workers. Occupational toxicology serves the aims of occupational risk management to protect workers from physical agents and chemical substances and to make their work environment safe. Occupational hazard and risk assessment rely on human and environmental toxicology, but the environment is restricted to the workplace in this case: the air inside an industrial plant, the risk of dermal or eye contact with chemical substances at work, as well as the development of occupational diseases in association with the chemical substances used or produced in the technologies.

*Regulatory toxicology* gathers and evaluates existing toxicological information, develops uniform, standardized, comparable methods for testing and evaluating not only the effects but also the fate and behavior of chemical substances both in the environment and in organisms. Regulatory toxicology means the interpretation and use of toxicity results for the establishment of effect-based quality criteria for food, drinking water, other water uses (e.g. irrigation), animal feed, for all environmental compartments such as air, surface waters, sediments, subsurface waters, soils, depending on their use and users (land use) as well as for waste utilization (e.g. sewage sludge utilization on soil). Regulatory toxicology tries to control hazardous substances and materials in a safe manner, ensuring an acceptable risk level, or in other words, a safe exposure, an exposure level which does not pose unacceptable risk to the ecosystem and humans. It supports the authorization, restriction or ban of chemicals, licensing of technologies using chemicals, and ensures a globally harmonized system for classification and labeling of chemicals for their safe use.

*Food toxicology* aims at providing a supply of safe and edible food to consumers, assessing poisonous or allergenic food components and the adverse effects of food processing additives and pesticide residues.

*Air toxicology* measures indoor and outdoor air quality and compares it to quality criteria. It also deals with the atmosphere as a source of contaminants that reach soil and water.

*Aquatic toxicology* is based on the response of the aquatic ecosystem (marine and freshwater), the response of ecosystem diversity or the response of representative ecosystem components in laboratory tests. Exploring the fate and behavior of chemical substances in water is an important part of aquatic toxicology.

*Terrestrial toxicology* concentrates on the terrestrial ecosystem as well as on the fate and behavior of chemicals in soil. The terrestrial ecosystem is understood to a lesser degree than the aquatic one. Diversity of terrestrial species, first of all soil microflora and fauna is difficult to investigate using traditional methods. In the case of soil and other solid phase compartments such as sediments, solid wastes and sludge, the mobility, partitioning, accessibility and bioavailability of toxic substances play an extremely important role in producing adverse effects and posing an actual risk to the ecosystem and humans.

*Clinical toxicology* is concerned with diseases and illnesses associated with short- or long-term exposure to toxic chemicals.

*Forensic toxicology* is the science that looks into the cause and effect relationships between exposures to a drug or chemical substance and the resulting toxic or lethal effects.

*Epidemiology* is closely related to toxicology and its key aim is finding connections between human health and contamination in food or drinking water or the environment. Epidemiology is the study of patterns of health and illness and associated factors at the population level. It supports evidence-based medicine with proper statistical data on risk factors for diseases including the risk of environmental pollution or the accidental exposure to toxicants. Epidemiology is based on good study design, data collection, statistical analysis and documentation. The methods and concepts of epidemiology provide suitable tools for finding relationships between GIS-based pollution maps and the health quality of human, animal or plant populations in the future. It is very important all over Europe and the world to harmonize and utilize the collection and processing of available data.

*Human toxicology* aims to specify the dose–response relationship between hazardous chemical substances and human responses. Since these relationships cannot be tested on a well-designed, statistically relevant human population (the human toxicity of chemical substances is mainly based on the results of animal toxicity tests), the extent of toxic effects on humans is usually an estimate. Extrapolation from animal data to humans is possible assuming that the animal species has been properly selected, i.e. its response is analogous to the human body's response and the test method and scenario applied truly model real human exposure. The main extrapolation methodology – for example from rat to man – applies a safety factor based on experience. The default value for the interspecies safety factor is $ED_{50}$ (human)/$ED_{50}$ (animal) $= 0.1$ because drugs and toxic chemical substances are generally 10 times more potent in humans as indicated by available pharmacological and toxicological data, therefore a 10-times smaller dose suffices to cause the same effect.

Animal data are suitable for establishing the dose or concentration of the chemical substance that would cause an adverse effect, damage or death to 10, 20, 50 or 90% of the treated animals, or for determining the lowest effect and the highest no-effect concentrations or doses. These end points represent manageable limit values.

Animal tests have many subclasses discerned according to the applied animal taxon (fish, bird, mouse, rat, dog, monkey, etc.), type of exposure (acute, repeated or chronic exposure), exposure routes (inhalation, oral, cutaneous and mixed routes) and aim of the test (testing of toxicity, mutagenicity, reprotoxicity, neurotoxicity, sensitization and irritation, endocrine disruption, immunotoxicity, phototoxicity or photoallergy). Depending on end points, tests can be based on death or minor deterioration of the whole organism such as immobilization of daphnids, or behavioral changes (e.g. contaminated soil avoidance by insects), irritation or corrosion of the skin or eye, changes in the metabolism. The end point can be

- organ-specific: cardiac, ophthalmic, cutaneous, muscle, bone- or hepatotoxic, etc.
- cellular: cell death, mitochondrial or peroxisome irregularities, cellular tight junctions, etc. and
- molecular: reactive oxygen species, glutathione and glutathione transferase, metabolomics, DNA changes, chemokines and other genomics.

Impairments in the reproduction can be studied through fertility, quantitative and qualitative characterization of the offsprings and contaminant analysis of the milk of mammals.

*Pharmacology* is the closest related branch of applied science to human toxicology, whose aim is the study of drug actions. The methodology is very similar to human toxicity assessment, the difference is that drugs are applied for their beneficial effects on human or other organisms, and the main effect has a special target, e.g. to kill pathogens or harmful cells. More specifically, pharmacology is the study of the interactions that occur between a living organism and the chemical substance, i.e. the drug that affects certain biochemical function(s). If the substances have medicinal properties, they are considered pharmaceuticals and the assessment is qualified as a pharmacological assessment. If the drug is a toxicant, it is toxicology that deals with its effects. As all drugs are toxicants, even though they have a dose range which leads to medicinal applications, the methodologies are very closely related. It is mainly the target that distinguishes pharmacology from toxicology.

*Ecotoxicology and human toxicology* represent the two main pillars of environmental toxicology. The subject of the study of ecotoxicology is the potential for physical, chemical and biological agents – also called stressors – to affect ecosystems. Such stressors might disrupt natural genetics, biochemistry, physiology, behavior, structure, form and interactions of living organisms that comprise the ecosystem if they occur in a concentration, level or density which causes adverse effects on microbial, plant, animal and human constituents as well as the complex structure they form in the ecosystem.

Truhaut coined the term ecotoxicology in 1969 defining it as 'the branch of toxicology concerned with the study of toxic effects, caused by natural or synthetic pollutants, to the constituents of ecosystems, animal (including human), vegetable and microbial, in an integral context' (Truhaut, 1977). Ecotoxicology covers not only the constituents of the ecosystem but also their forms and structures.

Ecotoxicology is the integration of toxicology and ecology or, as Chapman (2002) suggested, 'ecology in the presence of toxicants'. It aims to quantify the effects of stressors upon natural populations, communities, or ecosystems. Chapman (2002) tries to distinguish ecotoxicity and environmental toxicity, saying that environmental toxicity is more anthropocentric, but Landis and Yu (1999) consider environmental toxicity as impacts of chemicals and other stressors on ecological systems, their function and structure, including man. Our approach in this book, similarly to the development of environmental toxicology over the last few decades, corresponds to the latter: environmental toxicology includes both human toxicology and ecotoxicology.

## 1.2  Regulatory toxicology for chemical substances and contaminated land

Regulations are often discussed in this book in the context of environmental toxicology and environmental risk management of chemicals.

*Regulatory toxicology* is the study of the adverse effects of chemicals, not just to humans, but also to all living organisms including plants, animals or fungi. The integration of metabolism, toxicity, pathology and effect mechanisms plays a much greater role today than ever before. A better understanding of the interactions and relationships is essential for proper regulation of chemical substances and drugs and every other material, product or waste which contains hazardous chemical substances.

It should be emphasized that the origin of chemical risk is not only the hazard of a substance, but also the abnormal concentration or presence of a chemical substance in the wrong place at the wrong time. A large number of major pollution events are caused, for example, by petroleum products, metals, mine wastes, agricultural nutrients and various waste materials, which are natural products such as organic waste, but emerge in the improper quantity and quality or in the wrong place.

The complex approach of regulatory toxicology has the advantage that the results of environmental toxicity can be utilized for the regulation of activities which pose a risk to the environment. It is of utmost importance for regulation to be based on science and the decisions of environmental management should be effect-based. What does this mean? This means that decision making, risk management, environmental monitoring and the implementation of risk reducing measures should be based on scientific evidence concerning the real adverse effects of the pollutant in a generic area or on a real site. Science-based regulation reflects the state of the art of scientific and engineering knowledge, the continuous development in science and technology and in environmental risk management. This kind of dynamism in the regulation process makes possible the integration of new scientific knowledge and innovative risk management tools, which provide improvement vis-à-vis former BATs (Best Available Technologies).

Another aspect of regulatory toxicology is that regulations require information, data, methodology, standardized analytical and test methods as well as interpretations. Environmental toxicology has to fulfill the needs of regulation and fill gaps in our knowledge.

There is a close interaction between the two sides: environmental toxicology should serve regulation, while scientists are supposed to know the concepts of regulation and the way to fill the gaps with new methodologies and information. Science should provide advice about the need for the integration of scientific knowledge into regulatory decision making.

The concepts and problems involved with the generation, evaluation and interpretation of experimental ecotoxicological, animal and human data should be placed by regulatory toxicology in the wider perspective of societal considerations towards protecting human health and the environment.

Regulatory toxicology and risk management are closely related, and both of them should continuously include new scientific developments to slowly reduce or eliminate methodological shortcomings and data gaps impairing the quality of decisions and regulations. Even after upgrading the scientific grounds, many uncertainties will remain due to natural variabilities, statistical uncertainties and subjective errors and bias.

The importance of regulatory toxicology is well characterized by the regulations on pesticides, biocides, food additives, cosmetics and the regulation of hazardous chemical substances and materials all over the world. In Europe, hazardous chemical substances and materials are regulated by the REACH (Registration, Evaluation, Authorization and Restriction of Chemicals, EC No 1907/2006) (REACH, 2006) and CLP (Classification, Labelling and Packaging, 1272/2008) (CLP, 2008a) regulations. Additionally, specific regulations exist for specific families of products such as pesticides, biocides, fertilizers, detergents, explosives, pyrotechnic articles, and drug precursors.

In principle, REACH applies to all chemical substances: not only chemicals used in industrial processes but also those in our daily lives, for example, in cleaning products, paints and in articles such as clothes, furniture and electrical appliances.

The aims of REACH are:

–   improving the protection of human health and the environment from the risks posed by chemicals;
–   enhancing the competitiveness of the EU chemical industry, a key sector of the EU's economy;
–   promoting alternative methods for the assessment of hazards of substances;
–   ensuring the free circulation of substances on the EU's internal market.

Further details can be found in Chapter 2 in Volume 1 not only about REACH and CLP but other European and world-wide regulations related to chemical substances, such as the Globally Harmonized System (GHS, 2007), which ensures the uniform classification and labeling of chemical substances to alert the users of chemicals in industry, commerce, transport or households to be cautious and apply preventive measures or means.

Currently, chemical substances are classified according to GHS into the following 15 categories:

–   Explosive substances and preparations;
–   Oxidizing substances and preparations;
–   Extremely flammable substances and preparations;
–   Highly flammable substances and preparations;
–   Flammable substances and preparations;
–   Very toxic substances and preparations;
–   Toxic substances and preparations;
–   Harmful substances and preparations;
–   Corrosive substances and preparations;
–   Irritant substances and preparations;
–   Sensitizing substances and preparations;
–   Carcinogenic substances and preparations;
–   Mutagenic substances and preparations;
–   Substances and preparations which are toxic for reproduction;
–   Substances and preparations which are dangerous for the environment.

REACH, CLP and GHS regulations need a uniform and standardized analysis and testing methodology as well as globally harmonized evaluation, interpretation of the results such as the classification and labeling of chemical substances and the physico-chemical, ecotoxicological and toxicological methods. A standard and uniform methodology is a fundamental principle of regulatory toxicology, otherwise the results would not be comparable, and differences due to methodology, laboratory, country, etc. would limit the efficiency of regulation to a large extent.

The European Council regulation No 440/2008 of 30 May 2008 (EC, 2008) lays down the test methods pursuant to REACH.

OECD (OECD, 2013a) (Organisation for Economic Cooperation and Development) has standardized the guidelines in support of the regulation and uniform management of chemical substances (OECD, 2013b). OECD developed, validated and published a collection of the most relevant internationally agreed test methods used by government, industry and independent laboratories to determine the adverse effects, the hazards and safety of chemicals and chemical preparations, including pesticides and industrial chemicals. They cover tests for the physical-chemical properties of chemicals, human health effects, environmental effects, degradation and accumulation in the environment.

The European Union (EU) and OECD initiated the development of alternative test methods which either do not use animals at all, or use only a much smaller number of animals compared to traditional toxicological tests.

The European Centre for the Validation of Alternative Methods (ECVAM, 1991) was created by a 'Communication from the Commission to the Council and the Parliament' in October 1991 (COM, 1991), pointing to a requirement in Directive 86/609/EEC (EEC, 1986) on the protection of animals used for experimental and other scientific purposes, which requires that the Commission and Member States should actively support the development, validation and acceptance of methods which could reduce, refine or replace the use of laboratory animals.

ECVAM was established in 1992 at the Joint Research Centre (JRC, 2013), and is now the part of the Institute for Health and Consumer Protection (IHCP, 2013) located in Ispra, Italy. Its mission is to:

–   Promote the scientific and regulatory acceptance of non-animal tests that are of importance to biomedical sciences, through research, test development and validation and the establishment of a specialized database service;
–   Coordinate at the European level the independent evaluation of the relevance and reliability of tests for specific purposes, so that chemicals and products of various kinds, including medicines, vaccines, medical devices, cosmetics, household products and agricultural products, can be manufactured, transported and used more economically and more safely, whilst the current reliance on animal test procedures is progressively reduced.

## 1.3   Future of environmental toxicology

The protection of test animals and animal welfare is not the only reason for searching for alternatives, but also because of the developing trend in toxicology to find the molecular fundamentals of the effects of chemical substances on an organism's body, organ, cell or molecule. The methodologies slowly changed from the whole body level and drastic end points such as lethality, to earlier and milder indicators of the same effect. Both the protection of animals and a refined understanding of the mechanisms of effects guided events in the same direction: the use of molecular tools, metabolites, enzymes, RNA and DNA molecules as indicators or the heat production (Gruiz et al., 2010) of living organisms. MacGregor (2003) summarized the future possibilities of environmental toxicology, emphasizing that the molecular biology revolution and the development of genomic and proteomic technologies support the transformations in toxicological and clinical practice. Traditional biomarkers of cellular integrity, cell and tissue homeostasis and morphological alterations that stem from cell damage or death can be supplemented or replaced with DNA or other molecular end points and can become a realistic alternative.

Molecular and cell-based technologies are discussed in detail in the following sections; the application of QSAR-models – aiming at predicting the physicochemical and biological properties of molecules based on their structure and their application in environmental toxicology – was introduced in Chapter 1 in Volume 1.

### 1.3.1  Molecular technologies

*Molecular technologies* permit simultaneous monitoring of many hundreds or thousands of molecules (OMICS, 2009) and promise to allow functional monitoring of key cellular pathways simultaneously. New biomarkers based on molecular responses to functional perturbations and cellular damage are foreseen. Responses that can be monitored directly in humans should provide bridging biomarkers that may eliminate much of the current uncertainty in extrapolating from laboratory models to human outcome. Another aspect of genomics is the currently enhanced ability to associate DNA sequence variations with biological outcomes and individual sensitivity. Genetic approaches, for example, the correlation of genetic variants with human diseases and adverse reactions from exposure to toxicants proved that individual variations play an important role in sensitivity to environmental agents, disease susceptibility, and therapeutic responses. Regulatory toxicological practice is likely to be shaped in the future by the combination of conventional pathology, toxicology, molecular genetics, biochemistry, cell biology and computational bio-informatics – resulting in the broad application of molecular approaches to monitoring functional disturbances. The forecast of MacGregor (2003) has become a reality in the past few years.

The AltTox (2009) definition of *omics* is as follows: omics refers to the collective technologies used to explore the roles, relationships and actions of the various types of molecules that make up the cells of an organism (OMICS, 2009). Omics technologies are increasingly used in different fields of science, mainly in system biology, which is defined as biology focusing on complex interactions in biological systems, and using holism instead of reducing the scope of investigation, which might be particularly important in environmental sciences and management.

The main users of omics-based technologies are:

- Systems biology;
- Computational systems biology;
- Medicinal systems biology;
- Cell signaling and networking;
- Drug discovery and development;
- Synthetic biological systems;
- Environmental toxicology.

*Omics technologies* include

- *Genomics*: the study of genes and their function such as in the Human Genome Project (HUGO, 2012; DOE, 2003). The first of the "omics" technologies to be developed, genomics has resulted in massive amounts of DNA sequence data requiring great amounts of computer capacity. Genomics has progressed beyond sequencing of organisms (structural genomics) to identifying the function of the encoded genes (functional genomics). As well as functional genomics, other areas include nutrigenomics, gene expression profiling, micro-array technologies, drug metabolism and toxicology.

- *Transcriptomics*: the study of the mRNA as in genomics.
- *Proteomics*: the study of proteins, characterizing the identity, function, regulation, and interaction of all of the cellular proteins of an organism (the *proteome*). With proteomics, it is possible to identify the changes in cells and tissues exposed to toxic materials, and to understand the mechanisms of toxicity. Data collected in databases can be utilized in a predictive *in silico* tool in environmental toxicology. The most advanced approaches and technologies are: functional and computational genomics, molecular and cellular proteomics, gel-based proteome profiling, mass spectrometry-based and gel electrophoresis-based proteomics, and protein arrays.
- *Metabolomics*: the study of molecules involved in cellular metabolism. It provides information on the metabolic state of a cell, organ or organism to identify biochemical changes that are characteristic of environmental stresses such as exposure to chemical substances or environmental contaminants, which alter the metabolic pathways in cells, and result in a specific metabolite profile suitable for assessing and identifying toxic exposures and consequent responses.
- *Glycomics*: the study of cellular carbohydrates, using methods such as glycan array technology, structural and chemical glycobiology and microbial glycomics.
- *Lipomics*: the study of cellular lipids, such as cellular and membrane lipidomics.


The molecular profiles can vary with cell or tissue exposure to chemicals or drugs and thus have potential use in toxicological assessments. These new methods have already facilitated significant advances in our understanding of the molecular responses to cell and tissue damage, and of perturbations in functional cellular systems (Aardema & MacGregor, 2002).

The large amount of data provided by molecular methodologies should be managed and analyzed by bioinformatics which uses advanced computing techniques. Bioinformatics tools include computational tools that can mine information from large databases of biological data such as genomics or proteomics data.

The integrative approach of generating maps of cellular and physiological pathways and responses using bioinformatics and molecular databases is called **computational biology**.

The aim of **computational system biology** is the creation of computational models of the functioning of the cell, multicellular systems, and ultimately the whole organism. These *in silico* models will provide virtual test systems for evaluating the toxic responses of cells, tissues, and organisms to the effect of toxic chemical substances.

Innovative molecular methods such as **microarrays** have been developed in the last decade. Microarrays consist of DNA or protein fragments placed as small spots onto a slide, which are then used as 'miniaturized chemical reaction areas' (DOE, 2003). The studies typically involve looking for changes in gene or protein expression patterns from the effect of drugs or environmental chemicals. Thousands of genes or proteins can be investigated simultaneously. The evaluation by a robotic system, called HTS, i.e. High Throughput Screening, is another innovation.

*Toxicogenomics* investigates the genes expressed in organisms that have been exposed to a chemical substance, a drug, or a toxin and the results are compared to unexposed (untreated) organisms (negative controls). The goal of toxicogenomics is to identify patterns of gene expression related to specific chemical substances or chemical classes of substances to use these expression patterns as end points for toxicity assessments. Until today, toxicogenomics has been useful in refining animal experiments and identifying mechanisms of toxicity in laboratory tests. The new toxicogenomics approach can be utilized for cell cultures exposed to toxicants, and to develop a predictive tool for *in vivo* toxicity testing (Aardema & MacGregor, 2002).

### 1.3.2  Cell-based technologies

Cultured animal and human cells in so-called 'cell cultures' are widely used for screening and deeper exploration of the effects, including the toxic effects, of chemical substances. Many alternative test methods have been developed to replace animals for testing of chemical substances or reduce their number as much as possible. Replacing animal organisms with cellular models has many advantages, for example, creating a 'clear' model with good reproducibility by concentrating on the main exposure routes and end points. These *in vitro* cellular models are very useful in regulatory toxicology where uniformity and reproducibility have priority in generic risk assessment. A further advantage may be that the individual- or species-specific complexity of the answer can be excluded from the model. However, when the realistic response of a specific organism is required, a simplified model can hardly simulate the organism's complexity and its interactions with the real environment. The (general) model may be a disadvantage when a certain species, strain or individual is the target of testing. In this case simulation models or microcosms should be applied to approach reality as much as possible.

Simplified cellular models such as single layers of cells living on a nutrient medium in a culturing vessel are too much reduced, and they are not able to simulate living organisms. Part of the new developments aims to increase the complexity of the cellular models, for example, by three-dimensional (3D) cellular or tissue models that can improve replication of the *in vivo* tissue.

### 1.3.3  Computational toxicology

A vast quantity of measured toxicological data is available in databases because molecular toxicology, microarrays and their robotic evaluation disgorge abundant data. New mathematical tools have to be developed and introduced for the efficient evaluation and interpretation of these data (Kavlock *et al.*, 2005). Computational toxicology is the overall term for these tasks which deal with:

–   The construction and curation of large-scale data repositories necessary to save the information from new technologies (Richard, 2006).

– The introduction of virtual and laboratory-based high-throughput screening (HTS) assays on hundreds to thousands of chemicals per day and the development of high-content assays with hundreds to thousands of biological end points per sample for the identification of toxicity pathways (Inglese *et al.*, 2006; Dix *et al.*, 2007).

– The latest advances in computational modeling that provide the tools needed to integrate information across multiple levels of biological organization for the characterization of chemical hazard and risk to individuals and populations (Di Ventura *et al.*, 2006).

The mathematical, molecular and cellular models are often integrated into test systems or test batteries and used for risk assessment and decision making in environmental risk management. No doubt, in the future, these innovative molecular and cellular methods will play an important role in predictive and regulatory toxicology and in toxicological research, not only by replacing animals, but also because of the opportunity to increase the number of tests and experiments, which are highly limited when laboratory animals are used.

## 1.4    What environment means in the context of toxicology

Environment is much more than just natural environment: it includes – according to a holistic view – all those parts of the earth that create and influence the habitats of ecosystem members and human beings and those parts of the ecosystem that provide services for them. This means that waste in the earth's stratosphere, deep-sea oilfields, workplaces and technological rooms are part of our environment due to their direct or potential contacts with other parts of the ecosystem or living organisms!

Environment includes air, waters, both subsurface and surface waters as well as their sediments, rocks and soils. These compartments and their three main physical phases (gaseous, liquid, and solid) are in close connection and interaction with each other; soil gas with soil solid, pore water (partition) and atmosphere (diffusion); sediment with free water and with soil (flood); soil with sediment (erosion); groundwater with surface waters (convective transport in both directions), precipitation with all compartments (seepage, runoff), etc. One compartment additional to the non-living ones is the biota and its local habitat, which can be differentiated as a water-based environmental compartment and others, where the solid phase is dominant (biofilms, rhizosphere, activated sludge, soil and sediment etc.). These habitats are locations with intensive interactions between all environmental compartments and phases.

The ecosystem itself is difficult to define: it is a widespread view that the ecosystem is limited to the natural ecosystem, but in our context 'natural', agricultural, recreational and fully artificial ecosystems cannot be handled separately because of interactions and free transfer and also because of the participation of humans. Chemical substances in this complex system may have a very invasive effect, even though they are natural substances but occurring in abnormal concentrations, or xenobiotics which are *ab ovo* xenon, i.e. alien to the environment. As they are unknown invaders, their behavior and interactions with ecosystems or humans are not fully predictable. The best example is the problem of pesticides: producers, users and regulators are well aware of the target effect of pesticides. It was realized very soon that exploring and measuring the non-target effects on non-target ecosystem constituents is also necessary and it has become an obligation to minimize the adverse effects of pesticides on non-target ecosystem members. Today, we know that many of the pesticides have long-term effects on the endocrine system causing hormone and immune system disruption as well as neurotoxic effects. It is a painful fact that science is not ready to measure these emerging effects and to manage the problem of the so-called emerging pollutants.

## 1.5 Environmental toxicology *versus* human toxicology

The aim of human and environmental toxicologists was very different in the beginning: human toxicologists intended to characterize the acute risk to the human body due to chemicals through the usual exposure routes: inhalation, digestion or dermal contact and using the traditional pharmacological tools. As the epidemiological data sets are typically not complete, the quality of existing epidemiological information is still questionable and individual differences between the members of the human population may be significant. The toxicologists usually rely on animal tests whose results are then extrapolated to humans. Animal models have a number of advantages but even more disadvantages: similarities and differences in metabolic pathways, high deviation in references, and the small number of animals tested and, as a consequence, low statistical power and high cost. In addition, there are ethical problems because of welfare concerns of the experimental animals. End points of animal tests have also changed in the last 20 years: lethality and the number of progeny as end points have been refined by molecular and cellular indicators and metabolomics or genetic analyses. New branches of science were born such as neurotoxicology, toxicokinetics, toxicogenomics, hormone and immune system responses and many other sub-areas.

On the other hand, ecotoxicology, which was the sole concern in the beginning and dealt with changes in the structure and constituents of the ecosystem and their comparison with healthy ecosystems, has also changed: the extrapolation from individual species to the whole environment has become more widespread in comparison to more complex ecosystem assessments (which remains a simplification, in spite of going into details) due to its easier implementation and better statistical power. The use of microorganisms and other small organisms made it possible to increase the numbers of tested organisms, simultaneous tests and repetitions. Increased sensitivity of the test methods and early indication have moved to the fore as prime metabolic end points instead of lethality. Metabolomics and DNA markers have become part of the practice of ecotoxicity testing.

Finally, the importance and significant effects of interactions between human receptors and the ecosystem through food chains, agroecosystems and commonly used air and water have made it clear that environmental toxicology must find a common denominator between human toxicology and ecotoxicology.

In the end, a significant part of both ecotoxicology and toxicology can be reduced to the same cellular and molecular methods. Although toxicity and ecotoxicity tests may use the same methods and model systems, the extrapolation to human or ecosystem is different. Many of the omics are the same or very similar in biological systems. The genetic code (genomics), the enzymes (proteomics), the metabolites (metabolomics), the stress mechanisms and end products need the same methodology whether they are daphnids, fish, rats, human cells or organs. Uniform molecular methods which, of course, cannot estimate the response of a whole organism, can indicate almost all of the adverse effects. A suitable combination of a number of molecular markers can give a refined picture of the possible adverse effects and the mechanisms. At the other extreme *in vivo* tests on live animals, e.g. rats are also models with limitations: a rat can never replicate the complex response of humans. An efficient assessment strategy, preferably a tiered assessment, has to be applied to obtain the correct answer concerning the extent and quality of adverse effects and the subsequent risks. An efficient assessment based on optimized algorithms can provide a better and more precise answer at lower cost compared to traditional tests. For example, applying non-expensive and sensitive screening tests prior to animal tests makes it possible to exclude negative substances from further testing. *In silico* QSAR models (Quantitative Structure–Activity Relationship) are advantageous in tiered assessments as a screening tool both in human- and ecotoxicology. Toxicology and ecotoxicology are ready to fertilize each other, and their methodological tools are available to establish the uniform science of environmental toxicology. However, traditions, current routines and the different trends of the professional branches may raise obstacles which must be dealt with.

## 1.6  Animal studies

*Animal experiments* are common in physiological studies for the development of new chemicals or medicines, for studying environmental effects or testing new food additives. The protection and welfare of animals is covered by a number of international treaties and other legal instruments. These refer to wildlife, zoo and farm animals, animals in transport and also to animals used in scientific experiments. EU legislation on the protection of animals used for experimental and other scientific purposes is covered by Directive 86/609/EEC (EEC, 1986).

The Treaty of Amsterdam (ToA, 1997) requires the EU and its Member States to take animal welfare considerations into account in a number of policy areas and also in practice by developing and using alternative test methods instead of animal experiments.

US legislation on laboratory animals is based to a large extent on the international guidelines about the use of humans in research and testing. The basis of these guidelines is the Nuremburg Code, a list of criteria developed for judging if the tools applied by the Nazis in World War Two constitute crimes 'against humanity'.

The most pragmatic approach to reduce experiments on animals is the replacement of animal tests with non-animal ones. Whenever replacement is not possible, all efforts should be made to apply those methods which use a smaller number of animals and cause least harm to them. Animal tests can be replaced by *in vitro* methods which use living cells or tissue cultures, or with mathematical models such as QSAR.

The EU started a program in 2010 for the minimization and possible exclusion of higher animals from toxicity and other adverse-effect tests and for the development of alternative test methods, using microbes, single-cell animals, tissue cultures, isolated organs from slaughterhouse animals, or applying *in vitro* techniques and *in silico* methodologies. Many of these methods are identical to or very closely related to the methods used in environmental toxicology.

The European Chemicals Agency (ECHA, 2013) prepared a Practical Guideline entitled *How to avoid unnecessary testing on animals* (ECHA, 2010). Clearly, it takes time to replace all animal tests with new alternative tests which completely avoid the use of animals. According to the Practical Guideline, there are some other options which can reduce animal testing for the purposes of regulation of chemicals in the transitional period:

- Data sharing: the REACH test data can be shared between registrants to avoid unnecessary tests on animals. This means that use of studies conducted by one registrant on vertebrate animals can be shared by all registrants for that substance.
- Gathering information before test planning, looking for existing data in all possible sources, acquiring data on analogous substances if 'read across' is possible, using QSAR and SAR estimates; weight-of-evidence approach to fill data gaps for particular end points if this is appropriate.
- Strategies to avoid unnecessary tests on animals:

  o *In vitro* methods: *in vitro* tests performed in a controlled environment such as a test tube or Petri dish, without using a higher living organism. A test performed *in vivo* is one that is carried out on a living organism. Results obtained from suitable *in vitro* methods may indicate the presence of certain dangerous properties or may be important in relation to understanding the mode of action of the substance. In this context, 'suitable' means sufficiently well-developed according to internationally agreed test development criteria.
  o Grouping of substances and read across: animal tests on a substance can be avoided if there is enough evidence on similar substances and can be read across between the two substances. Substances whose physicochemical, toxicological and ecotoxicological properties are likely to be similar or follow a regular pattern as a result of structural similarity, may be considered as a group, or category of substances. Applying the group concept means that the physicochemical properties, human health effects and environmental effects or environmental fate may be predicted from data for one substance within the group by interpolation to other substances in the group; this is the read-across approach. This avoids the need to test every substance in the group for every hazard end point. Preferably, a category should include all similar substances.
  o Quantitative Structure–Activity Relationship (QSAR) models: animal tests can be avoided if the hazardous properties of a substance can be predicted using computer models. The (QSAR and SAR approach seeks to predict the intrinsic properties of chemicals by using various databases and theoretical models, instead of conducting tests. Based on knowledge of the chemical structure, QSAR quantitatively relates the characteristics of the chemical to a measure of a particular activity. QSAR should be distinguished from SAR, which makes qualitative conclusions about the presence or absence of a property of a substance, based on a structural feature of the substance.
  o Weight-of-evidence approach: animal tests can be avoided if there is a weight of evidence which points to the likely properties of a substance. This approach may be applied if there is sufficient information from several independent sources leading to the conclusion that a substance does or does not exhibit a

particular dangerous property, while the information from each single source alone is regarded as insufficient to support an assertion.

– Another possibility to reduce animal testing is eliminating duplication and utilizing scientific evidence from other sources. An obvious negative case is where no irritation or corrosion has been observed in an acute dermal toxicity test. In such cases no additional dermal irritation/corrosion test should be carried out. The positive evidence can also be utilized; a known irritant/corrosive substance, e.g. a strong acid or base, or a substance flammable in air at room temperature should not be tested for skin irritation/corrosion, but classified without animal testing. Low $K_{ow}$ organic substances can also be classified as non-bioaccumulative without animal testing.

The NTP Interagency Center for the Evaluation of Alternative Toxicological Methods (NICEATM, 2013) and the Interagency Coordinating Committee on the Validation of Alternative Methods (ICCVAM, 2013) define the alternatives to animal tests in a simpler way, which is called the 3R approach, including Reducing, Refining and Replacing animal use in toxicity testing (NICEATM/ICCVAM, 2013):

– Reducing the number of animals used to the minimum number required to obtain scientifically valid data.
– Refining procedures to lessen or eliminate animal pain and distress.
– Replacing animals with non-animal systems or one animal species with a less highly developed one (for example, replacing a mouse with a fish).

## 1.7 *In vitro* contra *in vivo*: alternative test methods

*ACuteTox* (2005–2010) was an integrated project in the EU between 2005 and 2010, which aimed to develop and pre-validate a simple and robust *in vitro* test strategy for the prediction of acute human toxicity.

The main objectives – as indicated on the website of the project (ACuteTox, 2005–2010) – are 'the compilation, evaluation and generation of high-quality *in vitro* and *in vivo* data on a set of reference chemicals for comparative analyses, and the identification of factors that influence the correlation between *in vitro* (concentration) and *in vivo* (dose) toxicity, particularly taking into consideration biokinetics, metabolism and organ toxicity (liver, central nervous system and kidney). Moreover, innovative tools (such as cytomics) and new cellular systems for anticipating animal and human toxicity were explored. Ultimately, the goal was to design a simple, robust and reliable *in vitro* test strategy amendable for robotic testing, associated with the prediction models for acute oral toxicity.'

Collected data and studies by ACuteTox have shown good correlation between *in vitro* basal cytotoxicity data and rodent $LD_{50}$ values and, in addition, a good correlation (around 70%) was found between *in vitro* basal cytotoxicity data and human lethal blood concentrations. Based on these results, the ACuteTox project is divided into the following scientific work packages (Anonymous, 2006):

– Generation of high quality *in vivo* and *in vitro* databases (which currently contain $LD_{50}$ values from 2206 animal studies and human data from 2902 case reports);

- Iterative amendment of the test strategy, including adapting various cell lines to commercially available high-throughput robotic platforms;
- New cell systems and new end points, such as *in vitro* production of cytokines in whole human blood cultures;
- Role of kinetics (ADE = absorption, distribution and elimination), including further evaluation of *in vitro* and *in silico* models for gut and blood–brain-barrier passage;
- Role of metabolism, including further evaluation of a variety of metabolically active cell systems;
- Role of target organ toxicity, with an emphasis on neuro-, nephro- and hepato-toxicities;
- Technical optimization of the amended test strategy;
- Pre-validation of the test strategy.

At the same time, the pharmaceutical acute toxicity working group expressed its opinion on acute (animal) toxicity results: 'the information obtained from conventional acute toxicity studies is of little or no value in the pharmaceutical development process'. Based on this, the ACuteTox project completed quantitative surveys and summarized the required expert opinion (Seidle, 2007; Chapman & Robinson, 2007; NC3Rs, 2007):

- 100% of respondents found data from acute toxicity studies of little or no use and only used the information in dose setting for other studies in exceptional circumstances;
- 100% of respondents agreed that they would not carry out acute toxicity testing if it were not a regulatory requirement;
- 100% of respondents agreed that acute toxicity studies were not used to identify target organs;
- 100% of respondents never use acute toxicity data to help set the starting dose in man;
- 81% of respondents thought the data obtained from acute toxicity studies was of no use to regulators or clinicians.

In addition to the above opinions, the time requirement, the costs and the ethical issues together urge the scientific community to find a solution for the replacement of animal testing with *in vitro* and *in silico* methods, and develop such a testing strategy where animal tests can be reduced to a minimum. Part of this work should be education and development and convincing experts that the traditional toxicity tests, in spite of their long history, are not the best possible methods for toxicity and, in particular, not for acute toxicity testing.

AltTox summarized arguments against animal tests as follows (AltTox, 2009):

- *Testing methods have not kept pace with scientific progress*: today, there has been a revolution in biology and biotechnology and these developments and opportunities are not reflected by the traditional animal testing methods. Emerging technologies such as bioinformatics, genomics, proteomics, metabolomics, systems biology, and *in silico* (computer-based) systems offer more potential alternatives to animal use.

*In vitro* methods are recommended to use cells, cell lines, or cellular components, in the case of human toxicology, preferably of human origin. The new approach would generate more relevant data, expand the number of chemicals that could be scrutinized, and reduce time, money, and animals involved in testing.

– *Questionable reliability and relevance of current testing methods*, using animals. It is recognized that different species may respond differently to the same substance. It is impossible to know whether the results of testing on rodents, rabbits, or dogs will provide an accurate prediction of toxic effects in humans. There is also much debate concerning the relevance of extrapolating from high doses administered to animals to realistic human or environmental exposure levels. The chronic testing, which goes on partly on elderly animals, may lead to an undue overestimate of toxicity. Despite efforts to standardize procedures, the results of some animal tests can be highly variable and difficult to reproduce.

– *Animal welfare considerations* also play an important role when stressing alternatives. Some conventional toxicity test methods consume hundreds or thousands of animals per substance examined. Toxicity testing causes pain, mainly because evaluation is to be done near death.

– *Time and cost* of conventional tests. These take months or years to conduct and analyze (4–5 years, in the case of carcinogenicity studies) at a huge cost.

– *Legal obligations* prohibit testing on animals where alternative methods are reasonably and practicably available (EEC, 1986).


*Computational toxicology* is a realistic means; its validity depends on the number and quality of existing data. A number of (quantitative) structure–activity relationships (QSAR and SARs), expert systems, and other *in silico* models have been developed for the prediction of various toxicity values. For example, the TOPKAT (Accelrys Inc., Cambridge, UK) models of oral $LD_{50}$ and inhalation $LC_{50}$ in rats and mice have been constructed based on published experimental toxicity data for thousands of substances. Regulatory authorities in Canada, Denmark and the US have reported extensive use of QSAR and SAR models as a basis for interim regulatory classifications of existing and/or new chemicals (Zeeman *et al.*, 1995; Danish EPA, 2001; Cronin *et al.*, 2003; Meek, 2005). The Danish EPA has published statistical results on the degree of accuracy of QSARs of 70–85%, depending on the model used (Danish EPA, 2001).

The OECD also has a project on QSAR development, for estimating the properties of a chemical from its molecular structure and has the potential to provide information on hazards of chemicals, while reducing the time, monetary cost and animal testing currently needed. The aim of the project is to facilitate practical application of QSAR and SAR approaches in regulatory contexts and to improve their regulatory acceptance. The OECD QSAR and SAR project has developed various outcomes such as the principles for the validation of QSAR and SAR models, guidance documents as well as the QSAR Toolbox.

The QSAR Toolbox is a software application designed for governments, industry and other stakeholders to fill the gaps in (eco)toxicity data needed for assessing the hazards of chemicals. This new document specifically provides guidance on how to use the Toolbox to group chemicals to fill data gaps for genotoxicity and carcinogenicity of chemicals (OECD Assessment of Chemicals, 2013c).

## 1.8   Evidence-based toxicology

The new and quick alternative test methods result in numerous toxicological data in addition to the already existing huge amount. A toolbox for the validation of toxicological methods and their results is essential for further development to reach a more dynamic toxicity measuring tool both in human and environmental fields (Hartung, 2010; 2012). This validation concept is also called *evidence-based toxicology* (along the lines of evidence-based medicine) and may be defined as all available data and information is collected from different sources and evaluated and reviewed in a systematic way. It makes possible both the retrospective validation of traditional animal studies and innovative non-animal test methods (OECD, 2005). It makes testing much quicker and less expensive than the traditionally used methodology for measuring and evaluating toxicity. Evidence-based toxicology can be applied for regulatory purposes and for the evaluation of actual environmental exposures endangering humans and the ecosystem.

The first summary of evidence-based toxicology was established at a workshop titled "21st Century Validation for 21st Century Tools" (Rudacille, 2010). It was followed by the formation of the Evidence-Based Toxicology Collaboration (EBTC, 2014). The EBTC's EU branch was officially opened during the 2012 Eurotox conference (Hoffmann, 2012).

# 2   ADVERSE EFFECTS TO BE MEASURED BY ENVIRONMENTAL TOXICOLOGY

Adverse effects of chemical substances are effects which endanger the health and integrity of living organisms and their communities over the short or long term.

Environmental toxicology is the main pillar of environmental policy, and regulates the production, transport and use of any hazardous chemical substances all over the world, whether they are industrial, agricultural or pharmaceutical substances, pesticides, cosmetics or food additives. But adverse effects are much more numerous and complex than those that are covered by regulations today.

Chemical substances with a specified acute effect – which directly affects the members of an ecosystem – may have the same or different chronic effects arising only over the long term. It can appear in the same member of the ecosystem, in other members of the ecosystem or in the whole ecosystem, including humans. This means that quantitative and qualitative changes are going on in the ecosystem: change in the structure, in species diversity, in food chains or food webs. The final output can be the change in the function and productivity of the whole ecosystem, influencing the environment regionally or globally.

Which of the many complex, often unknown ecosystem characteristics should be measured, which changes should be monitored? How can the most characteristic ones be found? Who knows the key points? The answer is that our knowledge is very moderate in this field, the situation is somewhat similar to climate change: we do not understand the system in its totality and the changes over the long term. We cannot decide whether or not short-term or longer-term trends have the same direction, are they part of the normal global ecological trends, and, in addition to the normal trends, does the scale of the changes endanger global equilibrium? Which changes are 'good' and which are 'bad'? On which time scale? And for whom? Human beings tend to place themselves on the top as the most important species, but the local and short-term interest of humans and mankind is not the same as the long-term interest of the world, the habitat of humans and ecosystem. The answer to these questions is as complicated as finding the answer to the main question of humanity: why are humans on the earth? We can only hope that we do not destroy it.

## 2.1 Hazardous effects of chemical substances

In spite of the European REACH and all other regulations such as pesticide, biocide, cosmetics, and pharmaceuticals regulation all over the world, there are many chemical substances used in large quantities without any control. None of the petroleum products are managed using a risk-based approach, and in most of the relevant regulations and standards, no composition is specified for oils, fuels and other petroleum-based products in terms of their chemical composition and hazardous properties.

Wastes are also managed without knowing their chemical composition, in spite of the well-known fact that adverse effects originate from certain chemical substances. Industrial waste is managed in view of its origin, without taking into consideration its exact composition. A good example is the red-mud catastrophe in Hungary in 2010: red mud, the residue of bauxite processing was not considered hazardous waste because there was no differentiation between wet storage of the highly alkaline sodium hydroxide (NaOH) solution and dry storage after neutralization and desiccation. If the chemical composition had been the basis of evaluation, red mud should have been considered as a mixture of sodium hydroxide, a highly corrosive chemical substance and the residue from aluminum ore, with a specific, in many cases harmless, mineral composition.

Household waste is also misjudged when treated as chemically neutral and handled as something with low chemical hazard (i.e. only its hygienic hazard is being considered). What a misunderstanding! Batteries, paints, metal tools, electronic equipment, plastics and everything which is collected non-selectively is accumulated in household garbage. The best recycling performance from households is 30% in the most environmentally conscious countries.

From the unduly underestimated risk of non-regulated or not risk-based regulated hazardous substances, we jump now to substances with known hazards. The necessary use of hazardous substances needs adequate control to manage and reduce their risk to an acceptable level.

The hazards of chemical substances due to their fate and behavior in the environment and their adverse effects on ecosystems and humans will be summarized here. Hazardous chemical substances and their environmental risks have been reviewed in Chapter 2 in Volume 1.

Pesticides and biocides, aimed at killing a target organism, and cosmetics, which are in regular and intensive contact with humans, will be discussed from the point of view of environmental risk. Food ingredients and additives are also important chemical substances and regulated outside REACH by food-centered European regulations.

GHS classifies the adverse effects into 15 classes whose most general and typical toxic effects in or through the environment will be discussed here:

- Acute toxicity;
- Skin corrosion;
- Skin irritation;
- Eye effects: irritation and corrosion;
- Sensitization;
- Mutagenicity;
- Carcinogenicity;
- Reproductive toxicity;
- Target organ systemic toxicity: single exposure and repeated exposure;
- Aspiration toxicity;
- Hazardous to the aquatic environment;
- Acute aquatic toxicity;
- Chronic aquatic toxicity.

REACH and GHS do not aggregate all existing hazards due to the adverse effects of chemical substances. Some others, which may play an important role in environmental and human health risk, are listed here:

- Acute toxicity on the terrestrial ecosystem;
- Chronic toxicity on the terrestrial ecosystem;
- All kinds of toxicity on a single species of wildlife;
- Phototoxicity;
- Endocrine disruption;
- Immune system disruption;
- Nervous system disruption.

## 2.2  Toxic effects of chemical substances

Toxicity or toxic effect is an adverse effect on the health of humans, wildlife or entire ecosystems. Chemical substances with adverse effects may be synthetic xenobiotics or natural chemical substances of mineral, organic or living origin.

The term toxic is used in a broad sense for most of the adverse effects, including acute and chronic toxicity, reprotoxicity, immunotoxicity, neurotoxicity, evolving through inhalation, digestion, dermal contact or whole body contact.

Human toxicity (discussed in Chapter 3) is traditionally based on animal tests. Ecotoxicity deals with the effects of toxic chemicals on ecosystem health and on single ecosystem members. Environmental toxicity integrates both human and ecotoxicity characterized by qualitative and quantitative tools and considers the interactions between the two.

According to Paracelsus' theory, toxicity is a dose-related phenomenon, and up to the present time, we define toxicity and the scale of toxicity based on limit doses or concentrations.

The consequences of classification of a chemical substance as hazardous are greater than just a hazard label; it also has a direct effect on the management of associated risks.

*Table 1.1* Toxic classes according to $ED_{50}$ and $EC_{50}$ upper limit values (GHS, 2007).

| Exposure route | Category 1 | Category 2 | Category 3 | Category 4 | Category 5 |
|---|---|---|---|---|---|
| Oral: $LD_{50}$ mg substance/kg of bw | 5 | 50 | 300 | 2000 | 5000 |
| Dermal: $LD_{50}$ mg substance/kg of bw | 50 | 200 | 1000 | 2000 | 5000 |
| Gas inhalation: $LC_{50}$ mg/kg (in volume) | 100 | 500 | 2500 | 20,000 | No |
| Vapor inhalation: $LC_{50}$ mg/L | 0.5 | 2.0 | 10 | 20 | No |
| Dust and mite inhalation: $LC_{50}$ mg/L | 0.05 | 0.5 | 1.0 | 5.0 | No |

No: no classification; bw: body mass

Classification and labeling of chemicals is regulated globally based on their hazard, in a harmonized way (GHS, 2007).

According to the definition of REACH, a substance fulfills the toxicity criterion when:

- an acute lethal effect may arise following oral, dermal or inhalation exposure, and is classified as toxic;
- a substance is classified as carcinogenic, mutagenic, or toxic for reproduction;
- there is evidence of chronic toxicity; its short-term or long-term toxicity for marine or freshwater organisms is under the threshold of $EC_{50} = 0.1$ mg/L and the NOEC $= 0.01$ mg/L respectively.

Table 1.1 shows the doses and concentrations of toxic substances, which cause an $LD_{50}$ or $LC_{50}$ value, namely the dose or concentration that kills half of the affected organisms. Table 1.2 shows the toxicity criteria for the water ecosystem.

As can be seen in Table 1.1, toxicity is split into five categories of severity where Category 1 requires the least amount of exposure and Category 5 requires the greatest exposure to be lethal, making it the least dangerous category.

$EC_{50}$ is determined from short-term aquatic toxicity data of key indicator species such as fish, crustaceans and algae or aquatic macroplants, and the lowest-effect concentration is selected and compared with the concentrations in Table 1.2. Fish are exposed for 96 h and crustaceans for 48 h and the results for algae are given in $E_rC_{50}$, based on the reproduction (growth) curve of the algal culture. NOEC or $EC_{10}$ values from long-term toxicity tests are used to determine chronic aquatic toxicity for classification purposes.

The European CLP criteria take also degradability into account when categorizing aquatic toxicity: chronic exposure to a readily biodegradable substance is categorized less strictly, but for those substances for which adequate chronic toxicity data are not available the categorization is stricter: $EC_{50} \leq 1$ mg/L belongs to Cat.1 (similar to categorization based on acute toxicity).

Photodegradation and hydrolysis are generally known physicochemical phenomena, but biodegradability should be measured in standardized test methods. Substances

*Table 1.2* Toxic to aquatic ecosystem: categories based on $EC_{50}$, $E_rC_{50}$ or NOEC values (CLP, 2008).

| Exposure | Category 1 | Category 2 | Category 3 |
|---|---|---|---|
| Acute (mg/L) | $\leq 1.0$ | – | – |
| Chronic, non-rapidly degradable (mg/L) | $\leq 0.1$ | $\leq 1.0$ | $\leq 10.0$ |
| Chronic, rapidly degradable (mg/L) | $\leq 0.01$ | $\leq 0.1$ | $\leq 1.0$ |
| Chronic, adequate chronic toxicity data are not available (mg/L): non-rapidly degradable, BCF > 500, acute toxicity results are available | $\leq 1.0$ | $\leq 10$ | $\leq 100$ |

are considered rapidly degradable in the environment if at least the following level of degradation is achieved within 10 days of the start of degradation in a 28-day ready biodegradation study:

– 70% in tests based on dissolved organic carbon;
– 60% of the theoretical maximum in tests based on oxygen depletion or carbon dioxide generation.

Bioaccumulation of substances within aquatic organisms can give rise to toxic effects over longer time scales even when actual water concentrations are low. For organic substances the potential for bioaccumulation will normally be determined by using the octanol–water partition coefficient, $K_{ow}$. A cut-off value of $\log K_{ow} \geq 4$ identifies the substances with a real potential to bioconcentrate. Of course an experimentally determined BCF provides a better measure. A BCF of $\geq 500$ in fish is indicative for classification purposes.

The European CLP introduces a "safety net" classification (category Chronic 4) for use when available data do not allow classification but there are some grounds for concern.

Category 1 substances with acute or chronic aquatic toxicity contribute as components of a mixture to the toxicity of the mixture significantly. This increased weight is considered by a multiplying factor, called M-factor.

$LC_{50}$ and NOEC values of chemical substances and their hazard-based classification can be used for the risk management of chemicals when their environmental concentration is also known.

When testing *environmental samples* and not pure chemical substances, the qualitatively measured toxicity directly indicates the extent of the risk of the environmental sample to the test organism, even if the chemical composition and concentration of the contaminants is not known. Extrapolation from single species to the whole ecosystem is also possible by using uniform testing and extrapolation methodologies.

Although the actual contaminants have not been identified and their concentration in environmental samples is not known, the no-effect dose or $ED_{50}$ value can be determined for water, soil or sediment samples, which gives the amount of sample in grams or liters that caused 50% lethality or a 10%, 20%, or 50% decrease in any measured end points. This measured effective or no effect dose is methodology-dependent because only 1 g of sample is used in one test, but 5 or 50 g is used in another. These kinds of test results need further interpretation for decision making. The basis of interpretation can be a kind of 'calibration' of the test method by comparing the toxicity of the environmental sample with the toxicity of known substances. As a result of this kind of quantitative calibration, the effective water or soil dose (ED) can be expressed as a substance-equivalent toxicity in concentration and as such, can better fit into a conventional risk assessment procedure. For example, the toxicity of soil contaminated with a mixture of metals is given in copper equivalents as if copper caused the entire toxicity. Another type of interpretation for toxicity results of environmental samples is verbal characterization of toxicity according to the scale of inhibition, for example, very toxic, toxic, moderately toxic and non-toxic. A third plausible mode for characterizing the measured toxicity of environmental samples is the necessary rate for achieving a "no effect" dilution. This interpretation is a good indication for the scale of the necessary risk reduction to achieve a "no risk" situation in the environment.

*Xenobiotics* are substances foreign to an entire biological system. Their interaction with living organisms may lead to the impairment of the organisms' normal function and health. Xenobiotics can be metabolized and secreted from the body, but they may be bound to essential molecules, enzymes or receptors of the cells and organs, resulting in damage to the genetic, biochemical, physiological, or behavioral soundness of the exposed organism. Organisms can adapt to xenobiotics; they (first of all the microbes) can learn to utilize, eliminate, tolerate or resist xenobiotics. The mechanism of xenobiotic metabolism is important both in medical practice and in the environment, and indicators such as glutathione S-transferases can be used for the exploration of both drug effect mechanisms and pesticide resistance.

*Mixtures of chemical substances* may interact with each other either in the environment or in the body. Their interaction may result in changes in the common adverse effect which may be additive, synergic or antagonistic. When estimating the risk of mixtures, it is necessary to know the type of interaction of the chemicals present.

Not only the effect of different chemicals but also the intakes through different *exposure routes* may be added together. A person who is exposed to contaminated air, contaminated water and soil at the same time, may inhale contaminating chemical substances, take them in orally or through his skin. Terrestrial and aquatic plants and animals live in the contaminated environmental matrix and are in contact with it through their whole body, digestive system and inhalation. In this case the response includes all exposures via different routes.

The details of the realization of the toxic effect are explored by toxicology, ecotoxicology and toxicokinetics.

*Toxicokinetics* studies the uptake and the further fate and interactions of the toxic substances in the organism's body, organs or cells, while toxicology mainly deals with the result, the end point of adverse effects and the dose–response relation. The steps of toxicokinetics are absorption, distribution, biotransformation and excretion. Absorption describes the entrance of the chemical substance into the body, through the air, water, food, or soil. Once a chemical is inside a body, it is distributed to other areas of the body through diffusion or biological transport. The chemical substance may be biotransformed into other chemicals; these are metabolites. These metabolites can be more toxic than the parent compound. After biotransformation, the metabolites may leave the body, may be transformed into other compounds, or continue to be stored in the body compartments. In many cases, the presence of the toxic substance is not a simple storage, but it may act biologically, evoking a false biological response of the endocrine, immune or nervous systems. Toxicokinetics tries to find relationships between exposures in animal toxicity tests and the corresponding exposures in humans. Some other branches of toxicology, such as genetic toxicology, reproductive, endocrine, immune and neurotoxicology will be investigated in detail under human toxicology.

*Toxicogenomics* is a new branch of toxicology, dealing with the collection, interpretation, and storage of information about the response of genes and proteins within the cell, tissue or organ of an organism when affected by a toxic substance. It combines toxicology with genomics, and other closely related new molecular technologies such as transcriptomics, proteomics and metabolomics. Toxicogenomics endeavors to explore the molecular mechanisms of toxic effects and to find connections between toxic effects and molecular biomarkers, which would indicate toxic effects at an early stage, making early warning possible in both medical and environmental practice.

*Ecotoxicogenomics* is the challenge of integrating genomics into aquatic and terrestrial ecotoxicology. Snape *et al.* (2004) formulated ecotoxicogenomics and its importance in understanding the response of ecosystems and proposed the term ecotoxicogenomics to describe the integration of genomics (transcriptomics, proteomics and metabolomics) into ecotoxicology. Ecotoxicogenomics is defined as the study of gene and protein expression in non-target organisms that is important in responses to environmental toxicant exposures. Instead of whole-organism responses (e.g. mortality, growth, reproduction), we can use a much more sensitive indicator for identifying chemicals of potential toxicants. To find the proper molecule to indicate the toxic potential of contaminating substances, we have to understand the effect mechanisms and the interactions between ecological levels and in the population. Snape *et al.* (2004) hope that ecotoxicogenomic tools may provide us with a better mechanistic understanding of aquatic ecotoxicology.

*Microbial ecology* and handling the microflora of water, soil, sediment or sludge as a community with a metagenome may help in understanding and following the differences and changes due to environmental conditions and differentiate spatial, climatic and seasonal changes from the adverse effects of contaminants.

Microbial species and communities have specific functions in ecosystems, in nutrient cycling and contaminant elimination. Identification of microorganisms by traditional microbiological methods is often difficult because only a minority of the species within the system will grow on artificial media. DNA techniques help to overcome this problem and to extend our knowledge on species diversity because non-cultivable species are also detected.

Soil microbial ecology has already found some interesting relationships, for example, the relative stability of the metagenome of soil microflora was found to be unrelated to temperature or latitude, which strongly determines plant or animal diversity. The most influential factors in soil are pH, redox potential, available energy substrates and contaminants (Jeffery *et al.*, 2010).

The study of the metagenome needs special molecular techniques, particularly DNA techniques, an adequate amount of historical or background data and a suitable statistical tool for the evaluation.

*Metagenomics*, in short, is the science of biological diversity. Technically, it is the study of nucleotide sequences, structure, regulation and function within the complex ecosystem. An environmental system such as the soil can be characterized by the DNA from all of the organisms living in the community. We can extract total DNA and study it directly or after cloning it into a host cell, for example *Escherichia coli*, to express the information coded in the cloned metagenome in the form of DNA (during propagation of *E. coli*) or in the form of RNAs and proteins (during the life of the cells). We can create a genetic library from the metagenome of the soil, and analyze it as a whole (characteristic DNA fingerprint), based on similarity (comparison with known fingerprints) or for the identification of specific genes, their presence and frequency.

Metagenomics represents a powerful tool to qualify and quantify the biodiversity of native environmental samples. It can characterize the genetic diversity in samples regardless of the availability of laboratory culturing techniques, or knowledge of the microorganisms present. The metagenomic DNA from the library, prepared for example from soil, can be analyzed by PCR techniques, using DNA probes and DNA chips or by the newly developed technique of DNA-SIP (Stable Isotope Probing). The problem of evaluation and interpretation may be the same as in the case of diversity assessments, namely what should be considered as reference. The advantage of metagenomics is that the number of tests is not limited, so one can generate temporal and spatial series for identifying concentration-dependence, trends and gradients.

The **DNA–SIP method** is based on the utilization of a substrate labeled with stable isotope of carbon ($^{13}$C). Those cells that are active and able to utilize the $^{13}$C-labeled substrate and proliferate, build the carbon isotope into their DNA, which is easily detectable after isolating the metagenome. The food chains or other trophic communities can also be mapped by this method.

**Metatranscriptomics** are the gene products of the metagenome, namely DNAs or RNAs synthesized from genes and intergenetic regions of DNA. Metatranscriptomics offers the opportunity to reach beyond the community's genomic potential as assessed in DNA-based methods, towards its in situ activity, meaning not only the presence but also the activity (transcription) of the genes.

Analyses by Shi *et al.* (2009) showed that metatranscriptomic data sets can reveal new information about the diversity, taxonomic distribution and abundance of small RNAs (sRNAs) in naturally occurring microbial communities and indicate their involvement in environmentally relevant processes including carbon metabolism and nutrient acquisition. In their published research, a large fraction of cDNA sequences (DNA synthetized *in vitro* from messenger RNA) were detected, which are partly identical to well-known sRNA, and partly to new groups of previously unrecognized putative sRNAs (psRNAs). These psRNAs mapped specifically to intergenic regions of microbial genomes recovered from similar habitats, displayed characteristic conserved secondary structures and were frequently flanked by genes that indicated potential regulatory functions. Depth-dependent variation of psRNAs generally reflected known depth distributions of broad taxonomic groups, but fine-scale differences in the psRNAs within closely related populations indicated potential roles in niche adaptation.

**Metaproteomics**, also called environmental proteomics, or community proteogenomics, is based on the analyses of expressed proteins by the metagenome. The proteins are extracted from the soil, then analyzed by mass spectrometry after electrophoretic or chromatographic separation. Metaproteomics supports the understanding of the microbial communities and the relation of community composition to its function.

Schulze *et al.* (2005) applied mass spectrometry-based proteomics to analyze proteins isolated from Dissolved Organic Matter (DOM) from soil. The focal question was to identify extracellular enzymes important in the carbon cycle. Proteins were classified using the National Center for Biotechnology Information (NCBI, 2013) protein and taxonomy database. They found that 78% of proteins in a lake but less than 50% in forest soil DOM originated from bacteria. In a deciduous forest, the number of identified proteins decreased from 75 to 28 with increasing soil depth and decreasing total soil organic carbon content. The number of identified proteins and taxonomic groups was 50% higher in winter than in summer. Cellulases and laccases were found among proteins extracted from soil particles, indicating that degradation of soil organic matter takes place in biofilms on particle surfaces. These results demonstrate that the 'proteomic fingerprint' is suitable to prove the presence and activity of organisms in an ecosystem.

## 2.3   Carcinogenic effects

*CMR substances* are chemical substances with carcinogenic, mutagenic or toxic effects to reproduction. The abbreviation CMR is used for the characterization of chemical substances with these properties within the REACH regulation.

A *carcinogenic effect* may be caused by a substance or a mixture of substances that induces cancer or increases its incidence and/or malignancy, or shortens the time to tumor occurrence. Cancer may stem from the ability to damage the genome or to disrupt the cellular metabolic processes. Carcinogenic chemicals are conventionally divided into two categories according to the assumed mode of action. Non-genotoxic modes of action include epigenetic changes, i.e. effects that do not involve alterations in DNA but which may influence gene expression, altered cell–cell communication, or other factors involved in the carcinogenic process.

*Cancer* is a disorder of the cells, characterized by the lack of programmed cell death, which is responsible for destroying damaged cells. If the pathway is damaged, the cell cannot prevent itself from becoming a cancer cell. Carcinogens induce the uncontrolled, malignant division of cells, ultimately leading to the formation of tumors. The objective of investigating the carcinogenicity of chemicals is to identify potential human carcinogens, their mode(s) of action, and their potency.

Once a chemical has been identified as a carcinogen, there is a need to elucidate the underlying mode of action, i.e. whether the chemical is directly genotoxic or not. For genotoxic carcinogens, it is assumed that usually there is no discernible threshold and that any level of exposure carries a risk. For non-genotoxic carcinogens, no-effect thresholds are assumed to exist and to be discernible. Human studies are generally not available to make a distinction between the above-mentioned modes of action, and a conclusion on this, in fact, depends on the outcome of mutagenicity testing and other mechanistic studies. In addition to this, animal studies may also inform on the underlying mode of carcinogenic action.

GHS classifies carcinogens into two categories, of which the first may be divided again into subcategories if so desired by the competent regulatory authority:

- Category 1: known or presumed to have carcinogenic potential for humans;

  o   Category 1A: the assessment is based primarily on human evidence;
  o   Category 1B: the assessment is based primarily on animal evidence;

- Category 2: suspected human carcinogens.

The cancer hazard and mode of action may also be highly dependent on exposure conditions such as the route of exposure. Therefore, all relevant effect data and information on human exposure conditions are evaluated.

A list of carcinogenic substances is given in Chapter 1.2 in Volume 1.

## 2.4  Mutagenic effects

*Mutagenic substances* are those which induce mutations in living cells. Mutagenicity refers to the induction of permanent transmissible changes in the amount or structure of the genetic material of cells or organisms. These changes may involve a single gene or gene segment, a block of genes or chromosomes.

*Alterations to the genetic material* of cells may occur spontaneously or be induced as a result of exposure to ionizing or ultraviolet radiation or genotoxic chemical substances. In principle, human exposure to substances that are mutagens may result in increased frequencies of mutations above the baseline. Heritable damage to offspring, and possibly to subsequent generations, of parents exposed to substances that are mutagens may follow if mutations are induced in parental germ cells (reproduction cells). Mutations in somatic cells (cells other than reproduction cells) may be lethal or may be transferred to daughter cells with deleterious consequences for the affected organism. There is considerable evidence of a positive correlation between the mutagenicity of substances *in vivo* and their carcinogenicity in long-term studies on animals. The aims of testing for mutagenicity are to assess the potential of substances to induce effects, which may cause heritable damage in humans or lead to cancer.

## 2.5  Reprotoxicity

*Reprotoxics* are chemical substances which may cause reproductive toxicity. Reproductive toxicity by the definition of REACH includes adverse effects on sexual function and fertility in adult females and males as well as developmental toxicity in the offspring:

– Adverse effects on sexual function and fertility;
– Adverse effects on development;
– Adverse effects on or via lactation.

*Reproductive toxicity* is clearly of great concern because the continuance of the human species is dependent on the integrity of the reproductive cycle. It is characterized by multiple diverse end points, such as impairment of female and male reproductive functions or capacity (fertility), induction of non-heritable harmful effects on the progeny (developmental toxicity) and effects on or mediated via lactation.

The objectives of assessing reproductive toxicity are to establish:

– Whether exposure of humans to the substance of interest has been associated with reproductive toxicity;
– Whether, on the basis of information other than human data, it can be predicted that the substance will cause reproductive toxicity in humans;
– Whether the pregnant female is potentially more susceptible to general toxicity;
– The dose–response relationship for any adverse effects on reproduction.

A complete list of the EU-classified substances is given in Annex VI of the CLP regulation (CLP, 2008b).

## 2.6  Persistent and very persistent substances

*Persistent and very persistent substances* are substances that persist in the environment for a long time. The reason for persistence is that the substance cannot be degraded by light or other radiation, heat, oxygen, water or moisture or by biological effects. Many of the xenobiotics are designed to be persistent in the environment (similar to drugs in the body), otherwise the amount to be applied and, as a consequence, the cost of the substance would be very high, and the efficiency, due to too short a contact time and unstable concentration, would be limited.

According to the REACH regulations, persistent substances fulfill the following criteria:

– The half-life in marine water is higher than 60 days or
– the half-life in fresh- or estuarine water is higher than 40 days or
– the half-life in marine sediment is higher than 180 days or
– the half-life in fresh- or estuarine water sediment is higher than 120 days or
– the half-life in soil is higher than 120 days.

Very persistent substances are those chemical substances that fulfill the following criteria:

– The half-life in marine, fresh- or estuarine water is higher than 60 days or
– the half-life in marine, fresh- or estuarine water sediment is higher than 180 days or
– the half-life in soil is higher than 180.

Hazard and persistency together increase the risk of the substance, as ecosystems or humans are exposed for longer times to the substance, so it has higher chance to cause harm.

## 2.7  Bioaccumulative and very bioaccumulative substances

*Bioaccumulative and very bioaccumulative substances* are those which are able to concentrate in the body of living organisms of microbial cells, plants or animals, including man. Bioconcentration is measured relative to the environment and is quantitatively characterized by the bioconcentration factor (BCF), which is the ratio of two concentrations, the concentration in the organism or organ and the concentration in the environmental compartment.

BCF plant is $C_{plant}/C_{soil}$, and BCF fish is $C_{fish}/C_{water}$. Bioaccumulation of certain substances, e.g. hydrophobic organic substances in liver or fat tissue leads to the corruption of the food chain. Bioaccumulation of inorganic substances such as toxic metals – lead, cadmium, or mercury – in plant shoots and leaves poisons the consumers, and may cause biomagnification along the food chain. The mode and rate of bioaccumulation along food chains and within food webs depends not only on the characteristics and environmental concentration of the chemical substance, but also on the genetic potential and the biochemical mechanism developed by the living organisms mainly as a defensive reaction (low accessibility deposits outside the cell membrane, in the cell membrane or inside the cell).

According to REACH, a substance fulfills the bioaccumulative criterion when its bioconcentration factor (BCF) is greater than 2000, and very bioaccumulative when greater than 5000. According to the REACH Regulations, the assessment of bioaccumulation shall be based on measured data on bioconcentration in aquatic species. Data from freshwater as well as marine water species can be used. In CLP regulation, a cut-off value of $\log K_{ow} \geq 4$ is intended to identify only substances with a real potential to bioconcentrate. If a BCF value e.g. in fish is available, a BCF $\geq 500$ is indicative for classification purposes. Relationships can be observed between chronic toxicity and bioaccumulation potential, as toxicity is related to body burden.

It can be seen from the hazard classes that chemical substances are not classified according to their chemical structure and properties, but according to their effects. Nevertheless, the grouping of hazardous chemical substances in environmental science and practice is a mixed system: sometimes the hazard has priority such as in the REACH regulations, which is a hazard- and risk-based system, but in other cases, physico-chemical character and behavior are more important, so that some groupings are based on chemistry (chlorinated solvents, PAH, PCB, etc.), origin (petroleum products, non-ferrous ore mining, etc.) or on the use of the chemical substance (plant protection products).

## 2.8  Emerging pollutants

*Emerging pollutants* are hazardous chemical substances that have recently been discovered in the environment, such as endocrine or immune disruptors, sensitizers and allergic compounds. Strictly speaking these are contaminants of emerging concern, also abbreviated as CECs. Their presence, frequency of occurrence, and sources are generally not known. Many pharmaceuticals, personal care products (PPCPs) and well-known contaminants are amongst them due to newly recognized adverse impacts different to their target effects. Emerging pollutants are chemicals or materials of interest that can be characterized by:

–   Perceived or real threat to human health or the environment;
–   Lack of a clear dose–response relationship;
–   Significant individual differences in sensitivity and the realization of effects;
–   Their effects depend on age, life-stage, season and other environmental conditions;
–   Lack of published health standards or an evolving standard;
–   Contaminants may also be 'emerging' because of the discovery of a new source, a new pathway to humans, or a new detection method or technology.

These chemical substances are discussed in detail in Chapter 2 of Volume 1, only a summary on the more frequent endocrine disruptors is given here:

–   An *endocrine disruptor* interferes with the synthesis, secretion, transport, binding, action, or elimination of natural hormones in the body, which are responsible for the maintenance of homeostasis, reproduction, development and/or behavior.

*Table 1.3* Some 'emerging pollutants' in the Hungarian course of the Danube: screening based on substance quantities and hazards (Molnár et al., 2013).

| Ranking | Chemical substance | Hazard-based score | Type of substance |
|---|---|---|---|
| 1 | Bisphenol A | 57 | Industrial chemical |
| 2 | Sulfamethoxazole | 55 | Antibiotic drug |
| 3 | Metamizole (sodium salt) | 53 | NSAID |
| 4 | Diuron | 52 | Pesticide |
| 5 | Nonylphenol | 51 | Surfactant |
| 6 | Urethane/ethyl carbamate | 51 | Fermentation product |
| 7 | Carbamazepine | 51 | Antidepressant |
| 8 | Doxorubicin | 50 | Anticancer antibiotic |
| 9 | Bis(tributyltin) oxide | 50 | Pesticide |
| 10 | Paracetamol | 50 | NSAID |
| 11 | Aminophenazone | 49 | NSAID |
| 12 | Nicotine/cotinine | 49 | Luxury drug |
| 13 | Verapamil | 46 | Calcium-antagonist drug |
| 14 | Carboplatin | 46 | Anticancer drug |
| 15 | Gemfibrozil | 46 | Anti-cholesterol and triglyceride lowering drug |
| 16 | Diclofenac | 45 | NSAID |
| 17 | 17ß-estradiol | 44 | Hormone: contraceptive and therapeutic agent |
| 18 | Ethinylestradiol | 44 | Hormone: contraceptive and therapeutic agent |
| 19 | Di(2-ethylhexyl)-phthalate | 42 | Plasticizer |
| 20 | Dibutyl phthalate | 42 | Plasticizer |
| 21 | Trifluralin | 41 | Pesticide |
| 22 | Progesteron | 40 | Hormone: therapeutic agent |
| 23 | Ketoprofen | 39 | NSAID |
| 24 | Naproxene | 38 | NSAID |
| 25 | Ibuprofen | 36 | NSAID |

NSAID: non-steroidal anti-inflammatory drug

- *Immune disruptors* affect the immune system via endocrine disruption or directly.
- *Sensitizers and allergens* are chemical substances that may develop an allergic reaction in normal tissue generally after repeated exposure. Sensitization is a causeless immune response.

Table 1.3 shows the results of a preliminary score-based risk assessment on emerging pollutants in the Hungarian course of the Danube. The risk score was created based on produced and used amounts; physical, chemical and environmental fate properties (partitioning, (bio)degradability, bioaccumulative potential); known or suspected adverse effects on the environment and human health (toxicity, mutagenicity, reprotoxicity, endocrine and immune disruption) as well as occurrence/measured concentrations in the Danube or in the waste water discharged to Danube. The available quantitative results were converted into scores by assigning ranges/classes created by experts.

Nearly one hundred suspected chemicals were assessed and ranked by the scores created from mixed information. The second tier, the quantitative environmental risk assessment of the priority chemicals modified the picture. For example, the risk posed on the Danube ecosystem ($RCR_{aq}$) by bisphenol A is 0.009, far below the threshold of $RCR = 1$. On the other hand diclofenac, based on several sampling, chemical analysis and ecotoxicity tests carried out by the authors, has an aquatic $RCR_{aq} = 1$–5, indicating high ecological risk.

## 3    INTERACTION OF A CHEMICAL SUBSTANCE WITH LIVING ORGANISMS

The effect of a chemical substance and its measurement is based on the interaction between the chemical substance and the living organism.

*Testing in the lab* makes the test conditions controllable, i.e. quality and quantity of the materials, time and temperature. The test can be performed at an equilibrium state or the changes can be followed over time.

*Field assessment* or monitoring is accompanied by uncertainties due to environmental circumstances which are uncontrolled and uncontrollable by the assessor, for example, the continuous transport and fate processes, the interactions between

– a few different contaminants;
– the contaminant and the matrix of the medium;
– the contaminant and the biota.

Non-equilibrium conditions in the environment make the evaluation indeterminate because the instantaneously measured values cannot be placed on the time curve of the ongoing processes. It must be decided whether laboratory testing under controlled conditions but low relevance for the environment or environmental testing with high relevance but low reproducibility and low interpretability fit better with the actual case or problem. A suitable micro- or mesocosm can unify the advantages of the two test concepts.

Regardless of the testing scenario used – laboratory bioassay, microcosm or field testing – the cause of the effect of a chemical substance to a living organism is the same: the interaction between the chemical substance and the test organism. The consequences of this interaction yield the end points needed for environmental toxicity testing, as shown in Figure 1.3.

The *first stage* in the process is the fate, transport and distribution of the material in the environment after its emission into the environment and before meeting the organism. The degradation or other modification of the chemical structure influences the effect which arises later on when meeting the living organism. Photolysis, hydrolysis, or enzymatic degradation of the chemical substance may take place in the environment without or before reaching the ecosystem members. Xenobiotics often mimic natural substances, which leads to their biodegradation in the environment by water-, soil- or sediment-living microorganisms.

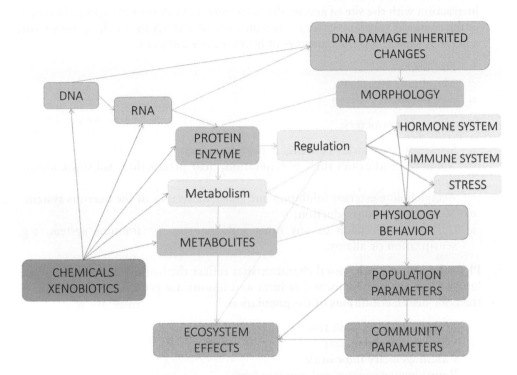

*Figure 1.3* Ecosystem effects and indicators/biomarkers of environmental toxicology.

The **second stage** is when the chemical substance reaches the site of action, i.e. a certain part of the living organism: a molecule, a receptor site or any biochemical compartment of the organism.

The **last stage** is the consequence of the second, and this is the interaction between the chemical substance and the site of action. The impact of this interaction can materialize at molecular, cellular, organ, organism, population, community or ecosystem levels.

The indicators of the interaction between chemical substances and the biota are summarized using the system of Landis and Yu (1999). These indicators can be used both for biomarkers and end point measurement when a toxicity test is being designed. Indicators of the interaction between a chemical substance and the ecosystem:

–  Biotransformation after introduction of the chemical substance/xenobiotic into the environment, and before meeting the site of action:

o  Induction of hydrolase enzymes;
o  Oxidation in the presence of mixed-function oxidases;
o  DNA repair enzymes;
o  Biotransformation;
o  Biodegradation;
o  Bioaccumulation.

–   Interaction with the site of action: this is the molecular-level interaction (bonding) of the chemical substance with the receptor site of a DNA molecule, a membrane, a nerve synapsis, specific receptors of hormones or antigens.

   o   DNA/RNA;
   o   Membrane receptors;
   o   Key enzymes.

–   Biochemical parameters:

   o   Stress proteins such as adrenalin, noradrenalin;
   o   Metabolic indicators such as respiration, heat production, substrate utilization;
   o   Acetylcholine esterase inhibition and other molecules of the nervous system;
   o   Metallothionein production;
   o   Immune suppression or any other influence on the immune system, e.g. sensitization or allergy.

–   Physiological and behavioral characteristics reflect the healthy state of the population. Tumors, developmental failures and lesions are typical markers showing the poor health conditions of the population.

   o   Mutagenic effect indicators;
   o   Chromosomal damage;
   o   Carcinogenicity indicators;
   o   Reproductive success and reprotoxicity;
   o   Developmental toxicity indicators;
   o   Lesion and necrosis;
   o   Mortality;
   o   Behavioral alterations, movement;
   o   Compensatory behaviors, migration, escape.

–   Population parameters such as size and structure, including age-structure of the population are closely related to environmental conditions and stress.

   o   Population density;
   o   Productivity;
   o   Mating success;
   o   Alterations in genetic structure measured by DNA techniques;
   o   Competitive alterations, e.g. benefit from accepting a xenobiotic substance as a substrate.

–   Community parameters, particularly species composition and community functioning, are characterized and compared. The problem in community characterization is that we do not know what is optimal for a certain locality and time.

   o   Community structure;
   o   Diversity: number of species and their relative abundance characterized by diversity indices;
   o   Existence and role of minor components of the community;

o  Energy transfer efficiency;
o  Dynamic character of the community;
o  Stability;
o  State of succession;
o  Chemical parameters.

All of these parameters together constitute the effect of the chemical substance on the ecosystem as a whole. The combined effect of chemical substances or other stress factors changes the composition and metabolism of an ecosystem and results in the endangerment and extinction of species and consequently the complete change of food chains and food webs, or the dominance of certain species, as in the case or eutrophication, acid rain or oil spills. These factors together with climate change and the accelerated evolution of the microflora in the environment undeniably increase the rate of ageing of the earth's ecosystem. Many of the changes in the ecosystem are irreversible and the constantly moving target of the ecosystem towards reaching equilibrium and maintaining homeostasis makes the effects non-proportional to the causes, and the system cannot be recognized in full detail. This is the reason for the simplification of ecotoxicity to the toxicity of hazardous substances on certain representative organisms or on a well-defined microcosm system. The extrapolation from this result to the whole ecosystem may not be able to truly model and predict the long-term history of the ecosystem in detail, but is no worse from the point of view of statistics than a field assessment and could be a useful tool for the management of the environment and hazardous chemicals.

## 3.1  Dose–response relationship

The well-known and widespread chemical models used for the management of the environment are based on the connection between the dose or concentration of the chemical substance and its effect. An increasing dose mixed into an experimental rat's feed causes an increase in the effect. An increasing concentration series of a tested substance in water shows gradually stronger effect on fish, plant or crustacean living in that water.

Paracelsus (1493–1541), the renaissance physician, botanist, alchemist, astrologer, and general occultist, published his experimental results that every chemical substance is poison, and the 'dosage makes the distinction between poison and remedy'. Based on this knowledge, he used low concentrations of mercury to cure syphilis. The same is true for many of the environmental contaminants, e.g. metals, semimetals and other natural chemical compounds, and man-made pharmaceutical and biocide substances.

Plotting any of the biomarkers, relevant for the actual measurable effect, against the dose or concentration, an S-shaped sigmoid is obtained in each case, as shown in Figure 1.4.

In Figure 1.4, the measured end point is the luminescent light of the test organism, the light emitting bacterium, *Vibrio fischeri*, which gives a concentration-dependent response by emitting less light in the presence of toxic chemicals than normal. In the experiment shown in the graph, the toxic chemical substance is copper, and the light emission of the bacterial suspension was measured in the presence of different copper concentrations between 5 and 200 mg/kg.

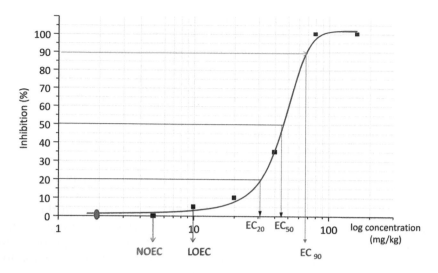

*Figure 1.4* Concentration–response curve: inhibition of light emission (H%) of luminobacterium as a function of Cu concentration in a dilution series.

The end point of the measurement – depending on the test organism and its response – can be any of the formerly listed biochemical, physiological, behavioral, population or community parameters. In the turning point (inflection point) on the curve, the response changes from a proportionally increasing phase to a saturation phase, where the concentration increase of the toxic agent cannot trigger a proportional response, but only a smaller one due to the damage (inactivation, death) of the test organism.

If the test organism is a bacterial strain or a plant or small insect, our goal is rarely to protect this particular species based on the results, but instead to try to extrapolate from the single species test results to the entire, more complex, ecosystem. Based on experience and known correlations, the rule is that the test results from species at a minimum of three trophic levels should be used, and depending on the number of tested species, test type and duration, safety factors or a probabilistic estimation from species sensitivity distribution (SSD) should be applied. The prediction of the effect (or no-effect) concentrations of the chemical substances for the ecosystem in question can be used for decision making or as an environmental quality criterion. The predicted no-effect concentration (PNEC) of a chemical substance e.g. for an aquatic ecosystem can be estimated based on the result of one algal, one crustacean and one fish species using factorial extrapolation. No-effect estimates for terrestrial ecosystems require test results on bacterial and plant species, and on the members of soil-living micro- or mesofauna. The value of the assessment factor depends on the type and duration of the tests: having three acute test results the assessment factor is 1000, but having one, two or three chronic test results instead of acute ones, the factor goes down to 100, 50 and 10, respectively.

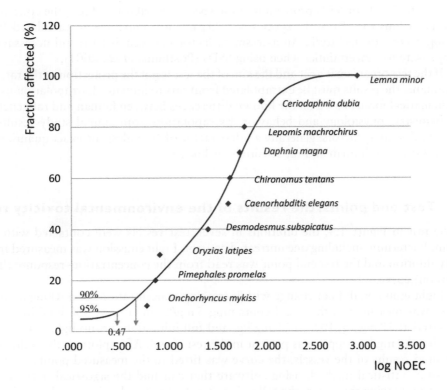

*Figure 1.5* SSD curve drawn from nonylphenol NOEC data (Nonylphenol, 2013).

The species sensitivity distribution (%) can be plotted if results from a minimum of 6–8 (preferably more) representative species are available: one should plot the proportion of species affected as a function of the logarithm of stressor concentration or dose (Posthuma *et al.*, 2002; Solomon, 2008; SSD, 2013). The log–normal distribution of the measured ecotoxicological end points is a prerequisite for this kind of extrapolation. The predicted no-effect, or protective concentrations (arbitrary protective level of 95%, 90%, 50%, etc.), can be read from the SSD curve as shown by the example of nonylphenol aquatic toxicity in Figure 1.5.

SSD curves are generally plotted based on $EC_{50}$, a rather strong end point. NOEC or LOEC values would serve the protective concept better. Laboratory ecotoxicity tests are based on the deterioration – on a response of the adversely affected organisms – while a real ecosystem assessment is based on the survival/presence of species. Species actually endangered and lowered in number cannot be accounted for by a field assessment. To harmonize the laboratory-based SSDs with the results of field assessments, for the latter case one should derive the difference of the expected (reference) and the counted species numbers and their distribution. This difference is comparable with the "affected proportion" of species read from the SSD curve based on laboratory test results. An SSD curve prepared from 10 chronic end points of nonylphenol NOEC is shown in Figure 1.5. One can read from the curve the desired target concentration assigned to 50, 90 or 95% protection of the species distribution. $HC_5$, the concentration hazardous for 5% of the species is 2.93 µg/L for nonylphenol, calculated from the log NOEC of 0.47 µg/L. An assessment factor between 5–10 is still necessary to compensate for uncertainties when using SSDs (Posthuma *et al.*, 2001).

If the test organism is a rat and the aim of the test is not the protection of the rat, but of humans, the results must be extrapolated from rats to humans. Extrapolation uses a mathematical model based on the known differences between human and rat genetics, biochemistry, physiology and behavior. Extrapolation from animal study results to man applies safety factors: generally a safety factor of 10, unless the exact quantitative relation between human and rat sensitivity is known.

## 3.2  Test end points: the results of the environmental toxicity test

In the test in Figure 1.4, the dilutions in seven test vessels were contacted with the luminobacterium, including one untreated control. Light emission was measured from every dilution and the test end point was read from the concentration–response (light emission) curve.

Light emission did not change with increasing concentration for a while; no inhibition was measured in the vessel containing 5 mg/kg of copper; 5% inhibition was measured at 10 mg/kg, 10% at 20 mg/kg, and full inhibition, i.e. no light emission at all, when 80 mg/kg copper was present in the test vessel. After plotting the individual measured results of the vessels, the curve was fitted to the measured points with the help of a statistical method, using software that can find the statistical optimum of the fitting. Having this curve, the following so-called test end points can be read from the concentration–response curve:

– $EC_{20}$ and $EC_{50}$: the concentrations that have the effect of a 20% and 50% decrease in the measured end points: luminescence inhibition in the example, but it can be respiration rate, enzyme activity, etc. in other cases. The abbreviation EC stands for effective concentration. These values are always estimated using graphical or computational means.
– $ED_{20}$ and $ED_{50}$: the doses that have the effect of a 20% and 50% decrease in the measured end point value. The abbreviation ED stands for effective dose. The difference from EC is that the amount of the effective material is given as a dose. This is the case when animal tests are used and the toxicant administered to the test animal (rat, mouse, rabbit) is measured, used and plotted on the graph in mass units such as µg, mg or g. Dose is also used when the amount or even the identity of the contaminants in an environmental sample are not known. In this case, the mass of water or soil which results in 20% or 50% inhibition in growth, respiration, light emission, etc. is determined.
– $LC_{20}$ and $LC_{50}$: the concentrations that cause 20% and 50% mortality in the test organisms – estimated by graphical or computational means. The difference from the previously introduced $EC_{20}$ and $EC_{50}$ is that the measured end point is not optional, but fixed: it is lethality and LC stands for lethal concentration.
– $LD_{20}$ and $LD_{50}$: the doses that cause 20% and 50% mortality in test organisms – estimated by graphical or computational means. Lethal dose is the amount of the

effective chemical substance or environmental sample given in mass units (μg, mg or g).

- NOEC and NOEL: No Observed Effects Concentration and Level. This is the highest applied concentration or dose in the test, which did not show any effect when tested, compared to the non-treated control.
- NOAEC/NOAEL: No Observed Adverse Effects Concentration or Level, the highest applied concentration or dose in the test that did not show an adverse effect. It has to be emphasized that stimulation as an effect is excluded from the evaluation.
- LOEC and LOEL: Lowest Observed Effects Concentration or Level, the lowest applied concentration that has caused an effect. When our concentration or dose series is of too large a scale, the difference between LOEC and NOEC can be significant.
- MATC: Maximum Allowable Toxicant Concentration, determined by graphical or statistical methods from NOEC and LOEC: NOEC < MATC < LOEC.


These end points of the test are uniformly used for the quantification of adverse effects in environmental toxicology. These are objective measures, and can be directly used in environmental management and applied in decision making.

It is necessary to clarify again: *dose*, being an amount of mass or volume, is mainly used in the practice of toxicity testing when the exposure routes of ingestion, injection or dermal contact are applied to the test animals. The concept is that the only exposure to the toxicant is the controlled intake and from this input dose, knowing the body mass of the test organism, the body burden can be calculated, which is very useful when extrapolating from test animal results to humans.

Concentration is the input parameter when the representative ecosystem members are tested in water or in soil/sediment. The concept is that the entire body of these organisms is in contact with water or soil and, consequently, more simultaneous exposure routes are responsible for interacting and taking up contaminants from the habitat medium.

Animal testing of exposures from air also applies the concentration of the toxicant in the air as an input parameter.

When contaminated land, site-specific effects or the quality of water and soil are tested, we do not know whether contaminants are present or not. Even if we know from clear signals that contaminants are present, they are not identified and quantified, so that the only thing we can test is the dilution rate or mass (dose) of the tested environmental sample (water or soil), which does not affect the test organisms adversely under standard conditions. The interpretation of the results given in dilution rates or sample doses is based on the experience of the assessors. This kind of interpretation may be different according to the test organism and test method.

A more sophisticated solution is the application of the 'calibration' method and giving the scale of effect not as a dilution rate or dose of the tested water or soil sample, but as a toxicity equivalent value, compared to a well-known and easy-to-test chemical substance. An example is introduced in Chapter 9.

The toxicity test end points are functions of the duration of the test. Acute (short-term) tests use the end points of EC/ED and LC/LD causing 20%, 50%, sometimes 10% or 90% inhibition.

NOEC/NOEL and LOEC/LOEL are applied for long-term (chronic) tests. Duration of the tests is discussed in Section 3.3 below, dealing with the classification of the test methods.

## 3.3   Classification of environmental toxicological tests

Environmental toxicology is the science and practice of the adverse effects – mainly of chemicals and other man-made agents – in and through the environment. The targeted receptors of these adverse effects may be both ecosystems and humans.

According to this definition, we consider physico-chemical analyses, biological tests and DNA techniques as environmental toxicity tests, which are capable of characterizing the fate and transport of the chemical substance, and its effect on living organisms, and their organizational systems, including man. As part of the environmental toxicity measuring tool battery, all those model test systems are considered that are feasible using extrapolation from the measured effect on individual organisms, ecosystem and man.

### 3.3.1   Test type according to the aim of the test

Depending on the aim and targeted end point of the test, the types of environmental toxicological tests vary from simple biotests to ecosystem assessments.

– Bioassays are simple, single species laboratory methods for testing:

   o   acute toxicity;
   o   chronic toxicity;
   o   mutagenicity;
   o   carcinogenicity;
   o   teratogenicity;
   o   reprotoxicity.

– Bioassays cannot simulate reality in detail, but they represent one single organism, they work with one single chemical substance or a known mixture. Environmental samples can be tested by bioassays: the samples may derive from one physical phase (filtrated water, soil gas, etc.) or from a complex compartment of the environment (whole soil with three phases). Bioassays need small amounts of sample, but still are statistically relevant. Their controllability is high, so that they are easy to standardize. Multiple numbers of replicates may be tested at the same time; they are reproducible and comparable.
– Microcosms and mesocosms are multispecies toxicity tests modeling the real ecosystem. Micro- and mesocosms are described in detail in Chapter 9 in Volume 3. Their main characteristics are environmental reality and their own history based on their individual evolution.
– *In situ* biomonitoring means the observation of indicator organisms existing naturally (passive biomonitoring) or placed by the assessor into the environment (active monitoring). Natural organisms show large deviation in size, age, sex, etc., causing large scale uncertainty in the results. Homogeneous (e.g. synchronized cultures) and well controlled, lab-grown test organisms may improve test statistics. The measured end points from molecular to population level are optional.

–   Diversity is the statistics of species, chiefly species richness and relative abundance of the species in the ecosystem. The observed species richness in an ecosystem is usually referred to as *species density*. The species evenness is the *relative abundance* or proportion of individuals among the species. By statistical means, we can obtain an index of diversity, such as species richness, Shannon index, Simpson's index or the Berger-Parker index.

–   Biodegradation tests can measure the biodegradability of a chemical substance using a standardized biodegradation test. Real biodegradation in a certain environment can be tested by biodegradation tests modeling the real, site-specific conditions (see also Chapter 2).

–   Bioaccumulation tests provide information on the bioaccumulative potential of a chemical substance or about the real biodegradation under site-specific conditions.

### 3.3.2 Test organisms

The type of test organisms can be chosen from a wide range of selections. Practically all organism types can be used when they fulfill the criteria for organisms for environmental toxicity testing. They should be suitable and feasible for providing a useful answer in the form of the organism's response to the questions of environmental toxicologists, decision makers, managers and regulators.

*General requirements* toward the test organism are summarized below:

–   *Availability*: the test organism should be widely available in the nature or in commerce.

   o   Laboratory cultures are the most widely used test organisms because, under controlled conditions, their stability and good quality can be ensured. The culturing lab can be the same as the testing one or it can be a specialized lab for culturing the test organisms and guaranteeing high quality.

   o   Other culture facilities may also be used, e.g. hatcheries for crustaceans, fish, clams or water plants. Test organs, tissues and blood can be collected from slaughterhouses.

   o   Collection from the field may also be a good solution, mainly in those cases where test organisms are not easy to culture or maintain in the laboratory: marine organisms, plankton, freshwater clams and higher water plants, terrestrial species such as insects, or mites.

–   *Maintenance* of the test cultures in the laboratory can be successful when the species requirements relating to food, space and stress are well known. It is very important to save the sensitivity and all of the required properties of the test organism during cultivation and maintenance, and have a sufficient supply for testing.

–   *Genetics* of the test organism and history of the culture are essential to be able to follow the changes and reach the required statistical quality of the test. The genomes of *Escherichia coli*, *Saccharomyces* yeasts, *Vibrio fischeri* luminobacterium, *Tetrahymena* the single-cell animal, *Drosophila* or some of the *Nematoda* have been fully mapped. The culture collections describe the main genetic and physiological characteristics of the organism species, subspecies or strain. This controlled origin may ensure the conservation of these characteristics over the long term by going back to the original culture any time. Amongst the higher test organisms, there are many species/subspecies which have been used for a long time and are well known from the genetic and physiologic point of view: rat, mice, guinea pig, birds or rabbits.

– *Sensitivity* of the test organism is closely related to the aim of the tests and this is why it is a significant parameter.

    o   Relative sensitivity means that a test organism shows different sensitivity to different toxicants/contaminants. The user should have this information, otherwise any additional chemical substance, its metabolites or contaminants may cause an effect which is comparable to the main toxicant's effect, and makes the response of the test organism disproportionately strong and non-linear in terms of the concentration. Another reason why information on sensitivity is needed is that the test organism must match the problem or the substance to be tested.

    o   Special sensitivity towards one or a few toxicants.

    o   Sensitivity towards a broad number of toxicants.

– *Representative* information about the ecosystem or ecosystem constituents is needed for devising the environmental tests.

– *Sensitivity should be representative* for a class or phyla to protect certain taxa, in this case, additional information is needed as to which families or phyla are represented by the test organism.

    o   The most sensitive ecosystem member must be used for early warning.

    o   The results of tests on organisms with higher (but not much higher) sensitivity can be integrated into a conservative risk assessment/management system.

    o   Average sensitivity is the best for risk managers since it provides a response that is characteristic of the entire ecosystem, without implementing complicated and costly monitoring.

    o   Lower sensitivity than that of the average test organisms can be used for screening hot spots or the most risky elements of a complex system.

    o   Some families or phyla, or minor components of a complex ecosystem are generally not represented by any of the test organism types.

– *Concentration/dose–response relationship* has multiple requirements:

    o   Connection between the amount of the toxicant/contaminant and the response of the test organism.

    o   Proportional response to the concentration/dose of the toxicant.

    o   The effective concentration/dose range should be as broad as possible.

*Reproducibility, statistics*: it is one of the most important requirements when environmental testing should be integrated into a quantitative risk assessment procedure where objective and quantitative data are needed, and evaluated together with physico-chemical analytical data. When this integrated application is understood, it becomes evident that environmental toxicity data cannot be of lower quality than physico-chemical data, otherwise poor statistics and validity dominate the entire procedure. The reproducibility and statistics of a test method also depend on the practice of the laboratory. Toxicity measuring laboratories prepare their own statistics and give the historical averages and deviations of control and reference tests as additional information to the test results.

*Organisms* that give some kind of response to toxicants/contaminants are numerous. The selection of the test organism is often determined by the historical expertise of the laboratory or organization. DNA, microbiology, veterinary labs or human toxicologists generally select those test organisms which they have the methods, tools and equipment to work with and the knowledge needed to truly interpret the results. Accordingly, all possible organisms and living systems of organisms, which are or may be used for environmental tests, can be listed:

- Bacterial cells;
- Algae;
- Fungi;
- Plants: micro- and macrophytes;
- Animals from single-cell through few-cell and micro-size animals to macro-size animals, viz. crustaceans, fish, clams, insects, rodents, or other mammals;
- Multispecies systems such as soil microflora, sewage sludge, rhizosphere, prey–predator systems, food chains or food webs;
- The whole ecosystem can also be tested in microcosms or mesocosms or in the field. Methodologies of testing aquatic and terrestrial ecosystems are distinguished owing to practical reasons and differences in test arrangements.

*Number of species* used in the test is a basic design parameter and the evaluation and interpretation needs different methodologies.

- Single-species tests apply one single species for testing the effect of chemical substances. These species are well-known organisms, deriving from controlled cultures. Single species are used in most of the laboratory bioassays and toxicological tests.
- Multispecies tests involve more species in the same test. In the field of microbiology, the competition test of two bacterial species uses the competing bacterial strains grown together in the test medium. A special relationship is tested in the prey–predator tests. Food-chain effects can be tested using the members of the food chain. Microcosms and mesocosms are multispecies tests where – like the real ecosystem – environmentally relevant species in representative numbers and distribution are tested together. In these models any ecosystem characteristic can be measured or monitored: the number of organisms, number of species, relative distribution of species, respiration or any other metabolic activity of the whole microcosm, independent of the contribution of the individual species or organisms. Mesocosms are larger in size and longer in time than microcosms, consequently, sampling and monitoring have fewer limitations than is the case for microcosms, therefore diversity and its changes, food chain and food web characteristics can also be traced and measured.

### 3.3.3 Test design

Exposure scenario, test duration and arrangement are manifold and their combination results in endless options in test design.

*Exposure Scenarios* describe the conditions how a toxicant comes into contact with the test organisms' body. Depending on the circumstances, the following exposure routes can be distinguished:

–   Full-body test is when the test organism is immersed in the tested water, sediment or soil and there is direct contact between the contaminant/contaminated environment and the test organism. In such cases, the whole body, the skin and all dermal surfaces, eye, gill, hair, etc. are exposed and in many cases, the internal mucosal surfaces in the mouth, (trachea, digestive system, etc.) as well.
–   Feeding studies aim at testing the eating of food or toxicants mixed into food or drinking water to model the uptake and effect through the digestive system. The problem with this kind of testing is food intake; when the test animals feel the presence of toxic material in the food, they eat much less or not at all. In proper tests, the amount of water drunk and food eaten are measured individually. This needs special drinkers and feeders.
–   Placement of a controlled amount into the stomach by a tube (gavage) makes the amount of food or water delivered more precise and controlled.
–   Injection of a controlled amount (intramuscular, intravenous) of toxicant into body-tissues or blood forces contact with cell membranes and uptake by the metabolic apparatus of the cells.

### Test Duration

–   Short-term: acute tests cover a relatively short time period compared to the life span of the test organism. 48-h *Daphnia* tests or 70-h fish tests are acute tests, and the result is given as $EC_{20}/ED_{20}$ or $EC_{50}/ED_{50}$ values. The testing time for animals with longer life spans (dog, monkey) is only a small portion of their whole life.
–   Long-term: chronic tests may cover one or more generations. The duration of chronic tests on test organisms with longer life span takes a significant proportion of their life, including the gestational period of females and spermatogenesis of male test organisms. Chronic test results are given as NOEC/NOEL or LOEC/LOEL values.
–   Growth tests of microorganisms are special cases because of their generation time, which is very short. The growth curve provides the best method for the characterization of the adverse effects of the chemicals, inhibiting metabolism and growth of the population. In growth tests, the microbes are cultured in a series of growth media containing increasing concentrations of the chemical substance to be tested. The number of cells or the biomass at a certain time as well as the whole growth curve of the culture is plotted and evaluated. Any parameter of the growth curve such as the length of the lag phase, the rise of the growth curve, or the cell concentration at the point of inflexion can be used as a response. In practice, the $ErC_{20}$ and $ErC_{50}$ values are applied for the characterization of toxicity where the growth rate can be read from the curve and used for drawing the concentration–response curve.

*Test arrangement* guarantees a constant toxicant concentration, nutrient supply and other parameters in the test medium. Test arrangement, size, medium, material fluxes, should be fitted to the concept and aim of testing.

– Static tests: the test medium which contains the chemical substance to be tested is the same solution in the same vessel during the whole test. The problem with this method is that, during the test, gradients in aerobicity, pH, distribution of the toxicant and concentration of metabolites develop both from normal substrates and the toxicant. Biodegradation of the tested toxicant may also be a problem which must be solved during testing. Most of the acute tests and microcosms are static tests.

– Static renewal tests: in this type of test, the medium is renewed from time to time. This can hinder the decrease or depletion of toxicants in the test solution and the accumulation of harmful or otherwise effective metabolites, thus abating their influence on the test results. Static renewal tests are often used for chronic aquatic tests.

– Recirculation may ensure a continuous toxicant supply, the removal of metabolites and keeping the whole test system in a steady state. This arrangement needs more sophisticated equipment and the costs are also higher compared to the static or the renewal tests. It may have certain disadvantages and technical problems which are caused by the limitations of the techniques used for the treatment of the recycled water: selective filters, supplies, additives, etc.

– Flow-through tests continuously ensure fresh test medium with the same nutrient concentration and unchanged toxicants. In addition to pumps and flow meters, it needs a mixing apparatus which refreshes the test medium. Of course this is the most expensive, but the most reliable test arrangement.

– Flow-through test chambers can be placed in the environment to establish a monitoring system. Flow-through chambers for *Daphnia*, fish, and clams are placed in rivers, lakes, and inlets of sensitive surface waters for effluent monitoring.

### 3.3.4  Most commonly measured end points

The measurement end points detected can be the characteristics of an organism, the components of the test medium, or any substrate, product or metabolite resulting from the activity of the test organism.

– Toxicity test end points may be:

  o Growth, in terms of cell number, mass production, root lengths, chlorophyll content, nitrogen content of the cell mass, reproduction, etc.;
  o Survival or mortality, sometimes immobilization;
  o Respiration by monitoring $O_2$ consumption, $CO_2$ production or measuring the enzyme activities of the respiratory (electron transport) chain, as well as ATP production;
  o Luminescence;
  o Other enzyme activities as well as the decrease in the substrate of the enzyme or increase in the product concentration;

- o   Metabolites of biochemical processes;
- o   Gene products such as RNAs and proteins.

– Mutagenicity test end points are:

- o   Number of mutants;
- o   Number of revertants;
- o   Chromosome abnormalities.

– Carcinogenicity end points are mainly tumors.
– Reprotoxicity can be quantitatively characterized by the following end points:

- o   Reproductive success;
- o   Cytogenetic characteristics of the offspring;
- o   Morphological characteristics of the offspring.

– Biodegradation tests generally use the end points of:

- o   $O_2$ consumption;
- o   Substrate consumption;
- o   Production of end products such as $CO_2$ or metabolites from the tested substance.

– Evaluation of bioaccumulation tests is based on the chemical analysis of the quality and quantity of accumulated substances and the medium tested (water, soil, food/feed, etc.).

### 3.3.5   Environmental compartments and phases to test

The design of the test methods for gaseous, liquid or solid aggregates is different depending on the tested environmental compartments and phases. Laboratory bioassays for the testing of chemical substances are mainly carried out in dissolved form in the liquid phase, but in some cases in artificial or real sludge or soil. Samples taken from the environment can be tested in their original form or as an extract or leachate. The most frequently tested environmental samples are:

– Liquid samples

- o   water from freshwater or marine environment;
- o   subsurface waters, runoffs and groundwater;
- o   pore water from soil and sediment;
- o   leachate or seepage from soil or waste;
- o   waste waters;
- o   liquid phase extracts, eluate or leachate from solid samples.

– Solid phase samples

- o   whole soil;
- o   whole sediment;
- o   solid waste.

– Slurries and sludges can also be tested: mainly waste sludge, including sewage sludge; liquid and solid phase can be separated before testing.

### 3.3.6 Aims of environmental toxicity tests

The aims of testing chemicals for their environmental toxicity are numerous. Regulation and managing chemical substances, contaminated land and waste needs the quantitative information on adverse effects listed below as a basis for decision making and managing the environment.

– Characterizing the environmental fate and behavior of chemical substances;
– Measuring toxicity, mutagenicity, carcinogenicity and reprotoxicity of single chemicals;
– Measuring toxicity, mutagenicity and reprotoxicity of mixtures;
– Establishing effect-based environmental quality criteria;
– Environmental monitoring via biomonitoring and/or integrated monitoring;
– Early warning within environmental monitoring and management;
– Measuring toxicity, mutagenicity, carcinogenicity and reprotoxicity of environmental samples;
– Assessing and monitoring the ecosystem;
– Measuring toxicity, mutagenicity and reprotoxicity of waste materials, leachates of wastes;
– Direct, effect-based decision making to support environmental management.

## 3.4 Environmental toxicology in relation to hazard and risk assessment

Hazard is the intrinsic character of a chemical substance, its hazardousness (flammability, toxicity, mutagenicity, etc.) is anchored in its chemical structure. It is the basis of QSAR (Quantitative Structure-Activity Relationship) and the fundament of hazard-based legislation and management of chemical substances. Nevertheless, hazard is only an indicator for environmental risk. Hazardous chemical compounds are risky to the environment when they are produced and used in such an amount which, on entering the environment, adversely effects humans and the ecosystem or members of the ecosystem. This means that risk depends on the amount of the substance produced, used and emitted, as well as on the potential adverse effects and the exposed receptors. Hazardous substances can be controlled, handled and used in a safe way when their risk is reduced to an acceptable level, in spite of being hazardous.

### 3.4.1 Testing hazard or risk?

According to the above explained difference between hazard and risk, environmental toxicity tests can also be divided into hazard-testing and risk-testing methods. When assessing hazard, the test focuses on the toxicant, while the test organism and the test method are standardized. Testing with the aim of characterizing risk, focuses on the exposed receptor and tries to simulate the real environmental conditions of the impact. One part of the test methods is well standardized, with good statistics both at the intra- and interlaboratory level, suitable for legislation of chemicals and regional or global decision making. In these tests, the environment plays a secondary role, environmental specialties are integrated in a generic way as standard test conditions; the tests are not

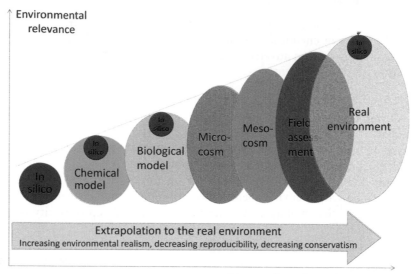

*Figure 1.6* Environmental realism of the assessment models.

intended for modeling the environment, rather ensuring a stable test scenario. Another part of the test methods is designed in such a way that they have close contact with the real environment and they can model the environment or at least its most important characteristics. If environmental reality of the tests were put on a scale, then the scale of realism of the test designs can be 'measured' (Figure 1.6).

The distinction between hazard and risk, their assessment and integration of one of them into an environmental management system with a certain target is the key feature of sustainable management of the environment.

A good example is the difference between degradability of a chemical substance and the degradation of the same substance as a pollutant in a generic or in a site-specific environment. Degradability is an intrinsic property of a chemical substance and depends on its chemical structure. However, one can speak about degradation in the context of interaction with the environment. Degradation may be estimated under theoretical or generic environmental conditions (average annual temperature, water flux, redox and pH conditions, average sediment characteristics, humidity, soil type and conditions in the soil, etc.). Degradation or especially biodegradation of the same contaminant in a contaminated site needs a site-specific approach, either by using site-specific environmental characteristics or by testing the biodegradation of the contaminant in the contaminated soil or water *in situ* or in a sample taken from the site in question. The prediction of the fate and behavior, e.g. degradation of a substance in the environment is a key issue in risk assessment because readily degradable substances have obviously less chance to meet receptors than persistent ones.

Photodegradability, readiness for hydrolysis and biodegradability of a substance can be calculated from its physico-chemical properties using physico-chemical functions or QSAR equations. There are test methods available for measuring degradability and these methods represent a uniform generic test environment. Certain generic environmental parameters such as radiation/light conditions, temperature, redox potential, pH, etc., which may significantly influence degradation depending on regions and climate, can be considered in the calculation or measurement of degradability of a substance. However, one must be aware of the fact that actual, site-specific degradation of the same substance can be higher or lower, and in some cases, significantly different from the calculated or measured degradability. Photodegradation of a photodegradable compound in particular can be zero in deep waters or in sediments, or in a northern region with little sunshine, but significant on a shallow seashore in the sunny south.

Biodegradability tests have been standardized by OECD. All regulations on hazardous chemical substances (REACH, pesticide, biocide, food, cosmetics, *etc.* regulations) require test results. Reproducibility and comparability of the test method is more important for regulatory purposes than its environmental reality, partly because the target environment, where the substance will be used and emitted, is not known beforehand. This is the reason why the potential biodegradability is tested in a standard test medium by using a standardized (?) biodegrading microflora. A key aspect, which weakens this concept, can be identified here: it is hardly possible to prepare, maintain and use a standard but still non-selective or non-adapted microflora for carrying out biodegradation tests. One or a few microorganisms are not good enough, they might not be able to degrade even a readily biodegradable substance or they may accidentally degrade something which is not easy to degrade in the environment. The use of a bacterial community, e.g. stemming from sewage sludge, may also be uncertain because its microbial composition depends on the history of the sewage sludge, its past opportunity of having met the same or similar substance and having been adapted to it or not. Experts should be aware that the microbiota of the sewage sludge are highly adapted to different, locally occurring contaminants and cannot represent a natural community in natural waters or sediments. The same applies to contaminated sites: biodegradability in general and the real biodegradation at a long-term acclimated site with a very likely adapted microflora may differ greatly for the same contaminant.

Due to the large-scale deviations biodegradability test results are loaded with high uncertainty, and, even though one would like to, this test cannot be made independent of the environment. In hazard assessment cases, it may be better to apply a mathematical model, especially if abundant historical data are available which the mathematical function or the QSAR equation can be based on.

The other extreme option is the site-specific assessment of biodegradation at a real site, at the source of a contaminant in a micro- or mesocosm, simulating the existing conditions. Managing risk at a contaminated site and making decisions on biodegradation-based risk reduction cannot be supported by biodegradability results; it can only be done with measured site-specific biodegradation results.

### 3.4.2  Standardized or customized test methods?

The dilemma of selecting the right test method, a standardized or a site- and problem-specific one for measuring adverse effects is severe and frequent. Standardized test methods have many advantages; they are uniform, their results are comparable, and less skilled personnel are capable of carrying them out. All test conditions are defined in the test description. Data collection, evaluation, interpretation and the necessary documentation are also specified. In most cases, the optional species or subspecies are listed as test organisms, from which the laboratory can select the proper one.

Before choosing and using the standardized methods, the key decision point, and the question to be answered by the test must be clearly identified. This also means that all standardized test options and the user's requirements towards the results must be known to be able to correctly select the best suiting method. In plain English: a non-suitable standard method will fail to provide the correct answer. Sometimes none of the standardized methods can provide the proper answer; in this case, a problem-specific test should be developed for the actual case. Standardized tests answer the most frequently asked questions, but it does not make any sense to develop standard methods for rarely occurring or very special problems.

Some regulations, e.g. REACH specify the required test method for all decision points and criteria of the regulation. The tests which serve the regulation can yield comparable results, and, based on them, priorities can be set. Another concept of decision making is to compare test results with legal criteria. Most of these standardized tests can measure the hazard due to adverse effects but not the actual risk. In order to obtain a risk-related answer from a test, a model must be created for the real situation, which represents the most important (even if not all) parameters and factors in an environmental problem, which contribute to risk generation.

When a contaminant is immobile in the environment, strongly bound to the solid matrix of an environmental compartment, e.g. soil, and the question to be answered is, how much threat this contaminant poses to the soil ecosystem (plants and soil-dwelling organisms), the proper way to assess the effect-based risk of the soil is not analyzing or testing the extract from this soil, contrary to the wrong practice. Rather, the representative terrestrial/soil-living test organism should be placed in the contaminated soil, let them interact with each other, simulating real situations. The response measured in such a direct contact scenario has high environmental relevance. But if the groundwater is the receptor, i.e. the assessment's target, a leachate from the soil column irrigated with real or model precipitation should be tested to be able to decide whether or not the groundwater is endangered by the immobile contaminant.

Characterizing site-specific risk by effect testing, the test design should simulate or model the real situation. The test design must be able to rely on proper knowledge of the site and the risk components. A conceptual risk model must be created which characterizes the source(s), transport routes, the contaminated environmental compartments and their dimensions, as well as the land uses and the users of the polluted environment.

Standardized tests, even complex ones, for example, long-term pesticide testing mesocosms (EPA, 1996/2008), are not capable of answering other questions related to, for example, sediments of special texture, of extreme organic or inorganic content, or to the presence of species other than specified by the standard. In conclusion, a simplified rule is that standardized tests give proper results for hazard-based or generic risk-based, mainly regulatory environmental management. On the other hand, site- and problem-specific simulations and models are useful in solving concrete risk management tasks.

### 3.4.3  Testing or modeling? – QSAR and environmental toxicology

The large number of existing chemicals makes the detailed testing of every single case impossible. Every year, several thousand new chemicals are added to the about 100,000 chemical substances already produced and used. This vast amount of substances, regardless of the quantities produced and used and whether or not they are hazardous, cannot be tested alone due to the work and tremendous costs involved. Additionally, the low environmental reality of many of the tests must be considered since the forecast for the real environment has a very poor statistical value.

Toxicity data are not available for all kinds of chemicals. Missing ecotoxicity data for existing chemicals cannot be reinstated by myriads of tests, but proper mathematical models enable the environmental toxicologist to estimate the toxicity of non-tested chemicals based on their chemical structure or any intrinsic physico-chemical characteristics, which are associated with chemical structures such as melting point, volatility, water solubility and octanol–water partitioning. Modeling is a valuable method and the proper place of modeling in environmental management must be found. However, even long-term monitoring or environmental measurements and tests performed over many years are just models, perhaps more realistic ones than mathematical models, but still models, which greatly simplify the real environment. A detailed ecological survey is another model because only a small part of the environment, which is considered representative, is being assessed. The acceptance of monitoring results as 'true' values may be misleading. One must interpolate and/or extrapolate from the information obtained from measurements at discrete points, which is tantamount to considering the collected information as a model. The results of both modeling and testing/monitoring should be validated. Table 1.4 displays a comparison between modeling and testing/monitoring to help assess their advantages and disadvantages.

#### Creating QSAR models

The QSAR (Quantitative Structure–Activity Relationship) is based on the fact that chemically similar substances have similar physico-chemical properties (solubility, partition between physical phases, mobility), environmental fate and behavior (degradation, concentration and accumulation) and biological effects such as toxicity, mutagenicity, carcinogenicity, reprotoxicity, neurotoxicity, hormone- and immune-disrupting, sensitizing or corrosive effects.

Based on the connection between chemical structure, behavior and effects of chemical substances and assuming that abundant data on the toxicity of chemically similar substances are available, the relationship between structure and toxic effect can be described using mathematical equations. In order to formulate the proper equation, a QSAR model must be established.

Data of good quality can be collected from the literature and existing databases, and a frame of reference with the coordinates of a certain toxic effect (e.g. fish or *Daphnia* toxicity) and the structure of any related physico-chemical property of the chemical substance can be created. Environmental behavior and biological/toxic effects of chemicals are closely associated with their $K_{ow}$ value (octanol–water partition coefficient). In our example, plotting literature results of fish/*Daphnia* toxicity as a function of $K_{ow}$ value of many substances, a linear relationship can be found between a large number of points, thus a straight line can be fitted to the data. The linear equation

*Table 1.4* Comparison of modeling and testing/monitoring (ECOFRAM, 2011).

| Modeling | Testing/monitoring |
|---|---|
| **Advantages** | |
| Cost effective | Provides an actual measurement of chemical |
| Time requirement: days to months | concentration, hydrologic response, etc. |
| Simple model scenario: what happens if ... | Avoids multiple conservatism due to compounding |
| Able to measure the effect of risk | conservative assumptions |
| reduction measures | Accounts for the inherent heterogeneity of the system |
| Able to predict concentrations in a | There is a greater acceptance of measured data |
| continuum over space and time | There is public confidence in monitoring data |
| Able to simulate concentrations below | |
| the analytical limits of quantification | |
| Comparative assessments are possible: | |
| both exposures and effects | |
| **Disadvantages** | |
| Environmental reality: low | Costly |
| Uncertainty in model predictions | Time requirement: weeks to years |
| Input of data may be uncertain and | Sampling difficulties: planning, selection of sampling |
| not feasible | points, statistics, interpolation and extrapolation |
| Simplifications required | Requires a long time to evaluate effectiveness |
| General public reluctance to accept | Handling non-detects is difficult |
| predicted data | Sampling represents discrete points |
| Needs calibration to see how closely | Sample represents one unique combination of |
| predicted values match reality | conditions |
| | Constrained by analytical precision and the level of detail |
| | Results can be misleading (1 event occurring in 100 years) |
| | Difficult to interpret results in a probabilistic fashion |
| | Cause and effect may be difficult to assign |

of the relationship can be used for the calculation of fish/*Daphnia* toxicity from $K_{ow}$. Most of the data will be close to the line, their deviation from the line shows how close the correlation is between $K_{ow}$ and fish toxicity. Some of the literature data will be far from the line, and in these cases, the reason for the variance must be investigated: it is highly probable that the chemical type of the substance should also be taken into account, and different equations must be established for aromatic, aliphatic, chlorinated, polycyclic or other chemicals. There are some individual substances that have special modes of action, and, as a consequence, do not fit to any of the group-specific lines. These identified outliers can be expunged from the training set of data used for creating the model, but an acceptable and scientifically reasonable explanation should be given for their removal.

A good-quality QSAR model needs several hundreds or thousands of data, their quality must be screened and they must be grouped according to chemical structure, molecular size, mode of action, etc. Series within the structure may fit well to the line within a certain range, but outside this range, the equation may not fit any more, which means that QSAR is applicable only under certain conditions and within certain ranges. It is therefore very important to publish the accuracy together with the equation. The definition of some terms, based on the Technical Guidance Document applied by the

REACH regulation (EU-TGD, 2003), should be given here for a better understanding of QSAR and the methodology:

- *QSAR method* is the theory underlying a QSAR, including the adequacy of the descriptor variables, the form of the model and the description of the activity represented by the model.
- *QSAR model* is the quantification of the QSAR method, for example, through the derivation of a mathematical equation describing the activity for a specific class of substances.
- *Domain* of a QSAR is the group of substances for which the model is valid. This group of substances can be defined by structural rules, mechanistic information and/or parameter ranges.
- A QSAR is considered *reproducible* if it can be applied by all assessors independently and leads to the same results.
- *Training set* is the set of data used to construct a QSAR model.
- *Validation set* is the set of data used to validate the QSAR model. The data in this set should not be included in the training set and should be chosen in the domain of the model, but independently of the training set (EU-TGD, 2003).
- *Accuracy*: it should be checked whether the correlation coefficient and the overall validity and accuracy of the model have been given. These statistics should include the estimated standard deviation of the prediction errors, the significance of the model as a whole, its variables and parameters.

The Technical Guidance Document (TGD) of the EC prepared for the risk assessment of new and existing chemical substances (EU-TGD, 2003) recommends the QSAR methodology for the prediction of environmental fate, behavior and effects of chemical substances on aquatic ecosystem members:

- Henry's Law Constant (H);
- Octanol–water partition coefficient (log $K_{ow}$);
- Partitioning between soil/sediment organic matter and water ($K_{oc}$);
- Photolysis ($k_{deg\,air}$);
- Hydrolysis ($k_{hydr\,water}$);
- Biodegradation (not-ready biodegradability);
- Acute toxicity to fish (96-h $LC_{50}$);
- Acute toxicity to *Daphnia* (48-h $EC_{50}$);
- Acute toxicity to algae (72–96-h $EC_{50}$);
- Long-term toxicity to fish (NOEC, 28-day study);
- Long-term toxicity to *Daphnia* (NOEC, 21-day study);
- Bioconcentration of fish and worms.

This branch of QSARs also underlines the fact that any adverse effect of chemicals depends to a large extent on their fate and behavior in the environment. This means that the environmental effect of a chemical substance can only be reliably predicted based on laboratory tests if additional data on partitioning, degradation and accumulation of the substance in the environment are gathered, from which the actual or forecasted environmental concentration of the substance can be established. The adverse effect of this environmental concentration, estimated as described, is the basis of the risk assessment.

### Available QSARs

In the case of aquatic toxicity, the guideline differentiates between polar and non-polar toxic substances as they cannot be handled by the same QSAR equation due to their different behavior and effect mechanisms in the organisms: polar substances are hydrophilic, while non-polar ones are hydrophobic. This is why biological availability of the latter is poor, mainly over the short term, but if their accessibility increases due to environmental effects, e.g. in the presence of co-solvents or tensides, their adverse effects can be triggered. Organism-specific active transport may occur and as a result, the non-polar substance can reach adipose tissues, nerves or the liver, where it can exert its toxic effect or accumulate. If no specific toxic mechanisms occur, the internal effective concentrations are almost constant.

Currently, reliable QSARs are available for chemicals that act through a non-specific mode of action such as non-polar narcosis (caused by inert compounds) as well as polar narcosis (caused by substituted phenols, anilines, pyridines and mononitrobenzenes). Regarding non-polar narcosis, QSARs are recommended for fish (short and long term), *Daphnia* (short and long term) and algae (short term). With respect to polar narcosis, QSARs are recommended for fish (short term) and *Daphnia* (short term). No QSARs have been recommended for substances that act through more specific modes of action.

Some typical applications of QSAR models should be introduced here (EU TGD, 2003). Many of the models recommended today have been selected and recalculated for ecotoxicity estimation by Verhaar *et al.* (1995) ($n$ is the number of data, $r^2$ is the correlation coefficient, $Q^2$ is the cross-validated $r^2$ and s.e. is the standard error of estimate).

### Biodegradability of chemical substances

Primary biodegradability of phthalate esters
$RC = -24.308 \cdot \log K_{ow} + 394.84$
$RC$ = primary biodegradability, $n = 12$, $r^2 = 0.87$ (Boethling, 1986)

Biological oxygen demand of alcohols, phenols, ketones, carboxylic acids, ethers, sulfonates
$BOD = 1105 \cdot \Delta/\delta/_{x-y} + 1.906$
$BOD$ = Biological Oxygen Demand, $\Delta/\delta/_{x-y}$ = difference in the modulus of atomic charge across a functional group bond x–y
$n = 112$, $r^2 = 0.98$ (Dearden & Nicholson, 1987)

Biological oxygen demand of alcohols
$BOD = 0.093 \cdot S_E - 3.163$
$S_E$ = electrophilic superdelocalizability
$n = 19$, $r^2 = 0.96$ (Dearden & Nicholson, 1987)

### Adverse effects

*Fish toxicity* of non-polar substances
*Pimephales promelas* 96-h $LC_{50}$, mol/L

$\log LC_{50} = -0.85 \cdot \log K_{ow} - 1.39$
$n = 58$, $r^2 = 0.94$, $Q^2 = 0.93$, s.e. $= 0.36$ (Verhaar *et al.*, 1995)
*Brachydanio rerio and P. promelas* 28- to 32-day NOEC, ELS test, mol/L
$\log NOEC = -0.90 \cdot \log K_{ow} - 2.30$
$n = 27$, $r^2 = 0.92$, $Q^2 = 0.91$, s.e. $= 0.33$ (Verhaar *et al.*, 1995)

*Daphnia toxicity* of non-polar substances
*Daphnia magna* 48-h $EC_{50}$ immobilization test, mol/L
$\log EC_{50} = -0.95 \cdot \log K_{ow} - 1.32$
$n = 49$, $r^2 = 0.95$, $Q^2 = 0.94$, s.e. $= 0.34$ (Verhaar *et al.*, 1995)
*Daphnia magna* 16-day NOEC, growth, reproduction, mol/L
$\log NOEC = -1.05 \cdot \log K_{ow} - 1.85$
$n = 10$, $r^2 = 0.97$, $Q^2 = 0.95$, s.e. $= 0.39$ (Verhaar *et al.*, 1995)

*Algae toxicity* of non-polar substances
*Selenastrum capricornutum* 72–96-h $EC_{50}$ growth, mol/L
$\log EC_{50} = -1.00 \cdot \log K_{ow} - 1.23$
$n = 10$, $r^2 = 0.93$, $Q^2 =$ n.d., s.e. $= 0.17$ (Van Leeuwen *et al.*, 1992)

*Fish toxicity* of polar substances
*Pimephales promelas* 96-h $LC_{50}$, mol/L
$\log LC_{50} = -0.73 \cdot \log K_{ow} - 2.16$
$n = 86$, $r^2 = 0.90$, $Q^2 = 0.90$, s.e. $= 0.33$ (Verhaar *et al.*, 1995)

*Daphnia toxicity* of polar substances
*Daphnia magna* 48-h $EC_{50}$ immobilization test, mol/L
$\log EC_{50} = -0.56 \cdot \log K_{ow} - 2.79$
$n = 37$, $r^2 = 0.77$, $Q^2 = 0.73$, s.e. $= 0.37$ (Verhaar *et al.*, 1995)

*Bioconcentration*

QSAR for BCF of substances with $\log K_{ow} < 6$
$\log BCF = 0.85 \cdot \log K_{ow} - 0.70$
$n = 55$, $r^2 = 0.90$ (Veith *et al.*, 1979)

QSAR for BCF of substances with $\log K_{ow} > 6$
a) Polynomial equation $\log K_{ow} > 6$
$\log BCF = 6.9 \cdot 10^{-3} \cdot (\log K_{ow})^4 - 1.85 \cdot 10^{-1} \cdot (\log K_{ow})^3 + 1.55 \cdot (\log K_{ow})^2 - 4.18 \cdot \log K_{ow} + 4.79n = 45$, $r^2 =$ n.a. (Connell & Hawker, 1988)
b) Parabolic equation $\log K_{ow} > 6$
$\log BCF = -0.20 \cdot (\log K_{ow})^2 + 2.74 \cdot \log K_{ow} - 4.72$
$n = 43$, $r^2 = 0.78$ (recalculated from Connell & Hawker, 1988)

Bioconcentration of aromatic compounds by *Daphnia magna*
$\log BCF = 0.898 \cdot \log K_{ow} - 1.315$ (Calow, 1994)

Bioconcentration in earthworms
$\log BCF = 1.0 \cdot \log K_{ow} - 0.6$
$n = 100$, $r^2 = 0.91$ (Connell & Markwell, 1990)

The toxicity or other $K_{ow}$-dependent parameters of a chemical substance can be calculated using a good QSAR, but this is still inaccurate compared to its environmental fate and behavior as well as the effects brought about in the real environment, in surface waters and sediments, in soils and sludges, and in the organism of individuals. In the real environment and individual bodies, interactions between the chemical substances and matrix materials, partition between physical phases, interactions of contaminants with each other and with members of the biota, and, after reaching the receptor, with the individuals' biochemical system will to a large extent modify the measurable behavior, characteristics and effects of the substance. One should be aware that QSAR and other models, e.g. chemical models, are suitable for the estimation and prediction of behavior and effect of chemicals in the case of incomplete data, but the uncertainty of these estimates may be high.

## 3.5   Statistical evaluation of ecotoxicological tests

When testing the adverse effects of chemicals, finding a good compromise between statistical values and practical factors such as time requirement, workload and cost is a key task of risk managers. Apart from expense, labor and time requirements, an excessive number of replicates may cause a time-shift, and thus the replicates are processed at different times, corrupting the statistics.

To be able to select the proper statistical tool, the environmental problem must be understood, the conceptual risk model devised and then the most suitable test methodologies and statistical tools must be found.

In the statistics of an acute test – given that the $EC_{50}$ point is on the steepest part of the concentration–response curve – much smaller errors are expected than is the case for NOEC or LOEC values where the dose–response curve just starts increasing and this increase is slow, related to the applied concentrations.

Statistical methods are supported by software tools, which are available as free or commercial products.

### 3.5.1   Evaluation of acute toxicity tests

Evaluation of an acute toxicity test implies the reading of the test end point from the dose–response curve.

*Graphical interpolation* requires plotting the concentration–response curve and reading the concentration which causes 50% or 20% (in special cases 10% or 90%) inhibition. From the plotted curve, the experienced environmental toxicologist knows whether the shape of the curve is typical or atypical. If the curve is atypical, the effect mechanism behind it is probably not a simple one, and the evaluation cannot be performed mechanically. This kind of subjectivity is both an advantage and a disadvantage. The best way to utilize the advantage is to use graphical interpolation as an exploratory or control means. The disadvantage is that the confidence interval cannot be calculated.

The *probit method* is the most popular statistical tool for the evaluation of typical S-shaped concentration–response curves. The original data are processed by probit-transformation. Probit is a binary response model that employs a probit link function. This model is most often estimated using a standard maximum likelihood procedure, such an estimation being called a probit regression. In the probit model, the inverse standard normal distribution of the probability is modeled as a linear combination of the predictors (PROBIT, 2013).

The confidence interval can be easily calculated using the probit method for the evaluation of acute toxicity test-end points. The disadvantage is that it needs two sets of partial kills, which differs from 0 and 100%.

A great number of computerized programs are available such as

- STATA-PROBIT (Data Analyses and Statistical Software) (STATA, 2013);
- SAS-PROBIT (Statistical Analysis System) (SAS, 2013);
- STATISTICA (STATSOFT, 2013), SPSS-PROBIT (original name: Statistical Package for the Social Sciences, but today SPSS is used without any explanation) (SPSS-PROBIT, 2013);
- Mplus-PROBIT (Mplus, 2013);
- R-PROBIT (R-project for Statistical Computing) (R-PROBIT, 2013).

Some specified methods are recommended for toxicity evaluation such as

- ASTM-PROBIT (developed by ASTM International, originally called: American Society for Testing and Materials) (ASTM, 2013);
- TOXSTAT (TOXSTAT, 2013);
- DULUTH-TOX (US-EPA), etc.

Roberts (1988) compared five readily available computer programs for probit analysis. The methods were evaluated for input/output options and reliability of output.

- DULUTH-TOX by Charles Stephan of the Environmental Protection Agency's Environmental Research Laboratory, Duluth;
- ASTM-PROBIT from the ASTM Guide for Probit Analysis, Draft 2;
- UG-PROBIT, which was written by statisticians at the University of Guelph, Canada;
- SPSSx as different from the commercially available software;
- SAS, part of the SAS software package.

According to Roberts' (1988) results, except for UG-PROBIT, the programs yielded essentially identical median lethal concentration ($LC_{50}$) values for the 20 data sets tested. DULUTH-TOX and ASTM-PROBIT include an objective evaluation of the validity of input data and calculated results, but they lack graphical output. SAS-PROBIT provides graphical output superior to that from SPSSx-PROBIT, which yields inappropriate fiducial limits in some cases.

UG-PROBIT, SPSSx-PROBIT and SAS-PROBIT purport to consider control or natural responses. Only SPSSx-PROBIT and SAS-PROBIT actually adjusted observed treatment responses for the control response.

DULUTH-TOX and ASTM-PROBIT, in all implementations, have a major advantage that objective tests are included to determine the validity of input data and to guide the interpretation of output. The commercial statistical programs have the advantages of graphical output and a method for handling control mortality (Roberts, 1988).

*Logit transformation* of data and fitting the curve based on the maximal likelihood method can also be applied for calculating EC or LD values. Similar to the probit method, in the case of lack of partial kill data, some assumptions are required for the calculations.

The *moving average method* is similar to graphical interpolation, but the concentration–response curve should be properly linearized. The confidence interval can be established only when partial kill data are available.

### 3.5.2    Data analysis for chronic toxicity tests

The standard method for analyzing chronic toxicity data is ANOVA, which is the abbreviation of Analysis of Variance. This determines the concentrations that are significantly different in effect from the untreated control.

In *ANOVA*, the observed variance in a particular variable is partitioned into components attributable to different sources of variation. In its simplest form, ANOVA provides a statistical test of whether or not the means of several groups are all equal, and therefore generalizes the t-test to more than two groups.

The fixed-effects model of variance analysis applies to situations in which the experimenter applies one or more treatments to the subjects of the experiment to see if the response variable values change. This allows the experimenter to estimate the ranges of response variable values that the treatment would generate in the population as a whole.

In the case of chronic toxicity, the equivalence of the control and treated is tested. Analysis of variance is performed on the treatment group. By multiple comparisons between the treatment groups, those groups can be identified which are different from control.

In the first step, ANOVA calculates the distance between all treated and control groups. If the F-score is not statistically significant, the treatment has the same effect, there is no difference between the groups. If the F-score is significant, data are examined in a second step to find the groups different from the control. Using multiple comparisons, the groups different from each other can also be identified.

The goal of the ANOVA-evaluation is to find the LOEC or NOEC values, namely the lowest contaminant concentration which is different from control showing a significant adverse effect or the highest no-effect concentration which is identical to the control but different from LOEC. This is a so-called hypothesis-testing model.

### 3.5.3    Data analysis of multispecies toxicity tests

Multispecies tests need multivariate techniques for the exploration of patterns within ecological data sets and microcosm and mesocosm data.

Multispecies toxicity tests – as discussed in detail in Chapter 8:

- They are similar to natural ecological systems in complexity and
- In the consequences of being historical: having their own evolution (the parallel tests may also differ from each other);
- They include trophic structures, food chains or food webs;
- Environmental parameters such as sunshine, precipitation, nutrient supply, interaction with and between environmental compartments and phases play an important role.

The main paradox of multispecies testing is the loss of statistical power with increasing ecological relevance.

In view of the complexity and unique history of micro- and mesocosms as well as natural structures and their dynamic nature, the evaluation and interpretation of measured data and the correspondence of changes to the treatments require carefully planned and statistically evaluated test design and monitoring. When designing the number of replicas, the unique history of these simultaneous microcosms must be taken into account, and the detected indicators should be multivariate and associated with the treatments as much as possible. The suitable statistical methods are also multivariate because univariate methods such as simple ANOVA cannot handle the temporal dependence of variables or the community level changes (which cannot be characterized by single species responses) and, as a consequence, the risk of the false zero hypothesis increases.

Evaluations of these multivariate tests need a concept that lets the variables explore the ecological patterns. Multivariate statistics is able to examine all data, but it is the test designer's responsibility to decide which data should be measured or gathered.

Multivariate statistical methods differ from each other to a large extent: the evaluators must know the relationships among variables and find the most suitable statistical means. Multivariate statistical evaluation should be coupled with association analyses to identify the result of the treatments. Multivariate statistical techniques commonly used for the evaluation of ecological structures are:

- PCA: principal components analysis (assumption: linearity);
- DPC: detrended principal components (a polynomial is used to remove non-linearity);
- NMDS: non-metric, multidimensional scaling (non-linearity is considered using ranks);
- RDA: PCA coupled with redundancy analysis;
- Clustering: grouping by similarities: algorithm has no knowledge about treatment groups;
- Divergence between treatment groups;
- NCAA: non-metric clustering and association analysis: a multivariate derivative of artificial intelligence (Landis & Yu, 1999).

## 3.6   Standardization and international acceptance of newly developed toxicity tests

After finding the relationship between the concentration or dose of a hazardous chemical substance and a measurable end point, there are many tasks until an internationally agreed, standardized test method is established.

- *Research & development*: exploring effect mechanisms and developing technical details.
- *Prevalidation*: an approximately 2-year process that aims to standardize and optimize the test protocol and evaluate within-lab variability and define a 'prediction model' or 'data interpretation procedure', which articulates the process by which test results can be used to predict toxicological end points in vivo.

- *Validation*: an approximately 1-year process which means the evaluation of the test's transferability to other laboratories and measuring between-labs variability and reproducibility (minimum four external laboratories).
- *Peer reviewing* by independent peer review body (ECVAM Scientific Advisory Committee, or ESAC).
- *Regulatory acceptance* in Europe means ESAC endorsement, in the US, ICC-VAM formulates recommendations. This process can take 2 years or more at the national/regional level and longer in the case of international consensus-driven bodies such as OECD (OECD, 2013a), the International Conference on Harmonization (ICH, 1991), and the Veterinary International Co-operation on Harmonization (VICH, 1996).

Environmental toxicology, this increasingly important tool in environmental decision making, should be better understood and fully integrated into the tool batteries of environmental management. The integration of direct toxicity testing of our environment and the application of their full models in the form of micro- and mesocosms bring environment (nature) and its inhabitants closer to the increasingly mechanically thinking and functioning humanity, which is one of the main obstacles in evolving a holistic approach necessary for really efficient environmental management. Environmental toxicology demonstrates and helps to understand and solve many of the pitfalls of environmental management and policy: e.g. handling uncertainties due to the variability of natural systems and the wide range of errors in sampling, measuring, evaluating and interpreting the acquired information, the importance of the references and benchmarks, the temporal and spatial extension of the impacts and the objective and subjective factors behind uncertainties. Understanding the dynamic nature of the environment and adapting the management concept to it, is a basic requirement. A future vision of the most efficient environmental management is a simple preventive tool, just based on a high level of respect for the environment. But before reaching this level of harmony with the environment, one has to learn more about the environmental responses to anthropogenic exposures by measuring and evaluating these impacts and responses. The general overview in this Chapter of the aims, approaches and developments in environmental toxicology has provided the basis to introduce the methodological details of human, aquatic and terrestrial toxicology in Chapters 2–6, and the methods for acquiring information from environmental toxicity data in Chapter 9.

## REFERENCES

Aardema, M.J. & MacGregor, J.T. (2002) Toxicology and genetic toxicology in the new era of 'toxicogenomics': impact of '-omics' technologies. *Mutation Research*, 29(499), 13–25.

ACuteTox (2005–2010) *An in-vitro test strategy for predicting human acute toxicity*. An integrated project within the sixth framework programme. [Online] Available from: http://www.acutetox.eu/. [Accessed 8th October 2013].

AltTox (2009) *Toxicity testing overview*. [Online] Available from: http://alttox.org/ttrc/tox-test-overview/. [Accessed 8th October 2013].

Anonymous (2006) *A CuteTox – research project for alternative testing*. [Online] Available from: http://www.acutetox.eu/. [Accessed 8th October 2013].

ASTM (2013) *American Society for Testing and Materials*. [Online] Available from: http://www.astm.org. [Accessed 8th October 2013].

Boethling, R.S. (1986) Application of molecular topology to quantitative structure – biodegradability relationships. *Environmental Toxicology and Chemistry*, 5, 797–806.

Calow, P. (1994) *Handbook of ecotoxicology*. Blackwell Science Ltd.

CAS (2013) Chemical Abstracts Service. [Online] Available from: http://www.cas.org/index.html. [Accessed 8th October 2013].

CAS (2014) *Database counter*. [Online] Available from: http://www.cas.org/cgi-bin/cas/regreport.pl. [Accessed 1st May 2014].

Chapman, K. & Robinson, S. (2007) *Challenging the requirement for acute toxicity studies in the development of new medicines*. [Online] Available from: http://www.nc3rs.org.uk/downloaddoc.asp?id=559&page=22&skin=0. [Accessed 8th October 2013].

Chapman, P.M. (2002) Integrating toxicology and ecology: putting the 'eco' into ecotoxicology. *Marine Pollution Bulletin*, 44(1), 7–15.

CLP (2008a) *Classification, Labelling and Packaging of Substances and Mixtures*. Regulation (EC) No 1272/2008 of the European Parliament and of the Council of 16 December 2008. Amending and Repealing Directives 67/548/EEC and 1999/45/EC, and Amending Regulation (EC) No 1907/2006. *Official Journal of the European Union*, L 353/1, 1–1355. [Online] Available from: http://eur-lex.europa.eu/LexUriServ/LexUriServ.do?uri=OJ:L:2008:353:0001:1355:en:PDF. [Accessed 8th October 2013].

CLP (2008b) *Classification, Labelling and Packaging of Substances and Mixtures*. Regulation of the European Parliament and of the Council. Amending and Repealing Directives 67/548/EEC and 1999/45/EC, and Amending Regulation (EC) No 1907/2006, Annex VI. [Online] Available from: http://echa.europa.eu/legislation/classification_legislation_en.asp or http://ecb.jrc.ec.europa.eu/classification-labelling/clp/. [Accessed 8th October 2013].

COM (1991) Communication from the Commission to the Council and the Parliament Establishment of the European Centre for the Validation of Alternative Methods. Sec 91:1794, 1–6. [Online] Available from: http://ecvam.jrc.ec.europa.eu/ft_doc/SEC%2092%201794%20final%2019911029.pdf. [Accessed 8th October 2013].

Connell, D.W. & Hawker, D.W. (1988) Use of polynomial expressions to describe the bioconcentration of hydrophobic chemicals by fish. *Ecotoxicology and Environmental Safety*, 16(3), 242–257.

Connell, D.W. & Markwell, R.D. (1990) Bioaccumulation in the soil to earthworm system. *Chemosphere*, 20, 91–100.

Cronin, M.T.D., Jaworska, J.S., Walker, J.D., Comber, M.H.I., Watts, C.D. & Worth, A.P. (2003) Use of QSARs in international decision-making frameworks to predict health effects of chemical substances. *Environmental Health Perspectives*, 111, 1391–1401.

Danish EPA (2001) *Report on the advisory list for self-classification of dangerous substances*. [Online] Available from: http://www2.mst.dk/common/Udgivramme/Frame.asp?http://www2.mst.dk/udgiv/publications/2001/87-7944-694-9/html/kap01_eng.htm. [Accessed 8th October 2013].

Dearden, J.C. & Nicholson, R.M. (1987) Correlation of biodegradability with atomic charge difference and superdelocalizability. In: Kaiser, K.L.E. (ed.) *QSAR in environmental toxicology – II*. Dordrecht, Reidel. pp. 83–89.

Di Ventura, B., Lemerle, C., Michalodimitrakis, K. & Serrano, L. (2006) From *in vivo* to *in silico* biology and back. *Nature*, 443(5), 527–533.

Dix, D.J., Houck, K.A., Martin, M.T., Richard, A.M., Setzer, R.W. & Kavlock, R.J. (2007) The ToxCast program for prioritizing toxicity testing of environmental chemicals. *Toxicology Science (Forum)*, 95, 5–12.

DOE (2003) *Genome glossary*. US Department of Energy (DOE). [Online] Available from: http://www.ornl.gov/sci/techresources/Human_Genome/glossary/glossary.shtml.   [Accessed 8th October 2013].

DSD (1967) *Classification, packaging and labelling of dangerous substances*. Council Directive 67/548/EEC of 27 June 1967 on the approximation of laws, regulations and administrative provisions. [Online] Available from: http://eur-lex.europa.eu/LexUriServ/LexUriServ.do?uri=CELEX:31967L0548:EN:HTML. [Accessed 8th October 2013].

EBTC (2014) *Evidence-Based Toxicology Collaboration*. [Online] Available from: http://www.ebtox.com. [Accessed 8th January 2014].

EC (2008) *Registration, Evaluation, Authorisation and Restriction of Chemicals (REACH)*. Council Regulation (EC) No 440/2008 of 30 May 2008 laying down test methods pursuant to Regulation (EC) No 1907/2006 of the European Parliament and of the Council(text with EEA relevance). *Official Journal L* 142:31/05/2008:0001–0739.EC 440. [Online] Available from: http://eur-lex.europa.eu/LexUriServ/LexUriServ.do?uri=CELEX:32008R0440:EN:HTML. [Accessed 8th October 2013].

ECHA (2010) *Practical Guide 10: How to avoid unnecessary testing on animals*. PG10. ISBN-13: 978-92-9217-402-6, ISSN: 1831-6727, ECHA-10-B-17-EN. [Online] Available from: http://echa.europa.eu/doc/publications/practical_guides/pg_10_avoid_animal_testing_en.pdf. [Accessed 8th October 2013].

ECHA (2013) *European Chemicals Agency*. [Online] Available from: http://echa.europa.eu/. [Accessed 8th October 2013].

ECOFRAM (2011) *Progress Report of the Ecological Committee on FIFRA Risk*. The ECOFRAM Aquatic Exposure and Aquatic Effects Subgroups. [Online] Available from: http://www.epa.gov/oppefed1/ecorisk/ecofram/aqexpeff.htm#assess. [Accessed 8th October 2013].

ECVAM (1991*) European Centre for the Validation of Alternative Methods*. [Online] Available from: http://ecvam.jrc.ec.europa.eu/. [Accessed 8th October 2013].

EEC (1986) *Protection of animals used for experimental and other scientific purposes*. Council Directive 86/609 EEC 1986 of 24 November 1986 on the approximation of laws, regulations and administrative provisions of the Member States. [Online] Available from: http://ec.europa.eu/food/fs/aw/aw_legislation/scientific/86-609-eec_en.pdf. [Accessed 8th October 2013].

EPA (1996/2008) *Federal Insecticide, Fungicide, and Rodenticide Act*. [Online] Available from: http://agriculture.senate.gov/Legislation/Compilations/Fifra/FIFRA.pdf. [Accessed 8th October 2013].

EU TGD (2003) Technical Guidance Document on risk assessment in support of Commission Regulation (EC) No 1488/94 on risk assessment for existing substances, Commission Directive 93/67/EEC on risk assessment for new notified substances and Directive 98/8/EC of the European Parliament and of the Council concerning the placing of biocidal products on the market, Part 3, Chapter 4, European Commission, Joint Research Centre, Eur 20418 EN/3. [Online] Available from: http://ihcp.jrc.ec.europa.eu/our_activities/public-health/risk_assessment_of_Biocides/doc/tgd. [Accessed 8th October 2013].

GHS (2007) *Globally harmonized system of classification and labelling of chemicals*. 2nd revised edn. United Nations. New York, Geneva. [Online] Available from: http://www.unece.org/trans/danger/publi/ghs/ghs_rev02/English/00e_intro.pdf. [Accessed 8th October 2013].

Gruiz, K., Feigl, V., Hajdu, Cs. & Tolner, M. (2010) Environmental toxicity testing of contaminated soil based on microcalorimetry. *Environmental Toxicology*, 25(5), 479–486.

Hartung, T. (2010) Lessons learned from alternative methods and their validation for a new toxicology in the 21st century. *Journal of Toxicology and Environmental Health*, 13, 277–290.

Hartung, T. (2012) Evidence-Based Toxicology – the Toolbox of Validation for the 21st Century? *Altex*, 27(4), 253–263.

Hoffmann, S. (2012) Kick-off of the Evidence-based Toxicology Collaboration Europe. *Altex*, 29, 456.

HUGO (2012) *Human Genome Project.* [Online] Available from: http://www.ornl.gov/sci/techresources/Human_Genome/home.shtml. [Accessed 20 May 2012].

ICCVAM (2013) *Interagency Coordinating Committee on the Validation of Alternative Methods.* [Online] Available from: http://iccvam.niehs.nih.gov/docs/about_docs/Milestones.pdf. [Accessed 8th October 2013].

ICH (1991) *International Conference on Harmonisation.* [Online] Available from: http://www.ich.org/home.html. [Accessed 8th October 2013].

IHCP (2013) *The Institute for Health and Consumer Protection.* [Online] Available from: http://ihcp.jrc.ec.europa.eu/. [Accessed 8th October 2013].

Inglese, J., Auld, D.S., Jadhav, A., *et al.* (2006) Quantitative high-throughput screening: a titration-based approach that efficiently identifies biological activities in large chemical libraries. *Proceedings of National Academy of Sciences USA*, 103, 11473–11478.

IPPC (2008) *Integrated pollution prevention and control.* Directive 2008/1/EC of the European Parliament and of the Council of 15 January 2008. (Codified version). *Official Journal of the European Union*, L 24/8, 1–22. [Online] Available from: http://europa.eu/legislation_summaries/environment/air_pollution/l28045_en.htm. [Accessed 8th October 2013].

Jeffery, S., Gardi, C., Jones, A., *et al.* (2010) *European atlas of soil biodiversity.* European Commission, Publications Office of the European Union, Luxembourg. ISBN 978-92-79-15806-3, ISSN 1018-5593, DOI: 10.2788/94222. [Online] Available from: http://eusoils.jrc.ec.europa.eu/library/maps/Biodiversity_Atlas/. [Accessed 8th October 2013].

JRC (2013) *European Commission Joint Research Centre.* [Online] Available from: http://ec.europa.eu/dgs/jrc/index.cfm. [Accessed 8th October 2013].

Kavlock, R., Ankley, G.T., Collette, T., *et al.* (2005) Computational toxicology: framework, partnerships, and program development. *Reproductive Toxicology*, 19, 265–280.

Landis, W.G. & Yu, M.H. (1999) *Introduction to environmental toxicology: impact of chemicals upon ecological systems.* New York, Boca Raton, Florida, CRC Press LLC.

MacGregor, J.T. (2003) The future of regulatory toxicology: impact of the biotechnology revolution. *Toxicology Science*, 75(2), 236–248.

Meek, B. (2005) *Tiered evaluation strategies – Canadian experiences under CEPA/DSL.* [Online] Available from: http://isrtp.org/nonmembers/Alternative%20Tox%20Methods%20Nov%202005/14.%20Meek%20Bullets.pdf. [Accessed 8th October 2013].

Mplus (2013) *Latent variable modeling program.* [Online] Available from: http://www.statmodel.com. [Accessed 8th October 2013].

Nagy, Zs.M., Molnár, M., Fekete-Kertész, I., Molnár-Perl, I., Fenyvesi, É. & Gruiz, K. (2014) Removal of emerging micropollutants from water using cyclodextrin, *Science of the Total Environment*, 485–486, 711–719. DOI: 10.1016/j.scitotenv.2014.04.003. [Online] Available from: http://www.sciencedirect.com/science/article/pii/S0048969714004975. [Accessed 18th July 2014].

NC3Rs (2007) *News: Challenging the requirement for acute toxicity studies* – workshop report published. [Online] Available from: http://www.nc3rs.org.uk/news.asp?id=512. [Accessed 8th October 2013].

NCBI (2013) *National Center for Biotechnology Information.* Available from: http://www.ncbi.nlm.nih.gov/. [Accessed 8th October 2013].

NICEATM (2013) *The NTP Interagency Center for the Evaluation of Alternative Toxicological Methods.* Available from: https://www.niehs.nih.gov/research/atniehs/dntp/assoc/niceatm. [Accessed 8th October 2013].

NICEATM/ICCVAM (2013) *The NTP Interagency Center for the Evaluation of Alternative Toxicological Methods* (NICEATM) and the Interagency Coordinating Committee on the Validation of Alternative Methods (ICCVAM). Available from: http://iccvam.niehs.nih.gov/. [Accessed 8th October 2013].

Nonylphenol (2013) *Nonylphenol and nonylphenoletoxylates in textiles.* Annex XV Restriction report. Proposal for a restriction. Version number 3. 29 July 2013. Swedish Chemicals Agency. [Online] Available from: http://echa.europa.eu/documents/10162/f28b5c79-11e0-4ce2-91db-e53f7daa4d5a. [Accessed 8th January 2014].

OECD (2005) Guidance document on the validation and International acceptance of new or updated test methods for hazard assessment. ENV/JM/MONO(2005)14. *OECD Series on Testing and Assessment* 34, 96 pp. Paris, France, OECD. [Online] Available from: http://www.oecd.org/officialdocuments/displaydocumentpdf?cote=env/jm/mono(2005)14&doclanguage=en. [Accessed 8th January 2014].

OECD (2013a) *Organisation for Economic Co-operation and Development.* [Online] Available from: http://www.oecd.org/home/. [Accessed 8th October 2013].

OECD (2013b) *Series on testing and assessment.* Adopted guidance and review documents. [Online] Available from: http://www.oecd.org/document/30/0,3746,en_2649_37465_1916638_1_1_1_37465,00.html. [Accessed 8th October 2013].

OECD (2013c) *Assessment of chemicals.* [Online] Available from: http://www.oecd.org/department/0,3355,en_2649_34379_1_1_1_1_1,00.html. [Accessed 8th October 2013].

OMICS (2009) *Omics, bioinformatics & computational biology.* [Online] Available from: http://www.alttox.org/ttrc/emerging-technologies/-omics/. [Accessed 8th October 2013].

Posthuma, L., Suter, G.W. II & Traas, T.P. (eds.) (2002) *Species sensitivity distributions in ecotoxicology.* Florida, US. CRC Press LLC. [Online] Available from: http://www.scribd.com/doc/90143799/Posthuma-Et-Al-2002-SSD-in-Eco- Toxicology and http://books.google.hu/books?id=67JZSpcDiEIC&printsec= frontcover&hl=hu#v=onepage&q&f=false. [Accessed 28th December 2013].

PROBIT (2013) *Probit Regression.* [Online] Available from: http://www.ats.ucla.edu/stat/stata/dae/probit.htm. [Accessed 8th October 2013].

REACH (2006) *Registration, Evaluation, Authorisation and Restriction of Chemicals (REACH),* establishing a European Chemicals Agency. Regulation (EC) No 1907/2006 of the European Parliament and of the Council of 18 December 2006. *Official Journal of the European Union,* L 396/1, 1–849. [Online] Available from: http://eur-lex.europa.eu/LexUriServ/LexUriServ.do?uri=OJ:L:2006:396:0001:0849:EN:PDF. [Accessed 8th October 2013].

Richard, A. (2006) The future of toxicology – Predictive toxicology: an expanded view of 'chemical toxicity'. *Chemical Research in Toxicology,* 19(10), 1257–1261.

Roberts, M.H. (1988) Comparison of several computer programs for probit analysis of dose-related mortality data. *ASTM Special technical publication,* 1007, 308–320.

R-PROBIT (2013) *The R Project for Statistical Computing.* [Online] Available from: http://www.r-project.org. [Accessed 8th October 2013].

Rudacille, D. (2010) 21st Century Validation Strategies for 21st Century Tools. Summary of the July 2010 Workshop. *Altex,* 27(2), 279–284.

SAS (2013) *Software Solutions.* [Online] Available from: http://support.sas.com. [Accessed 8th October 2013].

Schulze, W.X., Gleixner, G., Kaiser, K., Guggenberger, G., Mann, M. & Schulze, E.D. (2005) A proteomic fingerprint of dissolved organic carbon and of soil particles. *Oecologia,* 142(3), 335–343.

Seidle, T. (2007) *Opportunities and barriers to the replacement of animals in acute systemic toxicity testing.* [Online] Available from: http://www.alttox.org/ttrc/toxicity-tests/acute/way-forward/seidle/. [Accessed 8th October 2013].

Shi, Y., Tyson, G.W. & DeLong, E.F. (2009) Metatranscriptomics reveals unique microbial small RNAs in the ocean's water column. *Nature*, 459, 266–269.

Snape, J.R., Maund, S.J., Pickford, D.B. & Hutchinson, T.H. (2004) Ecotoxicogenomics: the challenge of integrating genomics into aquatic and terrestrial ecotoxicology. *Aquatic Toxicology*, 67(2), 143–54.

SPSS-PROBIT (2013) *Statistics software*. [Online] Available from: http://www.spss.com; http://www.spss.com.hk/statistics and http://www.hearne.com.au/Software/SPSS-Statistics-Family-by-IBM/Editions. [Accessed 8th October 2013].

Solomon, K.R., Brock, T.C.M., De Zwart, D., Dyer, S.D., Posthuma, L., Richards, S., Sanderson, H., Sibley, P. & van den Brink, P.J. (eds.) (2008) *Extrapolation Practice for Ecotoxicological Effect Characterization of Chemicals*. Florida, US, CRC Press and SETAC.

SSD (2013) *Species Sensitivity Distributions*. CADDIS, Data Analyses. US EPA. [Online] Available from: http://www.epa.gov/caddis/da_advanced_2.html. [Accessed 8th December 2013].

STATA (2013) *Data Analysis and Statistical Software*. [Online] Available from: http://www.stata.com. [Accessed 8th October 2013].

STATSOFT (2013) Developer of Statistica Softwares. [Online] Available from: http://www.statsoft.com. [Accessed 8th October 2013].

ToA (1997) *Treaty of Amsterdam*. Amending the Treaty on European Union, the Treaties Establishing the European Communities and Related Acts. *Official Journal*, C 340, 10. [Online] Available from: http://eur-lex.europa.eu/en/treaties/dat/11997D/htm/11997D.html. [Accessed 8th October 2013].

TOXSTAT (2013) Downloading TOXSTAT. [Online] Available from: http://www.soft32.com. [Accessed 8th October 2013].

Truhaut, R. (1977) Eco-toxicology – Objectives, principles and perspectives. *Ecotoxicology and Environmental Safety*, 1(2), 151–173.

Van Leeuwen, C.J., Van der Zandt, P.T.J., Aldenberg, T., Verhaar, H.J.M. & Hermens, J.L.M. (1992) Application of QSARs, extrapolation and equilibrium partitioning in aquatic assessment: I. narcotic industrial pollutants. *Environmental Toxicology and Chemistry*, 11, 267–282.

Veith, G.D., Defoe, D.L. & Bergstedt, B.V. (1979) Measuring and estimating the bioconcentration factor of chemicals in fish. *Journal of the Fisheries Research Board of Canada*, 36, 1040–1048.

Verhaar, H.J.M., Mulder, W. & Hermens, J.L.M. (1995) QSARs for ecotoxicity. In: Hermens, J.L.M. (ed.) *Overview of structure-activity relationships for environmental endpoints, Part 1: General outline and procedure*. Report prepared within the framework of the project 'QSAR for prediction of fate and effects of chemicals in the environment', an international project of the Environmental Technologies RTD Programme (DGXII/D-1) of the European Commission under contract number EV5V-CT92-0211.

VICH (1996) *International Cooperation on Harmonisation of Technical Requirements for Registration of Veterinary Medicinal Products*. [Online] Available from: http://www.vichsec.org. [Accessed 8th October 2013].

Zeeman, M., Auer, C.M., Clements, R.G., Nabholz, J.V. & Boethling, R.S. (1995) US EPA regulatory perspectives on the use of QSAR for new and existing chemical evaluations. *SAR and QSAR in Environmental Research*, 3, 179–201.

Chapter 2

# Fate and behavior of chemical substances in the environment

*K. Gruiz, M. Molnár, Zs. M. Nagy & Cs. Hajdu*
*Department of Applied Biotechnology and Food Science,*
*Budapest University of Technology and Economics, Budapest, Hungary*

## ABSTRACT

The production and use of chemical substances is accompanied with certain emissions to the environment. Depending on the physico-chemical characteristics of contaminants they are primarily emitted to air, water or the soil. The fate and behavior of chemicals in the environment is determined in addition to their physico-chemical characteristics, to their interactions with the ecosystem, with the living and non-living components as well as the users of the environment. The fate and behavior of chemical substances are well known under '*in vitro*' (laboratory) conditions; however, our knowledge about the interactions between chemical substances and environmental compartments, phases and their inhabitants is limited. The assessment, refinement and the pitfalls of the characterization of the behavior of chemicals in the environment is the topic of this chapter to support understanding the role and the use of fate properties in hazard and risk assessment, as well as in the risk-based management of chemicals.

## I INTRODUCTION

Environmental behavior, fate, mobility, availability and degradability of a contaminant or a mixture of contaminants in the environment have a strong influence on their actual toxicity and environmental risk. Furthermore, information about the contaminants' behavior and fate such as extreme partitioning between physical phases, photodegradation, hydrolysis or biodegradation may provide key data for the remediation of contaminated land.

Environmental *risk assessment* attributes a high priority to environmental fate characteristics for the assessment of chemical substances and contaminated sites. The real toxicity of chemical substances is determined by their effective concentration, accessibility and availability to the potentially impacted receptors. A rapidly hydrolyzing compound does not pose a high risk to the aquatic ecosystem because it has limited opportunity, i.e. a very short time period, to meet and interact with the receptors. A highly sorbable toxic element or molecule is bound to the solid phase of the sediment and does not interfere with aquatic organisms. However, sediment-dwelling organisms may be affected by the same sorbed substances. These examples explain the difference between 'actual' and 'potential' adverse impacts, or more generally, the hazard and the risk of chemical substances. The potential occurrence of chemical substances in the environment and their potential adverse effects are associated with their production, use and emission to the environment as well as with their intrinsic material properties. The hazard of a substance can be characterized based on intrinsic properties. Risk is based on the real interaction of the chemical substance and the actors of the environment: it is the actualization of the potential. The potential is not closely related to existence in the environment: a molecule is known as hazardous already on the design board. In addition to hazard and risk, regulators created a third category, called generic risk, indicating the adverse impact of chemicals under generic environmental and land-use conditions.

Along the parallel lines of hazard and risk, one can speak of environmental fate properties and actual environmental behavior, about toxicity of chemical substances and actual adverse/toxic effect of the same toxicant under certain conditions. For example, toxicity in freshwater in general differs from the toxic effect in the river Danube; the adverse effect of a toxic metal in soil in general may greatly differ from the same effect in a tropical sandy soil in the desert, or the same soil in a kindergarten, etc. Soil type, land use and the users of the land influence the realization of the contaminant's toxic potential.

The effect of the chemical substance can be measured in standardized water or soil samples by standard bioassays, using standard test organisms, but the actual impact should be measured *in situ* or in an artificial test system, which works with a representative sample and truly simulates the environmental conditions of the case being assessed. The extrapolation from hazard to environmental risk is possible, but is loaded with multiple uncertainties. For example, the absence of short-term risk in an acute test does not mean that no risk will occur over the long term in the presence of hazardous substances in the environment; some negative aquatic toxicity tests do not imply that the substance is harmless in the natural environment.

Similar to actual toxicity, 'actual concentration' is also an estimate based generally on the simplified model of a dynamic process. The potentially emitted amount of a chemical substance and its potential partitioning among physical phases in the environment can be calculated with relatively low uncertainties. On the other hand, real interactions with the living and non-living components of the environment moving away from the source in time and space are less and less predictable because of the contaminant properties. The concentration of the contaminant depends on its fate and transport in the environment, and the instantaneous concentration at a specific time and location highly depends on the environmental conditions (climatic, seasonal, meteorological parameters, hydrological and hydrogeological properties, soil type, geochemistry, heterogeneities, etc.). The effective concentration at the site of action, which can produce an effect on the receptor, is determined by the accessibility and availability of the contaminant for the targeted receptor.

Environmental *risk reduction* should involve and utilize the environmental fate characteristics of pollutants. Volatile substances should be volatilized in order to remove them, water-soluble pollutants should be flushed out, and the risk of sorbable substances should be reduced by a technique using absorption or adsorption. Technology can handle the sorbent in a separate reactor volume, which can be a proper engineering device or a specified environmental volume (reactive soil zone).

Strong sorption, without removing the pollutant, may decrease the risk by reducing the contaminants' mobility and availability to receptors (see also Chapter 2.4 in Volume 4). If the contaminant is biodegradable or highly reactive to chemical components or other reactive agents of the environment, this property is worth being utilized for its elimination from the environment.

*Natural attenuation* of contaminants in the environment – their transport performance, degradation or transformation – may serve as a basis for site remediation. Bioaccumulation can also be the basis of remediation, and in this case, the technologist should separate the accumulating organisms from the biologically 'extracted' medium, e.g. by filtrating microorganisms from water or by harvesting plants from soil. Blocking or changing the transport pathway itself provides an easy and commonly used solution for risk reduction of water and soil pollution, for example, collecting leachate from the contaminant-containing wastes instead of letting it enter surface waters or infiltrate into soil. A natural cap may reduce contaminant discharge into the air and water, etc.

The suggestion to consider environmental fate in risk assessment and utilize it for environmental risk reduction appears self-evident. However, the history of environmental management shows the opposite: environmental risk assessment is still being performed based on emitted amounts or measured concentrations of chemicals without taking into account degradation, elimination or low bioavailability. Limiting values are often based on pure chemical models, and not on a risk value which, for example, would take into account the strong bonding capability of certain elements and compounds to environmental matrices, where their adverse effects cannot be activated because they are not accessible/available for the receptors.

Risk reduction by remediation, the other main task of risk management, does not utilize sufficiently the information about transport and fate of contaminants either. 'Washing out' contaminants from soil by groundwater, pumping them to the surface and treating them (pump and treat) is still practiced at a poor technology level. It has been applied for decades at very high cost and little success, e.g. for solid-bound contaminants with limited water solubility or subsurface lenses of dense non-aqueous phase liquids (DNAPLs) which are a long-term source for dissolved contaminant plumes.

These examples elucidate how important the environmental fate and behavior of chemical substances are and how important it is that environmental managers acquire and use this information. Ignorance of fate characteristics deteriorates the efficiency of risk management and leads to risk overestimation, resulting in unnecessary costs.

Site-specific risk primarily depends on the environmental fate of contaminants, highly influenced by the quality of the environment. The transport and behavior of the same contaminant are different in a sandy or loamy soil to those in a river or a lake. The true characterization of the actual fate and effects as well as the consequent risk of a contaminant requires an integrated approach, which includes a joint assessment of the contaminant's physico-chemical and environmental fate properties as well as the characteristics of the environment.

The two main facets of the environmental fate of chemicals – partitioning among physical phases and biodegradation – have not been extensively studied by chemical science. The need for this kind of information emerged with the development of environmental toxicology and environmental risk assessment. The hazard of chemicals mainly depends on their own characteristics and their predicted interaction with the environment. The behavior of the chemicals under standardized conditions is measured and put into databases as transport and fate characteristics such as partitioning between physical phases, degradability, or their (bio)accumulative potential.

The mobility and bioavailability of a contaminant depend on its intrinsic properties and represent its potential to move on and affect biological entities. Mobility and availability can only be interpreted when interactions with other environmental agents and matrices are considered. Mobility of a chemical only makes sense when its interaction with the medium is taken into account. Availability should reflect the interactions with radiation/light, water, or the members of the ecosystem. The characteristics of the environment determine the receiving capability, the resistance toward the behavior and impacts of the contaminant. For example, transport of the contaminant in soil will be determined by its mobility and the retention capacity of the soil. Retention capacity of the soil depends on its texture, i.e. sandy, loamy sandy, loamy or clayey. The sorption capacity of the solid phase and the partitioning of the contaminant between physical phases determines the scale of volatilization, leaching, desorption or sorption, and highly influences the contaminant's availability for water and living organisms. Magnitude of the toxic effect depends both on the ecotoxicity of the chemical compound and the sensitivity of the organisms living on the site of exposure, as well as the mode or interactions between them. Transport and availability of contaminants, e.g. in soils can properly be characterized only by the integration of chemical analytical, biological and ecotoxicological information.

Partitioning of toxicity in soil among the solid, aqueous and gaseous phases results in risk to groundwater, soil air and, as a consequence, to the surrounding atmospheric air and surface waters. Partitioning in sediments between the solid phase and pore water determines water quality. Strong binding and low biodegradability may lead to the development of a chemical time bomb, i.e. the latent presence of a contaminant without any symptoms for a while, and the potential for a sudden risk increase due to a change in environmental conditions such as redox potential, pH, temperature or pressure, chemical composition or any other factors influencing mobility and toxicity.

When measuring the toxicity of solid environmental matrices, actual toxicity can be better characterized by contact tests that simulate a realistic scenario, where mutual interactions occur. The results of interactive bioassays include the results of the possible interactions between all participants: contaminant, contaminated medium and test organism.

## 2  INTERACTION OF THE CONTAMINANTS WITH ENVIRONMENTAL PHASES

Interactions between chemical substances and environmental compartments can be classified as physico-chemical or biological processes. Another classification of interactions is based on the compartment: interactions within the atmosphere, in surface waters and sediments or soils. The biota is considered as an additional compartment linked to aquatic and terrestrial ecosystems. All compartments have three physical phases, but one phase generally dominates: gas in the atmosphere, liquid in aquatic compartments and solid in soils and sediments. Other phases are always present such as liquid in the form of vapor and solid in the form of dust in air, suspended solid and dissolved gas in waters, pore water in soil and sediment, and soil gas in the three phases of soil. We shall discuss the interactions of chemicals with the environment in the following three groups:

- Transport and fate processes, including partitioning;
- Physico-chemical interactions;
- Biological interaction, primarily biodegradation and bioaccumulation.

## 2.1  Transport and partitioning

Transport processes and partitioning of a chemical substance between the phases play a major role in the contaminant's spread in the environment. These transport processes alter the distribution and concentration of contaminants, but do not change the absolute amount present in the environment.

*Transport* refers to the way chemicals move in the environment or in organisms to their ultimate destination and also the way they reach their final target. Transport of chemicals in the environment integrates advection, diffusion, sorption and desorption, solution and dissolution, infiltration, drainage, dilution, partitioning, etc. of the contaminant on the pathway from the source to the receptor.

*Partitioning* is a complex transport process which includes the distribution of a chemical substance among the environmental phases. The two main partitioning scenarios occur between air and water, and water and solid. In these situations opposing processes seek to balance evaporation and condensation; dissolution and precipitation; sorption and desorption; uptake and release, integration and disintegration, mobilization and immobilization, etc.

*Environmental* transport and fate – this term covers all physical, chemical and biological processes linked to the movement (changing position and phases) and transformation (physical alteration, chemical and biological reactions, including degradation processes) of the chemical agent after it enters the environment. The fate of contaminants in the environment allows for prediction of impacts and risk calculations for the ecosystem and humans.

Equilibrium partitioning of a chemical substance between physical phases can be calculated from the basic physico-chemical parameters of the chemical substance (vapor pressure, water solubility, etc.) and the environmental parameters (temperature, pH, etc.). Given that there is never an equilibrium state in the environment, measured distribution – preferably as a function of time – should be used to reflect the real situation. In the following paragraphs the calculation of equilibrium partitioning will be introduced.

### 2.1.1  Partitioning between air and water

Partitioning between air and water is characterized by Henry's law constant quantitatively, giving the equilibrium ratio of the concentration of the chemical substance in air to that in water.

$$Henry = p \cdot M/c,$$

where

- *Henry*: Henry's law constant (Pa·m$^3$/mol);
- *p*: vapor pressure (Pa);
- *M*: molecular mass (g/mol);
- *c*: solubility in water (mg/L).

Partitioning between any liquid phase and gaseous phase can be characterized by the dimensionless Henry's law constant:

$$K_{\text{liquid-gas}} = Henry/R \cdot T,$$

where

- $K_{\text{liquid-gas}}$: partitioning of the chemical substance between liquid and gas;
- *R*: gas constant (Pa·m$^3$/mol·K);
- *T*: temperature (K).

### 2.1.2   Partitioning between solid and water

Partitioning between solid and water is quantified by the ratio of the concentration in the soil and sediment solid to that in the liquid phase. When calculating partitioning, one should distinguish between organic and inorganic contaminants. The equilibrium partitioning can be expressed as

$$K_p = C_{\text{solid}}/C_{\text{water}}$$

The value of $K_p$ for organic substances depends on the organic matter content in the solid phase, given that organic contaminants are bound to the organic matrix (e.g. the humus fraction of the soil)

$$K_p = F_{\text{oc}} \cdot K_{\text{oc}},$$

where

- $F_{\text{oc}}$: the organic fraction in the soil or sediment;
- $K_{\text{oc}}$: partitioning of the contaminant between the soil's organic carbon content and water content (i.e. pore water or soil moisture).

$$K_{\text{oc}} = a \cdot K_{\text{ow}}/1000,$$

where

- *a*: a constant (a = 0.411);
- $K_{\text{ow}}$: octanol–water partition coefficient.

Cations, the mobile inorganic atoms or molecules with an electrical charge, are distributed between liquid and solid phases in the soil. $K_d$, the equilibrium constant, characterizes their partitioning. The easiest way to determine this value is the analytical measurement of the cations in both phases after equilibrium setting:

$$K_d = C_{\text{cation water}}/C_{\text{cation solid}}$$

The value of $K_d$ depends on the size and strength of the cation and the sorption capacity of soil. When using these mathematical relationships, it has to be considered that there is no equilibrium (or steady state) in the environment, in spite of the fact that the processes tend towards equilibrium. This means, for example, when contaminants are transported by groundwater which flows through various volumes of the solid soil, a part of the contaminant is sorbed on the solid soil depending on contact time, but the concentration in the water is always higher than the equilibrium concentration. Using equilibrium concentration in the calculations may lead to underestimation of the risk.

### 2.1.3 Transport models

For the purpose of generic risk assessment, for example, authorization or restriction of chemicals, the European REACH regulation proposes the use of transport models, which should uniformly be used. The models indicate all possible transport routes and the risk assessor should decide about the priority risk components and adapt the generic model to the specific chemical substance and problem. The same transport models can be used for site-specific purposes, but only after a careful study and creation of the conceptual risk model for the site identifying the key transport routes and risk components.

The main difference between the generic and site-specific use of transport and fate models is that, for regulatory purposes, so-called generic environmental parameters are included as default values, which are valid for the whole of Europe (i.e. for no particular site). In the site-specific version, one has to apply the characteristic parameters of the particular site, which are not readily available and should be gathered or measured by the user of the transport model.

Currently, many fate and transport models are available for chemicals in the environment. When using a transport model to estimate the predicted environmental concentration (PEC) of a chemical substance, the environmental concentration in all environmental compartments must be determined, taking into consideration partitioning, degradation and other physical, chemical and biological changes and interactions (Figure 2.1). The measured and calculated parameters and the pre-chosen data have to be discussed and documented in the risk assessment study.

Figure 2.1 shows the scheme of transport from the source of chemical substance to environmental compartments on single or multiple transport pathways. Environmental compartments themselves are complex systems with up to three physical phases. They also contain organic and inorganic constituents in different chemical forms and the community of the members of the ecosystem. The transport itself includes every kind of displacement or movement of the contaminant by precipitation, infiltration, surface runoff, leaching, diffusion, dilution, erosion, sedimentation, partition, chemical reactions (photodegradation, hydrolysis, oxidation and reduction), interaction with the biota (biotransformation, uptake and biodegradation), etc. The end points of the transport are the environmental compartments which are themselves endangered by the contaminating chemical substance, in addition to their ecosystem and the land users.

Generally, the following parameters are given as input data in exposure models:

– Physical-chemical characterization of the substance:
  o molecular mass;
  o vapor pressure;

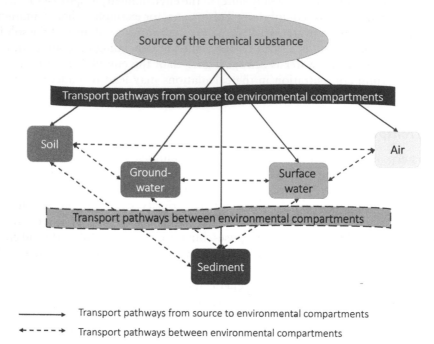

*Figure 2.1* Transport from source to environmental compartments.

  o  boiling point;
  o  Henry or Henry's law constant
  o  octanol–water partition coefficient;
  o  water solubility.

– Known amounts or fluxes of the chemical substance, taking into consideration their use patterns:

  o  amount produced;
  o  amount used;
  o  amount emitted;
  o  measured or calculated flux in the environment.

– Environmental parameters:

  o  environmental compartments concerned;
  o  environmental phases;
  o  geochemical properties;
  o  hydrogeological properties, etc.

The concentration of substances found in the environment can be construed on local ($PEC_{local}$), regional ($PEC_{regional}$) and continental ($PEC_{continental}$) scales. These PEC values have an effect on each other because the continental area contains the regional scale and the regional scale contains the local area. The boundaries of the site to be modeled should be exactly defined and delineated.

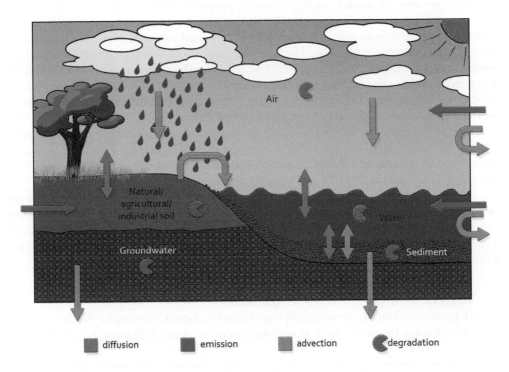

*Figure 2.2* Simple Box concept: environmental compartments and transport routes.

The most frequently used conceptual model for environmental purposes is the Simple Box model (Mackay *et al.*, 1992; Van de Meent, 1993; Brandes *et al.*, 1996) which enables the calculation of both regional and local exposure concentrations in all environmental compartments. The main facet of a box model is that it does not resolve the spatial heterogeneities and the distribution of the contaminant concentration inside the box. It is assumed that the box is "well-mixed" and from the difference of input and output contaminant fluxes one can calculate a removal rate or residence time.

The Simple Box model incorporates the fate properties of direct and indirect emissions, biotic and abiotic degradation in all compartments, diffusive transport, advective transport and partitioning between phases. Substance input to the model is considered continuous and equivalent to continuous diffuse emission. The results provided by the model are steady-state concentrations which can be considered long-term average exposure values. In the model, the substance released is distributed among the compartments according to the properties of the substance and the modeled environment. The concentration of a substance at the border of the modeled region must be taken into account as background concentration.

The Simple Box model (Figure 2.2) can be tailored to concrete site-specific uses, modified or simplified by excluding any of the environmental compartments, phases and non-typical processes in a certain case, for example, evaporation and partitioning between air and water can be excluded when the transport of a non-volatile chemical substance is modeled in a two-phase system of solid and water.

Another type of model, used for generic purposes, such as regulation of chemicals or catchment-scale environmental management, having an important role in the estimation of environmental concentrations of chemical substances in aquatic systems, describes wastewater treatment. The basis of these models is the SimpleTreat model shown in Figure 2.3 (Struijs *et al.*, 1991) which provides a quantitative description of the processes that take place in an average-sized sewage treatment plant based on aerobic degradation using activated sludge. The SimpleTreat model enables computation of the steady-state concentrations in a sewage treatment plant.

The SimpleTreat model provides information about the amounts of chemical substances entering and leaving a sewage treatment plant. The removal of a substance is influenced by the physical-chemical and biological characteristics of the substance and the operational conditions of the sewage treatment plant.

The revised version of the SimpleTreat model (Mikkelsen, 1995) incorporates an improved process formulation for volatilization from the aeration tank. More specific information on the biodegradation of a substance may be available in the modified version of the SimpleTreat model at a higher tier of the risk assessment process. The following scenarios are optional:

- Temperature dependence of biodegradation;
- Degradation kinetics according to the Monod equation, created for the description of biological reactions, such as enzyme reactions, microbial growth rate, etc.;

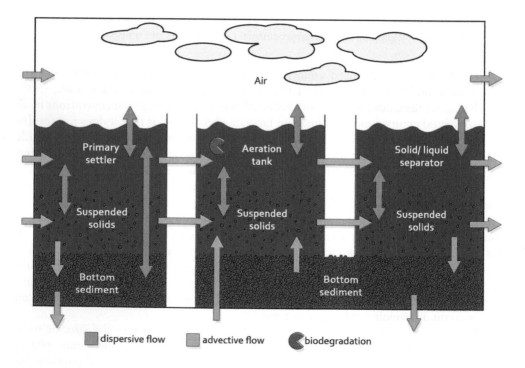

*Figure 2.3* SimpleTreat model for characterizing the fate and transport of contaminants in a wastewater treatment plant using activated sludge.

- Degradation of the substance in the adsorbed phase;
- Variation in sludge retention time;
- Exclusion of a primary settler.

The Simple box model is a strong simplification suitable for local transport modeling with local parameters or for regional transport modeling with generic parameters, supposing homogeneous environmental distribution in the "box". A different concept is required for site-specific transport modeling, e.g. for the calculation of flux or access time, for example from the contamination source of an industrial facility to the next drinking water well. This kind of calculation needs (partly) numerical models, including the site-specific environmental parameters, their distribution, heterogeneities and interactions between the contaminant and the medium. When regional, e.g. watershed-scale transport is to be modeled by real parameters or well established generic ones, a GIS-based approach should be applied, using meteorological, topographical, hydrological, geological, geochemical, soil typological parameters, surface coverage, etc. of the area in the form of 3-dimensional maps (see also Chapter 10 in Volume 1).

## 2.2 Chemical interactions between chemical substances and the environment

Physico-chemical interactions between contaminants and the environment take place in the form of photolysis, hydrolysis, sorption, desorption and chemical reactions such as oxidation, reduction, radical reactions and chemical modifications. Certain interactions with air or water cause a very abrupt or very intensive reaction or produce dangerous chemicals, for example, flammable solids, spontaneously combustible solids (dangerous when wet) or explosives.

Some of the chemical reactions in the environment result in a decrease in toxicity or other adverse effects of chemicals, but the opposite may also occur: a chemical transformation may produce a more dangerous form (oxidation of sulfides into sulfuric acid), or a more risky situation (desorption from soil or sediment particles and thus contaminating sensitive waters). Risk-reducing chemical interactions, such as precipitation or reduction in an insoluble form can be utilized in risk-reduction measures (waste (water) treatment, reduction of technological discharge, etc.) and in remedial technologies applied to contaminated environments.

### 2.2.1 Photolysis

*Photolysis,* also called photodissociation or photodecomposition is a chemical reaction in which a chemical compound is broken down by photons. Photodissociation is not limited to visible light: electromagnetic waves or ultraviolet light, X-rays and gamma rays are usually involved in these radical photoreactions, given that their energy is higher than that of visible light.

*Photolysis in the atmosphere* plays an important role in eliminating many atmospheric pollutants, such as hydrocarbons (e.g. methane) or nitrogen oxides. The formation of the ozone layer in the stratosphere is also a result of a photoreaction of oxygen.

The degradation rate by photolysis in the *atmosphere* can be described by the equation:

$$k_{\text{deg air}} = k_{\text{OH}} \times C_{\text{OH air}} \times 24 \times 3600,$$

where

- $k_{\text{OH}}$: specific degradation rate constant with OH radicals ($cm^3$/molecule/s);
- $C_{\text{OH air}}$: concentration of OH radicals in the atmosphere (molecules/$cm^3$)[1];
- $k_{\text{deg air}}$: pseudo first-order rate constant for degradation in air (1/day).

*Photochemical degradation in water* is important for such chemical substances that are not degradable in any other way i.e. by hydrolysis or biodegradation. Photolysis in natural waters is strongly limited by the intensity of light (seasonal and geographical differences) and the density of water (dissolved and suspended solids in water) which absorbs most of the light.

The value of half-life for photolysis in water (if known) can be converted into a pseudo first-order rate constant:

$$k_{\text{photo water}} = \ln 2 / DT_{50 \text{ photo water}}$$

where

- $DT_{50 \text{ photo water}}$: half-life for photolysis in surface water (days);
- $k_{\text{photo water}}$: first-order rate constant for photolysis in surface water (1/day);
- ln: natural logarithm (to the base e).

### 2.2.2  Hydrolysis

*Hydrolysis* is the degradation of chemical compounds by water. It can be a chemical process, including the hydrolysis of salts into acids and alkaline products, degradation of starch or cellulose into sugars, and the hydrolysis of esters and fatty acids. For many substances, the rate of hydrolysis depends to a large extent on the specific environmental pH and temperature and, in the case of soil, on moisture content as well. The rate of hydrolysis always increases with increasing temperature.

In addition to chemical hydrolysis, enzymatic hydrolysis is also common in the environment. Enzymatic hydrolysis can proceed in water and soil triggered by free enzymes bound to soil or sediment particles or by enzymes within living organisms. The free ones are exoenzymes produced and secreted by living organisms, or otherwise derived from them, e.g. following their death and decomposition. Most of the enzymatic processes are connected to or take place within living organisms.

Hydrolysis is quantified by the half-life ($DT_{50}$) of hydrolysable chemical substances. $DT_{50}$ or the degradation rate of chemical substances can be determined in standardized tests. $DT_{50}$ can be converted into degradation rate. The first-order rate constant $k_{\text{hydr water}}$ is:

---

[1]The global annual average OH radical concentration can be assumed to be $5 \times 10^5$ molecules/$cm^3$ (EU-TGD 2003).

$$k_{\text{hydr water}} = \ln 2 / DT_{50 \text{ hydr water}},$$

where

- $DT_{50 \text{ hydr water}}$: half-life for hydrolysis in surface water (days);
- $k_{\text{hydr water}}$: first-order rate constant for hydrolysis in water (1/day);
- ln: natural logarithm (to the base e).

### 2.2.3  Chemical oxidation and reduction

*Oxidation* is the loss of electrons or an increase in oxidation number. *Reduction* is the gain of electrons or a decrease in oxidation number.

Both oxidation and reduction may strongly influence the contaminant's chemical form and behavior in the environment. Oxidation or reduction of a substance in the environment depends on its oxidation state and the redox potential of the environment. The change in redox potential generally modifies the fate and transport properties of chemical substances as well as their availability to and effects on living organisms. For example, mercury is highly toxic in elemental or methylated forms, but not at all in low redox-potential wetlands where it is present in the chemical form of sulfides. Chromium VI loses its toxicity when reduced to chromium III, due to the alternative respiration of facultative anaerobic microbes that lower redox potential in the soil. Toxic organic compounds degrade and lose their toxicity because of the oxidizing effect of chemicals or microorganisms. They can be degraded into $CO_2$ and water which are not toxic.

As there is a redox gradient proportional to depth both in water and soil, this controls the form and rate of oxidation for the contaminants and other components of the environment.

OECD has issued Test Guidelines for testing fate properties and REACH made them obligatory and uniformly applicable. The key tests for the characterization of transport, fate and behavior of chemical substances in the environment are shown in Table 2.1.

Table 2.1  OECD guidelines for the measurement those physico-chemical properties of chemicals that influence their environmental fate and behavior.

| Test Guideline | Test name |
| --- | --- |
| OECD TG 101 (1981) | UV-VIS Absorption Spectra |
| OECD TG 104 (2006) | Vapor Pressure |
| OECD TG 105 (1995) | Water Solubility |
| OECD TG 106 (2000) | Adsorption–Desorption Using a Batch Equilibrium Method |
| OECD TG 107 (1995) | Partition Coefficient (n-octanol–water) – Shake Flask Method |
| OECD TG 108 (1981) | Complex Formation Ability in Water |
| OECD TG 111 (2004) | Hydrolysis as a Function of pH |
| OECD TG 112 (1981) | Dissociation Constants in Water |
| OECD TG 113 (1981) | Screening Test for Thermal Stability and Stability in Air |
| OECD TG 116 (1981) | Fat Solubility of Solid and Liquid Substances |
| OECD TG 117 (2004) | Partition Coefficient (n-octanol–water) – HPLC Method |
| OECD TG 120 (2000) | Solution/Extraction Behavior of Polymers in Water |
| OECD TG 121 (2001) | Estimation of the Adsorption Coefficient ($K_{oc}$) on Soil and on Sewage Sludge using High Performance Liquid Chromatography (HPLC) |
| OECD TG 123 (2006) | Partition Coefficient (n-octanol–water) – Slow-stirring Method |

# 3 INTERACTIONS OF CHEMICAL SUBSTANCES WITH THE BIOTA

Interactions of chemicals with the biota reflect both the ability of the chemical substance to become available and accessible to living organisms in the environment and the genetic, metabolic, physiological and behavioral characteristics of the living organism to respond to it. This response can be detrimental, lethal or adaptive, determining the survival of the individual, the population or community.

Adverse effects such as toxic, mutagenic, reprotoxic, hormone and immune system disrupting and sensitizing effects are discussed in detail in Chapter 3. In the course of interactions with the contaminants, the sensitivity of an organism and the adaptive behavior of a community in the environment are essential: for example, emergence and dispersal of one single gene in a microbial community can reverse the response from lethality to survival.

Adaptation, a change in the genome of individual organisms and its rapid spread to the whole population and the community, is a basic property of microorganisms which typically live in soils, sediments, slurries and sludges, in liquid and solid wastes. Adaptation plays an essential role in environmental biodegradation, biotransformation, bioaccumulation, bioleaching and biostabilization in general, and, in particular, related to contaminants and xenobiotics.

A chemical's three specific properties used to describe its potential hazard to the environment and to humans through the environment are summarized here.

- Toxicity and other adverse effects: the hazard posed by a substance to living organisms, based on toxicity measured by aquatic and terrestrial organisms, bacteria, fungi, plants, animals and humans.
- Degradability: persistence of the substance in the environment, based on its molecular structure or test results.
- Bioaccumulation/bioconcentration: the accumulation of a substance in living organisms, and toxication of humans and other top predators causing biomagnification through the food chain.

Toxicity and its testing are discussed in Chapters 3, 4 and 5, biodegradation and bioaccumulation in the following two sections (3.1 and 3.2).

## 3.1 Biodegradation and biotransformation

Biodegradation of chemical substances, mainly xenobiotics, is a key process of the elimination of chemicals from the environment both spontaneously and by engineering solutions using biological and ecological technologies. The risk of a chemical substance is primarily determined by its biodegradability which, together with its toxicity and bioaccumulative potential, determines the type and scale of risk, the risk management methods, the urgency of intervention and the legal obligations.

### 3.1.1 Classification of environmental fate of chemicals for regulatory purposes

Biodegradation and bioaccumulation play key roles in the categorization of the most dangerous chemicals, the so-called PBTs (Persistent, Bioaccumulative and Toxic chemicals). Classification is based on the combination of persistency, bioaccumulative potential and toxicity within the REACH, and is described in Annex XIII of the regulation (PBT REACH, 2008). The REACH criteria are the following:

– Persistent (P) if bioaccumulation half-life is:

  o $t_{1/2} > 60$ days in marine water or
  o $t_{1/2} > 40$ days in fresh- or estuarine water or
  o $t_{1/2} > 180$ days in marine sediment or
  o $t_{1/2} > 120$ days in fresh- or estuarine sediment or
  o $t_{1/2} > 120$ days in soil.

– Bioaccumulative (B), characterized with a bioconcentration factor of

  o $BCF > 2000$ L/kg or $K_{ow} \geq 4$.

– Toxic (T), and fulfilling the following chronic toxicity criteria:

  o NOEC (long-term) $<0.01$ mg/L;
  o substance is classified as carcinogenic (category 1 or 2), mutagenic (category 1 or 2), or toxic for reproduction (category 1, 2 or 3) or
  o there is other evidence of chronic toxicity, as identified by the classifications: T, R48, or Xn, R48 according to Directive 67/548/EEC.

Another priority group within the REACH Regulation is formed by PBTs or vPvBs:

– Very persistent (vP) characterized by the biodegradation rate of:

  o $t_{1/2} > 60$ days in marine, fresh- or estuarine water or
  o $t_{1/2} > 180$ days in marine, fresh- or estuarine sediment or
  o $t_{1/2} > 80$ days in soil.

– Very bioaccumulative (vB) is a substance which fulfills the biodegradation criterion of:

  o $BCF > 5000$ L/kg or $K_{ow} > 5$.

Both PBTs and vPvBs are subjected to close scrutiny in the EU as they may have a long-term impact on the environment.

Quantitative criteria for persistence, bioaccumulation and toxicity in other countries are also strictly regulated, in Canada, for example, the following applies (GM CEPA, 1999):

– Persistence: half-life values expressed in days

  o air $\geq 2$
  o water $\geq 182$
  o sediment $\geq 365$
  o soil $\geq 182$

– Bioaccumulation:

  o bioconcentration factor, $BCF \geq 5000$ or
  o log $K_{ow} \geq 5$

– Toxicity (mg/L)

  o acute hazardous effect $L(E)C_{50} \leq 1$
  o chronic hazardous effect $NOEC \leq 0.1$.

The regulations also suggest that, in addition to toxicity, the environmental fate properties, first of all biodegradability, determine the environmental risk of chemicals and their categorization as PBT.

### 3.1.2 Biodegradation – definitions

Before introducing the biodegradation measurement methods, some terms must be clarified in connection with biodegradation of chemicals.

*Biodegradability* or more precisely a chemical's potential to be degraded by biological systems is an intrinsic material characteristic, which depends on the chemical structure and molecular mass of the chemical substance.

*Biodegradation* is the biological decomposition or breakdown of a substance. The term is mainly used for the action of environmental microorganisms such as bacteria or fungi. Biodegradation is affected by both the chemical substance's structure and resistance to biodegradation and the biological activity of microorganisms, or the community of microorganisms in the environment, which can be highly diverse depending on climate, season and the adaptive character and behavior of the degrading community. The presence of other chemicals is another influencing factor. Mineralization is distinguished by biodegradation: it denotes the natural process of complete biodegradation in which organic matter is fully degraded under aerobic or anaerobic conditions. As a result, gaseous or water-soluble inorganic residues of $H_2$ or $H_2O$ are produced from all reduced organic molecules, C or $CO_2$ from organic carbon, $NH_3$ or $NO_3^-$ from nitrogen-containing organic molecules, $PO_4^{3-}$ from phosphorous- and $S^{2-}$, $SH^-$ or $SO_4^{2-}$ from sulfur-containing biomolecules, and positive or negative ions from all kinds of metals, originally contained in living or dead organic matter. These inorganic forms are available and can be taken up by plants and other primary producers in soil and sediment. Mineralization is responsible for the biogeochemical element cycles i.e. the transition from biologically controlled forms to abiotic forms in the Earth's hydrosphere, lithosphere, and the atmosphere. Many of the xenobiotics are only partially biodegraded, but not mineralized, so they become excluded from the closed element cycles for a long time.

*Primary biodegradation* changes the identity of the parent chemical as a first step toward more extensive mineralization, whereas *ultimate biodegradation* results in the complete mineralization of the chemical to water, carbon dioxide and inorganic compounds.

*Biotransformation* of chemical substances plays a role in higher-order organisms so that they can be more readily eliminated whereas simpler organisms such as bacteria and fungi are responsible for most of the extensive biodegradation and elimination of substances from the environment as part of the mineralization process. Biotransformation can also be aimed to reach a stable form of the toxicant which is not available for the organisms during their normal lives. The stable form may include reserve nutrients, defender compounds or any kind of harmful material stored after translocation and sequestration in organelles. These biochemically and genetically established protective mechanisms ensure reduced risk for the species, but not for consumers that come next in the food chain.

### 3.1.3 Biodegradation – the process

Biodegradation of contaminants in the environment as part of the mineralization process means that the community of microorganisms tries to utilize its existing genetic and metabolic tools to produce energy from any form of a chemical substance that contains energy (molecules in reduced form). If their existing metabolic potential is not enough to acquire this chemically rapidly available energy, they try to adapt themselves to the 'new', less accessible, often toxic energy source by genetic modifications. These include:

– Switching on formerly switched-off (reserve) genes;
– Increasing the mutagenic rate to multiply the number of mutations and selecting the beneficial mutants for replication through normal reproduction;
– Creation of genetic recombinants by horizontal gene transfer, meaning the delivery (handover) of genes between the cohabiting members of the population or the whole community without any sexual event generating offspring.

For the mineralization of a chemical, several enzymes must act sequentially to transform or break down the chemical substance to molecules that enter the pathways of intermediary metabolism. Synthetic chemicals that are likely to be substrates for enzyme(s) produced by microorganisms can enter these pathways and become mineralized. Xenobiotics, not being complementary to any existing enzymes, have no chance of entering any of the active metabolic pathways and become degraded.

The rate of biodegradation in soil, as an example is characterized by the following equation (EU-TGD, 2003):

$$k_{\text{biodeg soil}} = \ln 2 / D_{50 \text{ biodeg soil}}$$

where

– $k_{\text{biodeg soil}}$: first-order rate constant of biodegradation (1/day);
– $DT_{50}$: half-life of the chemical substance (days);
– ln: natural logarithm (to the base e).

Biodegradation of chemical substances in water and soil is classified according to their biodegradation rate. The classes proposed by the European guidance are shown in Table 2.2.

The rate of biodegradation in surface water, soil and sediment is related to the chemical structure and other environmental properties, such as $K_{ow}$ of substances,

Table 2.2 Biodegradation classes, relevant rate constants and half-lives.

| Classification of biodegradation | k (1/day) | $DT_{50}$ (days) |
|---|---|---|
| Readily biodegradable | $4.7 \times 10^{-2}$ | 15 |
| Readily, but failing 10-day window | $1.4 \times 10^{-2}$ | 50 |
| Inherently biodegradable | $4.7 \times 10^{-3}$ | 150 |
| Not biodegradable | 0 | $\infty$ |

microbial activity, and environmental conditions such as temperature, pH, redox potential, moisture content, etc. These properties vary from site to site: the random adaptation of the microbiota yields biodegradation of a site-specific character. Therefore the best practical solution is to directly measure biodegradation as developed at a certain location in soils, sediments or wastes.

For generic and regulatory purposes, the notion of 'biodegradability' of a chemical substance is being used, which is a function based on quantitative structure–activity relationship (QSAR) models or standard bioassays and is the intrinsic property of a chemical substance. QSAR applies the knowledge of structure and biodegradability relationships to create a mathematical model. The already existing information, i.e. test results, are collected and statistically evaluated and a generally applicable QSAR equation is established, which can be used to calculate biodegradability of most chemical substances when their chemical structure and/or $K_{ow}$ is known.

Laboratory bioassays measuring biodegradability are not well established because the test cannot be applied to a stable, uniform, standardized microflora, given that such a microflora does not and will not exist. Even if it were established, its stability could not be ensured (unchanged metagenome) for a long time. It cannot be guaranteed either that specific genes, responsible for certain biodegradation pathways, would not be present, or would not appear suddenly. So, one must be aware of all these shortcomings when applying a standardized biodegradability test for the characterization of the chemical substances' intrinsic ability to be biodegraded.

Precise estimation is hampered by the simplifications of the models that assume that the kinetics of biodegradation is of pseudo first-order, which is strictly speaking not true in the real environment. Another simplification is that the dissolved portion of the substance is considered as available for biodegradation, which is of course incorrect in a dynamic system (shift in balance, mobilization and recharge, interactions with microorganisms) and is refuted by the results of those interactive test setups which let the interactions develop between solid-matrix-bound contaminants and living organisms. The microbes produce biotensides, exoenzymes and mucilage to control mobility and increase biological access to contaminants of strong tendency to be bound to soil matrix.

Site-specific biodegradation tests may provide a realistic and adequate answer to the elimination rate of a contaminant at a certain locality under natural conditions.

Biodegradation reduces the concentration of chemicals in the environment and influences their adverse effects. If the chemical substance is degradable via photolysis, hydrolysis or biodegradation, the amount that reaches the receptors and is capable of exerting an adverse effect decreases in time. Emissions from the same amount of a readily biodegradable and a persistent chemical substance – even if they have the same level of toxicity – result in very different environmental concentrations after a certain time. For example, the biodegradable one will be completely eliminated within two weeks reducing its concentration to 0%, while 100% of the persistent one will remain in the environment.

### 3.1.4   QSAR for biodegradation

QSAR, based on a database that contains a number ($n$) of empirical data on biodegradation, establishes the mathematical relationship between the structure of a chemical substance and its biodegradability. One set of results, serving as the basis of a QSAR equation, should be derived from the same standardized test method.

Qualitative information is available for many biodegradation pathways. Major sources of empirical data are the University of Minnesota Biocatalysis/Biodegradation Database (Wackett & Ellis, 1999), Biodegradability Evaluation and Simulation System (Punch et al., 1996a,b; BESS, 2013), the Syracuse Research Corporation's BIODEG database (BIODEG, 2013), which contains over 5800 records of experimental results on biodegradation studies for approximately 800 chemicals, and the MITI database of the Japanese Ministry of International Trade and Industry (METI, 2013). Most of these data are derived from laboratory studies. These databases have been used in model development and can also be used in generic risk assessment. However, the established models are not considered directly applicable for estimating environmentally relevant biodegradation rates for a wide range of chemicals.

Amongst the many software developed for the estimation of biodegradation in nature and biodegradability of a chemical for regulatory purposes, we emphasize the Biodegradation Probability Program (BIOWIN, 2013), which estimates the probability for the rapid biodegradation of an organic chemical in the presence of mixed populations of environmental microorganisms (Boethling & Sabljic, 1989; Raymond et al., 1999; Howard et al., 1992; Boethling et al., 1994, 2003; Tunkel et al., 2000; Howard et al., 2005; Pavan & Worth, 2006).

BIOWIN was developed by the Syracuse Research Corporation (SRC, 2013) and the US EPA as part of the Estimation Program Interface Suite (EPISUITE™, 2013) model package and is freely available from the US EPA website (US EPA, 2004). BIOWIN includes six different models:

– Linear probability BIODEG;
– Non-linear probability BIODEG;
– Expert survey ultimate biodegradation model (USM);
– Expert survey primary biodegradation model (PSM);
– Japanese MITI linear and
– Japanese MITI non-linear (Arnot et al., 2005).

These are generally referred to as BIOWIN 1–6. Estimates are based on fragment constants and molecular mass and require only a chemical structure. BIOWIN models have been developed and tested for a range of chemical substances for assessing the biodegradation potential of chemicals by regulatory agencies to exclude the uncertainty due to the natural variability and adaptive character of degrading microflora. So one can directly estimate biodegradation from the structure of the chemical substance, which is the cheapest way to obtain useful data.

Some QSAR equations are introduced here, for the biodegradability of selected chemical substances using the octanol–water partition coefficient ($K_{ow}$) and the charge of chemical bonds or electrophilicity of the molecules for the determination of biodegradability.

– Phthalate esters:
   $RC = -24.308 \times \log K_{ow} + 394.84$
   RC: primary biodegradability $n = 12$ $r^2 = 0.87$ (Boethling, 1986);
   log: common logarithm (to the base 10);

– Alcohols, phenols, ketones, carboxylic acids, ethers, sulfonates:

$BOD = 1105 \cdot \Delta/\delta/_{x-y} + 1.906$
BOD: Biological Oxygen Demand  $n = 112$  $r^2 = 0.98$ (Dearden & Nicholson, 1987)
$\Delta/\delta/_{x-y}$: differences in the modulus of atomic charge across a functional group bond $x-y$;

– Alcohols:

$BOD = 0.093 \times S_E - 3.163$
$S_E$: electrophilic superdelocalizability  $n = 19$  $r^2 = 0.96$ (Dearden & Nicholson, 1987).

### 3.1.5  Aims of testing biodegradation

The principle of **biodegradability** testing is the same in any applied test method: degradation of a chemical substance as the only carbon source. In biodegradability tests a standardized culture medium containing microflora or artificial/natural water or soil is used to indicate and monitor biodegradation. The most widespread way is measuring $O_2$ depletion and/or $CO_2$ production during an aerobic biodegradability test, but any other actor of the energy production and the respiration chain in the cells can be selected and detected, most frequently the substrate, metabolites or enzymes. Using alternative electron acceptors to indicate electron transport – the concomitant of energy production – makes the evaluation easy, i.e. by producing a proportionate amount of simply detectable colored product. These types of tests reveal the tendency of a chemical substance to become biodegraded as its intrinsic property. In spite of careful standardization, the composition of the microorganisms used in these standardized tests cannot be controlled completely: this may cause high uncertainties and makes biodegradability of chemicals a fiction.

The **simulation of biodegradation** of a certain chemical substance under water, sediment or soil conditions is a test type which combines the chemicals' intrinsic biodegradability with typical environmental conditions existing in water, sediment or soil. Testing biodegradability in soil, a standard soil type with constant texture, organic matter content, soil moisture, bulk density, pH and redox potential should be ensured. Simulation tests provide useful results for the evaluation of the generic risk of chemicals on aquatic or terrestrial ecosystems, but the uncertainties are high and the differences between the generic and the site-specific ecosystem may deviate – theoretically in both directions jeopardizing the conservative (pessimistic) approach to be applied in an environmental risk assessment.

When extrapolating from generic biodegradability test results to actual biodegradation at a real site, deviation mainly occurs in favor of biodegradation in a real environment, due to ecosystem flexibility. The real biodegradation rate may be much higher compared to the results of biodegradability or simulation tests due to the adaptation of the biodegrading microbiota at a contaminated site, especially if it has been contaminated for a long time.

**Testing local biodegradation** in contaminated water, sediment or soil reflects the complex situation and the progress of degradation. The outcome of site-specific biodegradation tests includes the results of interactions between contaminants, contaminated matrices and the degrading community. The intrinsic biodegradability of the chemical substance is combined with its actual accessibility and bioavailability, and with the genetic and biochemical potential of the exposed microbiota. Based on biodegradation test results, the site-specific risk can be precisely assessed and adequate risk reduction measures can be implemented. The biodegradability measurement methods cannot be standardized similar to biodegradability or generic simulation tests: the main rule in site-specific testing is to copy or mimic real conditions as much as possible, e.g. by using microcosms or mesocosms.

The most frequently used test set-ups for measuring biodegradation are the closed and open bottles and the flow-through systems. These test systems are suitable for testing water and soil under aerobic, anoxic and fully anaerobic conditions with or without additive injection, component extraction or sampling.

### 3.1.6  Measurement end points for characterizing biodegradation

Biodegradation tests may detect the following end points:

– Substrates used for energy consumption or co-metabolism;
– The degrading cells' density and activity;
– Enzymes produced playing a role in the biodegradation process;
– The metabolites and end products of the biodegradation.

The chemical analysis methods which are based on chemical substance abatement measure the residue of the chemical substance and its possible metabolites.

Figure 2.4 shows the enzymatic reaction of biodegradation with the components suitable as test end points. Symbols of the equation are:

– S: substrate which produces energy by aerobic/anaerobic respiration: the chemical substance to be degraded.
– E: enzyme or enzyme system: proteins produced by degrading microorganisms to break down the substrate into smaller units and oxidize it for energy extraction, including the enzymes of aerobic/anaerobic respiration and the electron transport chain (terminal oxidation).
– $I_n$: intermediate, a temporary enriched product, whose utilization varies;
– P: product, end product, e.g. $CO_2$ as end product of aerobic mineralization;
– $E_p$: a measurable or easy-to-detect consequence of the presence of the product, e.g. pH of an acidic product.

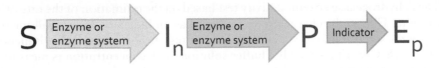

*Figure 2.4* Enzyme reaction of biodegradation and measurable endpoints.

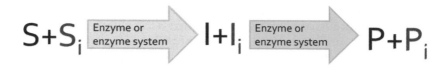

$$S+S_i \xrightarrow{\text{Enzyme or enzyme system}} I+I_i \xrightarrow{\text{Enzyme or enzyme system}} P+P_i$$

*Figure 2.5* Enzymatic reaction applying an indicator substrate.

In addition to the natural players of the substrate utilizing reaction, one may apply artificial alternative substrates or intermediates, which results in special products utilizing the enzymes active in the biodegrading cells.

Figure 2.5 shows the concept of using an assistant molecule in the test to facilitate the detection of the selected measuring end point of $S_i$, $I_i$ or $P_i$. Symbols of the equation are:

–   $S_i$: indicator substrate, added to the test in addition to the chemical substance (S) for easy indication. It can be a labeled substance or an alternative electron acceptor, producing a color reaction as the effect of the enzymes.
–   $I_i$: Intermediate from the indicator substrate.
–   $P_i$: product of $S_i$, an easy-to-indicate substance due to the radioactive label or color reaction.

Respiration-based methods work in such a way that the chemical substance to be tested for biodegradability is added to the 'standard' degrading microflora in artificial or natural water, sediment or soil. Another concept suggests testing biodegradation in the contaminated environmental sample itself by measuring the biodegradation potential of its own microflora. End points to be measured can be $O_2$ depletion or $CO_2$ production for aerobic biodegradation, gaseous products of $H_2$, $CH_4$, $NH_3$, $H_2S$ for anaerobic degradation or any other molecule which plays a role in energy production and the respiratory chain of the cells. The use of alternative electron acceptors is the most widely applied solution to indicate energy production from the contaminant. Isotope-labeled chemicals make the evaluation easy and selective.

The measurement of one or more enzyme activities in the respiratory chain can be used as an index for the total oxidative activities of the cell; therefore, dehydrogenase enzyme activity has been used as a measure for overall microbial activity (Alef & Nannipieri, 1995). The assay of dehydrogenase can be used as a simple method to investigate microbial activities and possible inhibitory effects of the contaminants on the microorganisms. Nearly all microorganisms reduce 2,3,5-triphenyl-tetrazolium-chloride (TTC), this artificial electron acceptor, to triphenyl formazan (TPF), which can be measured colorimetrically.

The dehydrogenase enzyme activity test based on the estimation of the rate of TTC reduction to TPF can be characterized and used as an indicator for overall microbial activity of the soil (Molnár *et al.*, 2009a). For this test, field-moist soil (5 g) is incubated for 24 h at 28°C in TTC–tris–HCl buffer solution. TPF concentration is measured by colorimetry after extraction with acetone at 546 nm. For the interpretation of the dehydrogenase activity, the TPF concentration is given in μg TPF/g soil.

Table 2.3 OECD Guidelines for testing the biodegradability of chemical substances.

| Test guideline | Test name |
|---|---|
| OECD TG 301 (1992) | Ready Biodegradability |
| OECD TG 302A (1981) | Inherent Biodegradability: Modified Semi-Continuous Activated Sludge (SCAS) Test |
| OECD TG 302B (1992) | Inherent Biodegradability: Zahn-Wellens/EVPA (EMPA) Test |
| OECD TG 302C (2009) | Inherent Biodegradability: Modified MITI Test (II) |
| OECD TG 303 (2001) | Simulation Test – Aerobic Sewage Treatment – A: Activated Sludge Units; B: Biofilms |
| OECD TG 306 (1992) | Biodegradability in Seawater |
| OECD TG 307 (2002) | Aerobic and Anaerobic Transformation in Soil |
| OECD TG 308 (2002) | Aerobic and Anaerobic Transformation in Aquatic Sediment Systems |
| OECD TG 309 (2004) | Aerobic Mineralization in Surface Water – Simulation Biodegradation Test |
| OECD TG 310 (2006) | Ready Biodegradability – $CO_2$ in Sealed Vessels (Headspace Test) |
| OECD TG 311 (2006) | Anaerobic Biodegradability of Organic Compounds in Digested Sludge: by Measurement of Gas Production |
| OECD TG 314 (2008) | Simulation Tests to Assess the Biodegradability of Chemicals Discharged in Wastewater |

Soil respiration can also be determined on the basis of $CO_2$ *production* or $O_2$ *consumption* in closed static jars and also in a dynamic flow-through system. $CO_2$ is usually trapped in NaOH and determined by HCl titration or by measuring the electrical conductivity of the NaOH solution. Manometric indication can be applied in a closed apparatus, and a direct $CO_2$ analysis in flow-through systems, e.g. by infrared spectroscopy, due to the asymmetrical stretch of $CO_2$ giving a strong band in the IR spectrum at $2350\,cm^{-1}$.

As a microbiological end point of reproduction/growth of a microbial culture, the number of cells or their measurable structural and functional compartments or products are suitable (cell number, growth curve, nitrogen content, chlorophyll content, light emission, etc.), when directly associated with biodegradation (Gruiz *et al.*, 2001; Hyman & Dupont, 2001).

### 3.1.7 Standardized biodegradability test methods for chemical substances

Biodegradability tests aim at the generic characterization of the fate and behavior of chemical substances in the environment, but their objective also includes determination of the site-specific biodegradation rate.

The OECD test guidelines – prepared to measure the generic biodegradability of chemical substances – are shown in Table 2.3. Simulating aquatic ecosystems or wastewater, the test mainly applies to microorganisms from the environment or simply natural water, sediment or wastewater.

The biodegradability of chemical substances provides priority information for authorization, classification and restriction of chemicals in European or other regional regulations. Based on the test results, decisions are taken throughout Europe and biodegradation results of the listed OECD tests are applied as generic values. They can be applied in the conservative approach for generic environmental risk assessment

*Table 2.4* Standardized ISO tests for testing biodegradation in the environment.

| Test guideline | Test name |
|---|---|
| ISO 7827:2010 | Evaluation of the 'ready', 'ultimate' aerobic biodegradability of organic compounds in an aqueous medium – Method by analysis of dissolved organic carbon (DOC) |
| ISO 14592-1:2002 | Evaluation of the aerobic biodegradability of organic compounds at low concentrations – Part 1: Shake-flask batch test with surface water or surface water/sediment suspensions |
| ISO 14593:1999 | Evaluation of ultimate aerobic biodegradability of organic compounds in aqueous medium – Method by analysis of inorganic carbon in sealed vessels ($CO_2$ headspace test) |
| ISO 14592-2:2002 | Evaluation of the aerobic biodegradability of organic compounds at low concentrations – Part 2: Continuous flow river model with attached biomass sealed vessels ($CO_2$ headspace test) |
| ISO 9408:1999 | Evaluation of ultimate aerobic biodegradability of organic compounds in aqueous medium by determination of oxygen demand in a closed respirometer |
| ISO 9439:1999 | Evaluation of ultimate aerobic biodegradability of organic compounds in aqueous medium – Carbon dioxide evolution test |
| ISO 10634:1995 | Guidance for the preparation and treatment of poorly water-soluble organic compounds for the subsequent evaluation of their biodegradability in an aqueous medium |
| ISO 10707:1994 | Evaluation in an aqueous medium of the 'ultimate' aerobic biodegradability of organic compounds – Method by analysis of biochemical oxygen demand (closed-bottle test) |
| ISO 10708:1997 | Evaluation in an aqueous medium of the ultimate aerobic biodegradability of organic compounds – Determination of biochemical oxygen demand in a two-phase closed-bottle test |
| ISO 16221:2001 | Guidance for determination of biodegradability in the marine environment |

because real biodegradation rates tend to be higher than those in the tests due to greater diversity and adaptation in the environment.

Table 2.4 shows some ISO biodegradability test standards.

### 3.1.8 Measuring biodegradation in soil

Testing biodegradation in a defined environment needs *in situ* measurement or the simulation of the real situation under laboratory conditions. This can be done by standardized or non-standardized test methods. Some of the possibilities are introduced here, with the focus on soil, where many of the contaminants are finally eliminated by myriads of microorganisms.

The direct chemical analysis of contaminant residues in the soil or groundwater determines overall losses and thus may provide equivocal evidence for biodegradation (Crawford, 2002; Alvarez & Illman, 2006), provided that abiotic degradation does not occur. A decrease in concentration or a change in composition may be a consequence of abiotic mechanisms such as volatilization, leaching and complexation to soil organic matter, or may be due to biotic processes such as biodegradation. A variety of quantitative and qualitative techniques are available to measure organic contaminant losses in the environment (in soil, sediment, groundwater). The choice of a particular method usually depends both on the type of the contaminant and the matrix. For example, the biodegradation of a single and easy-to-analyze biodegradable contaminant should be tested by a substance-specific chemical analysis, but biological end points (e.g. production of $CO_2$ or degrading enzymes) should have priority for the biodegradation of complex mixtures such as petroleum products (e.g. TPH or PAHs), or mixed and unknown contaminants.

The analysis of the organic contaminants usually requires separation from the matrix by extraction with organic solvents, assisted by ultrasound or microwave digestion. Sometimes, further separation may be necessary before determining the concentration of the components of mixtures. The measured end point is a concentration value in the extract. Solvents of different polarities (water, fluid state $CO_2$, DMSO, hexane, acetone, n-pentane, dichloromethane, etc.) are used for extraction (separation of the contaminant from the matrix) depending on the chemical nature of the contaminant to be analyzed. Mixtures of contaminants are further separated into components or fractions before their quantitative determination using fractionated extraction, gas chromatography, high performance liquid chromatography, capillary electrophoresis or any other suitable separation method. The contaminant residues may be quantitatively determined by gravimetry, conventional chemical analysis, mass spectrometry or atomic and molecular spectrometric methods, such as UV or visible light (UV-VIS), infrared (IR) and fluorescence spectroscopy. FID (flame ionization detection), based on the detection of ions formed during combustion of organic compounds in a hydrogen flame, became a frequently used detector, mainly coupled with gas-chromatography.

Biological methods play an important role in measuring aerobic and anaerobic biodegradation in the environment, in contaminated surface waters or wastewaters, dredged or living sediments, or soils and groundwater.

### 3.1.9 Soil respiration, biodegradative activity of the soil – problem-specific applications

Measuring soil respiration is one of the oldest and the most frequently used methods for quantifying microbial activity in soils (King et al., 1992; Alef & Nannipieri, 1995; Crawford, 2002; Alvarez & Illman, 2006).

Soil respiration in a 'closed bottle' is applied for (Molnár et al., 2009a):

– Testing the biodegradability of chemicals under standard conditions;
– Measuring the actual biodegradation in contaminated soil;
– Testing enhancement of biodegradation in soil and sediment;
– Testing toxicity in soil.

#### Manometric respirometry: the closed-bottle technique

Closed bottles filled with real soil can be considered as a special soil microcosm as discussed in Chapter 9. Manometric respirometry carried out in a Sensomat system applies a piezoresistive pressure sensor technology. Test conditions can be properly standardized, and the soil used can be 'generic' or site-specific. Since the continuously decreasing oxygen concentration is unrealistic, the method is of limited applicability for the assessment of realistic biodegradability. However, proper interpretation of the

Cap with pressure sensor

Bottle with soil to be tested

Remote data reading

*Figure 2.6*   Closed bottle for testing the respiration rate in soil based on manometry. The pressure decrease is detected by the sensor in the cap and read remotely.

results can yield a useful and sound indicative tool, similar to BOD (biological oxygen demand) used in wastewater tests.

The Sensomat System for respiration and biodegradation testing in contaminated soil (Figure 2.6) was characterized and compared to other test methods by Molnár *et al.* (2009a). The pressure that develops in the closed bottle filled with moist soil due to microbial activity and gas exchange is measured and logged with the pressure sensor on top of the jar. A vial containing NaOH is placed in each vessel to trap $CO_2$. If oxygen is consumed in the closed vessels at a constant temperature, negative pressure develops. The pressure difference (decrease) due to microbial activity in closed vessels is generally measured for 5 days at 25°C and read remotely. For the interpretation of the results, the linear part of the respiration curve (pressure–time) can be used from which the respiration index can be determined by linear regression.

Advantages of the manometric method are:

–   simplicity: no dilution of sample required, no seeding, no blank sample;
–   direct reading of pressure, $CO_2$ production or $O_2$ consumption values;
–   continuous display of any of the end point values at the current incubation time.

Figure 2.7 illustrates the biodegradation of crude oil in the closed-bottle system. The blue curve shows the freshly contaminated soil's respiration in the first 24 hours, which is negligible, indicating no oxygen consumption, no biodegradation. Testing after a two-day adaptation period – demonstrated by the pink curve – the soil microbiota was able to utilize the hydrocarbon mixture as an energy source, indicated by the decline of the pink curve due to oxygen consumption. A steepening negative slope of the curve indicates ongoing adaptation and increasing respiration.

As another example, Figure 2.8 shows the respiration curves of the biodegradation of transformer (PCB-free) oil in soil. Transformer oil was added to an average-quality forest soil in concentrations of 10, 20 and 30 g/kg. The pressure decrease (increase in

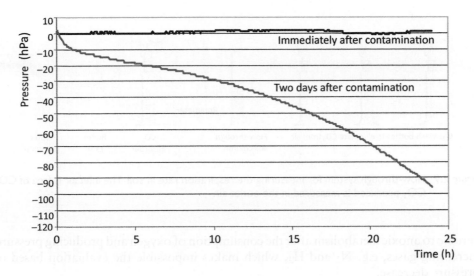

*Figure 2.7* Respiration curve of crude oil in soil immediately and 2 days after contamination (MOKKA Project, 2004–2008).

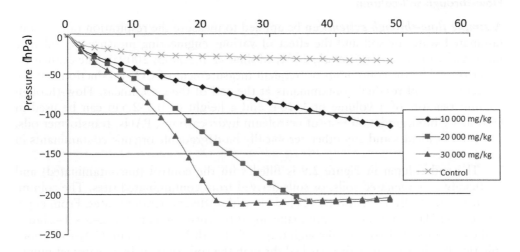

*Figure 2.8* Transformer oil biodegradation in soil; tested in a closed bottle by measuring the pressure decrease due to oxygen consumption (MOKKA Project, 2004–2008).

the negative direction) is proportional to $O_2$ consumption. Respiration rate is proportional to biodegradation in the soil containing 10 g/kg contaminant, and is inhibited after 35 and 21 hours in soils contaminated with 20 and 30 g/kg oil. The inhibition is due to the decreasing oxygen content in the bottle.

The closed-bottle soil respiration test faces two problems: the continuously decreasing oxygen concentration and the microbiota's capability of switching from

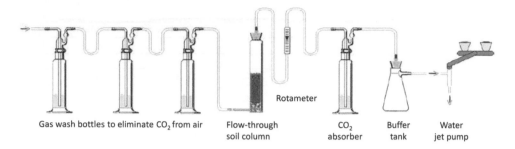

Rotameter

Gas wash bottles to eliminate CO₂ from air       Flow-through       CO₂       Buffer       Water
                                                 soil column       absorber   tank       jet pump

*Figure 2.9*   Flow-through system for measuring the respiration rate in soil. The inlet air is free of $CO_2$. $CO_2$ produced is measured in the outlet air.

aerobic to anoxic metabolism after the consumption of oxygen and producing pressure-increasing gases, e.g. $N_2$ and $H_2$, which makes impossible the evaluation based on pressure decrease.

### Flow-through soil-column

A *simple flow-through system* can be applied to measure the respiration rate of contaminated water or soil and the effect of various engineering measures to enhance biodegradation rate (Gruiz *et al.*, 2001; Molnár *et al.*, 2005; 2009a). The evaluation is performed by the continuous or frequent measurement of $CO_2$ production and the concentration of residual contaminants at the end of the experiment. Flow-through column reactors of a volume of 1–20 L and a height of 0.5–2.5 m can be used in laboratory biodegradation tests of petroleum hydrocarbons, PAHs, transformer oils, creosote, pesticides and any other aerobically biodegradable organic contaminants in soil or water.

The soil column in Figure 2.9 is filled with the control (uncontaminated) and artificially contaminated soils, or soils derived from contaminated sites. The column reactor is generally aerated for 2–24 h daily with different aeration rates. From experience, 24 h/day aeration does not make any difference compared to 2–3 × 1–2 h/day in the case of soil. Two flow-through traps filled with NaOH ensure $CO_2$-free atmospheric air. The $CO_2$-free air is sucked through the soil columns by a water jet pump, ventilation or a small vacuum pump. Air flow is measured by a gas meter. The $CO_2$ produced by the soil microorganisms during the aeration period is trapped in 1 N NaOH solution and determined by HCl titration or measured by $CO_2$ meters based on infrared spectrography.

The cumulated $CO_2$ production by the microorganisms in the soil column is proportional to the biodegradation/mineralization. The contaminant to be tested is not the only carbon source (unlike in water or in liquid phase test solutions). Therefore, $CO_2$ production in the uncontaminated soil is simultaneously measured, and the biodegradation rate of the tested contaminant is calculated from the difference. Figure 2.10 shows the $CO_2$ production in two soils contaminated with 10,000 mg/kg diesel (blue) and 10,000 mg/kg crude oil (brown), compared to the uncontaminated control soil

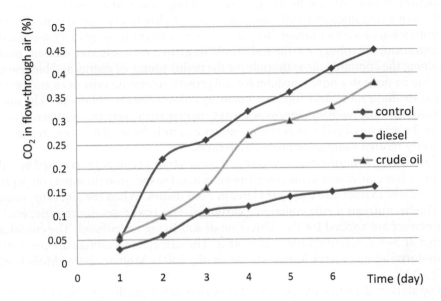

*Figure 2.10*   Diesel oil and crude oil biodegradation in soil: $CO_2$ production compared to control soil (Molnár, 2006).

(green). The curves indicate that diesel oil started to degrade on the second day, while crude oil only after the third. A certain fraction of diesel oil was rapidly degraded on the second day, but the degradation rate from the third day onwards is similar to that of the crude oil: the two curves rise in a parallel fashion. The residual contaminant mass or the number of cells can be used to validate the test results of $CO_2$ production. 82% of diesel oil was removed by biodegradation, and 66% of crude oil. The cell number of oil-degrading microorganisms on the 7th day was $3 \times 10^7$ in the soil contaminated with diesel oil and $2.5 \times 10^7$ in the one contaminated with crude oil.

The use of radioactive isotope-labeled chemical substances in biodegradation studies and the application of radioactivity as the measured end point increase the specificity and selectivity of the biodegradation tests, both in chemical biodegradability testing and soil biodegradative activity testing. Labeled substrates can also be used as an internal standard in the biodegradation studies.

### Cell counting for contaminant-utilizing microorganisms

Microbial enumeration is an indirect method for characterizing microbial activity in the environment; it provides data for the estimation of microbial vitality and forecasting biodegradation potential. It is important to note that uncertainties may result in real soil from other available energy sources out of the contaminant and by the isolation and growth methods, which may only capture a small percentage of the total microbial community. Cell counting by direct microscopy and viable plate counts of organisms able to grow on agar media are enumeration methods that are widely applied.

The quantitative metrics of soil microbiota (cell concentration) may be closely related to the biodegradation of a contaminant if it is the only substrate for energy production or biosynthesis. In average soils with high nutrient and humus content, the numbers of living microbial cells in general are not closely associated with the microbial utilization of a contaminant. To obtain a contaminant biodegradation-related cell concentration, microbes from the soil should be grown on a selective nutrient medium containing the contaminant as the only (or the main) source of energy and biosynthesis. This may be done in a liquid medium for cell growth (counting cells as an end point) or on agar medium for colony forming (counting colonies as an end point). The number of colony-forming microorganisms is always smaller than their real number in the soil so that the measured cell concentration values can only be used for comparison of the differently treated samples or processes in time.

The *aerobic heterotrophic bacterial cell concentration* can be determined by colony counting after cultivation of microorganisms in a soil suspension (in water) on peptone–glucose–meat-extract nutrient agar plates in Petri dishes. When the aim is to count the fungal cells, nutrient media containing carbohydrates (molasses, saccharose, etc.) and yeast extract are applied for the cultivation of soil-living microfungi. The colonies are counted at 3–5 different dilutions after 48 h. The calculated averages are presented as colony-forming units (CFU/g soil) (Gruiz *et al.*, 2001; Molnár, 2006; Molnár *et al.*, 2005; 2009b).

Population densities of specialized contaminant-degrading bacteria can be best estimated by applying a serial dilution to the point of extinction in numbers of cells. The method is called limiting dilution analysis and is used to measure the abundance of cells present in a mixed population and able to perform a particular function: degrading a specific contaminant in this case.

Performing 5–9 simultaneous tests, the statistical method of *Most Probable Number (MPN)* can be applied based on cell density or appearance of color of an alternative electron acceptor in the wells, which is identical to the presence (+) or absence (–) of microorganisms. The calculation of cell concentration is possible without an actual count of single cells or colonies (Alef & Nannipieri, 1995). The selective nutrient broth-containing wells are inoculated with aliquots of the serial dilution. For the propagation (growth) of the contaminant-degrading cells, a nutrient medium is prepared containing the organic contaminant (petroleum products, pesticides, etc.) as the only energy source. The dilution series prepared from the contaminated soils (minimum three replicates) contains all of the soil microorganisms (see Figure 2.11), but only those that can utilize the contaminant as an energy source will be able to grow (propagate) (Gruiz *et al.*, 2001; Molnár *et al.*, 2005). The liquid medium is supplemented with inorganic salt solution, trace elements and with an artificial electron acceptor of 2-(p-iodophenyl)-3-(p-nitrophenyl)-5-phenyl tetrazolium chloride (INT). The Most Probable Number (MPN) is calculated from the red color (+/–) using probability tables.

The example in Figure 2.12 demonstrates the correlation of respiration rate and oil degrading cell concentration in a soil contaminated with transformer oil after treatment with a bioavailability enhancing, solubilizing agent (SA). The transformer oil in the soil is difficult to access for oil degrading microorganisms, so that it was treated with the oil solubilizing agent (SA) in increasing concentrations (SA 0.1, 0.5 and 1.0%). The additive increased pollutant bioavailability, which resulted in higher cell concentrations and respiration rates in the flow-through reactors. Cell concentration was measured in a nutrient medium containing transformer oil as the only carbon source, using the above described limiting dilution method.

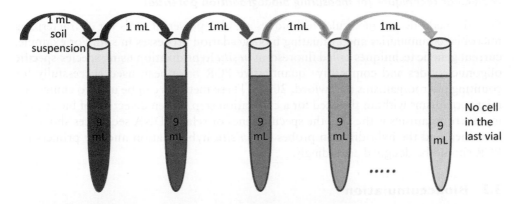

Figure 2.11  Dilution series for determining the most probable cell concentration.

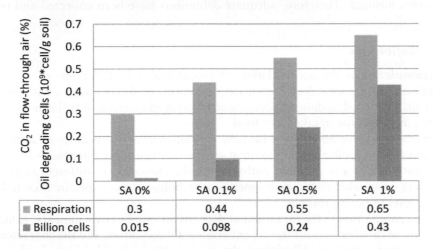

| | SA 0% | SA 0.1% | SA 0.5% | SA 1% |
|---|---|---|---|---|
| ■ Respiration | 0.3 | 0.44 | 0.55 | 0.65 |
| ■ Billion cells | 0.015 | 0.098 | 0.24 | 0.43 |

Figure 2.12  Increase in transformer oil-degrading cell concentration and respiration rate due to treatment with a bioavailability-enhancing agent (Molnár, 2006).

### ATP in the cells – a biochemical endpoint

The biodegradative activity of the biota can be quantitatively characterized by measuring the adenosine triphosphate (ATP) concentration in the microorganisms. ATP is the central reactant in both the energy-producing (catabolic) and biosynthetic (anabolic) reactions within the cells and is proportional to energy production. One of the most straightforward methods is based on use of the luciferin reaction (Paul & Clark, 1996). The substrate luciferin can react with ATP and in the presence of Mg and $O_2$ luciferase breaks down to produce free adenosine monophosphate, inorganic P and light. This emitted light can be measured by a photometer.

The prerequisite for using the ATP-proportional light for the characterization of the biodegradation of chemical substances is that energy is produced from, and only from, the chemical substance concerned in the test.

### Molecular techniques for measuring biodegradation potential

Genetic methods and molecular techniques are also powerful tools for characterizing microbial communities and evaluating biodegradation processes in soil. For example, current genetic techniques called fluorescent *in situ* hybridization using species-specific oligonucleotides and competitive quantitative PCR have been used successfully for counting microorganisms (Crawford, 2002). These methods can be used to enumerate microorganisms without the need for a cultivation step. When detection of biodegrading microorganisms is the aim, the specific genes or related DNA sequences should be identified and the hybridization probes for *in situ* hybridization and the primers for PCR should be designed accordingly.

## 3.2   Bioaccumulation

Bioaccumulation, bioconcentration, and biomagnification are relatively new terms, sometimes misused. Therefore, adequate definitions have been collected and posted here.

### 3.2.1   Definitions

*Bioaccumulation* is the accumulation of contaminants in the tissue of organisms through any route, including respiration, ingestion, or direct contact with contaminated air, water, soil, sediment, pore-water, or dredged material. It is the most general process, including the uptake from food.

*Bioaccumulative potential* is the inherent potential of a substance to accumulate in organisms. Its application in a general sense – for regulatory purposes – bears high uncertainty due to the different pathways of the chemicals in different organisms, depending on uptake, elimination kinetics, metabolism, organ-specific accumulation and the level of bound residues.

*Bioaccumulation in the environment* depends on specific environmental scenarios, the nature of the chemical substance, environmental conditions and the characteristics of the receptor organisms. All of these play a role in the realizable bioaccumulation, in the bioaccumulated amount and in the mode of bioaccumulation. We know of defense mechanisms inhibiting uptake and/or accumulation, and hyperaccumulation, e.g. those plants which are capable of extracting toxic metals from soil. Bioaccumulation is the result of the following processes:

– Absorption: chemical transport across a biological membrane into the organism;
– Distribution after absorption by circulation throughout the body;
– Metabolism by enzymes (biotransformation);
– Excretion: elimination of the chemical.

*Bioavailability greatly influences bioaccumulation.* The bioavailable fraction of the total concentration (e.g. in water or soil) is that which can be absorbed by the organisms via a specific route of intake. It does not depend on the rate at which the chemical is absorbed and accumulated.

*Bioconcentration*: this term has been monopolized for the process which produces a net accumulation of a chemical directly from water into aquatic organisms resulting from simultaneous uptake (e.g. by gill or epithelial tissue) and elimination

(Rand & Petrocelli, 1985; Rand, 1995). In a more general sense, the term bioaccumulation is used for all kinds of uptake and accumulation of chemicals from air, water, soil and sediment into the body of an organism. In the real environment the accumulation via inhalation, oral and dermal routes are added together, resulting in an aggregated body burden.

*Bioconcentration factor* (BCF) is used in the EU (EU-TGD, 2003), and is also known as the *bioaccumulation factor (BAF)*, mainly by the US EPA (Kravitz, 1998), but both have the same meaning: the ratio of a substance's concentration in the tissue of an aquatic organism to its concentration in ambient water, in situations where the organism is exposed to the substance through the water only, and the ratio does not change substantially over time. This is the static bioconcentration factor, the ratio of the concentration in the organism to the concentration in water in a steady-state (dynamic equilibrium) situation. When uptake and depuration kinetics are measured, the dynamic bioconcentration factor can be calculated from the quotient of the uptake and depuration rate constants:

$$BCF_{fish} = C_{fish}/C_{water} = k_1/k_2,$$

where

- $C_{fish}$ : concentration in fish (mg/kg);
- $C_{water}$: concentration in water (mg/L);
- $k_1$: uptake rate constant from water (L/kg/day);
- $k_2$: elimination rate constant (1/day);
- $BCF_{fish}$: bioconcentration factor (L/kg).

Others than the equilibrium model based on $K_{ow}$ are also used for the determination/calculation of *BCF*. The most widespread *BCF* calculation models are:

- Equilibrium partitioning models (depending on physico-chemical properties of the chemical substance).
- Correlation with $K_{ow}$ (see aquatic and terrestrial bioaccumulation).
- Correlation with water solubility or other chemical properties.
- Physiological models (relying on physico-chemical properties of the chemical and biological attributes of the target animal).
- FGETS (Food and Gill Exchange of Toxic Substances) simulation model, for example, uses fish's gill and intestinal morphometry, the body weight of the fish, and fractional aqueous, lipid, and structural organic composition (EPA Food Chain Model, 2013).
- PBPKs, physiologically based pharmacokinetic models describe the chemical's kinetics in an animal species using literature and measured parameters. The model can be applied for simulating time trends.

*Body burden* is the result of bioaccumulation from the point of view of the organisms: it means the amount of chemicals which are not readily excreted and so can remain for years in the blood, adipose (fat) tissue, semen, muscle, bone or brain tissue, or in certain organs of the organisms.

*Biomagnification* is the result of the process of bioconcentration and bioaccumulation by which tissue concentrations of bioaccumulated chemicals increase as the chemical passes up through two or more trophic levels. The term implies an efficient transfer of a chemical from food to consumer, so that residue concentrations increase systematically from one trophic level to the next, resulting in an increase in the internal concentration in organisms at higher levels in the trophic chain.

*Secondary poisoning* is concerned with toxic effects in the higher members of the food chain living either in the aquatic or terrestrial environment, and which results from ingestion of organisms from lower trophic levels that contain accumulated substances.

*Biomagnification factor (BMF)* is defined as the relative concentration in a predatory animal compared to the concentration in its prey:

$$BMF = C_{predator}/C_{prey}.$$

The concentrations used to derive and report BMF values should, where possible, be lipid normalized.

*Biota-sediment accumulation factor (BSAF)* is the relative concentration of a substance in the tissues of an organism compared to the concentration of the same substance in the sediment (US EPA and US ACE, 1998).

*Bioaccumulative chemicals of concern (BCC)* in sediment: chemicals identified as a concern for sediment quality assessment because of their ability to accumulate in the tissue of organisms through any route, including respiration, ingestion, or direct contact with contaminated water, sediment, pore-water, or dredged material (US EPA & US ACE, 1998).

### 3.2.2  Bioaccumulative potential of chemicals

The most important and widely accepted indication of bioaccumulation potential is a high value of the n-octanol–water partition coefficient. It is accepted that values of log $K_{ow}$ greater than or equal to 3 indicate that the substance may bioaccumulate (EU-TGD, 2003). For certain types of chemicals, for example, surface-active agents and those which ionize in water, log $K_{ow}$ values may not be suitable for the calculation of a BCF. Also, metal concentration cannot be calculated from partitioning, because $K_{ow}$ models only the lipid fraction of organisms' tissues.

The theoretical correlation between $K_{ow}$ and bioaccumulation is modified by or completely changed by the active transport of the chemical in the organism, metabolic pathway-specific interactions with tissue components, receptors or enzymes, uptake and depuration kinetics. Further conditions influencing bioaccumulation are: adsorption and uptake, hydrolysis, degradation, and the size of the molecule.

#### Aquatic bioaccumulation

For substances with a log $K_{ow}$ of 2–6, the following linear relationship can be used as developed by Veith *et al.* (1979) for aquatic organisms:

$$\log BCF_{fish} = 0.85 \times \log K_{ow} - 0.70,$$

where

- $K_{ow}$: octanol–water partition coefficient (−);
- $BCF_{fish}$: bioconcentration factor (L/kg wet fish);
- log: common logarithm (to the base 10).

Table 2.5 Biomagnification factors (BMF) for organic substances estimated from bioconcentration in fish.

| Log $K_{ow}$ of substance | $BCF_{fish}$ | BMF |
|---|---|---|
| <4.5 | <2000 | 1 |
| 4.5–5 | 2000–5000 | 2 |
| 5–8 | >5000 | 10 |
| 8–9 | 2000–5000 | 3 |
| >9 | <2000 | 1 |

Knowing the *BCF* values, future bioaccumulation can be estimated from environmental concentrations: $C_{fish} = C_{water} \times BCF$.

Factors influencing a substance accumulation have an impact on *BCF* both in the environment and during tests:

– Slow uptake;
– Steric hindrance or other obstacles to transfer the chemical into the body;
– Reduced bioavailability of the chemical substance;
– Organism growth and other changes, e.g. in lipid content;
– Ongoing metabolism of the chemical substance;
– Variation in the concentration of the chemical substance in water over time;
– Solubilization and mobilization of the contaminant by biotensides or other contaminating tensides.

*The biomagnification factor* (BMF) should ideally be based on measured data. However, the availability of such data is at present limited. Therefore in simple cases, biomagnification can be estimated from bioaccumulation which originates from $K_{ow}$. Table 2.5 shows the recommendation of the guidance of REACH regulation (EU-TGD, 2003).

### Terrestrial bioaccumulation

Bioconcentration can be described as a hydrophobic partitioning between the pore water and the phases inside the organism and can be modeled according to the following equation as described by Jager (1998):

$$BCF_{earthworm} = \frac{0.84 + 0.012 K_{ow}}{RHO_{earthworm}}$$

where

– $RHO_{earthworm}$: earthworm density, by an average of 1 ($kg_{wwt}/L$) can be assumed.

Using this equation, the concentration in the full worm can be calculated as:

$$C_{earthworm} = \frac{BCF_{earthworm} \times C_{pore\ water} + C_{soil} \times F_{gut} \times CONV_{soil}}{1 + F_{gut} \times CONV_{soil}},$$

where

- $F_{gut}$: fraction of gut loading in worm $(kg_{dwt}/kg_{wwt})$;
- $CONV_{soil}$: conversion factor for soil concentration wet-dry weight soil $(kg_{wwt}/kg_{dwt})$;

$$CONV_{soil} = \frac{RHO_{soil}}{F_{solid} \times RHO_{solid}},$$

where

- $F_{solid}$: volume fraction of solids in soil $(m^3/m^3)$;
- $RHO_{soil}$: bulk density of wet soil $(kg_{wwt}/m^3)$;
- $RHO_{solid}$: density of solid phase $(kg_{dwt}/m^3)$.

### 3.2.3   QSAR for bioaccumulation

The QSAR models of bioaccumulation of organic contaminants in the cells and tissues of organisms are based on the $K_{ow}$ value, which is proportional to the binding of the chemicals to any organic, e.g. lipid phases within the organism (Mackay & Fraser, 2000). This simple relationship is written as:

$$BCF = a \times K_{ow},$$

When $K_{ow}$ is less than 6, the following linear equation is recommended by Veith *et al.* (1979):

$$\log BCF = 0.85 \times \log K_{ow} - 0.70$$

where

- log: common logarithm (to the base 10),
- $n = 55$, $r^2 = 0.90$

Meylan et al. (1999) refined the correlation between $BCF$ and $\log K_{ow}$ for nonionic and ionic compounds as follows:

- Nonionic organic compounds:

  o  $\log K_{ow} < 1 \rightarrow \log BCF = 0.50$;
  o  $\log K_{ow}$ 1–7 $\rightarrow \log BCF = 0.77 \log K_{ow} - 0.70 + \Sigma F_i$,
     $(F_i = $ structural correction factor$)$;
  o  $\log K_{ow} > 7 \rightarrow \log BCF = -1.37 \log K_{ow} + 14.4 + \Sigma F_i$;
  o  $\log K_{ow} > 10.5 \rightarrow \log BCF = 0.50$.

- Ionic organic compounds:

  o  $\log K_{ow} < 5 \rightarrow \log BCF = 0.50$;
  o  $\log K_{ow}$ 5–6 $\rightarrow \log BCF = 0.75$;
  o  $\log K_{ow}$ 6–7 $\rightarrow \log BCF = 1.75$;
  o  $\log K_{ow}$ 7–9 $\rightarrow \log BCF = 1.00$.

Fu *et al.* (2009) recommended the following equation for ionizable chemicals:

o   $\log K_{ow}(\text{ion}) = \log K_{ow} (\text{neutral}) - 3.5$

Isnard and Lambert (1988) found linear correlation between the logarithms of BCF and water solubility, investigating 107 chemicals with diverse water solubility of 0.02–36 000 mg/L:

o   $\log BCF = -0.47^{*}\log S + 2.02$

The application of QSARs for the estimation of the bioaccumulative potential of chemicals is essential; otherwise, laboratory tests and the accompanying chemical analyses of the bioaccumulated contaminants in the organism make the determination very expensive. In addition, the biological model is a simplified model of reality, simulating the uptake by one or a few organisms only (Pavan *et al.*, 2006). Of course QSARs also have disadvantages: they cannot model a complex organism and complicated interactions, e.g. biochemical mechanisms and in the case of special chemicals, e.g. at high $K_{ow}$, the model's validity is questionable.

### 3.2.4 Testing bioaccumulation

Empirical determination of bioaccumulation is a simple way to collect information on the bioaccumulation of chemicals in aquatic and terrestrial organisms. Direct determination can be conducted using either laboratory-exposed or field-collected organisms, and generally this approach minimizes or eliminates many of the problems associated with QSAR models. Nevertheless, the number of chemicals and the cost of tests, as well as the ethics associated with particular animals used for testing, justify the importance of QSAR models.

The goal of testing bioaccumulation is to monitor existing bioaccumulation or to predict bioaccumulation under future exposure conditions. For the prediction of bioaccumulative risk, a generic approach can be used which characterizes chemical substances in general, creating a default bioaccumulation value which depends primarily on the chemical substance and secondarily on a default environment, e.g. the 'European environment', or the prediction can be based on the bioaccumulation of a chemical in a well-known locality of the environment. Bioaccumulation tests can be:

–   Laboratory testing of the bioaccumulation of chemical substances in standardized tests for regulatory purposes, e.g. the OECD tests for the purposes of REACH regulation in aquatic and terrestrial scenarios.
–   Simulation tests for modeling specific environmental scenarios and bioaccumulation under site-specific conditions. In this case, measured BCF is multiplied by an appropriate factor (for example, food chain multiplier), reflecting the difference between the laboratory organisms and the field organisms.
–   A site-specific bioaccumulation rate (BCF) is the result of field testing. Field assessment of bioaccumulation applies to plants and animals living in the local habitat. The contaminants can be followed through the whole food chain, but the best indicators are those animals which are at the top of the food chain. Passive monitoring of bioaccumulation means the collection of adequately selected species by sampling. We can sample plants and herbivorous insects, as well as predator

birds or mammals from the terrestrial ecosystem of contaminated land. In the aquatic ecosystem, algae, plankton, clams and fish as well as sediment-dwelling organisms are suitable as bioaccumulation indicators. In addition to sampling of indigenous organisms, adequately maintained and controlled cultures of test organisms can be placed into the environment and the accumulated contaminants in their body tissues can be measured after recollection. This method is called active biomonitoring.

It must be borne in mind that laboratory testing of bioaccumulation represents only one step in a complex process in the environment: the uptake of contaminants from the environment by direct contact with water and soil by one species. This means that many of the results of bioaccumulation tests need further extrapolation for the members of the food chain and the whole ecosystem. For example, data on the bioaccumulation of one fish species is only an indication for the secondary poisoning of other fish-eating organisms. In the case of terrestrial plants, the situation is better because we can compare their contaminant content to NOEC values or regulatory limit values for food or feed.

When planning risk assessment and selecting data for estimation/prediction of BCF or BMF, it is important to know the advantages and disadvantages of $K_{ow}$-based QSAR, laboratory tests and field assessments.

–   $K_{ow}$-based QSAR:

   o   Advantages: ready to apply, non-expensive;
   o   Disadvantages: depends on the accuracy of $K_{ow}$ data; metabolism and specific physiology of the animals are not covered, some types of chemicals cannot be properly modeled.
–   Laboratory tests:

   o   Advantages: standardized tests with good reproducibility, large databases are available on results, applicable for all kinds of chemicals and the mechanism and kinetics can also be investigated.
   o   Disadvantages: use of animals is less and less acceptable, expensive, food-chain and food-web effects are not covered.
–   Field tests:

   o   Advantages: applicable for all type of chemicals in the presence of all environmental phases, relevant for metabolism, kinetics, food-chain and food-web situations, applicable for mixtures, are site specific, and include bioavailability and other interactions, etc.
   o   Disadvantages: quality of data is often questionable, environmental concentration of the chemical is uncertain, sediment–water distribution and presence of other contaminants and surface active agents may influence the results, heterogeneities in environmental parameters and variability of the organisms in size, age, gender, etc. increase uncertainties.

The best concept for the estimation of environmental risk due to bioaccumulation is the tiered approach. The first tier is screening, using QSAR. In the case of a positive indication of the bioaccumulative potential by QSAR, chemicals are laboratory tested for bioaccumulation. As a third step, the most expected members at the top of the food chain are field sampled and chemically analyzed to validate the results of QSAR and biotests.

Databases with generally sufficient amount of information are available to start a first-tier prediction in the course of risk assessment. Some of the environmental fate databases containing *BCF* information are listed below:

–  Chemical Risk Information Platform Database, METI–NITE Japan (METI–NITE–CHRIP, 2013)
–  Hazardous Substances Database, US National Library of Medicine's (HSDB – Toxnet, 2013)
–  Environmental Fate Data Base, Syracuse Research Corporation (EFDB–SRC, 2013)
–  European Chemical Substances Information System, JRC (ESIS–JRC, 2013)

The worst situation for the risk manager is the problem of accumulation of multiple chemicals from contaminated water, soil, sediment or waste because QSARs cannot handle the interactions between contaminants (only additivity), and the chemical analyses cannot help in the case of an unknown mixture. These tests would considerably increase the cost of testing.

An interesting theory was suggested by Yoo *et al.* (2003) based on the critical body residue (CBR) theory that non-polar organic contaminants acting via non-polar narcosis (anesthesia) cause acute toxicity when the total tissue concentration of all organic compounds exceeds 2–8 mmol/kg (McCarty & Mackay, 1993). This approach assumes that the compounds acting together through non-polar narcosis exert their effect synergistically. Using DDE (dichlorodiphenyldichloroethane) as a challenging toxicant in addition to the already bioaccumulated chemicals provokes a higher response than without preceding bioaccumulation. The difference between the toxic effect of DDE without previous accumulation and after bioaccumulation gives the rate of bioaccumulation and the toxicity in one test series.

### 3.2.5  Standardized tests for measuring bioaccumulation

Standardized bioassays for bioaccumulation of chemicals aim to indicate the bioaccumulation potential of chemical substances or contaminants in an environmental sample.

#### Testing bioaccumulation of chemicals in water, sediment and soil

The OECD test guidelines are prepared to support the testing of chemicals for regulatory purposes such as for authorization, classification and restriction of chemicals under REACH regulation. The same or slightly modified methodology can also be applied for testing contaminated environmental samples for *in situ* assessment and monitoring (Table 2.6).

Testing the bioaccumulation potential of chemicals uses artificially contaminated water, sediment or soil. Sorption and binding pure chemicals after their addition to the test medium can never truly copy the much more complex and generally long-term environmental process. One can assume that the biological availability is greater in the

*Table 2.6* OECD Guidelines for testing bioaccumulation in aquatic and terrestrial organisms.

| Test guideline | Test name |
| --- | --- |
| OECD TG 305 (1996) | Bioconcentration: Flow-through Fish Test |
| OECD TG 315 (2008) | Bioaccumulation in Sediment-dwelling Benthic Oligochaetes |
| OECD TG 317 (2010) | Bioaccumulation in Terrestrial Oligochaetes |

soil or sediment artificially contaminated and the result of the test is not an underestimate. Mobility/bioavailability-enhancing agents can ensure certain overestimation in simplified tests.

### Bioaccumulation testing by sediment-dwelling organisms

OECD 315 (2008) measures the bioaccumulation of sediment-associated chemicals in endobenthic oligochaete worms. The test consists of two phases: uptake and elimination phases.

During the uptake phase, worms are exposed to sediment spiked with the test substance, topped up with reconstituted water and equilibrated as appropriate. Groups of control worms are held under identical conditions. The duration of the uptake phase is by default 28 days, unless a steady state has been reached before.

For the elimination phase, the worms are transferred to a sediment-water system free of the test substance. This second phase is terminated when either the 10% level of steady-state concentration, or of the concentration measured in the worms on day 28 of the uptake phase, is reached. Change of the concentration of the test substance in/on the worms is monitored throughout both phases of the test. The uptake rate constant ($k_s$), the elimination rate constant ($k_e$) and the kinetic bioaccumulation factor (BAFK = $k_s/k_e$) are calculated. *Lumbriculus variegates*, *Tubifex tubifex* or *Branchiura sowerbyi* can be applied. The worms are weighed, and the accumulated chemical is analyzed (OECD 315, 2008).

Other freshwater oligochaete species are recommended by Ingersoll *et al.* (1995) for testing bioaccumulation from sediment, such as *Stylodrilus heringianus*, *Limnodrilus hoffmeisteri*, and *Pristina leidyi*. Ingersoll *et al.* (1996) recommended the use of the amphipod *Hyalella azteca* and the midge *Chironomus tentans* for testing the bioavailability of contaminants in freshwater sediments.

### Sediment testing by fish

Bioaccumulation testing using fish has a number of advantages. Certain fish species resuspend sediments, thus increasing contaminant bioavailability. Fish can provide adequate tissue mass for chemical analysis because they have higher lipid mass compared to invertebrates. Fish can also experience sediment-associated contaminant exposure through several routes, including direct ingestion and accumulation through gills and skin, and their gills may act as a substrate for the dissociation of hydrophobic chemicals from sediment particles. Fish can be used in laboratory exposure tests as well as in *in situ* studies on caged or free-ranging animals for testing biomagnification and estimating human exposure via secondary poisoning (OECD 305, 1996).

### Bioaccumulation and bioavailability

The fate and effects of persistent, bioaccumulative and toxic (PBT) pollutants in soils and sediments need special attention, as also stated in the status report of US EPA (2000). The document discusses the basics of bioaccumulation, the fate and nature of PBTs and the factors that affect the bioavailability of sediment-associated contaminants. It also identifies how bioaccumulation data can be used for sediment management decisions. The factors influencing bioavailability in sediment are grouped under physical, chemical and biological factors.

– Physical factors:
   o   Rate of mixing;
   o   Rate of sedimentation;
   o   Diffusion;
   o   Resuspension.
– Chemical factors:
   o   Acid-volatile sulfide (AVS) concentrations for Cu, Cd, Pb, Ni, Zn, a possible predictive tool for divalent metals toxicity (cold acid extraction of hydrogen sulfide and iron sulfide);
   o   Redox conditions;
   o   pH;
   o   Interstitial water hardness;
   o   Sediment organic carbon content;
   o   Dissolved organic carbon content;
   o   Organic-water equilibration constants for organic compounds;
   o   Organic matter characteristics;
   o   Equilibration time with sediment.
– Biological factors:
   o   Biotransformation;
   o   Bioturbation;
   o   Organism size/age;
   o   Lipid content;
   o   Gender;
   o   Organism behavior;
   o   Diet, including sediment ingestion, feeding mechanism;
   o   Organism response to physicochemical conditions.

### Laboratory testing of bioaccumulation in environmental samples

Bioaccumulation in the environment is much more complex than in the laboratory because the environmental fate of chemical substances, food-chain and food-web effects always produces a unique combination, which is not easy to model or simulate. Decision makers should be cautious when extrapolating from laboratory test results to the real environment and take uncertainties into account.

Actual bioconcentration can be determined using simplified environmental models such as direct contact tests and microcosms, or field testing based on active or passive biomonitoring.

*Table 2.7* Rapid plant accumulation test: metal-containing mine waste before and after chemical stabilization using different fly ashes (Feigl, 2012).

| Bioaccumulated metal | Mine waste | Mine waste + FA 1 | Mine waste + FA 2 | Mine waste + FA 3 | QC forage |
|---|---|---|---|---|---|
| Cd (mg/kg) | 10.6 | 5.7 | 0.7 | 0.4 | 1 |
| Zn (mg/kg) | 884 | 130 | 80 | 90 | Non-existent |
| Pb (mg/kg) | 96.1 | 9.1 | 8.3 | 4.7 | 10 |
| As (mg/kg) | 15.1 | 2.0 | 1.9 | 1.3 | 2 |

QC forage: Hungarian quality criteria for animal fodder (in mg/kg)
FA: fly ash

An important application of bioaccumulation tests is the determination of the mobile and bioavailable fraction of chemically measurable contaminants in environmental samples, needing an integrated approach in the application and evaluation of physico-chemical and biological methods.

The fate of a chemical time bomb in an environmental compartment can be investigated by dynamic bioaccumulation tests applying a provoked increase in the mobility of known or often unknown contaminants (see also Chapter 9).

The authors of this chapter have developed and applied a rapid bioaccumulation test (Table 2.7) on plant seedling *Sinapis alba* (white mustard). In this test, 40 seeds were placed on 5 g of soil wetted with water to its water holding capacity and incubated at 20°C in darkness for 5 days. The soil was re-wetted on the third day. The seedlings were harvested after 5 days, their roots and shoots were separated, washed and air-dried. The metal content of the plants was analyzed by ICP-AES (inductively coupled plasma atomic emission spectroscopy) after a $HNO_3$ and $H_2O_2$ (2.5:1) digestion at 105°C for 3 hours.

### 3.2.6   Field determination of bioaccumulation

To measure bioaccumulation in field experiments, first the appropriate test species has to be selected, then it must be decided whether natural or transplanted populations should be used. Finding a suitable reference site may also be a problem.

For bioaccumulation field tests, test animals can be taken from resident populations on the site of interest or they can be transplanted from other sources or locations. Resident populations can be selected, collected, maintained and analyzed before carrying out the field test on them. Those stemming from other sources may be commercial or self-grown populations, maintained in a lab, on the original site or on the site to be tested. This means that there are many variations for the selection of test organisms. The concept and test assemblage may vary from caged test organisms through field micro- or mesocosms to gathering and analyzing free living organisms. The advantages and disadvantages of each approach should be determined in each case.

The US EPA (2000) summarized the advantages and disadvantages of alternative bioaccumulation tests in connection with the testing of sediment quality. According to their statement, the use of transplanted test animals confined to cages can simplify the field assessment of bioaccumulation in a number of ways. Test animals can often

*Table 2.8* Metal accumulation in field-grown and laboratory plants in the same contaminated soil before and after chemical stabilization.

| Accumulated metals | Field grown | | | | | | Laboratory rapid test | |
|---|---|---|---|---|---|---|---|---|
| | Sudan grass | | Sorghum | | Maize | | Sinapis alba | |
| | Untreated soil | Stabilized | Untreated | Stabilized | Untreated | Stabilized | Untreated | Stabilized |
| Cd (mg/kg) | 3.0 | 0.9 | 6.6 | 0.7 | 5.3 | 1.6 | 3.0 | 0.9 |
| Zn (mg/kg) | 348 | 104 | 503 | 108 | 665 | 301 | 743 | 218 |
| Pb (mg/kg) | 3.3 | 2.2 | 6.3 | 1.9 | 25.4 | 5.6 | 3.0 | 2.0 |
| As (mg/kg) | 0.8 | 0.4 | 0.5 | 0.5 | 4.7 | 1.0 | 0.6 | 0.4 |

be obtained from commercial vendors in large quantities or from uncontaminated reference locations. Because of this, the variability associated with exposure conditions and times for individual test animals can be eliminated. All test organisms can begin the exposure period with the same, or similar, background contaminant concentrations in their tissues, and they will receive essentially the same exposure during the test period. Test animals can be selected based on size and age as well as sex. Studies can also be devised so that water column and sediment exposures can be differentiated and uptake kinetics studied in situ. Finally, and perhaps most importantly, exposure concentrations or conditions can be accurately determined because the animals will be confined to a small, well-defined area. The exposure period is controlled by the investigator, permitting time to achieve steady state and other issues to be considered when designing a transplant study on caged animals. Transplant studies provide a balance between experimental control and environmental reality not usually attained in either standard laboratory bioassays or assessment of resident populations.

Testing plant bioaccumulation from soil is less problematic, given that plants grow naturally and in large quantities both on agricultural and natural areas. Plants used for assessing bioaccumulation are those which are cultivated or are planned to be cultivated on agricultural and natural land. Plants in contaminated areas are the primary toxic actors in food chains causing "secondary poisoning" to consumers, including man. Their accumulating capacity varies considerably. Using available information, cultivation can be managed aiming to reduce risk as much as possible. The example in Table 2.8 shows differences between plants growing in soil contaminated with metals and the effect of chemical metal stabilization in the same soil. Accumulation of field-grown and laboratory test plants are also compared in the table (Feigl, 2012).

## 3.3 Bioleaching

Bioleaching is a type of leaching where the extraction of metal from solid minerals into a solution is facilitated by the metabolism of certain microbes, e.g. thiobacilli, also called miner bacteria. Bioleaching is a process described as 'the use of microorganisms to transform elements so that the elements can be extracted from a material when water is filtered through it' (BioMineWiki, 2013). It is a risky process in the environment but it provides benefits under controlled circumstances, e.g. for mining.

OECD prepared a guideline for the testing of the leaching behavior of chemical substances – OECD TG 312 (2004) Leaching in Soil Column. The method described in this Test Guideline is based on soil column chromatography (OECD 312, 2004). Two types of experiments are performed:

1   Those on chemical substances for regulatory purposes: determining the leaching potential of the test substance and the leaching potential of transformation products in soils under controlled laboratory conditions.
2   The same test equipment is suitable for testing the leachates from disturbed or undisturbed solid environmental samples. Based on the assumption that the leachate flows into surface waters, the aquatic toxicity of the leachate is measured. When the target of the leachate in the environment is the groundwater, the tests may be of generic use or may simulate a specific groundwater use.

The test design of the OECD test guideline – duplicate leaching columns, packed with air-dried and sieved soil (<2 mm) up to a height of 30 cm, saturated and equilibrated with an 'artificial rain' solution and allowed to drain – can be applied or fitted to any environmental process related to the infiltration of precipitation and/or contaminants with the precipitation. The leachate can be collected in optional fractions. The leached soil can also be analyzed as a whole or in layers.

The authors of this chapter used soil 'minilysimeters' to monitor the stabilization of metals in microcosms because the sample volume was strongly limited. In a 31 mm diameter glass column, 200 g air-dried and sieved (<2 mm) soil was placed on a ceramic filter covered with a synthetic textile at the bottom of the column. The soil was wetted up to its water holding capacity and irrigated with 50 mL fractions of model precipitation (0.16 mmol/L $CaCl_2$ solution). The leachate was collected and analyzed for metal content, pH and electrical conductivity (EC). In another version, the stabilizing additive was layered between the ceramic filter and the soil. This provided a model for permeable reactive barriers.

Other problem-specific methods for modeling and assessing leaching in microcosms are described in Chapter 8.

## 4   AVAILABILITY OF CONTAMINANTS FOR ENVIRONMENTAL ACTORS

The environmental behavior of contaminants is primarily determined by the physico-chemical properties of the chemical substances, but their actual fate can be characterized and interpreted only as an interaction with the environment. Sorption of a contaminant to the soil's solid matrix is determined by the chemical's sorbability and the soil particles' sorption capacity. Biodegradation of a contaminant depends on its biodegradability and its bioavailability to potentially degrading microorganisms, their access to the chemical substance and their degrading activity. Partitioning between liquid and solid in soil, and leaching depend on the access of water to the contaminant.

Moreover, the result of chemical analyses is determined by the access of the extracting solvent to the contaminant, which can be bound either weakly or strongly to the matrix of the environmental compartment. The availability of the contaminant from the point of view of an organic solvent is completely different to the availability for water or for living cells in the soil. As a rough approximation we can say that an organic solvent can extract much more, while water extracts much less from the organic contaminant than the fraction available for living organisms. The models and tools to be used to assess a contaminated environment must take into account the differences between the contaminants' environmental behavior and their interactions with the environmental matrix and other components such as water or microorganisms.

Two case studies are introduced here to explain contaminants' availability for different actors in the environment and for the agents used by the assessors in analyses and tests. Traditional extracting solvents can never truly mimic the "extraction" by biological organisms: methods using organic solvents generally intend to extract "all" of the contaminants (with moderate success). Risk assessment results based on total extracts may be a massive overestimate. To bring a chemical model closer to the biological response in the environment, one direction of the methodological developments is to find so-called *biomimetic* extractants, which are able to mimic the access of organisms or biological systems to certain contaminants.

One of the pollution cases presented involves organic, the other inorganic contaminants. These pollution cases were analyzed in detail with regard to contaminant availability for different solvents, microorganisms and plants. The topic is discussed in greater detail in Chapter 7.

Accessibilities of transformer oil (a PCB-free mixture of hydrocarbons of petroleum origin) for different solvents (hexane–acetone and water dissolved cyclodextrins used as solubilizing agents) as well as for soil organisms were compared and the correlation of the results evaluated. Table 2.9 contains the values measured in the course of a soil remediation case (biodegradation-based clean-up of a transformer oil-contaminated soil) (Molnár *et al.*, 2009a)

Abbreviations in Tables 2.9 and 2.10:

- TO: Transformer oil, a commercial product, namely TO 40A;
- SEM: Solvent extractable mass, TO accessible for hexane:acetone 3:1. Measured by gravimetry;
- EPH GC: TO accessible for hexane:acetone = 3:1. Measured by gas chromatography;

*Table 2.9* Analyses and tests of soil sample series from a remediation experiment: different solvents and microorganisms under different conditions have different access to the contaminant.

| Measurement end points | 0 week | 2 weeks | 4 weeks | 6 weeks |
|---|---|---|---|---|
| SEM (mg TO/kg soil) | 34,000 | 33,400 | 30,500 | 26,200 |
| EPH GC (mg TO/kg soil) | 27,850 | 26,600 | 22,800 | 19,400 |
| REH GC (mg TO/kg soil) | 10,600 | 16,200 | 11,000 | 13,800 |
| HEH GC (mg TO/kg soil) | 7,400 | 11,200 | 3,500 | 2,600 |
| $CO_2$ (% in flow through air) | 0.27 | 0.34 | 0.25 | 0.21 |
| CB Respiration Index (1000*hPa/min) | 24 | 36 | 18 | 17 |
| DEH (μg TPF/g soil) | 87 | 62 | 34 | 47 |
| MPN of oil degrading cell conc. (cell/g soil) | 210,000 | 350,000 | 15,000 | 23,000 |

Table 2.10  Correlations between chemical analytical and biotest results; correlations highlighted in bold are significant at $p < 0.05$.

| | Biological end points | | | |
| | MPN | DEH | RESP CB | RESP AER |
|---|---|---|---|---|
| **Chemical end points** | | | | |
| SEM | 0.844 | 0.793 | 0.706 | 0.820 |
| | $p = 0.156$ | $p = 0.207$ | $p = 0.294$ | $p = 0.180$ |
| EPH GC | 0.902 | 0.864 | 0.711 | 0.826 |
| | $p = 0.098$ | $p = 0.136$ | $p = 0.289$ | $p = 0.174$ |
| REH GC | 0.343 | −0.306 | 0.681 | 0.540 |
| | $p = 0.657$ | $p = 0.694$ | $p = 0.319$ | $p = 0.460$ |
| HEH GC | **0.966** | 0.609 | **0.961** | **0.990** |
| | $p = 0.034$ | $p = 0.391$ | $p = 0.039$ | $p = 0.010$ |

- REH GC: TO accessible for randomly methylated beta-cyclodextrin, measured by GC-FID;
- HEH GC: TO accessible for hydroxypropyl beta-cyclodextrin, measured by GC-FID;
- GC-FID: Gas chromatography coupled with flame ionization detector;
- CB: Closed-bottle test;
- DEH: Dehydrogenase enzyme activity in contaminated soil;
- TPF: Triphenyl formazan (product of an alternative electron acceptor);
- MPN: Most probable number.

The correlation matrix in Table 2.10 includes the correlation coefficients and the significance levels (p). The correlation between chemical analytical and soil microbiological end points was investigated for transformer oil where the trends were correlated to each other.

One can learn from the results that the access of organic solvents and microorganisms to the transformer oil in this soil differs greatly: SEM and EPH GC have no good correlation with any of the biological end points. One of the two water-based cyclodextrin solutions is able to mimic the access of the degrading microorganisms: the HEH extract shows good correlation with the cell number and with the respiration both in closed and flow-through systems. This leads to the conclusion that a model can be created based on a hydroxypropyl-beta-cyclodextrin solution to mimic and extrapolate to the biodegradability of transformer oil. However, this does not mean that HEH can generally be applied for all kinds of contaminants.

In another case a toxic metal contaminant endangers agricultural soil at a former mining site. The site was exposed to floods over many years and became contaminated with cadmium and zinc. Table 2.11 shows the metal concentrations of the soil treated with water, a mildly acetic acetate solution, aqua regia, and test plants. Access of test plants to the metals in the soil was measured through plant toxicity (only the available part is effective) and plant bioaccumulation (only the metal available for the root is taken up and accumulated). For comparison, the same results for the chemically stabilized soil are also shown in the table.

*Table 2.11*  Availability of toxic metals in soil for different solvents and plants.

| Test method | Non-treated soil | Stabilized by fly ash |
|---|---|---|
| Cd accessible to aqua regia (mg/kg) | 5.2 | 5.2 |
| Cd accessible to acetate solution* (mg/kg) | 1.5 | 0.3 |
| Cd accessible to water (mg/kg) | 0.05 | <0.004 |
| Cd accumulated by plant (mg Cd/kg dry plant) | 6.6 | 0.7 |
| Cd extracted from soil by plant biomass (mg Cd/kg soil) | 0.33 | 0.13 |
| Zn accessible to aqua regia (mg/kg) | 1102 | 1102 |
| Zn accessible to acetate solution (mg/kg) | 237 | 48 |
| Zn accessible to water (mg/kg) | 4.1 | 0.3 |
| Zn accumulated in plant (mg Zn/kg dry plant) | 503 | 108 |
| Zn extracted from soil by plant biomass (mg Zn/kg soil) | 25 | 19 |
| Plant biomass in accumulation tests (g biomass/5 g soil) | 0.25 | 0.9 |
| Relative accessibility to plant (shoot growth in mm)** | 20 | 32 |

*Acetate solution: an ammonium-acetate buffer solution of pH = 4.5
**These results cannot be converted into available metal concentrations

## 5  UTILIZING FATE PROPERTIES OF CHEMICALS TO REDUCE THEIR RISK IN THE ENVIRONMENT

Many of the transport and fate processes in the environment may lead to a reduction of environmental risk.

The term **natural attenuation** is generally used for spontaneous biodegradation of organic contaminants in the groundwater. In this chapter, the term natural attenuation will be used in a wider sense, integrating all spontaneously occurring natural risk reducing processes such as *in situ* transformation and degradation processes (mainly biodegradation), irreversible immobilization by sorption or by incorporation into the solid matrix as well as dilution, dispersion and partitioning if they result in a decrease in environmental risk. It is advisable not to use the term 'decrease in the concentration or amount of a contaminant' instead of the decrease of risk because it is not necessarily proportional to the concentration or amount of the contaminant, but is strongly influenced by its physical and chemical forms as well by the characteristics of the matrix and the receptors.

The continuous change of a contaminant's risk due to natural processes in the environment results in a site-specific *risk profile* and the change of risk as a function of time. A continuous falling slope of the profile indicates natural attenuation of the risk which may run parallel to the decrease in contaminant concentration, but it may also be attributed to decreasing mobility and bioavailability. A short summary is given here about the topic, a detailed discussion can be found in Volume 4.

### 5.1  Environmental transport and fate processes change contaminant risk

Transport processes such as advection, dispersion, diffusion, dilution, volatilization and partitioning change the flux of dissolved contaminants. The gravitational transport of precipitation, groundwater and surface waters results in dispersion, increasing the area or volume of contaminated land (which may increase the risk), but at the same time, decreasing concentration and risk due to sorption or filtering out and dilution along the transport pathways.

Partitioning between air–water and solid–water may result in either increased or decreased risk, depending on the source of the contaminant and the direction of the process. Due to partitioning between air and water, the contaminant can be eliminated from the water (to the air) or the water can absorb the contaminant from the air. Contaminants transported by groundwater can be distributed by partitioning between water and solid soil phases. This process is also called sorption of the contaminant on solid particles or filtration of the contaminant by the solid fraction. The same amount of a chemical is less risky when bound to the solid phase of the soil: there is a three to four orders of magnitude difference between the effective concentrations of the same contaminant in the soil/sediment solid phase and in soil moisture, pore-water or groundwater. In which of the phases with the same amount of contaminant is the risk higher depends on many factors, but chiefly on the volume contaminated, the sensitivity and density of exposed receptors and exposure pathways.

The rearrangement of contaminants by partitioning may proceed between water and the living 'phase' as well as between solid and the living 'phase'. The harmful nutrients (N, P) from water can be taken up by algae and water plants, which leads to eutrophication when the processes are uncontrolled, but may be beneficial when the plants are removed from the water system after take-up of nutrients and before dying and nutrient recycling. Partitioning of metals between soil and plants may be harmful to the food chain, but it can be beneficial for the soil if the bioaccumulating plants are removed from the soil after the contaminant has been extracted. Natural ecological systems, e.g. the plants of aerobic or semianaerobic wetlands are able to filter nutrients and contaminants with the help of their rhizosphere. Bacteria settled in the rhizosphere are ready to mineralize trapped nutrients and contaminants. This is why the concentration of contaminants and their risk decrease in the water and soil or even in wastewaters or leachates that reach the wetland.

When living organisms interact with the toxicant, active transport, not only simple partitioning, may occur, and the contaminant may accumulate in the tissues of the living organisms.

The equilibrium of natural process pairs such as mobilization–immobilization, sorption–desorption, oxidation–reduction depends to a large extent on the environmental conditions such as pH, temperature and redox potential.

Mobilization may support the elimination and removal of contaminants from soil and sediment as well as solid waste, reducing the risk in the solid but increasing it in the liquid/water phase. The solution is again the controlled and isolated handling of the water. Immobilization may have a direct risk reducing effect: strong and irreversible sorption in or on the solid phase of environmental compartments brings the transport of water-soluble contaminants to a halt. This kind of stabilization lowers the risk in waters, but over the long term, may result in the enrichment of contaminants in the soil or sediment, which, after a while, will act as a chemical time bomb. Mobilization first appears to be a risk-increasing phenomenon, but if the mobilized and transported contaminant can be diluted and resorbed under less risky conditions (e.g. at a place isolated from the environment or at a less sensitive location), the result of dilution is satisfactory. When only the front part of a subsurface contaminant plume has been degraded by soil microbes, dispersion of the plume may intensify biodegradation but increase the size of area affected.

There is a continuous redox gradient in nature, e.g. a decreasing redox potential with depth in water and soils. The chemical form and, as a consequence, the mobility of chemicals are determined by the redox potential. Certain contaminants, such as dissolved metals can be immobilized by higher pH and lower redox potential, depending on the chemical species. Precipitation of metals in deeper soil layers, sediments and wetlands in sulfide form produces a highly stable and biologically inaccessible form, as long as the redox potential is low.

The pH is also responsible for chemical forms of different mobility, for example, limestone in nature may convert ionic metals to hydroxide compounds, which are less mobile. This means that the waters and the living organisms are protected from the risk, but the metals are still there.

The most efficient and final reduction of the risk due to chemical contaminants is their degradation and complete elimination from the environment. Hydrolysis, photolysis, chemical degradation by oxidation or reduction and biodegradation can eliminate organic contaminants. The utilization of risk reducing natural processes in remedial technologies is highly recommended. The vast genetic and biochemical potential of the microorganisms in the environment makes possible the elimination of any organic compound, even xenobiotics. Mankind should be grateful for this natural risk-reducing 'measure'. Of course, risk may also increase due to temporary toxic metabolites or end products, but this is not a typical case. However, this possibility, too, has to be checked and controlled, e.g. by using ecotoxicity tests for the monitoring of natural attenuation. Inorganic compounds, chiefly metals differ in this respect from organic substances: they can hardly be removed from the environment, so that the "background" concentration is continuously increasing. This gives all the more cause for concern as global reserves of some metals are already close to depletion.

The above summarized environmental processes may serve as the basis for bio- and ecotechnologies. Engineers can optimize the environmental conditions for the biological catalyst, i.e. the living organisms which can reduce or eliminate risk.

Not only are these nature-based *in situ* technologies suitable for the remediation of contaminated water, sediment and soil, but also for nature conservation and environmental health protection, in one word for the sustainable management of waters and land. The prevention of contamination and deterioration of the environment is more important; it can be the key tool of long-term precautionary environmental risk management in the future. These 'soft' technologies are eco-efficient, they can harmonize the good environmental/ecological status with land use and ensure sustainability. Ecoengineers have to select the beneficial, risk-reducing environmental transport and fate processes of the chemical substances and utilize them. In terms of engineering, there are no risk-increasing processes in the environment, only inadequate control and application! The same process can increase or reduce the risk, depending on whether it is isolated from, or is in contact with the environment. A properly selected technology, e.g. a simple 'barrier', which is able to isolate the process of sorption, accumulation or leaching from the environment, can turn the processes into beneficial ones. Another approach to increase efficacy by engineering tools is the intensification of the beneficial risk-reducing processes such as biodegradation, bioprecipitation or bioleaching with the help of redox potential control and nutrient circulation.

The risk profile of a contaminant due to naturally occurring processes provides information about the direction and velocity of the changes in its risk. Technologists may develop remediation technology on the existing risk profile without any change; they can speed it up, or change the direction of the risk profile from an increasing risk profile (in contact with the environment) into a decreasing risk profile by isolating the process from the environment, e.g. collecting the leachate of high contaminant concentration (see also Chapter 8 in Volume 4 on risk-reducing natural processes).

## REFERENCES

Alef, K. & Nannipieri, P. (1995) *Methods in applied soil microbiology and biochemistry.* London, Academic Press.

Alvarez, P.J.J. & Illman, W.A. (2006) *Bioremediation and natural attenuation.* NJ, Wiley-Interscience.

Arnot, J., Gouin, T. & Mackay, D. (2005) *Development and application of models of chemical fate in Canada – Practical methods for estimating environmental biodegradation rates.* Report to Environment Canada. CEMN Report No. 200503. [Online] Canadian Environmental Modelling Network, Peterborough, Ontario K9J 7B8, Canada, Trent University. Available from: http://www.trentu.ca/academic/aminss/envmodel/CEMNReport200503.pdf. [Accessed 10th October 2013].

BESS (2013) *Biodegradability evaluation and simulation system. Applying and Discovering Knowledge of Biodegradation Pathways.* [Online] Available from: http://www.cse.msu.edu/~punch/projects/biodeg.html. [Accessed 10th October 2013].

BIODEG (2013) Environmental Fate Data Base (EFDB). BIODEG Chemical Search. Syracuse Research Corporation (SRC) Available from: http://srcinc.com/what-we-do/databaseforms.aspx?id=382. [Accessed 10th October 2013].

BioMineWiki (2013) *Bioleaching.* [Online] Available from: http://wiki.biomine.skelleftea.se/wiki/index.php/Bioleaching. [Accessed 10th October 2013].

BIOWIN (2013) *Part of the EPI (Estimation Programs Interface) Suite™.* [Online] Available from: http://www.epa.gov/oppt/exposure/pubs/episuite.htm. [Accessed 10th October 2013].

Boethling, R.S. (1986) Application of molecular topology to quantitative structure–biodegradability relationships. *Environmental Toxicology and Chemistry,* 5, 797–806.

Boethling, R.S. & Sabljic, A. (1989) Screening-level model for aerobic biodegradability based on a survey of expert knowledge. *Environmental Science & Technology,* 23, 672–679.

Boethling, R.S., Howard, P.H., Meylan, W.M., Stiteler, W.M., Beaumann, J.A. & Tirado, N.F. (1994) Group contribution method for predicting probability and rate of aerobic biodegradation. *Environmental Science & Technology,* 28, 459–465.

Boethling, R.S., Lynch, D.G. & Thom, G.C. (2003) Predicting ready biodegradability of premanufacture notice chemicals. *Environmental Toxicology and Chemistry,* 22, 837–844.

Brandes, L.J., den Hollander, H. & van de Meent, D. (1996) *SimpleBox 2.0: a nested multimedia fate model for evaluating the environmental fate of chemicals.* RIVM Report 719101 029. Bilthoven, The Netherlands, National Institute of Public Health and Environmental Protection (RIVM).

Crawford, R.L. (2002) Biotransformation and biodegradation. In: Hurst, C.J., Crawford, R.L., Knudsen, G.R., McInerney, M.J. & Stetzenbach, L.D. (eds.) *Manual of environmental microbiology.* 2nd edition. Washington, DC, ASM Press.

Dearden, J.C. & Nicholson, R.M. (1987) Correlation of biodegradability with atomic charge difference and superdelocalizability. In: Kaiser, K.L.E. (ed.) *QSAR in environmental toxicology* – II, Dordrecht, Reidel. pp. 83–89.

EFDB–SRC (2013) *Scientific Databases*. [Online] Available from: http://www.srcinc.com/what-we-do/environmental/scientific-databases.html. [Accessed 10th October 2013].

EPA Food Chain Model (2013) *Food and Gill Exchange of Toxic Substances*. [Online] Available from: http://www2.epa.gov/exposure-assessment-models/fgets. [Accessed 10th October 2013].

EPISUITE™(2013) *Estimation programs interface*. Available from: http://www.epa.gov/oppt/exposure/pubs/episuite.htm. [Accessed 10th October 2013].

ESIS–JRC (2013) *European Chemical Substances Information System*. [Online] Available from: http://esis.jrc.ec.europa.eu/. [Accessed 10th October 2013].

EU-TGD (2003) Technical Guidance Document on risk assessment in support of Commission Directive 93/67/EEC on risk assessment for new notified substances; Commission Regulation (EC) No 1488/94 on risk assessment for existing substances; Directive 98/8/EC of the European Parliament and of the Council concerning the placing of biocidal products on the market – Part II. European Commission. [Online] Available from: http://ihcp.jrc.ec.europa.eu/our_activities/public-health/risk_assessment_of_Biocides/doc/tgd. [Accessed 10th October 2013].

Feigl, V. (2012) *Combined chemical and phytostabilization of toxic metal contaminated soil and mine waste*. Ph.D. Thesis. Budapest University of Technology. Budapest, Hungary.

Fu, W., Franco, A. & Trapp, S. (2009) Methods for estimating the bioconcentration factor of ionizable organic chemicals. *Environmental Toxicology and Chemistry*, 28, 1372–1379.

GM CEPA (1999) Guidance manual for the risk evaluation framework for Sections 199 and 200 of CEPA 1999. *Decisions on environmental emergency plans*. [Online] Available from: http://www.ec.gc.ca/lcpe-cepa/default.asp?lang=En&xml=BAE803CC-9ED3-3D10-9D8B-5C8637B45E4A#241. [Accessed 10th October 2013].

Gruiz, K., Horváth, B. & Molnár, M. (2001) *Környezettoxikológia (Environmental Toxicology)*. Budapest, Műegyetem Publishing Company (in Hungarian).

Howard, P.H., Boethling, R.S., Stiteler, W.M., *et al.* (1992) Predictive model for aerobic biodegradability developed from a file of evaluated biodegradation data. *Environmental Toxicology and Chemistry*, 11, 593–603.

Howard, P., Meylan, W., Aronson, D., *et al.* (2005) A new biodegradation prediction model specific to petroleum hydrocarbons. *Environmental Toxicology and Chemistry*, 24(8), 1847–1860.

HSDB–Toxnet (2013) *Hazardous Substances Data Bank*. [Online] Available from: http://toxnet.nlm.nih.gov/cgi-bin/sis/htmlgen?HSDB. [Accessed 10th October 2013].

Hyman, M. & Dupont, R.R. (2001) *Groundwater and soil remediation: Process design and cost estimating of proven technologies*. Reston, VA, ASCE Press.

Ingersoll, C.G., Ankley, G.T., Benoit, D.A., *et al.* (1995) Toxicity and bioaccumulation of sediment-associated contaminants with freshwater invertebrates: A review of methods and applications. *Environmental Toxicology and Chemistry*, 14, 1885–1894.

Ingersoll, C.G., Haverland, P.S., Brunson, E.L., *et al.* (1996) Calculation and evaluation of sediment effect concentrations for the amphipod Hyalella azteca and the midge Chironomus riparius. *Journal of Great Lakes Research*, 22, 602–623.

Isnard, P. & Lambert, S. (1988) Estimating Bioconcentration factors from octanol–water partition coefficient and aqueous solubility. *Chemosphere*, 17, 21–34.

Jager, T. (1998) Mechanistic approach for estimating bioconcentration of organic chemicals in earthworms (Oligochaeta). *Environmental Toxicology and Chemistry*, 17, 2080–2090.

King, R.B., Long, G.M. & Sheldon, J.K. (1992) *Practical environmental bioremediation*. Boca Raton, FL, Lewis Publishers.

Kravitz, M. (1998) *Briefing presentation*. Washington, DC, U.S. Environmental Protection Agency.

Mackay, D. & Fraser, A. (2000) Bioaccumulation of persistent organic chemicals: mechanisms and models. *Environmental Pollution*, 110, 375–391.

Mackay, D., Paterson, S. & Shiu, W.Y. (1992) Generic models for evaluating the regional fate of chemicals. *Chemosphere*, 24(6), 695–717.

McCarty, L.S. & Mackay, D. (1993) Enhancing ecotoxicological modeling and assessment. *Environmental Science & Technology*, 27, 1719–1727.

METI (2013) Japanese Ministry of Economy, Trade and Industry, before 2001 Japanese Ministry of International Trade and Industry (MITI). Available from: http://www.meti.go.jp/english/. [Accessed 10th October 2013].

METI–NITE (2013) National Institute for Technology and Evaluation. Available from: http://www.safe.nite.go.jp/english/index.html. [Accessed 10th October 2013].

METI–NITE–CHRIP (2013) Chemical Risk Information Platform. NITE. Available from: http://www.safe.nite.go.jp/english/db.html. [Accessed 10th October 2013].

Meylan, W.M., Howard, P.H., Boethling, R.S., Aronson, D., Printup, H. & Gouchie, S. (1999) Improved method for estimating Bioconcentration/Bioaccumulation Factor from Octanol/Water Partition Coefficient. *Environmental Toxicology and Chemistry*, 18, 664–672.

Mikkelsen, J. (1995) *Fate model for organic chemicals in an activated sludge wastewater treatment plant – Modification of SimpleTreat*. National Environmental Research Institute, Denmark. Prepared for the Danish EPA.

MOKKA Project (2004–2008) Innovative Decision Support tools for risk based environmental management. [Online] Available from: http://www.mokkka.hu/index.php?lang=eng&body=mokka. [Accessed 10th December 2013].

Molnár, M. (2006) *Intensification of soil bioremediation by cyclodextrin – From the laboratory to the field*. PhD thesis. Budapest University of Technology and Economics, Budapest, Hungary.

Molnár, M., Leitgib, L., Gruiz, K., Fenyvesi, É., Szejtli, J. & Fava, F. (2005) Enhanced biodegradation of transformer oil in soils with cyclodextrin – from the laboratory to the field. *Biodegradation*, 16, 159–168.

Molnár, M., Fenyvesi, É., Gruiz, K., Illés, G., Hajdú, Cs. & Kánnai, P. (2009a) Laboratory testing of biodegradation in soil: a comparative study on five methods. *Land Contamination & Reclamation*, 17(3–4), 495–506.

Molnár, M., Leitgib, L., Fenyvesi, É. & Gruiz, K. (2009b) Development of cyclodextrin enhanced soil bioremediation: from laboratory to field. *Land Contamination & Reclamation*, 17(3–4), 599–610.

Molnár, M., Fenyvesi, É., Gruiz, K., Lantos, N. & Fehér, Zs. (2010) Innovative remediation of groundwater contaminated by chlorinated hydrocarbons. In: UFZ-Deltares/TNO (ed.) *Proceedings of the 11th UFZ-Deltares/TNO Conference on management of soil, groundwater & sediments*, CD, Consoil 2010, Salzburg, Austria, September 22–24. Poster No. A3-48. ISBN 978-3-00-032099-6.

OECD 305 (1996) *OECD guidelines for testing of chemicals – Bioconcentration: flow-through fish test*. http://www.oecd-ilibrary.org/content/book/9789264070462-en. [Accessed 10th October 2013].

OECD 312 (2004) OECD guidelines for the testing of chemicals – leaching in soil columns. [Online] Available from: http://www.oecd-ilibrary.org/environment/test-no-312-leaching-in-soil-columns_9789264070561-en. [Accessed 10th October 2013].

OECD 315 (2008) *OECD guidelines for the testing of chemicals – Bioaccumulation in sediment-dwelling benthic oligochaetes*. [Online] Available from: http://www.oecd-ilibrary.org/content/book/9789264067516-en. [Accessed 10th October 2013].

Paul, E.A. & Clark, F.E. (1996) *Soil microbiology and biochemistry*. 2nd edition. San Diego, CA, Academic Press.

Pavan, M. & Worth, A.P. (2006) *Review of QSAR models for ready biodegradation*. European Commission Directorate General, Joint Research Centre, Institute for Health and Consumer Protection EUR 22355 EN. [Online] Available from: http://ecb.jrc.ec.europa.eu/documents/QSAR/QSAR_Review_Biodegradation.pdf. [Accessed 10th October 2013].

Pavan, M., Worth, A.P. & Netzeva, T.I. (2006) *Review of QSAR Models for Bioconcentration*. European Commission, Directorate – General Joint Research Centre, Institute for Health and Consumer Protection. [Online] Available from: http:// http://ecb.jrc.it/QSAR. [Accessed 10th October 2013].

PBT REACH (2008) *Guidance on information requirements and chemical safety assessment Part C: PBT assessment*. [Online] Available from: http://guidance.echa.europa.eu/docs/guidance_document/information_requirements_part_c_en.pdf?vers=20_08_08. [Accessed 10th October 2013].

Punch, B., Patton, A., Wight, K., Larson, R.J., Masscheleyn, P. & Forney, L. (1996a) A biodegradability evaluation and simulation system (BESS) based on knowledge of biodegradation pathways. In: Penijnenburg, W.J.G.M. & Damborsky, J. (eds.) *Biodegradability prediction*. Dordrecht, Kluwer. pp. 65–73.

Punch, B., Patton, A., Wight, K., Larson, R., Masscheleyn, P. & Forney, L.J. (1996b) *BESS, a system for predicting the biodegradability of new compound*. Presented at QSAR 96, Copenhagen, Denmark, June 1996.

Rand, G.M. (ed.) (1995) *Fundamentals of aquatic toxicology: Effects, environmental fate, and risk assessment*. 2nd edition. Washington, DC, Taylor & Francis.

Rand, G.M. & Petrocelli, S.R. (eds.) (1985) *Fundamentals of aquatic toxicology: Methods and applications*. New York, Hemisphere Publishing Corp.

Raymond, J.W., Rogers, T.N., Shonnard, D.R. & Kline, A.A. (1999) A review of structure-based biodegradation estimation methods (R825370C064). *Journal of Hazardous Materials*, American Chemical Society, Washington, DC. [Online] Available from: http://cfpub.epa.gov/si/si_public_record_Report.cfm?dirEntryID=78758. [Accessed 10th October 2013].

SRC (2013) Syracuse Research Corporation. Available from: http://srcinc.com/about/; http://www.syrres.com/; http://srcinc.com/what-we-do/category.aspx?id=62. [Accessed 10th October 2013].

Struijs, J., Stoltenkamp, J. & Van De Meent, D. (1991) A spreadsheet-based model to predict the fate of xenobiotics in a municipal wastewater treatment plant. *Water Research*, 25(7), 891–900.

Tunkel, J., Howard, P.H., Boethling, R.S., Stiteler, W.M. & Loonen, H. (2000) Predicting ready biodegradability in the Japanese Ministry of International Trade and Industry (MITI) test. *Environmental Toxicology and Chemistry*, 19, 2478–2485.

US EPA (2000) *Bioaccumulation testing and interpretation for the purpose of sediment quality assessment – Status and needs*. United States Office of Water (4305) EPA-823-R-00-001. Environmental Protection Agency, Office of Solid Waste (5307W). [Online] Available from: http://water.epa.gov/polwaste/sediments/cs/upload/bioaccum.pdf. [Accessed 10th October 2013].

US EPA (2004) *Estimation programs interface*. U.S. EPA Office of Pollution Prevention Toxics and Syracuse Research Corporation. [Online] Available from: http://www.epa.gov/oppt/exposure/pubs/episuite.htm. [Accessed 10th October 2013].

US EPA & US ACE (1998) *Evaluation of dredged material proposed for discharge in waters of the U.S. – Testing manual: Inland testing manual*. EPA-823-B-98-004. Washington, DC, U.S. Environmental Protection Agency and U.S. Army Corps of Engineers.

Van de Meent, D. (1993) *Simplebox: a generic multimedia fate evaluation model*. RIVM Report No. 672720 001. Bilthoven, The Netherlands, National Institute of Public Health and the Environment (RIVM).

Veith, G.D., Defoe, D.L. & Bergstedt, B.V. (1979) Measuring and estimating the bioconcentration factor of chemicals in fish. *Journal of the Fisheries Research Board of Canada*, 36, 1040–1048.

Wackett, L.P. & Ellis, L.B.M. (1999) Predicting biodegradation. *Environmental Microbiology*, 1, 119–124.

Yoo, L.J., Steevens, J.A. & Landrum, P.F. (2003) Development of a new bioaccumulation testing approach: the use of DDE as a challenge chemical to predict contaminant bioaccumulation. ERDC/TN EEDP-01-50. [Online] Available from: http://www.glerl.noaa.gov/pubs/fulltext/2003/20030006.pdf. [Accessed 10th October 2013].

Chapter 3

# Human toxicology

## K. Gruiz

*Department of Applied Biotechnology and Food Science,*
*Budapest University of Technology and Economics, Budapest, Hungary*

## ABSTRACT

Humans are exposed to chemical hazards through inhalation, consumption of water and food, and dermal contact. Human life today is highly non-natural, and it takes place predominantly in a man-made (built) environment which is different from the natural ecosystem. Human exposure is restricted to (mainly indoor) air, processed foods and beverages and to minor dermal contact. Occupational health protection gadgets and equipment prevent employees from absorbing hazardous chemical substances at the workplace. In addition to protective wear and equipment, the human organism is highly adaptive and capable of self-defense. But all of this is no help when people are highly exposed to chemical hazards, including new, emerging chemicals whose effects are not (yet) fully known. These effects are different from those of traditional toxicants and are not proportional to their concentrations or doses, and they cannot be measured using traditional toxicological methods.

This chapter gives an overview of traditional toxicology, including acute and chronic toxicity, corrosivity, mutagenicity, carcinogenicity and reprotoxicity, as well as newly recognized chemical hazards such as sensitization, allergization, and endocrine and immune system disruption. In addition to traditional epidemiological studies and animal tests, alternative methods using molecules, cells or tissues instead of living animals will be discussed. Extrapolation methods from animal or alternative end points to humans and the associated uncertainties are also discussed because they are key factors in the utilization of human toxicity data for environmental management.

## I INTRODUCTION

People are exposed to environmental chemicals when they are in contact with contaminated land. Exposure to air, surface waters and sediments, soil and groundwater can take the form of residential, recreational, agricultural, and industrial land uses, as illustrated in Figure 3.1. Humans are exposed to contaminants directly by inhalation, ingestion and dermal contact or through the food chain. Exposure of humans via plants, fish and livestock to environmental contaminants is also called secondary

*Figure 3.1*   Human exposures to chemical substances in the environment.

poisoning. The two main sources of secondary poisoning are crops and fish. The most sensitive exposure pathways are:

–   The inhalation of air, vapors and dust;
–   The consumption of water from subsurface or surface water resources;
–   The consumption of agricultural products such as cereals, fruits, vegetables, animal products (meat and milk);
–   Eye and dermal contact with air, water (suspended solids) and soil.

The adverse effects of environmental chemicals on humans cannot be tested directly on a statistically suitable number of test persons at different concentrations aiming to find the no-effect concentration or dose. Theoretically, we could use existing epidemiological information (mainly from occupational poisoning cases or chemical accidents) for extrapolation to an average population or to children, but since epidemiological data are neither available for every chemical substance nor are the data reliable, extrapolation is accompanied by a high level of uncertainty resulting in undue overestimation.

A more realistic estimate may be obtained from animal tests, based on the assumption that the applied test organism is a good model for humans. Traditional toxicological testing is largely based on the use of higher laboratory animals: rat, mouse, rabbit, dog, etc. The drawbacks of this approach are low throughput, high cost, and difficulties inherent to inter-species extrapolation.

Increasing knowledge of the differences between humans and the most widely used test organisms suggests that more advanced methods are needed to extrapolate e.g. from rat to man. Scientific toxicological knowledge and sophisticated tests are useless, if a safety factor of 10 must be applied to extrapolate from rat to man. Satisfactory results are difficult to obtain and they are not proportional to the investments in animal testing. Overestimates, which cannot always be justified, result in high risk reduction costs. In addition, animal testing is ethically questionable and is of limited use in evaluating the very large number of chemicals with regard to inadequate toxicological data.

These problems have led to a search for alternative test methods to replace laboratory animal tests with *in vitro* or *in silico* methods. Current developments in molecular biology and computer science open new possibilities to find the genes or other key-molecules and mechanisms responsible for the interaction with toxicants and effectuating toxicity of the chemical substances.

The tests must have clear aims and well-identifiable results in order to be able to develop and apply alternative methods. Generic testing of the chemicals' effects for regulatory purposes has different requirements than site-, age- or gender-specific assessments for managing specific, e.g. locally arising problems. Tests on pure chemicals or water, soil and agricultural products also have their own specific aims and methodology. Every task and every case needs its own optimal test method that requires the smallest number of animals, provides rapid results and is cheap yet informative. Tiering may help integrate the rapid screening methods into the risk assessment procedure. Rapid testing can exclude negative cases for a large number of chemicals or environmental samples at as early a stage as possible, thus minimizing expenditure. The expensive, detailed assessment with costly labor and animal requirements should only be used for the most risky cases.

## 1.1  Adverse effects of chemicals on humans

Adverse effects of chemicals on humans are classified into groups according to the time duration necessary for the appearance of an effect in the target of the impact:

- – *[1]Acute toxicity: adverse effects occurring following oral or dermal administration of a single dose or repeated dose within a short time (24 h) of a substance, or inhalation exposure (of 4 hours);
- – *Specific target organ toxicity with single exposure via any of the exposure routes. It is a specific, non-lethal toxicity arising from a single exposure resulting impaired function of any or the organs, reversible and irreversible, immediate and/or delayed. See also specific target organ toxicity with repeated exposure.
- – Neurotoxicity belongs to specific target organ toxicity, developmental neurotoxicity to reprotoxicity.
- – Repeated dose toxicity studies characterize the toxicological profile of a toxicant following repeated administration. The profile includes the identification of potential target organs, the exposure–response relationships and reversibility of toxic effects. It may play a role in the determination of the maximal tolerated dose (MTD).
- – Chronic toxicity is the ability of a chemical substance to cause harm to an organism as a result of repeated exposures for long periods of time, even a lifetime. Very

---

[1]The effects marked with * are GHS classes as well.

small chronic exposures to a potentially harmful chemical can be handled by the exposed organisms, but increasing exposures cannot be fended off. The turning point between no-effect and effect levels is the threshold for chronic exposures; it should be the regulatory governing value for chronic toxicants.

–   *Specific target organ toxicity which arises from a repeated exposure to a substance or mixture resulting in impaired function of any of the organs, reversible and irreversible, immediate and/or delayed. Typical signs are: morphological and functional damage of organs (necrosis, fibrosis or granuloma, cell death, etc.), hematotoxicity, serious damage or morbidity due to bioaccumulation of the substance, damage to the nervous system's function, etc.

–   Mutagenicity: the effect of mutagenic agents which are giving rise to an increased occurrence of mutations in populations of cells and/or organisms, including somatic cells and germ cell. The latter will lead to a heritable genetic change and form generally a separate group called genotoxicity or germ cell mutagenicity. *In vivo* mutagenic activity may indicate that a substance also has a potential for carcinogenic effects.

–   *Germ cell mutagenicity impacts the egg or sperm cells (germ cells) and therefore can be transmitted to the next generation.

–   Genotoxicity and genotoxic carcinogenicity mean alteration in the structure, information content, or segregation of DNA, including the normal replication processes. Genotoxicity test results are usually taken as indicators for mutagenic or carcinogenic effects.

–   *Carcinogenicity of a substance or a mixture of substances means that they induce cancer or increase its incidence. Carcinogenicity is traceable to genotoxicity or non-genotoxic origin.

–   *Toxicity for reproduction, also called reprotoxicity, includes adverse effects on sexual function and fertility of adults, the developmental toxicity on the fetus (also called teratogenic effect), as well as adverse effects via lactation. Genetically based heritable effects in the offspring are addressed in germ cell mutagenicity.

–   *Irritation: production of reversible damage to the skin or reversible changes in the eye;

–   *Corrosiveness: the production of irreversible damage to the skin or causing serious eye damage;

–   *Sensitization: respiratory sensitizers lead to hypersensitivity of the airways following inhalation of the substance, skin sensitizer to an allergic response following skin contact.

–   Endocrine disruption is an adverse effect of endocrine disruptor chemicals, which may interfere with the body's endocrine system and produce adverse developmental, reproductive, neurological, and immune effects in both humans and wildlife. It means that immune-system disruption and neurotoxicity may be a special case of endocrine disruption.

–   Immune-system disruption is a subcategory of endocrine disruption, meaning a long term adverse effect resulting in allergic and autoimmune diseases and their growing prevalence rate, e.g. in the case of type 1 diabetes.

These types of adverse effects are thoroughly defined and specific test methods are developed for providing end points for the identification and characterization of these groups of effects. The grouping includes overlaps in deterioration and the classification point of views are also combined with each other: the definition of the single adverse effects includes details on duration, routes of exposure, metabolism, reversibility and other characteristics of the evolution and consequences of the effect.

There are many important issues to be refined and clarified, only a few from them are mentioned below, just to draw the attention:

- Dose–response: all chemicals may pose adverse effects in abnormally large doses/concentrations and toxicants can be harmless under a certain threshold.
- The threshold is a turning point between no-effect and effect levels/concentrations: values measurable in tests and applicable as quality criteria in risk management.
- Threshold chemicals are those chemicals which do not show adverse effects after a certain case of dilution, and the opposite: show an effect with growing doses/concentrations. In contrast, some chemicals are thought to be non-threshold chemicals because a certain no-effect value for them could not be determined. Scientists do not agree, some of them state that all chemicals have a threshold (including mutagens, carcinogens. teratogens, endocrine disruptors), some others maintain that there are no thresholds e.g. for carcinogenic effects of chemicals, meaning that one molecule of a carcinogen is capable of starting a cancer. There is no scientific evidence for either the first or the second view.
- Bioaccumulation is a risk factor both in the ecosystem and in humans, but its interpretation is often oversimplified by considering as a one-way process with a static outcome. The storage of toxicants is a defense mechanism in the body, and is a dynamic process: the depots are increasing when inflow is larger than the outflow, and start to decrease when no resupply is coming in; it is a question of equilibrium, which of course depends greatly on the partitioning of the chemical substance between blood and the tissue of storage.
- Reversibility–irreversibility of effects may be an important issue when the risk of the chemicals is calculated. Reversible effect means that the exposed tissue recovers and the organism returns to normal, when exposure ceases, e.g. skin or eye irritation and non-lethal acute narcotic effects. Irreversible effects are final and permanent, no recovery takes place such as in the case of skin or eye corrosion, tissue lesion, developmental/congenital disorder or cancer, etc.
- Mixtures of chemicals and additivity of effects: legislation is based predominantly on hazard and risk of individual substances. In recent years several debates arose due to the large number of chemicals present in the environment in low concentration (under their individual threshold), and the consequent co-exposure of ecosystem and man. The exposure from several environmental compartments and secondary poisoning from water and food cannot be quantified by currently existing scientifically based methodologies. The additivity of the effects is also questionable in most of the cases: even the effects of chemicals of similar effect-mechanism are not additive because significant synergism or antagonism may occur. The cumulated effect of mixtures and coexisting chemicals in the environment should be considered, but a groundless overestimation by multiplying highly conservative risk scenarios for each should be avoided.
- Weight of evidence (WoE) is a term scientifically not well-defined. It covers the assessment of the relative weights of different pieces of the available information

on the topic to decide on. The weight given to the available evidence can be formalized by pre-established factors which characterize the quality of the data, the consistency of the results, the type and severity of the effect and the relevance of the information. REACH Regulation recognized the WoE as a useful tool in classification and labeling (CLP) as well as in the safety assessment of chemicals (CSA). It is a sufficient weight of evidence for a chemical substance having, or not having, dangerous properties when several independent sources of information conclude the same (e.g. well-documented human cases), or when newly developed test methods, not yet acknowledged by the regulations, demonstrate the effect or prove the "no effect".

"Where sufficient weight of evidence for the presence or absence of a particular dangerous property is available: i) further testing on vertebrate animals for that property shall be omitted; ii) further testing not involving vertebrate animals may be omitted. In all cases adequate and reliable documentation shall be provided" (REACH, 2006).

## 1.2  Testing the adverse effects of chemicals on humans

*In vivo* or *ex vivo* animal tests and *in vitro* non-animal models can be used to measure the adverse effects of chemicals on humans. *In silico* models can be created for any of the data sets that satisfies the statistical requirements. The most frequent human toxicity end points are dermal and eye irritation and corrosion, acute and repeated-dose systemic toxicity, organ-specific toxicity, chronic toxicity, carcinogenicity, mutagenicity, reprotoxicity, endocrine and immune toxicity, neurotoxicity, and sensitization. In this chapter, test organisms, existing standardized animal test methods and *in vitro* alternatives will be introduced.

The methods of extrapolating from test or model results to humans should be based on knowledge, experience and statistical analysis. Extrapolation from mathematical (*in silico*), molecular or biochemical and the more complex biological models bears high uncertainty due to the great standard deviation of the test methods and the differences between the models and the human body. Increasing knowledge on these similarities and differences may reduce uncertainties, but intraspecies differences within the human population still remain and represent a major unknown factor.

Figure 3.2 shows the relationship of models to the human population: the distance (D) from humans (due to differences and complexity) and the angle $\alpha$, $\beta$ (characterizing the error of the extrapolation method) together determine the deviation of the result of the model from real human values. The dotted line around the models illustrates their standard deviation. We can see that a smaller extrapolation error and smaller distance of animal models compared to molecular models cause smaller deviation from current human values, in spite of a higher standard deviation of animal testing. If an amended molecular model could reduce the angle "$\beta$" to "$\alpha$", i.e. to that of the animal tests, the extrapolated value would be closer to real human values (small deviation). Experts should choose the optimal tests to obtain the most reliable result with minimum effort. Depending on the initial data set, *in silico* methods can be either very close or very far from reality (human body or human populations). With an extremely large amount of data available, data mining and processing methods may help develop mathematical models which are perfectly fit for human risk assessment.

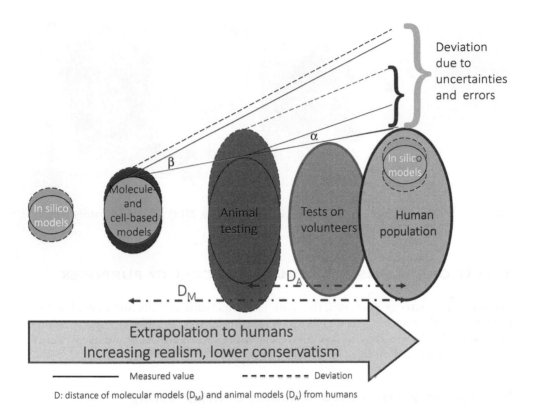

Figure 3.2 Modeling human exposures to hazardous chemical substances.

Nowadays the assessment of toxicological risk relies primarily on *in vivo* animal experiments that were designed decades ago and cost about $3 billion/year world-wide (Hartung, 2012). Uncertainties in animal representativeness (due to the differences in animal and human toxicity responses) are handled by applying safety factors when extrapolating from animal toxicity data to human risk (NRC, 2000; Kavlock *et al.*, 2009) which may lead to significant overestimates. Another shortcoming of animal testing is the use of extremely high doses which are much higher than those administered in reality. In these studies the highest dose gives mostly positive response which entails undue measures in risk management (Hartung, 2009). FDA (2004) statistics on drug development demonstrate that not only do overestimations occur, but more dangerous underestimations do too: 20% of the developed drugs fail due to toxic effects in humans not identified in pre-clinical animal testing (another 72% fails due to other reasons, totaling 92%).

Validated *in vitro* test methods (OECD, 2005) and evidence-based toxicology may help (see Chapter 1) by their systematic collection and validation of all existing data including the results of traditional and innovative alternative toxicity measurement methods, the knowledge on mechanisms and computer evaluations.

*Figure 3.3* Wistar rat (Stephens, 2014) and guinea pigs (Sandos, 2014), popular laboratory animals.

## 2   TEST ORGANISMS FOR HUMAN TOXICOLOGY PURPOSES

For human toxicology purposes, the main test organisms are rats and mice, but for a reduction of the number of test animals, isolated organs from cadavers, tissue cultures and isolated cells are also used in alternative tests. Some effects of chemicals on the genome are modeled on microorganisms.

### 2.1   Microorganisms used in human toxicity testing

–   *Salmonella* for Reverse Mutation Test (Ames Test) (Figure 3.3);
–   *Escherichia coli* for Reverse Mutation Assay;
–   *Escherichia coli* for UMU test;
–   *Saccharomyces cerevisiae* for Gene Mutation Assay;
–   *Saccharomyces cerevisiae* for Mitotic Recombination Assay;
–   *Saccharomyces cerevisiae* with cloned estrogen receptor in YES Assay.

### 2.2   Isolated cells, tissue cultures in human toxicology

–   BALB/c 3T3 – mouse fibroblast cell line for cytotoxicity testing;
–   Normal human keratinocytes for cytotoxicity;
–   LLC-PK1 kidney proximal tubule cell line for cell damage;
–   MDCK dog kidney epithelial cell line for cell damage;
–   HepG2 liver cell line (hepatoma) for cell growth and cytotoxicity;
–   HL-60 human acute promyelocytic leukemia cell line for energy production and metabolism;
–   Granulocytes/macrophages for acute neutropenia;
–   Mammalian chromosomes for *in vitro* Chromosome Aberration Test;
–   Mammalian cells for *in vitro* Gene Mutation Test;
–   Mammalian cells for *in vitro* Sister Chromatid Exchange Assay;
–   Mammalian cells for testing DNA Damage and Repair;

- Mammalian cells for testing Unscheduled DNA Synthesis
- Mammalian erythrocytes for Micronucleus Test;
- Human peripheral blood lymphocytes for Micronucleus Test;
- Syrian Hamster Embryo (SHE) for Micronucleus Test;
- Cell lines CHO, V79, CHL/IU and L5178\Y for Micronucleus Test;
- Human reconstructed skin for Micronucleus Test.

## 2.3 Lower animals in human toxicology

- *Drosophila melanogaster* (Drosophilidae, flies, Diptera), the common fruit fly or vinegar fly, is a historical species used in genetics and developmental research. Originally, its giant chromosomes made this little fly popular. Its use is still common due to easy maintenance, quick breeding, and their laying many eggs. It is used in alternative genotoxicity test methods.
- *Xenopus* genus of aquatic frogs, also known collectively as African Clawed Frogs or Platanna. They are a popular model for gene and protein expression and knock-down studies. *Xenopus* oocytes, 1 mm in diameter, are very large cells, which are easy for scientists to culture and use in experiments. *Xenopus laevis* is the most commonly used species for developmental toxicology studies. For example, the Frog Embryo Teratogenesis Assay on *Xenopus* (FETAX) is one such alternative bioassay that can be effectively used for toxicity testing.
- *Bufo marinus* (Anura, Bufonidae) is a large, terrestrial true toad, native to Central and South America. It is a popular species for toxicological studies of environmental pollutants, but it is also used in pregnancy testing and other laboratory research. It is easy and inexpensive to maintain and handle in the lab.
- *Bull frog* (*Rana catesbeiana*, Anura, Salientia) belonging to the tailless amphibians such as frogs and toads. It is an aquatic frog, a popular laboratory animal.
- *Mudpuppy* (*Necturus maculosus*, Caudata, Urodeles), belonging to amphibians, comprising newts and salamanders. Mudpuppies or waterdogs are aquatic salamanders of the family Proteidae. Their name originates from the misconception that they make a dog-like barking sound.

## 2.4 Birds

- Domestic chickens can be easily bred and housed so that they are increasingly used as experimental animals in many areas of scientific research and toxicology. Their main fields of use are breeding and genetics, embryology, anatomy, health, hygiene, toxicology and pharmacology, physiology, biochemistry, endocrinology and neurobiology.
- Other domesticated and wild birds such as

  o Dabbling duck, *Anas platyrhynchos;*
  o Rock dove, *Columba livia;*
  o Virginia quail, *Colinus virginianus;*
  o Japanese quail, *Coturnix japonica;*
  o Common pheasant, *Phasianus colchicus;*
  o Red-legged partridge, *Alectoris rufa.*

## 2.5  Mammals

Estimates of animals used globally for experiments range from tens of millions to 100 million or more, annually. In toxicology, the most popular mammal test organisms are rodents.

– *Norway rat* (*Rattus norvegicus*, Rodentia, Muridae) is the species most commonly used in research and human toxicology testing. In the USA alone, about 4 million rats are used annually in different laboratories. The adult rat ranges in weight from 250 to 500 g. Brown color is dominant and typical. The Wistar stock of *Rattus norvegicus* was established in Chicago in the Wistar Institute during the years 1906–1930. This lab began standardizing albino rat strains and today most of the laboratory rats are white. The Norway rat is easily maintained and relatively resistant to diseases. Rats are important in all kinds of toxicity testing and behavioral research.

– *Mice* (*Mus domesticus*, Rodentia, Muridae) are the most common laboratory animals. Laboratory strains of mice used today are descendants of the western European house mouse, *Mus domesticus*, with some genes from Asian species. *Mus musculus* is a composite taxonomic designation for several interbreeding species. A yellow mutant is used in studies of pigmentation, implantation, obesity, and sterility. BALB/c, the white laboratory mouse, is the most widely used test animal; 14 million are used in the US alone annually, and the total number used globally is close to 100 million. Mice are especially useful for cancer research because of their high tumor incidence. Their small size and rapid reproduction make them useful in all areas of research and risk management such as toxicity, radiobiology, cancer research, behavior research, nutrition, and genetic studies (FAU, 2013c).

– *Guinea pig* (*Cavia porcellus*, Rodentia, Caviidae) is used in antibody production, tumor genesis, nutrition, genetics, radiation research, and dental studies. It was an important species in the discovery of vitamin C and the diagnosis of tuberculosis. The guinea pig has been widely employed in biomedical research since 1780. Lavoisier used the cavy to measure heat production (FAU, 2013a).

– *Hamsters* (Syrian hamster, *Mesocricetus auratus*, Rodentia, Cricetidae) have a unique feature. Their reversible cheek pouch provides a site for normal and abnormal tissue transplants that have the virtue of visibility and ready access. In some studies, a tumor maintained in one cheek pouch was exposed to an experimental treatment while the control tumor was maintained in the opposite cheek pouch (FAU, 2013b).

– *Other rodents* such as the white-footed mouse (*Peromyscus leucopus*, Rodentia, Cricetidae), Mongolian gerbil (*Meriones unguiculatus*, Rodentia, Cricetidae) and chinchilla (*Chinchilla laniger*, Rodentia, Chinchillidae) are also used as laboratory animals in special cases, mainly for research.

– *Rabbit* (family Leporidae, order Lagomorpha) has been used in research studies in genetics, nutrition, and toxicology, also in legislative toxicology, physiology, immunology and reproduction. The pharmaceutical and cosmetics industries use the rabbit widely to test the toxic effects of cosmetics and pharmaceuticals. It is widely used for the production of antibodies and antiserums because of its relatively large size (FAU, 2013d).

- *Sheep, goat, cow,* and *pig* are also used as laboratory animals, primarily as transgenic animals or for immunological purposes; they have less importance in toxicology. *Ovis aries*, the domestic sheep, has been used in experiments in such diverse fields of study as endocrinology and reproductive physiology, cardiovascular physiology, fluid and electrolyte homeostasis, immunology, neurophysiology and neuroanatomy, thermoregulation, hematology, ingestive behavior, nutrition and gastrointestinal physiology. The first cloned mammal was also a sheep, called Dolly.

- *Dogs* and *cats* are often used in laboratories. Although all breeds and mixes are used, Beagles are the most popular test dogs because of their size and docile behavior. Dogs have been used to study maternal deprivation, the effects of smoking, chemical toxicity of many substances, and the effectiveness of medical devices. The laboratory cat (*Felis catus*) is mainly used in reflex studies, exposure to chemical stimuli, in neuropharmacology, particularly the testing of psychotropic drugs, in behavioral studies, toxicology, oncology and for the study of chromosomal abnormalities.

- *Non-human (NH) primates*: the best known laboratory primates are the squirrel monkey (*Saimiri sciureus*), baboon (*Papio hamadryas,*), rhesus (*Macaca mulatta*), Japanese or snow macaque (*Macaca fuscata*), cynomolgus or crab-eating macaque (*Macaca fascicularis*) and the chimpanzee (*Pan troglodytes*). The most famous laboratory monkeys are the Silver Spring macaque monkeys, having been used for behavioral research for a long time, but becoming the subjects of the first animal research case to reach the United States Supreme Court in 1991.

## 2.6   3R in animal testing

"3R" in animal testing means replacement, reduction and refinement. The ethical issue has priority in animal testing. Ethics may play a role in deciding the dilemma of testing on animals at all. The approach today is very much in favor of humanity and because of the lack of alternative methods of the same quality. The other component of the issue is how to keep and handle animals. In this topic, there is now a unified approach regarding animal welfare, humanized methods, suitable space and food for the animals, mitigating pain due to interventions and in the case of injury or illness, and not to mistreat animals.

Non-profit organizations such as 'Against Cruelty to Animals' work for animal welfare and against the use of animals in laboratories (Figure 3.3). Though animals cannot be removed from tests right away, a number of traditional toxicological tests have been replaced by alternative methods, and a stepwise reduction of animal use has already been decided and practiced all over the world. In Europe, the protection of animals used for experimental and scientific purposes is regulated by Directive 2010/63/EU (EU, 2010). Several organizations and funds such as FRAME in the UK (Fund for the Replacement of Animals in Medical Experiments) promote the development and application of sound scientific principles and methodology which can lead to the progressive reduction and replacement of laboratory animal procedures. They are dealing with new concepts, more efficient experimental design, *in vitro* test methods and animal welfare education. The 3R concept has been in use for over 50 years and is still timely, but is unlikely to meet future requirements. The attitude toward, and the basic fundamentals of, toxicology should be changed, i.e. animal tests results should be stripped of their golden standard role. Evidence-based toxicology is a step forward in this direction.

*Replacement* refers to the preferred use of non-animal methods over animal methods whenever it is possible to achieve the same scientific aim.

*Reduction* refers to methods that enable researchers to obtain comparable levels of information from fewer animals, or to obtain more information from the same number of animals using more adequate statistical methods.

*Refinement* refers to methods that alleviate or minimize potential pain, suffering or distress, and enhance animal welfare for the animals still used.

AltTox (2014) is dedicated to reducing the numbers and suffering of animals used in current toxicology assessments. Its website is designed to exchange information on *in vitro* and *in silico* methods for all types of toxicity tests.

The use of alternatives to testing on animals is on the agenda of ECHA too. ECHA published two comprehensive reports on alternative methods for the REACH Regulation (ECHA, 2011 and 2014).

## 3   TOXICITY END POINTS AND METHODS

In this section, the well-known toxicity end points are discussed in line with the regulatory requirements for chemical substances.

### 3.1   Acute toxicity

*Acute toxicity* is short-term toxicity, the adverse effects of chemical substances resulting either from a single exposure or from multiple exposures in a short period of time. In animal tests, toxicity is considered acute when the adverse effects occur within 14 days of administering the substance. In ecotoxicity tests, acute toxicity is defined as a period of time shorter than the generation time of the test organism. The $EC_{50}$, $LC_{50}$ or $ED_{50}$ and $LD_{50}$ values represent the end points used to quantitatively characterize acute toxicity.

Acute toxicity is distinguished from chronic toxicity: it describes the adverse health effects from repeated exposures, often at lower levels, to a substance over a longer time period (months or years).

*Acute Systemic Toxicity* means adverse effects occurring within a short time after administering a single, usually extremely high-dose substance via one or more of the exposure routes of oral way, inhalation and dermal contact or injection into the bloodstream or the muscles. Systemic effects require absorption and distribution of the toxicant to a site distant from its entry point. Systemic toxicity does not cause a similar degree of toxicity in all organs, but usually to one or two organs, which are called the toxic effect's target organs.

Acute systemic toxicity testing is the estimation of the human hazard potential of a substance by determining its systemic toxicity in an animal test system following an acute exposure. Its assessment has traditionally been based on the median lethal dose ($LD_{50}$) value – an estimate of the dose of a test substance that kills 50% of the test animals.

*Table 3.1*  OECD test guidelines (TG) for acute systemic testing (OECD, 2014).

| Test guideline | Test name |
| --- | --- |
| OECD TG 403 (2009) | Acute Inhalation Toxicity |
| OECD TG 436 (2008) | Acute Inhalation Toxicity – Acute Toxic Class Method |
| Draft OECD TG 433 (2004) | Acute Inhalation Toxicity – Fixed Concentration Procedure |
| OECD TG 420 (2002) | Acute Oral Toxicity – Fixed Dose Procedure |
| OECD TG 423 (2002) | Acute Oral Toxicity – Acute Toxic Class Method |
| OECD TG 425 (2008) | Acute Oral Toxicity: Up-and-Down Procedure |
| OECD TG 402 (1987) | Acute Dermal Toxicity |

### 3.1.1   Animal tests for acute systemic toxicity

In recent years, using death as an end point has been discouraged in all testing contexts, so the use of the reduction alternatives is now mandatory for acute systemic toxicity testing. Three reduction alternatives to the oral $LD_{50}$ tests have been developed and validated: the Fixed Dose Procedure, the Acute Toxic Class method, and the Up-and-Down Procedure as summarized in Table 3.1.

As a non-animal alternative, the EU Reference Laboratory for Alternatives to Animal Testing (EURLECVAM, 2014) proposed a tiered testing strategy for acute systemic toxicity (Siebert *et al.*, 1996; ECVAM, 2002a). In this testing strategy, a substance would be sequentially evaluated by (quantitative) structure–activity relationship and *in vitro* assays in the following way: QSAR or SAR → cytotoxicity testing → computational model for metabolism → biotransformation assays → cell-specific toxicity tests. When classified as very toxic at any step, the testing would end with the classification of the substance. A limited *in vivo* acute toxicity test would be performed only if all of the prior assessments indicated the substance to be not very toxic.

### 3.1.2   Non-animal, in vitro tests for acute systemic toxicity

Several organizations are involved in developing non-animal tests for human toxicity. In the US the Interagency Coordinating Committee on the Validation of Alternative Methods (ICCVAM, 2014), a permanent committee of the NIEHS under the National Toxicology Program Interagency Center for the Evaluation of Alternative Toxicological Methods (NICEATM, 2014), in Europe, the European Union Reference Laboratory for Alternatives to Animal Testing (EURL ECVAM, 2014) is responsible for the 3R in animal testing. ECVAM DB-ALM (2014) is a searchable database containing *in vitro* methods and QSAR models.

ICCVAM validated and endorsed two basic *in vitro* cytotoxicity assays: the Neutral Red Uptake (NRU) test with 3T3 rodent cells and the NRU test with normal human keratinocyte (NHK) cells as adjunct for determining starting doses. OECD established a technical guidance (TG for 3T3 NRU (2010) and a draft TG for NHK NRU (2010). EURL ECVAM also recommended the 3T3 NRU assay for supporting the identification of substances that do not require classification for acute oral toxicity.

Table 3.2 shows validated, and Table 3.3 not yet validated, alternative test methods. A summary on validated and non-validated alternative test methods is presented on the AltTox web site

*Table 3.2* Acute systemic toxicity testing by cell lines: validated *in vitro* test methods.

| Test guideline | Method | Test purpose | Validation authority |
|---|---|---|---|
| OECD TG (3T3 NRU, 2010) | BALB/c 3T3 Neutral Red Uptake Assay | Adjunct to *in vivo* acute oral toxicity tests for determining starting doses | ICCVAM |
| Draft OECD TG (NHK NRU, 2010) | Normal Human Keratinocyte (NHK) Neutral Red Uptake Assay | Adjunct to *in vivo* acute oral toxicity tests for determining starting doses | ICCVAM |
| ECVAM DB-ALM Protocol No.101 (CFU GM, 2006) | Colony Forming Unit-Granulocyte/Macrophage Assay | Hematotoxicity test for acute neutropenia | ECVAM DB-ALM |

*Table 3.3* Acute systemic toxicity testing by cell lines: some *in vitro* methods under development.

| Method | Test purpose | End point |
|---|---|---|
| LLC-PK1 kidney proximal tubule cell line | Transepithelial resistance (TER) and paracellular permeability | Barrier integrity/cell damage |
| MDCK dog kidney epithelial cell line | Transepithelial resistance (TER) and paracellular permeability | Barrier integrity/cell damage |
| HepG2 liver cell line (hepatoma) | Protein content | Cell growth |
| HL-60 human acute promyelocytic leukemia cell line | Adenosine triphosphate (ATP) content | Energy production and metabolism |
| Change of liver cell line | Morphology change followed by pH change | Cell growth/cytotoxicity |

AltTox (2014) is dedicated to advancing non-animal methods of toxicity testing through online discussion and information exchange.

## 3.2   Repeated-dose and organ toxicity testing

General toxicological effects occur as a result of repeated daily exposure to a substance (via oral, inhalation and/or dermal routes) for a portion of the expected life span (i.e., subacute or subchronic exposure) or for a major part of the life span (i.e., chronic exposure) (REACH, 2006).

These general toxicological effects include effects on body weight and/or body weight gain, absolute and/or relative organ and tissue weights, alterations in clinical chemistry, urinalysis and/or hematological parameters, functional disturbances in the nervous system as well as in organs and tissues in general, and pathological alterations to organs and tissues examined macroscopically and microscopically. In addition to this information on possible general toxicological effects, repeated-dose toxicity studies may also provide other information on reproductive toxicity or carcinogenicity or may identify specific manifestations of toxicity such as neurotoxicity, immunotoxicity or endocrine-mediated effects (REACH, 2006).

Table 3.4 OECD test guidelines of animal test methods for repeated-dose and organ toxicity (OECD, 2014).

| Test guideline | Test name |
|---|---|
| OECD TG 412 (2009) | Repeated-Dose Inhalation Toxicity: 28-Day or 14-Day Study |
| OECD TG 413 (2009) | Subchronic Inhalation Toxicity: 90-Day Study |
| OECD TG 408 (1998) | Repeated-Dose 90-Day Oral Toxicity Study in Rodents |
| OECD TG 409 (1998) | Repeated-Dose 90-Day Oral Toxicity Study in Non-Rodents |
| OECD TG 407 (2008) | Repeated-Dose 28-Day Oral Toxicity Study in Rodents |
| OECD TG 410 (1981) | Repeated-Dose Dermal Toxicity: 21/28-Day Study |
| OECD TG 411 (1981) | Subchronic Dermal Toxicity: 90-Day Study |
| OECD TG 422 (1996) | Combined Repeated-Dose Toxicity Study with the Reproduction/Developmental Toxicity Screening Test |

### 3.2.1  Animal test methods for repeated-dose and organ toxicity

Animal test methods are listed in Table 3.4.

OECD is updating TG 407, the repeated-dose 28-day oral toxicity test in rats, to validate a protocol suitable for detecting the endocrine disruption potential of test substances. OECD 422 makes a reduction in the number of animals possible.

The following tissues/organs can be studied histologically in a repeated-dose toxicity study:

Adrenal glands
Aorta
Blood smears
Brain
Epididymides
Eyes and optic nerves
Gall bladder
Gross lesions
Heart
Joint with bone
Kidneys and ureters
Large intestines
Larynx
Liver
Lungs with bronchi and bronchioles
Lymph nodes
Mammary glands
Esophagus
Ovaries
Pancreas
Peripheral nerves

Pituitary gland
Prostate
Salivary glands
Seminal vesicles (rodents)
Skeletal muscle
Skin and subcutaneous tissue
Small intestines
Spinal cord
Spleen
Sternebrae, femur or vertebrae
Stomach
Testes
Thymus
Thyroid/Parathyroid glands
Tissue masses of tumors
Tongue
Trachea
Urinary bladder
Uterus with uterine cervix and oviducts
Vagina

### 3.2.2 *Alternative methods for repeated-dose and organ toxicity testing*

Potential non-animal alternatives for chronic toxicity testing include human and animal perfused organs such as liver, kidney, lung, organ tissue slices, isolated, suspended cells, primary cultured cells, cultured cell lines, genetically engineered cell lines, reaggregating cell cultures, three-dimensional cell cultures and co-cultures, and computational systems for SAR and QSAR.

As yet, ICCVAM and ECVAM have not validated any non-animal methods for assessing chronic toxicity end points or repeated exposure target organ toxicity, but there are several potential candidates among the innovative *in vitro* test methods e.g. in the DB-ALM database:

- Alginate entrapped primary hepatocytes: cryopreserved, immobilized three-dimensional cell cultures, which can be used to study hepatotoxicity;
- Collagen-gel sandwich configuration culture of primary hepatocytes: a long-term hepatocyte culture system used to perform cytotoxic and metabolism-mediated toxicity studies;
- Culture of freshly isolated nonparenchymal liver cells and nonparenchymal cell lines (Kupffer cells, biliary epithelial cells, stellate cells) is used to study the mechanisms of hepatotoxicity;
- Genomics technologies are used to investigate hepatotoxicity *in vitro*, e.g. toxicogenomics (mRNA analysis) and proteomics (gene product analysis) for screening and mechanistic hepatotoxicity studies;
- Hepatocyte cultures on biosensor systems: direct examination of certain cellular processes following exposure to xenobiotics;
- Monolayer culture of genetically engineered liver cell lines are designed to express defined xenobiotic metabolizing enzymes. They may be used in short- and long-term metabolism and hepatotoxicity studies;
- Monolayer culture of liver cell lines can be used for basal toxicity, hepatotoxicity, genotoxicity and metabolism-mediated toxicity studies;
- Monolayer culture of primary hepatocytes is the most frequently used *in vitro* system for cytotoxic, genotoxic and mechanistic studies of hepatotoxicity and metabolism-mediated toxicity;
- Perfused two- and three-dimensional culture systems (bioreactors) with liver cells improve the viability and functionality of liver cells in culture and allow for short- and long-term studies of hepatotoxicity, metabolism-mediated toxicity and genotoxicity;
- Suspension culture of freshly isolated or cryopreserved hepatocytes derived from animals or humans are used for short-term studies on hepatotoxicity and metabolism-mediated toxicity;
- Avian (hen, turkey, quail, duck) embryonic tissue are applied for studies on metabolism and metabolism-mediated toxicity of chemicals in ovo;
- Freshly prepared or preserved liver slices from laboratory and domestic animals, humans, fish, birds and amphibians can be used for short- or long-term studies on hepatotoxicity;
- Stem or progenitor cell-derived hepatocyte-like cells may be used in studies on hepatotoxicity and metabolism-mediated toxicity;

– Subcellular fractions of liver tissue (such as microsomes and mitochondria) and liver homogenate can be used for short-term studies on hepatotoxicity and metabolism-mediated toxicity studies;
– Isolated Rat Glomeruli and Proximal Tubules isolated from the kidney show cytotoxic effect of chemicals by cell glucose and/or fatty acid oxidation and de novo protein synthesis.

## 3.3 Genotoxicity

*Genotoxicity* is chemically induced genetic mutations and/or other alterations of the structure, information content, or segregation of genetic material. Genotoxic chemicals are genotoxins, which cause heritable changes in the genetic material of spermatocytes or oocytes. Compared to mutagens which cause mutations in DNA, genotoxins may interact in a broader sense with DNA or non-DNA targets. Genotoxic carcinogens can lead to DNA mutations with the potential to cause cancer. Distinguishing the point at which exposure to a carcinogen increases mutation rates, is challenging. Currently, there is a general agreement that, for genotoxic carcinogens, no specific threshold can be identified. On the other hand, scientists have experienced in practice that a series of mutation events are needed before malignancy occurs and a single, small exposure may not result in disease. In addition, cells have their own defense mechanism to counter the effects of mutagens. All in all, many research and scientific activities deal with the question of thresholds for genotoxic chemicals, because a scientifically based NOAEL may play an important role in regulation and risk management of genotoxic and mutagenic substances (Greim, 2012).

A simple classification of the measurement end points for genotoxic substances and genotoxic mutagens is given below:

– *In vitro* gene mutation – Ames mutagenicity;
– *In vitro* chromosomal mutation – micronucleus assay;
– *In vivo* chromosomal mutation – micronucleus assay;
– *In vivo* genotoxic carcinogenicity – rodent assay.

The term genotoxicity is used in general, without specifying the type of interaction or the name of the test assay. There are *in vivo* and alternative *in vitro* test methods for assessing the potential of heritable genotoxicity on germ cells, on somatic cells, and on chromosomes, as the above list shows.

### 3.3.1 In vivo *animal tests for assessing potential heritable genotoxicity*

Table 3.5 shows the traditional animal test methods for genotoxicity.

*Table 3.5* OECD test guidelines of animal test methods for genotoxicity (OECD, 2014).

| Test guideline | Test name |
| --- | --- |
| OECD TG 478 (1984) | Genetic Toxicology: Rodent Dominan Lethal Test |
| OECD TG 479 (1986) | Genetic Toxicology: *In vitro* Sister Chromatid Exchange Assay in Mammalian Cells |
| OECD TG 484 (1986) | Genetic Toxicology: Mouse Spot Test |
| OECD TG 485 (1986) | Genetic Toxicology: Mouse Heritable Translocation Assay |
| OECD TG 483 (1997) | Mammalian Spermatogonial Chromosome Aberration Test |
| OECD TG 475 (1997) | Mammalian Bone Marrow Chromosome Aberration Test |

*Table 3.6* OECD guidelines for non-animal genotoxicity and mutagenicity testing methods (OECD, 2014).

| Test guideline | Test name |
| --- | --- |
| OECD TG 477 (1984) | Genetic Toxicology: Sex-linked Recessive Lethal Test in *Drosophila melanogaster* |
| OECD TG 479 (1986) | Genetic Toxicology: *In vitro* Sister Chromatid Exchange Assay in Mammalian Cells |
| OECD TG 480 (1986) | Genetic Toxicology: *Saccharomyces cerevisiae*, Gene Mutation Assay |
| OECD TG 481 (1986) | Genetic Toxicology: *Saccharomyces cerevisiae*, Mitotic Recombination Assay |
| OECD TG 482 (1986) | Genetic Toxicology: DNA Damage and Repair, Unscheduled DNA Synthesis in Mammalian Cells *In Vitro* |
| OECD TG 471 (1997) | Bacterial Reverse Mutation Test (Ames Test) |
| OECD TG 472 (1997) | Genetic Toxicology: *Escherichia coli*, reverse assay |
| OECD TG 473 (1997) | *In vitro* Mammalian Chromosome Aberration Test |
| OECD TG 474 (1997) | Mammalian Erythrocyte Micronucleus Test |
| OECD TG 476 (1997) | *In vitro* Mammalian Cell Gene Mutation Test |
| OECD TG 486 (1997) | Unscheduled DNA Synthesis (UDS) Test with Mouse Liver Cells *in vitro* |
| OECD TG 487 (2010) | *In vitro* Mammalian Cell Micronucleus Test (MNT) – Alternative to the *in vitro* chromosome aberration assay for genotoxicity testing |

### 3.3.2 OECD test guidelines for in vitro genotoxicity and mutagenicity testing

The non-animal mutagenicity tests may use bacterial cells, yeasts or isolated mammalian cells. Most of them are used as screening assays in tiered mutagenicity testing. Table 3.6 shows the OECD guidelines for *in vitro* genotoxicity test methods.

The *Ames test* is one of the most widely used screening tests for mutagenicity and carcinogenicity, and uses several different strains of histidine auxotrophic *Salmonella typhimurium* mutants that carry mutations in genes responsible for histidine synthesis, and essential for growth. If histidine synthesis is absent in the bacterial cell, it cannot grow on a histidine-free nutrient medium. Six strains are specially selected and mixed to have both frameshift and point mutations in the genes required to synthesize histidine, which allows for the detection of mutagens acting via different mechanisms.

A positive test indicates that the chemical substance might act as a mutagen or carcinogen (Figure 3.4). A substance that proves positive in the Ames test is not necessarily mutagenic or carcinogenic in tests performed on mammal cells or organisms. As the Ames test overestimates mutagenic risk, it is appropriate to use it as a screening method in a tiered assessment procedure.

### 3.3.3 New in vivo genotoxicity tests

The *Comet Assay* or single cell gel electrophoresis (SCGE) assay is a rapid, sensitive and relatively simple method for detecting DNA damage at the level of individual cells (Singh *et al.*, 1988). It combines the simplicity of biochemical techniques for detecting DNA single strand breaks (strand breaks and incomplete excision repair sites), alkali-labile sites, and cross-linking, with the single cell approach typical of cytogenetic assays (COMET, 2013).

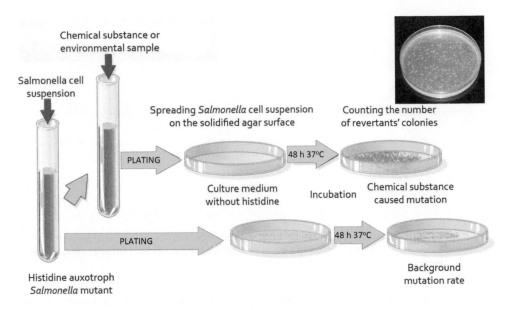

*Figure 3.4* Ames reverse mutation of *Salmonella typhimurium*: the number of revertants is proportional to the mutagenicity of the tested chemical substance.

Depending on the origin of the sample cells with which the chemical substances interact (toxic chemicals or drugs to be tested), the Comet Assay can be *in vivo* (cells from an *in vivo* test subject) or *in vitro* (cells extracted from an *in vitro* cell culture). As only a small numbers of cells are required for analysis, literally any tissue or organ is amenable to investigation. The only requirement is that a sufficient number of single cells (or nuclei) are obtained for analysis and that no or minimal damage is induced during tissue processing (Tice *et al.*, 2000).

The Comet assay is eventually gel electrophoresis of the DNA from whole cells under alkaline (pH > 13) conditions, resulting in an image similar to a 'comet' with a distinct head and tail (Figure 3.5). The head is composed of intact DNA, while the tail consists of damaged and broken pieces of DNA with double strand breaks, single strand breaks, alkali-labile sites, oxidative base damage, and DNA cross-linking with DNA or protein. Both animal and plant cells can be investigated using this method.

The thin agarose-gel together with treated cells is poured onto a microscope slide and cells are lysed by alkali, which lets the DNA unwind. Unwinding allows the broken and damaged DNA fragments to migrate away from the nucleus. Staining with ethidium bromide or other DNA-specific dyes allows the head and tail length to be evaluated based on fluorescence intensity. The extent of DNA liberated from the head and forming the tail is proportional to the rate of DNA damage.

The international validation studies of both the *in vitro* and *in vivo* genetic toxicity Comet Assays were ongoing in 2010, with the lead of JaCVAM (2013) – the Japanese Center for the Validation of Alternative Methods – in collaboration with NICEATM-ICCVAM and ECVAM.

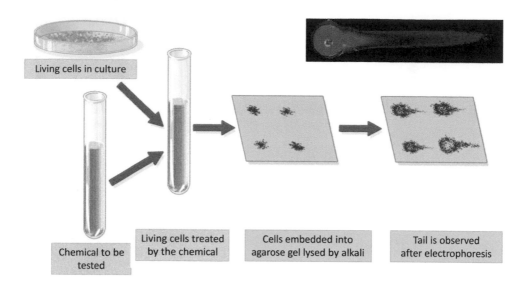

*Figure 3.5* Comet Assay: living cells are treated with the chemical substance to test, and then embedded into agarose which is poured on a slide. After electrophoresis it is stained.

The *in vivo transgenic mutation assay* uses live animals. This is a well-established assay employing transgenic rats or mice, which contain multiple copies of chromosomally integrated plasmid or phage DNA that harbor reporter genes for the detection of mutations. The mutations of the transgenes of *lac* repressor or *lacZ* (gene of beta-galactosidase) are of no consequence for the rodents as they are genetically neutral. This means that the mutations provide their host cell with neither an advantage nor disadvantage in either survival or proliferation. OECD concluded that there is more than sufficient evidence of validation of the *in vivo* transgenic mutation assay to support the establishment of a Test Guideline (Douglas, 2010).

An ECVAM panel proposed that total replacement of animal testing for genotoxicity/mutagenicity would require models for evaluating toxicokinetics and metabolism (Maurici *et al.*, 2005a). *In vitro* genotoxicity tests also need to be modified to use cell lines relevant to the target organs of interest, which would require standardization and validation of *in vitro* assays in mammalian germ cells for predicting heritable germ cell damage. Some other genotoxicity and mutagenicity test protocols from the ECVAM DB-ALM (2014) database include:

–   *Cytotoxicity and Genotoxicity in Primary Cultures of Human Hepatocytes* of the test compound by measuring cell viability, DNA damage and unscheduled DNA synthesis (DB-ALM Protocol No.16);
–   *Stable Cell Lines Expressing Cytochromes CYP cDNA* can be inserted into cell lines such as V79 Chinese hamster cells, which are used for *in vitro* toxicity testing. This means that the metabolites of xenobiotics are produced in the same cells in

which any toxic effect will be observed, thus negating the problems associated with co-cultures and the use of subcellular fractions (DB-ALM Protocol No.107);

– *DNA Binding in Bacteria* may be used to elucidate primary genotoxic mechanisms through the analysis of mutated bacterial DNA by high pressure liquid chromatography (HPLC) (DB-ALM Protocol No.8);

– *Alkaline Unwinding Genotoxicity Test* applies to mouse lymphoma cells cultured in the presence of test chemicals, with or without a metabolic activating system, and resultant DNA-strand breaks detected by alkaline unwinding and hydroxyapatite elution (DB-ALM Protocol No.19);

– *Prostaglandin H Synthase (PHS) mediated Genotoxicity of Xenobiotics*. This protocol describes the use of SEMV cells (a cell line derived from ram seminal vesicles) for studying prostaglandin H synthase-mediated metabolism of xenobiotics in intact cells (DB-ALM Protocol No.61).

### 3.3.4 QSAR for genotoxicity and genotoxic carcinogenicity

The QSAR approach is also applicable for genotoxic chemicals based on the covalent DNA binding as the molecular initiating event of genotoxicity and genotoxic carcinogenicity (Schultz, 2010). OECD prepared a QSAR Toolbox to support grouping of genotoxic and carcinogenic substances, on the basis of a common molecular initiating event, the ability of a chemical to bind covalently to DNA (the effect-mechanism of non-genotoxic mutagens includes many other molecular initiating events). Out of the chemical structure and other physico-chemical characteristics the following databases were used by the OECD QSAR Toolbox (OECD, 2012):

– Bacterial mutagenicity ISSSTY: 7367 chemicals' effect on gene mutation, measured by *in vitro Salmonella typhimurium* test;

– Carcinogenic potency database CPDB: 1530 chemicals' carcinogenic effect ($TD_{50}$), measured by *in vivo* animal tests (rats, mice, hamsters, dogs and non-human primates);

– Carcinogenicity ISSCAN: 1150 chemicals' carcinogenic effect ($TD_{50}$), measured *in vivo* on rats and mice;

– Cell Transformation Assay ISSCTA: 327 chemicals' non-genotoxic carcinogenicity, measured by *in vitro* cell lines (SHE cells, BALB/c 3T3, C3H/10HT1/2 and Bhas42);

– Genotoxicity OASIS: 7500 chemicals' gene mutation and chromosomal mutation, both (yes/no), tested *in vitro* by *Salmonella typhimurium* and *in vitro* by Chinese hamster lung cells and T-lymphoma cell lines;

– Micronucleus ISSMIC: 564 chemicals' chromosomal mutation (yes/no), measured *in vivo* by rats and mice;

– Micronucleus OASIS: 557 chemicals' chromosomal mutation, measured *in vivo*.

## 3.4  Chronic toxicity

Chronic toxicity describes the adverse health effects due to repeated exposures to a substance over a longer time period (months or years).

*Table 3.7*   OECD guidelines for testing chronic toxicity
(OECD, 2014).

| Test guideline | Test name |
| --- | --- |
| OECD TG 452 (2009) | Chronic Toxicity Studies |
| OECD TG 453 (2009) | Combined Chronic Toxicity/ Carcinogenicity Studies |

### 3.4.1   *Chronic toxicity testing methods on animals*

– The objectives of chronic toxicity studies include:
– The identification of the hazardous properties of a chemical substance
– The identification of target organs
– Characterization of the dose–response relationship
– Identification of a No Observed Adverse Effect Level (NOAEL) or point of departure for establishment of a Benchmark Dose (BMD)
– The prediction of chronic toxicity effects at human exposure levels
– Provision of data to test hypotheses with regard to mode of action (OECD, 2008a).

Currently, two OECD guidelines are in existence for testing chronic toxicity (Table 3.7). The updating of TG 452 has been carried out at the same time as the revisions of the Test Guidelines 451 on Carcinogenicity Studies and TG 453, with the objective of obtaining additional information from the animals used in the study and providing further details on dose selection.

The majority of chronic toxicity studies are carried out in rodent species, therefore Test Guideline 452 is intended to apply primarily to studies carried out on these species.

In a chronic test, the test substance is administered daily in graduated doses to several groups of experimental animals for a period of 12 months, although longer or shorter durations may also be chosen.

This duration is chosen to be sufficiently long to allow any effects of cumulative toxicity to become manifest, without the confounding effects of geriatric changes. Deviations from the exposure duration of 12 months must be justified, particularly for shorter durations. The test substance is normally administered by the oral route although testing by the inhalation or dermal routes may also be appropriate. The study design may also include one or more interim kills, e.g. at 3 and 6 months, and additional groups of animals may be included to accommodate this. During the period of administration, the animals are observed closely for signs of toxicity (OECD, 2008a).

## 3.5   Carcinogenicity

*Carcinogenicity* is the character of a chemical substance, which is able to induce cancer or is carcinogenic. This may occur through genotoxic (see Section 3.3) or non-genotoxic mechanisms. While covalent DNA binding has been identified as the molecular initiating event of genotoxicity and genotoxic carcinogenicity,

Table 3.8 OECD guidelines for testing carcinogenicity (OECD, 2014).

| Test guideline | Test name |
|---|---|
| OECD TG 451 (2009) | Carcinogenicity Studies |
| OECD TG 453 (2009) | Combined Chronic Toxicity/ Carcinogenicity Studies |

non-genotoxic carcinogenicity may be based on other molecular initiating events such as protein binding, non-covalent interactions with protein receptors, intercalation with DNA and the formation of free radicals. Carcinogenic substances are those that induce benign or malignant tumors, increase their incidence or malignancy, or shorten the time of tumor occurrence when they are inhaled, injected, dermally applied, or ingested.

### 3.5.1  Animal methods for carcinogenicity testing

The conventional test for carcinogenicity is the long-term rodent carcinogenicity bioassay described in OECD TG 451 (Table 3.8). The objective of this test is to observe test animals for a major portion of their life span for the development of neoplastic lesions during or after exposure to various doses of a test substance by an appropriate route of administration.

Two end points in animal bioassays, carcinogenicity and chronic toxicity, can be combined to reduce animal use, as described in OECD TG 453. Another guidance gives new methodological support for chronic toxicity and mutagenicity testing, for example, with proper dose selection (OECD, 2009).

### 3.5.2  Non-animal testing of carcinogenicity

Non-animal methods include *in vitro* cell-based assays and computational prediction models. The *cell transformation assay* (CTA) detects phenotypic changes induced by chemicals in mammalian cell cultures (Maurici *et al.*, 2005b). Different bioassays use different cell cultures:

- Syrian hamster embryo (SHE) assay, detects the early steps of carcinogenesis
- Low-pH SHE assay, detects the early steps of carcinogenesis
- Balb/c 3T3 assay, detects later carcinogenic changes
- C3H/10T1/2 assay, detects later carcinogenic changes (OECD, 2006a).

*The gap junction intercellular communication* (GJIC) method is based on the disruption of the intercellular exchange of low-molecular-weight molecules through the gap junctions of adjacent cells; this disruption can result in abnormal cell growth and behavior (Maurici *et al.*, 2005b). The assay appears to be a good candidate for screening for non-genotoxic carcinogens and tumor promoters, but it still needs to be standardized and validated.

*Mutagenicity/genotoxicity assays* are the most commonly used *in vitro* test systems to predict carcinogenicity. Mutagenicity refers to the induction of transmissible changes in the structure of the genetic material of cells or organisms. Mutations may involve a single gene or a group of genes. Genotoxicity is a broader term that refers to changes to the structure or number of genes via chemical interaction with DNA and/or non-DNA targets such as the spindle apparatus and topoisomerase enzymes. The term genotoxicity is generally used unless a specific assay is being discussed. In use for over 30 years, genotoxicity assays are employed in a tier-testing approach that starts with Tier I *in vitro* tests, followed by Tier II *in vivo* genotoxicity tests to determine the biological relevance of chemicals that are positive in the *in vitro* tests. Common genotoxicity testing batteries include assays that measure mutations as well as structural and numerical chromosome aberrations (Maurici *et al.*, 2005a).

*In vitro* genotoxicity test methods have been adopted at the EU level with OECD guidelines and, additionally, the old *in vitro* chromosome aberration assay has been replaced with the *in vitro* micronucleus test for genotoxicity testing (see Table 3.6 in Section 3.3.2).

Numerous other *in vitro* genotoxicity tests, including the *in vitro* Comet assay, are being developed but are not yet validated. ECVAM DB-ALM Database, (2014) comprises some protocols for carcinogenicity testing:

- *Lucifer Yellow Intercellular Exchange assay* for Tumor Promoters: the effect of the test substance on the transfer of the dye lucifer yellow between SV-40-transformed hamster fibroblasts is an indication of potential tumor-promoting activity (DB-ALM Protocol No.65);
- *DNA Binding Studies for Alkylating Compounds* using isolated perfused rat liver. This procedure uses an adaptation of the perfused rat liver technique to assess the capacity of directly alkylating compounds to induce DNA-binding and therefore mutagenicity (DB-ALM Protocol No.89);
- *GreenScreen HCTM Genotoxicity test* is a fast, quantitative genotoxicity assay *in vitro*. The assay uses the DNA damage-inducible "Growth Arrest and DNA Damage 45 alpha-Green Fluorescent Protein" reporter gene, expressed in p53-competent TK6 cell line. The response to a chemical insult leads to an increase in green fluorescence. GreenScreen HC + S9 allows for a detection of genotoxic potential of the test compound with and without metabolic activation. It is available in reagent kit form (DB-ALM Protocol No.132);
- *In vitro Syrian Hamster Embryo Cell Transformation* assay (SHE CTA) is a short-term assay recommended as an alternative method for testing of the carcinogenic potential of chemicals (both genotoxic and non-genotoxic). The assay is based on the change of the phenotypic features of cell colonies undergoing the first steps of the conversion from normal cells to neoplastic-like cells with oncogenic properties (DB-ALM Protocol No.136).

## 3.6  Reproductive and developmental toxicity

Chemically induced adverse effects on sexual function, fertility, and/or normal offspring development (resulting in, for example, spontaneous abortion, premature

*Table 3.9*  OECD guidelines for reproductive and developmental toxicity (OECD, 2014).

| Test guideline | Test name |
| --- | --- |
| OECD TG 415 (1983) | One-Generation Reproduction Toxicity Study, in Rats and Mice |
| OECD TG 421 (1995) | Reproduction/Developmental Toxicity Screening Test, with male and female rats, oral administration for 4–9 weeks |
| OECD TG 422 (1996) | Combined Repeated Dose Toxicity Study with the Reproduction/ Developmental Toxicity Screening Test |
| OECD TG 414 (2001) | Prenatal Development Toxicity Study, with female rats and rabbits |
| OECD TG 416 (2001) | Two-Generation Reproduction Toxicity, by dosing offspring |
| OECD TG 426 (2007) | Developmental Neurotoxicity Study |
| OECD TG 443 (2012) | Extended One-Generation Reproductive Toxicity Study |

delivery, and birth defects) are generally determined through the breeding of one or more generations of offspring.

*Reproductive toxicity* includes the toxic effects of a substance on the reproductive ability of an organism and the development of its offspring. It has been defined by UNECE (2003a; 2009a) as any effect of chemicals that would interfere with reproductive ability or capacity, including effects on lactation. The definition of *developmental toxicity* is very broad; GHS (Globally Harmonized System) considers the following UNECE (2003a) definition sufficient for classification purposes: adverse effects induced during pregnancy, or as a result of parental exposure, that can be manifested at any point in the life span of the organism.

### 3.6.1   Animal tests for reproductive and developmental toxicity

OECD standardized five animal tests, including rat, mice and rabbit one- and two-generation tests and developmental neurotoxicity. Animal tests are listed in Table 3.9.

### 3.6.2   In vitro methods for reproductive and developmental toxicity

The ECVAM Scientific Advisory Committee (ESAC) endorsed three *in vitro* methods for embryotoxicity testing as scientifically validated (ECVAM, 2001):

– Embryonic stem cell test for embryotoxicity
– Micromass embryotoxicity assay
– Whole rat embryo for embryotoxicity assay.

An ICCVAM-NICEATM meeting (2000) reviewed the *Frog Embryo Teratogenesis Assay: Xenopus* (FETAX) as a potential alternative for assessing developmental toxicants. The method was deemed not ready for validation, so recommendations were made for its development.

FETAX is a 4-day, whole embryo-larval developmental toxicity screening assay, which uses young embryos of the South African clawed frog, *Xenopus laevis* (Figure 3.6). The FETAX system is capable of monitoring acute, chronic, developmental, and behavioral toxicity for ecological and human health hazard assessment. The first

*Figure 3.6*  Albino *Xenopus laevis* (Kenpei, 2013) (left) and *Xenopus laevis* (Gratwitzke, 2012) (right).

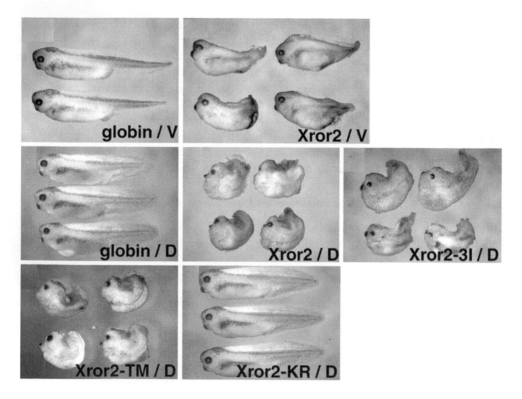

*Figure 3.7*  Malformations of *Xenopus* embrios due to gene mutations (Hikasa *et al.*, 2002).

96 h of embryonic development in *Xenopus* parallel many of the major processes of human organogenesis.

FETAX is able to measure embryolethality, embryonic malformation, embryonic growth reduction and embryonic developmental delay. Figure 3.7 shows the morphological consequences of the mutations of the globin gene and the Xror2 (timidin kinase receptor) gene (Hikasa *et al.*, 2002).

*Table 3.10*   OECD guidelines for *in vivo* and *in vitro* testing of dermal penetration (OECD, 2014).

| Test guideline | Test name |
| --- | --- |
| OECD TG 427 (2004) | Skin Absorption: *in vivo* method |
| OECD TG 428 (2004) | Skin Absorption: *in vitro* method |

As a screening test, a positive FETAX response would indicate a potential human hazard while a negative FETAX response would not indicate the absence of a hazard. In the role of a screening assay, a negative response would be followed by *in vivo* mammalian testing, while a positive response would require no further testing unless the investigator is concerned about a potential false positive response (ICCVAM, 2000).

*Xenopus* is a well-developed vertebrate model for biomedical research to understand fundamental biological processes and toxicity testing. The knowledge on *Xenopus* is collected and published in the bioinformatic database of Xenbase (2014).

## 3.7   Dermal penetration

Dermal penetration, also called *percutaneous penetration* or skin absorption (dermal uptake), measures the rate of penetration of a chemical substance through the dermal barrier and getting into the systemic circulation. Two mechanisms may play a role in the transport of chemicals: passive diffusion and biotransformation.

### 3.7.1   Animal testing of dermal penetration

Local or systemic effects on the rat, the test organism, follow dermal penetration as described in OECD TG 427. Table 3.10 includes one *in vivo* and another *in vitro* method standardized by OECD.

### 3.7.2   In vitro testing of dermal penetration

A variety of *in vitro* methods have been developed for dermal penetration testing. The chemical substance is applied to the surface of a skin sample separating the two chambers (a donor chamber and a receptor chamber) of a diffusion cell. Static and flow-through diffusion cells are both acceptable. Most of these methods use full or partial thickness human or animal skin mounted in a diffusion chamber (Figure 3.8). Although viable skin is preferred, non-viable skin can also be used. Pigskin is commonly used.

Human cell-based or reconstituted human skin cell models such as EpiDerm, EpiSkin, and SkinEthic and a rat keratinocyte culture model (ROC) are also being used to evaluate dermal penetration. These models are considered to be metabolically active, but in most cases, they are more permeable than *in vitro* human and animal skin preparations.

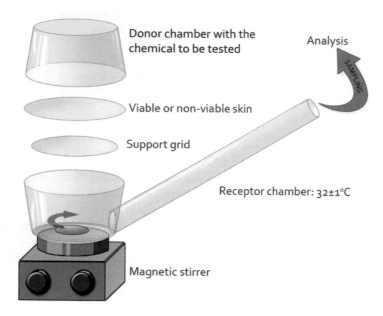

Figure 3.8 Testing skin penetration applying viable or non-viable skin.

## 3.8   Skin irritation and corrosion

Chemically induced skin damage may be reversible (called irritation) or irreversible (called corrosion).

GHS defines skin irritation as the production of reversible damage to the skin following the application of a test substance for up to 4 h. Skin corrosion is defined as the production of irreversible damage to the skin; namely, visible necrosis through the epidermis and into the dermis, following the application of a test substance for up to 4 h (UNECE, 2009c).

### 3.8.1   Animal testing of skin irritation and corrosion

One of the functions of the skin is to protect the body from environmental hazards such as toxic or corrosive chemicals. When the skin is exposed to a chemical, specific immunological and histological responses can occur. Four OECD test guidelines (Table 3.11) exist on acute, repeated-dose and subchronic dermal toxicity.

TG 402 Acute Dermal Toxicity test is also mentioned amongst acute systemic toxicity tests, 410 and 411 as repeated-dose and organ-specific test methods.

### 3.8.2   Alternative, non-animal test methods for skin irritation and corrosion

A number of alternative test methods have been developed for dermal irritation and corrosion as shown in Table 3.12. Most of them are included in the European legislation and used for classification and labeling (EU Test Methods, 2008; EU Test

*Table 3.11*  OECD guidelines for *in vivo* testing of skin irritation and corrosion (OECD, 2014).

| Test guideline | Test name |
|---|---|
| OECD TG 402 (1987) | Acute Dermal Toxicity |
| OECD TG 410 (1981) | Repeated-Dose Dermal Toxicity: 21/28-Day Study |
| OECD TG 411 (1981) | Subchronic Dermal Toxicity: 90-Day Study |
| OECD TG 404 (2002) | Acute Dermal Irritation/Corrosion, 4 h exposure and 14 day observation of exposed albino rabbit |

*Table 3.12*  OECD guidelines for *in vitro* testing of skin irritation and corrosion (OECD, 2014).

| Test guideline | Test name |
|---|---|
| OECD TG 439 (2013) | *In vitro* Skin Irritation – Reconstructed Human Epidermis Test Method for hazard identification of irritant chemicals and to identify non-classified chemicals<br>– EpiSkin SIT for dermal irritation<br>– EpiDerm SIT modified for dermal irritation |
| OECD TG 430 (2013) | *In vitro* Skin Corrosion – Rat Skin Transcutaneous Electrical Resistance (TER) |
| OECD TG 431 (2013) | *In vitro* Skin Corrosion – Reconstructed Human Epidermis (RHE) Test Methods:<br>– EpiSkin – standard model<br>– EpiDerm – skin corrosivity<br>– SkinEthic – for distinguishing corrosive from non-corrosive substances<br>– Epidermal Skin Test epiCS for distinguishing corrosive from non-corrosive substances |
| OECD TG 435 (2006) | *In vitro* Membrane Barrier Test Method for Skin Corrosion – Corrositex – non-cellular membrane |

Methods; 2009, B.4, 2008; and B.46, 2008). These *in vitro* test methods generally use *in vitro* grown human skin for testing. To make the penetration and response of the skin more realistic, 3D models have also been established.

*In vitro 3D skin models*, or reconstructed human epidermis (RHE) models, consist of human cells grown on a membrane at the air–liquid interface. This method of culturing induces the cells to grow in multilayers and to form junctions between the cells so that the cultures are similar to small pieces of human skin in the wells of a plate.

- EpiDerm Skin Irritation Test (SIT) (MatTek Corporation, 2014);
- EpiSkin SIT, Nice, France (EpiSkin, 2014);
- SkinEthic RHE, Nice, France (SkinEthic, 2014);
- LabCyte EPI-MODEL from J-TEC, Japan (Vitro-Life Skin, 2014);
- EST-1000 RHE (CellSystems, 2014).

The end points typically evaluated for skin irritation and corrosion testing are:

- Cytotoxicity (MTT assay) for both irritation and corrosion;
- Cytotoxicity (MTT assay) plus IL-1a (cytokine release) for irritation.

*Ex vivo models,* or skin explants, consist of pieces of skin from humans or animals for *in vitro* testing applications. These have been used in screening for skin irritants but are more useful for testing skin corrosion or dermal absorption (skin penetration). *Ex vivo* models for skin irritation, penetration and corrosion are:

- Mouse skin integrity function test (SIFT);
- Human cadaver skin;
- Human skin from surgery;
- Pig ear test (Zuang *et al.* 2005);
- Rat skin transcutaneous electrical resistance (TER) method has been validated for skin corrosion testing.

*A cell-free barrier model* called Corrositex® has also been validated for skin corrosion testing. The Corrositex assay is based on the time it takes for a chemical to penetrate an artificial biobarrier. In 1999, an ICCVAM review of Corrositex (National Toxicology Program, 1999) recommended its use as a stand-alone assay for evaluating acids, bases, and acid derivatives for the US Department of Transportation, and otherwise, as part of a tiered testing strategy.

## 3.9   Skin sensitization

The induction of allergic contact dermatitis following exposure to a chemical agent is the response of the immune system. This is the only validated immunotoxicity test that measures chemically induced adverse effects on the immune system through the skin.

GHS defines a skin sensitizer as a substance that will induce an allergic response following skin contact in a substantial number of persons or when there are positive results from an appropriate animal test (UNECE, 2003c). Human experiments to evaluate the skin sensitization potential of a material are not generally conducted, but human testing might be done to confirm a negative animal test result. Human data could also be the result of an epidemiological study or documentation of allergic contact dermatitis from more than one dermatology clinic (AltTox, 2008).

### 3.9.1   Skin sensitization: animal tests for regulatory requirements

The initial exposure is called the sensitization phase; it has no clinical symptoms. The delayed skin response from a later exposure to the allergen is called the elicitation phase. Clinical symptoms include erythema (redness), vesicles/bulla, papules, scaling, and pruritus (itching). Common examples of substances that can induce a skin sensitization reaction in certain individuals – also known as allergic contact dermatitis – include metals in jewelry and chemicals in cosmetics or in latex gloves.

Table 3.13 shows the traditional rat and mice tests for skin sensitization. Local Lymph Node Assay has two new versions, the non-radioactive and the reduced versions of the Local Lymph Node Assay.

### 3.9.2   Non-animal alternative methods

There are currently no validated *in vitro* or *in silico* methods to replace animal testing for the detection of skin sensitizers. However, many promising methods are in the

*Table 3.13* OECD guidelines for *in vivo* testing of skin sensitization (OECD, 2014).

| Test guideline | Test name |
| --- | --- |
| OECD Test TG 406 (1992) | Skin Sensitization – testing with rat and mice |
| OECD Test TG 429 (2010) | Skin Sensitization – Local Lymph Node Assay (LLNA) with rat or mice |
| OECD Test TG 442A (2010) | Skin Sensitization – a non-radiactive LLNA method, based on chemiluminescence of ATP |
| OECD Test TG 442B (2010) | Skin Sensitization – a non-radiactive LLNA method, based on thymidine analogue of 5-bromo-2-deoxyuridine (BrdU) content |

developmental stage. It is expected that a predictive method to totally replace animal testing will be a test battery composed of molecular, cell-based, and/or computational methods, as follows:

– *Peptide reactivity* (or depletion) assay measures binding to skin proteins.
– *Cell-based assays* with cultured cells to model the mechanism(s) of induction of contact allergy in the skin.
– *Quantitative structure–activity relationship* for classifying allergens and non-allergens on the basis of physico-chemical data and the reactivity parameters of functional groups.
– *Predictive Test Battery* integrates SAR or QSAR and molecular and/or cell-based assays to form predictive test batteries to identify human skin sensitizers and replace animal testing.

None of these methods have been validated yet.

## 3.10   Eye irritation and corrosion

*Eye irritation* is defined as the production of changes in the eye following the application of a test substance to the anterior surface of the eye, which are fully reversible within 21 days of application (UNECE, 2003b; UNECE, 2009b). *Eye corrosion* (serious eye damage) is defined as the production of tissue damage in the eye, or serious physical decay of vision, following application of a test substance to the anterior surface of the eye, which is not fully reversible within 21 days of application (UNECE, 2003b).

### 3.10.1   Animal testing of eye irritation and corrosion on rabbits

The Draize rabbit eye irritation test still remains the standard method (Table 3.14) for evaluating the ocular irritation/corrosion potential of a substance for regulatory purposes (Draize *et al.*, 1944). In this test, a material is instilled into one eye of albino rabbits (the other eye serving as the negative control), and the response of the animals is monitored using a standardized scoring system for injury to the cornea, conjunctiva, and iris.

Apart from the ethical issues, the Draize rabbit eye test has other disadvantages, for example, it overpredicts the human responses.

*Table 3.14* OECD guidelines for *in vivo* testing of eye irritation and corrosion (OECD, 2014).

| Test guideline | Test name |
| --- | --- |
| OECD TG 405 (2012) | Acute Eye Irritation/Corrosion |

*Table 3.15* *In vitro* ocular test methods considered valid for limited regulatory testing (OECD, 2014; DB-ALM, 2014).

| Test guideline | Method | Test purpose | Validation authority |
| --- | --- | --- | --- |
| OECD TG 438 (2013) | Isolated Chicken Eye (ICE) assay | Eye corrosion/ severe irritation | ICCVAM; ECVAM |
| OECD TG 437 (2013) | Bovine Corneal Opacity and Permeability (BCOP) assay | Eye corrosion/ severe irritation | ICCVAM; ECVAM |
| DB-ALM Protocol No.85 | Isolated Rabbit Eye (IRE) assay | Eye corrosion | Ongoing/CLP R41 |
| DB-ALM Protocol No.96 | Hen's Egg Test – Chorioallantoic Membrane assay (HET-CAM) | Eye corrosion | Ongoing/CLP R41 |

CLP R41: Although not formally endorsed as valid, positive outcomes can be used for classifying and labeling substances (CLP) as severe eye irritants (R41) in the EU.

A number of non-animal test methods have been developed in the search for a replacement for the Draize rabbit eye test (Table 3.15). These are not validated and accepted test methods yet. The most interesting question is, if a negative *in vitro* test result can be and will be accepted by regulatory agencies.

### 3.10.2   *Non-animal alternative methods for evaluating eye irritation and corrosion*

The ECVAM Scientific Advisory Committee (ESAC) statement (ECVAM, 2007) on the ICCVAM (2006) retrospective study of the four *in vitro* screening assays for ocular corrosive/severe irritants endorsed the validity of the BCOP and ICE methods for use in a tiered strategy as part of the weight of evidence approach. ESAC indicated that further work is needed for the IRE and HET-CAM methods. *In vitro* ocular test methods considered valid for regulatory application can be seen in Table 3.15. OECD TG 438 was amended in 2013 for the identification of chemicals not requiring classification for eye irritation or serious eye damage.

Although only two *in vitro* test methods have been endorsed with some conditions, many new developments are ready to be validated and accepted for standardized and uniform application (Table 3.16) (AltTox, 2007a).

Many of the raw materials for *in vitro* testing originate from slaughterhouses.

## 3.11   Toxicokinetics, pharmacokinetics and metabolism

The study of the absorption, distribution, metabolism, and elimination (usually abbreviated as ADME) of chemicals in the body is also called toxicokinetics or pharmacokinetics.

*Table 3.16* Alternative *in vitro* methods for testing eye irritation and corrosion.

| Test type | Name of the test |
| --- | --- |
| Isolated eye assays | Bovine Corneal Opacity and Permeability (BCOP) |
| | Porcine Corneal Opacity and Permeability (PCOP) |
| | Isolated Chicken Eye (ICE) assay |
| | Isolated Rabbit Eye (IRE) |
| | Isolated Mouse Eye |
| | Human Cornea (discarded from eye banks) |
| Chicken egg membrane assays | Chorioallantoic Membrane Vascularization Assay (CAMVA) |
| | Hen's Egg Test – Chorioallantoic Membrane (HET-CAM) assay |
| Reconstituted human cornea models | Human corneal equivalent |
| Reconstituted rabbit cornea models | 3D corneal tissue construct |
| Reconstituted bovine cornea models | Epithelium and stroma |
| 3D human corneal epithelial cell models | HCE-T human corneal epithelial cell model |
| | SkinEthic HCE model; CEPI |
| | Coty corneal epithelial model |
| 3D epithelial cell models | EpiOcular |
| | MDCK fluorescein leakage |
| | Tissue equivalent assay |
| 3D human conjunctival epithelial cell models | Human conjunctival model |
| Monolayer epithelial cell cultures | Human corneal epithelial cells |
| | Rabbit corneal cells; SIRC cell line |
| | Various cultured cells |
| | Epithelial and fibroblast cell lines |
| Red blood cells | Red cell hemolysis |
| Neural cell models (for detecting neurogenic ocular pain) | TRPV1-expressing neuroblastoma SH-SY5Y cells |
| Acellular models | EYTEX/Irritection test |
| | Hemoglobin denaturation |
| Computational models | SAR/QSAR |

*Pharmacokinetics or toxicokinetics* is defined as the study of the rates of absorption, distribution, metabolism, and excretion of toxic substances or substances under toxicological study (OECD, 2006b). Pharmacokinetics/toxicokinetics testing involves describing 'the bioavailability of a substance and its kinetic and metabolic fate within the body'. Pharmacokinetics is also the term used to describe the assessment of absorption, distribution and metabolism in the context of drug preclinical testing.

*Absorption* is the process of active or passive transport of a substance from the environment into the organism, across the cell membranes, through the lungs, the gastrointestinals or the skin. Absorbed dose of a substance is the mass of the substance transported into the organs of the organism. Ratio of the absorbed amount of a substance to the administered amount can be characterized by the absorption coefficient.

*Distribution* is a dynamic process following absorption, in which a chemical substance translocates throughout the body and reaches the site of action *via* blood circulation. Bioavailability is the fraction of an administered dose entering the systemic circulation. As the result of distribution, a chemical substance is able to reach the target site and evolve its effect, it can be transformed, eliminated or stored in depots. The rate of distribution depends on uptake and elimination rates, the blood flow, the metabolic activity of the organism, the affinity of the substance to specific tissues and its partitioning among blood and the target tissue. The substance in the blood can be unbound or bound to the blood cells or the plasma proteins. Some peripheral tissues for example the adipose tissues are preferred by fat-soluble (high $K_{ow}$) organic substances, or bone tissues by lead, cadmium or fluoride.

*The metabolism* of a substance distributed in the body (blood and peripheral tissues) is an enzyme-mediated change in the physico-chemical properties, e.g. chemical structure, size, configuration, polarity, reactivity of the toxicants and may result in: i) the activation of the chemical substance making it able for a physiologic effect and increased toxic potential at the target site of action; ii) deactivation/detoxication and producing a harmless product; iii) storage by accumulation in special tissues with high affinity to the substance; iv) elimination e.g. excretion from the body mainly via the kidneys and the digestive tract, but also sweat and tears. Hydrophobic (lipophilic) substances cannot partition into urine and faeces.

OECD has defined metabolism as all aspects of the fate of a substance in an organism; however, metabolism generally refers to the biotransformation of a substance (via an enzymatic or non-enzymatic process) within the body to other molecular species (metabolites). Two types of enzymes are involved in metabolism: phase 1 (cytochrome P450 enzyme family) and phase 2 enzymes.

An understanding of the metabolism of a substance in the body is critical to understanding its toxicity. For example biotransformation may result in a more toxic substance, or the lack of metabolism of a substance may result in its bioaccumulation in the body. Understanding a substance's metabolism can also facilitate identification of possible target organs and the route of clearance.

Pharmacokinetic/toxicokinetic data may be used to

- assist in the interpretation of other toxicological data,
- select doses for other toxicological studies, and/or
- extrapolate data from animals to humans (AltTox, 2007b).

*Elimination* is the disappearance of a substance in the body; it is the result of metabolism, involving excretion from the body or transformation to other substances. The rate of disappearance may be expressed by the elimination rate constant, biological half-time or residential time. *Clearance* is a measure of the body's ability to completely clear a toxicant from blood or plasma. Clearance is the rate of elimination by all routes (blood, renal, enzyme activities) relative to the concentration in a systemic biologic tissue.

### 3.11.1   Testing of toxicokinetics, pharmacokinetics and metabolism on animals

OECD accepts one animal test (TG 417) to study toxicokinetics, shown in Table 3.17. The dermal route of entry can play an important role in human exposure to chemicals, typically for personal care products or agrochemicals. OECD worked out *in vitro* and *in vivo* skin absorption models in 2004.

Table 3.17 OECD guidelines for *in vivo* and *in vitro* pharmacokinetic and metabolic tests (OECD, 2014).

| Test guideline | Test name |
| --- | --- |
| OECD TG 427 (2004) | Skin Absorption: *in vivo* method |
| OECD TG 428 (2004) | Skin Absorption: *in vitro* method |
| OECD TG 417 (2010) | Toxicokinetics using various animal species and routes of administration |

### 3.11.2  In vitro *dermal testing*

Dermal absorption methods are defined by an OECD Test Guideline TG 428, and are accepted by some regulatory authorities as non-animal pharmacokinetics and metabolism test methods.

ECVAM publications have reviewed *in vitro* and *in silico* approaches, including tiered testing strategies, mathematical models such as SAR, QSAR, METEOR, MetabolExpert, COMPACT, META or protein modeling and assessing metabolism by using microsomes, suspended liver cells, cultured monolayer and 3D liver cells, and precision-cut liver slice methods (ECVAM, 2002b; Coecke *et al.*, 2005). Mathematical Models aim to predict percutaneous absorption of a test compound based on its intrinsic properties (empirical models, using quantitative structure-permeability relationships, QSPR) or by describing skin structures and the flux of a chemical through these structures using mechanistic models (physiological mass transfer and diffusion models). Some protocols from the DB-ALM database are enlisted here:

- *Permeability assay* can be used to determine the permeability of the test compound across an *in vitro* model of the human intestine Caco-2 cells. The permeability is expressed as apparent permeability coefficient ($P_{app}$) (DB-ALM Protocol No.142);
- *Metabolic Stability assay* can be used to rank the test compounds with respect to their metabolic stability or biotransformation. The disappearance of the parent compound is measured during an incubation with human and rat liver microsomal fractions, S9 or hepatocytes (DB-ALM Protocol No.141);
- *Plasma Protein Binding assay* gives an estimate on the affinity of a test compound for plasma proteins by measuring the fraction unbound in pooled human plasma using equilibrium dialysis against an isotonic buffer (DB-ALM Protocol No.143).

## 3.12  Neurotoxicity

Neurotoxicology is the study of the adverse effects (such as deficits in learning or sensory ability) by chemical, biological, and certain physical agents on the nervous system (the brain, spinal cord, and/or peripheral nervous system) and/or behavior during development and in maturity. Many common substances are neurotoxic, including lead, mercury, some pesticides, and ethanol (Harry *et al.*, 1998).

Neurotoxicity testing is used to identify potential neurotoxic substances. Neurotoxicity is a major toxicity end point that must be evaluated for many regulatory

*Table 3.18* OECD guidelines for *in vivo* neurotoxicity testing (OECD, 2014).

| Test guideline | Test name |
| --- | --- |
| OECD TG 418 (1995) | Delayed Neurotoxicity of Organophosphorus Substances Following Acute (oral) Exposure on hens |
| OECD TG 419 (1995) | Delayed Neurotoxicity of Organophosphorus Substances: 28-day Repeated-Dose Study on hens |
| OECD TG 424 (1997) | Neurotoxicity Study in Rodents, acute (28 d), subchronic (90 d) and chronic (1 year) |
| OECD TG 426 (2007) | Developmental Neurotoxicity Study, *in utero* and early postnatal effects on offspring after exposure of pregnant rats |

applications. Sometimes neurotoxicity testing is considered as a component of target organ toxicity. Developmental neurotoxicity means *in utero* exposure to chemicals and drugs exerting an adverse effect on the development of the nervous system. Common exposure routes such as oral, dermal, or inhalation and both acute and repeated dosing (chronic) may cause neurotoxicity. OECD accepted four animal-using test methods, introduced in Table 3.18. None of the alternative methods have been validated yet.

### 3.12.1  Animal testing of neurotoxicity

The best alternatives encompass the most important neurotoxic end points organized into test batteries (ECVAM, 2002c). The first test tier would distinguish neurotoxicants from cytotoxic chemicals, and the second tier would consist of mechanism-specific tests. It was proposed that a minimum battery might consist of methods for assessing blood–brain barrier function, basal cytotoxicity (damage of the basic cellular function), and energy metabolism.

### 3.12.2  In vitro models for neurotoxicology studies and testing

Test systems used for neurotoxicological testings are:

- Primary cells: neurons and glia (microglia, oligodendrocyte, astrocyte) from different brain regions;
- Cell lines: neuroblastoma, astrocytoma, glioma, pheochromocytoma;
- Brain slices: hippocampus;
- Reaggregating neuronal and glial cell cultures;
- Organotypic (3D) cultures: usually co-cultures (more than one type of cell);
- Neural stem (progenitor) cells: primary cell cultures and cell lines.

ICCVAM and OECD have not reviewed or validated any non-animal methods or alternative testing strategies for assessing neurotoxicity. Regulatory authorities have not accepted any non-animal methods or alternative testing strategies for neurotoxicity testing until now (AltTox, 2009). ECVAM DB-ALM established a protocol titled: "Whole Rat Brain Reaggregate Spheroid Culture". This culture system (single cell

*Table 3.19  In vivo testing of endocrine disrupting effect of chemical substances (OECD, 2014).*

| Test guideline | Test name |
| --- | --- |
| OECD TG 440 (2007) | Uterotrophic Bioassay in Rodents: a short-term screening test for estrogenic properties |
| OECD TG 441 (2009) | Hershberger Bioassay in Rats: a short-term screening assay for (anti)androgenic properties |
| OECD TG 455 (2009) | The Stably Transfected Human Estrogen Receptor-alpha Transcriptional Activation Assay for Detection of Estrogenic Agonist-Activity of Chemicals |

suspension) allows the testing of neurotoxic compounds during development, differentiation and relative maturity of the brain reaggregate (DB-ALM Protocol No.11 (DB-ALM, 2014).

## 3.13  Endocrine toxicity and disruption

An endocrine disruptor is an exogenous agent that interferes with the synthesis, secretion, transport, binding, action or elimination of natural hormones in the body that are responsible for the maintenance of homeostasis, reproduction, development, and/or behavior. We gave a detailed overview of endocrine disruption in Chapter 7 in Volume 1. Here we discuss endocrine disruption only from the testing point of view.

### 3.13.1  Animal tests for screening endocrine disruption

The estrogenic and androgenic effects of chemical substances can be tested and evaluated from the results of animal tests (Table 3.19).

The uterotrophic bioassay evaluates the ability of a chemical to elicit biological activities consistent with agonists or antagonists of natural estrogens (e.g. 17β-estradiol). It is based on the response of the uterus to estrogens: an increase in weight due to water imbibition as an initial response, which is followed by a weight gain due to tissue growth.

The Hershberger bioassay is based on the changes in weight of five androgen-dependent tissues in the castrate-peripubertal male rat.

### 3.13.2  Validated non-animal alternatives for endocrine disruptor activity

Endocrine disruption is an emerging problem; screening and proving whether or not chemical substances have an endocrine disrupting potential is an urgent task for developers of new chemicals, authorities, regulators, managers or other decision makers. Fast and cheap *in vitro* methods are needed and a large number of such methods have been developed in recent years. Some of them have been validated and accepted or are under evaluation as screening methods for regulatory purposes. Table 3.20 contains the validated *in vitro* methods.

OECD has already accepted the *Stably Transfected Transactivation Assay* (STTA) using the hERα-HeLa-9903 cell line to detect estrogenic agonist activity for evaluating the ability of a chemical substance to function as an estrogenic receptor-α ligand and

Table 3.20  In vitro testing of endocrine disrupting potential of chemical substances (OECD, 2014; OCSPP, 2009).

| Test guideline | Test name |
|---|---|
| OECD TG 456 (2011) | Steroidogenesis Assay with H295R human cell line – Screening assay |
| OECD TG 455 (2012) | Performance-Based Test Guideline for Stably Transfected Human Estrogen Receptor-$\alpha$ Transcriptional Activation Assay for Estrogenic Agonist-Activity – Screening assay |
| OECD TG 457 (2012) | BG1Luc Estrogen Receptor Transactivation Test Method for Identifying Estrogen Receptor Agonists and Antagonists |
| 890.1150 (OCSPP, 2009) | Androgen receptor binding assay (rat prostate) – Endocrine screen |
| 890.1200 (OCSPP, 2009) | Aromatase inhibition assay (human recombinant) – Endocrine screen |
| 890.1250 (OCSPP, 2009) | Estrogen receptor binding assay – Endocrine screen |

activate an agonist response. It is used for screening and prioritization purposes but can also provide mechanistic information that can be used in a weight of evidence approach.

The *Steroidogenesis Assay* uses the H295R human carcinoma cell line to assess the effects of chemicals on testosterone and 17β-estradiol production.

### 3.13.3  The US EPA endocrine disruptor screening program

The EPA methods shown in Table 3.21 are part of a battery for screening the endocrine disruptor effect of chemicals in the Endocrine Disruptor Screening Program (EDSP, 2014). The Office of Chemical Safety and Pollution Prevention developed the harmonized Endocrine Disruptor Screening Program Test Guidelines – Series 890 (OCSPP, 2009) for the tier 1 and tier 2 screening. Tier 2 guidelines are not issued yet. The validation status of the tests under development is shown in the Assay Status Table (EDSP AST, 2014).

At the end of the *in vivo* tests, measurements are made that are reflective of the status of the reproductive endocrine system, including male secondary sex characteristics, gonadal histopathology, gonado-somatic index, and plasma concentrations of vitellogenin.

The *YES (Yeast Estrogen Screen) assay* is one of the promising test methods under development. It uses transgenic yeast to test the endocrine disrupting potential of chemical substances. The transgene built into the yeast cells is the human estrogen receptor.

*Saccharomyces cerevisiae* is chromosomally transformed by hER gene, coding the human estrogen receptor protein, which binds estrogen or other xenoestrogens with a complementary spatial structure. After binding to the estrogen responsive element on the yeast plasmid, the complex of hER and the estrogenic compound initiates the transcription of Lac-Z reporter gene, producing β-galactosidase, an enzyme capable of hydrolyzing the sugar from chlorophenol red-β-D-galactosidase (used as an indicator in the assay) and producing chlorophenol red with an easily detectable color. The concept is shown in Figure 3.9 after Routledge and Sumpter (1996).

Instead of *Saccharomyces cerevisiae*, other yeast species, e.g. *Blastobotrys adeninivorans*, have also been used for the construction of the YES assay with the

*Table 3.21*  Test guidelines for the Endocrine Disruptor Screening Program of US EPA (OCSPP, 2014).

| Tier/test number | Test name |
| --- | --- |
| **Tier 1 *In vitro*** | |
| 890.1150* | Androgen Receptor Binding – Rat Prostate |
| 890.1200* | Aromatase – Human Recombinant Microsomes |
| 890.1250* | Estrogen Receptor Binding – rat uterine cytosol |
| 890.1300** | Estrogen Receptor Transcriptional Activation – Human Cell Line HeLa-9903 |
| 890.1550** | Steroidogenesis – Human Cell Line H295R |
| **Tier 1 *In vivo*** | |
| 890.1600 | 15-Day Intact Adult Male Rat Assay Uterotrophic (rat) |
| 890.1400** | Hershberger (rat) |
| 890.1450 | Female Pubertal (rat) |
| 890.1500 | Male Pubertal (rat) |
| 890.1100** | Amphibian Metamorphosis – frog |
| 890.1350* | Fish – Short Term Reproduction |
| **Tier 2 *In vivo*** | |
| | Amphibian 2-Generation Development, Reproduction |
| | Avian 2-Generation |
| | Fish Lifecycle |
| | Invertebrate Lifecycle |
| | Mammalian 2-Generation |
| | *In utero* and through Lactation |

*in vitro* test, the same as in Table 3.20; **existing test methods as OECD TG.

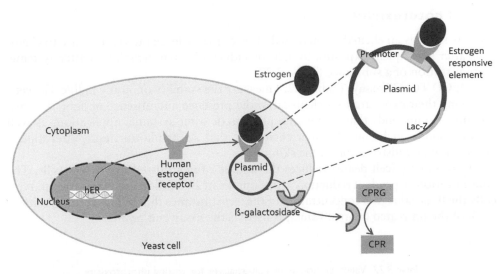

CPRG: chlorophenol red-ß-D-galactopiraniside
CPR: chlorophenol red

*Figure 3.9*  Schematic explanation of the YES assay.

reporter gene of phytase. Instead of estrogenic receptors, androgenic receptors can also be built in.

The advantages of the method are:

- Easy to cultivate test microorganism;
- Unlimited number of cells to use in the test;
- Well-known reporter genes ensuring an easily detectable response;
- Short time duration of the test makes high throughput screening possible;
- Able to test bioavailability of substances with estrogenic effect;
- Possibility to use radioactive substances.

Disadvantages of the YES assay are:

- Yeast cells differ from animal cells in their membrane composition, first of all in the sterol content and, as a consequence, in the binding of the tested chemical substances;
- Different yeast strains give different test results;
- The YES assay may significantly overestimate the estrogen disrupting effect;
- The YES assay may give false negative results.

Further development and finding the most fitting yeast strain may result in a fast and cheap *in vitro* method in the future for screening chemicals for endocrine disruption.

## 3.14  Phototoxicity

Phototoxicity is an elicited or increased (at lower dose levels) toxic response to chemicals after subsequent exposure to light or induced by skin irradiation after systemic administration of a substance (Table 3.22).

OECD TG 432 assays (Table 3.22) measure the viability of mouse Balb/c 3T3 cells following their exposure to a chemical in the presence and absence of light. The test identifies compounds that act *in vivo* phototoxic after systemic application, as well as compounds, including UV filter chemicals, that act as photoirritants after topical application and distribution to the skin.

Cytotoxicity (cell death) reduces the uptake of neutral red dye by the cells. The concentration-dependent reduction in dye uptake in 24 h following the treatment, typically the $IC_{50}$ value (the concentration of the test substance that reduces cell viability to 50% of the untreated control value), is used as the assay end point.

Table 3.22  Validated non-animal alternatives for testing phototoxicity (OECD, 2004; OECD/OCDE, 2004).

| Test guideline | Test name |
| --- | --- |
| OECD TG 432 (2004) | *In vitro* 3T3 Neutral Red Uptake Phototoxicity Test (3T3 NRU PT) |

# REFERENCES

3T3 NRU (2010) *BALB/c 3T3 Neutral Red Uptake Assay.* Guidance document on using cytotoxicity tests to estimate starting doses for acute oral systemic toxicity tests. OECD Series on Testing and Assessment No. 129. ENV/JM/MONO (2010) 20. [Online] Available from: http://ntp.niehs.nih.gov/iccvam/SuppDocs/FedDocs/OECD/OECD-GD129.pdf. [Accessed 20th October 2013].

Against Cruelty to Animals (2013) [Online] Available from: http://www.facebook.com/pages/Against-Cruelty-To-Animals/202320584952. [Accessed 20th October 2013].

AltTox (2007a) *Eye irritation/corrosion.* [Online] Available from: http://alttox.org/ttrc/toxicity-tests/eye-irritation/. [Accessed 20th October 2013].

AltTox (2007b) *Pharmacokinetics & metabolism.* [Online] Available from: http://alttox.org/ttrc/toxicity-tests/pharmacokinetics-metabolism/. [Accessed 20th October 2013].

AltTox (2008) *Skin sensitization.* [Online] Available from: http://alttox.org/ttrc/toxicity-tests/skin-sensitization/. [Accessed 20th October 2013].

AltTox (2009) *Neurotoxicity.* [Online] Available from: http://alttox.org/ttrc/toxicity-tests/neurotoxicity/. [Accessed 20th October 2013].

AltTox (2014) *Non-animal methods for toxicity testing.* [Online] Available from: http://alttox.org/ and http://alttox.org/ttrc/existing-alternatives/acute-systemic.html. [Accessed 20th January 2014].

ECHA (2011) The Use of Alternatives to Testing on Animals for the REACH Regulation, 2011 [Online] Available from: http://echa.europa.eu/documents/10162/13639/alternatives_test_animals_2011_en.pdf. [Accessed 20th December 2014].

ECHA (2014) The Use of Alternatives to Testing on Animals for the REACH Regulation. ECHA, 2014. [Online] Available from: http://echa.europa.eu/documents/10162/13639/alternatives_test_animals_2014_en.pdf. [Accessed 20th December 2014].

CellSystems (2014) [Online] Available from: http://www.lifelinecelltech.com/pdf/est-ast-application.pdf. [Accessed 20th May 2014].

CFU GM (2006) *Colony Forming Unit-Granulocyte/Macrophage Assay.* DB-ALM Protocol No 101. [Online] Available from: http://ecvam-dbalm.jrc.ec.europa.eu. [Accessed 20th October 2013].

Coecke, S., Blaauboer, B.J., Elaut, G. *et al.* (2005) Toxicokinetics and metabolism. *Alternatives to Laboratory Animals,* 33(1), 147–175.

COMET (2013) *Comet assay.* [Online] Available from: http://cometassay.com/. [Accessed 20th October 2013].

DB-ALM (2014) *DataBase service on ALternative Methods to animal experimentation,* EURL ECVAM. [Online] Available from: http://ecvam-dbalm.jrc.ec.europa.eu. [Accessed 20th January 2014].

Douglas, G.R. (2010) *Transgenic rodent gene mutation assays: current state of validation, validation of transgenic rodent gene mutation assay.* [Online] Available from: http://www.oecd.org/dataoecd/6/39/46161373.pdf. [Accessed 20th October 2013].

Draize, J.H., Woodward, G. & Calvery, H.O. (1944) Methods for the study of irritation and toxicity of substances applied topically to the skin and mucous membranes. *Journal of Pharmacology and Experimental Therapeutics,* 82, 377–390.

ECHA (2011) The Use of Alternatives to Testing on Animals for the REACH Regulation, 2011 [Online] Available from: http://echa.europa.eu/documents/10162/13639/alternatives_test_animals_2011_en.pdf. [Accessed 20th December 2014].

ECHA (2014) The Use of Alternatives to Testing on Animals for the REACH Regulation. ECHA, 2014. [Online] Available from: http://echa.europa.eu/documents/10162/13639/alternatives_test_animals_2014_en.pdf. [Accessed 20th December 2014].

ECVAM (2001) *ESAC statements: The use of scientifically-validated in vitro tests for embryotoxicity.* [Online] Available from: http://ecvam.jrc.it/publication/Embryotoxicity_statements.pdf. [Accessed 20th October 2013].

ECVAM (2002a) Acute lethal toxicity. *Alternatives to Laboratory Animals*, 30(1), 27–33.

ECVAM (2002b) Biokinetics. *Alternatives to Laboratory Animals*, 30(1), 55–70.

ECVAM (2002c) Target organ and target system toxicity. *Alternatives to Laboratory Animals*, 30(1), 71–82.

ECVAM (2007) ESAC statement on the conclusions of the ICCVAM retrospective study on organotypic *in vitro* assay as screening test to identify potential ocular corrosives and severe irritants as determined by US-EPA, EU (R41) and UN GHS classifications in a tiered testing strategy, as part of a weight of evidence approach. [Online] Available from: http://ecvam.jrc.it/publication/ESAC26_statement_Organotypic_20070510c.pdf. [Accessed 20th October 2013].

ECVAM DB-ALM (2006) *Colony forming unit – granulocyte/macrophage, CFU-GM assay. Haematoxicity*. ECVAM-DB-AL INVITTOX protocol. [Online] Available from: http://ecvam-dbalm.jrc.ec.europa.eu/public_view_doc.cfm?id=6E7E72104B2DEFD6BE979B3B139176C67180BB0BC12CB10496CDA74B54630A05A3291B895581F634. [Accessed 20th October 2013].

ECVAM DB-ALM (2013) *In vitro toxicity testing protocols*. [Online] Available from: http://ecvam-dbalm.jrc.ec.europa.eu/s_invitoxprot.cfm?idmm=6&idsm=7. [Accessed 20th October 2013].

ECVAM DB-ALM (2014) *DataBase service on ALternative Methods to animal experimentation* (DB-ALM), EURL ECVAM. [Online] Available from: http://ecvam-dbalm.jrc.ec.europa.eu. [Accessed 20th January 2014].

EDSP (2009) *Endocrine Disruptor Screening Program (EDSP)*. Tier 1 Screening Battery. US EPA. Online] Available from: http://www.epa.gov/endo/pubs/assayvalidation/tier1battery.htm. [Accessed 20th January 2014].

EDSP (2014) *Endocrine Disruptor Screening Program* (EDSP). US EPA. [Online] Available from: http://www.epa.gov/scipoly/oscpendo/index.htm. [Accessed 20th January 2014].

EDSP AST (2014) *Assay Status Table of the Endocrine Disruptor Screening Programme*. [Online] Available from: http://www.epa.gov/endo/pubs/assayvalidation/status.htm. [Accessed 20th January 2014].

EpiSkin (2014) *In vitro human skin for testing dermal corrosion*. [Online] Available from: http://www.invitroskin.com/_int/_en/episkin/home.aspx?tc=R_HOME_EPISKIN&cur=R_HOME_EPISKIN. [Accessed 20th January 2014].

EST-1000 (2014) *Epidermal Skin Test 1000 for testing dermal corrosion*. [Online] Available from: www.cellsystems.biz. [Accessed 20th January 2014].

EU (2010) *Protection of animals used for scientific purposes*. Directive 2010/63/EU of the European Parliament and of the Council of 22 September 2010. *Official Journal L*, 276(20.10.2010), 33–79. (revising Directive 86/609/EEC). [Online] Available from: http://eur-lex.europa.eu/LexUriServ/LexUriServ.do?uri=OJ:L:2010:276:0033:0079:EN:PDF. [Accessed 20th October 2013].

EU Test Methods (2008) *Test methods pursuant to Regulation (EC) No 1907/2006 of the European Parliament and of the Council on the Registration, Evaluation, Authorisation and Restriction of Chemicals (REACH)*. Council Regulation (EC) No 440/2008 of 30 May 2008. [Online] Available from: http://eur-lex.europa.eu/LexUriServ/LexUriServ.do?uri=CELEX:32008R0440:en:NOT. [Accessed 20th January 2014].

EU Test Method B.4 (2008) *Acute toxicity: dermal irritation/corrosion*. Commission Regulation (EC) No.440/2008/EC. [Online] Available from: http://eur-lex.europa.eu/LexUriServ/LexUriServ.do?uri=CELEX:32008R0440: en:NOT. [Accessed 20th January 2014].

EU Test Method B.46 (2009) *In vitro skin irritation: reconstructed human epidermis model test*. Annex to 440/2008/EC. Commission Regulation (EC) No 761/2009/EC. [Online] Available from: http://eur-lex.europa.eu/JOHtml.do?uri=OJ:L:2009:220:SOM:EN:HTML. [Accessed 20th January 2014].

EU Test Methods (2009) *Amending 440/2008 for the purpose of its adaptation to technical progress.* Commission Regulation (EC) No 761/2009 of 23 July 2009. [Online] Available from: http://eur-lex.europa.eu/JOHtml.do?uri=OJ:L:2009:220:SOM:EN:HTML. [Accessed 20th January 2014].

EURL ECVAM (2014) *European Union Reference Laboratory for Alternatives to Animal Testing.* [Online] Available from: http://ihcp.jrc.ec.europa.eu/our_labs/eurl-ecvam. [Accessed 20th January 2014].

FAU (2013a) *Guinea pig handout.* Florida Atlantic University Veterinary Services. [Online] Available from: http://www.fau.edu/research/vetservices/handout_gpig.php. [Accessed 20th October 2013].

FAU (2013b) *Hamster handout.* Florida Atlantic University Veterinary Services. [Online] Available from: http://www.fau.edu/research/vetservices/handout_hamster.php. [Accessed 20th October 2013].

FAU (2013c) *Mouse handout.* Florida Atlantic University Veterinary Services. http://www.fau.edu/research/vetservices/handout_mouse.php. [Accessed 20th October 2013].

FAU (2013d) *Rabbit handout.* Florida Atlantic University Veterinary Services. [Online] Available from: http://www.fau.edu/research/ovs/VetData/rabbit.php. [Accessed 20th October 2013].

FDA (2004) Challenge and opportunity on the critical path to new medical products. U.S. Food and Drug Administration (FDA). [Online] Available from: http://www.fda.gov/ScienceResearch/SpecialTopics/CriticalPathInitiative/CriticalPathOpportunities-Reports/ucm077262.htm. [Accessed 20th January 2014].

Gratwicke, B. (2012) *Xenopus laevis* – photo. [Online] Available from: http://en.wikipedia.org/wiki/File:Xenopus_laevis_02.jpg. [Accessed 20th January 2014].

Greim, H. (2012) *The Cellular Response to the Genotoxic Insult: The Question of Threshold for Genotoxic Carcinogens.* London, Royal Society of Chemistry.

Harry, G.J., Billingsley, M., Bruinink, A. *et al.* (1998) *In vitro* techniques for the assessment of neurotoxicity. *Environmental Health Perspectives*, 106(1), 131–158.

Hartung, T. (2009) Toxicology for the twenty-first century. *Nature,* 460, 208–212.

Hartung T. (2010. Lessons learned from alternative methods and their validation for a new toxicology in the 21st century. *Journal of Toxicology and Environmental Health*, 13, 277–290.

Hartung, T. (2012) Evidence-Based Toxicology – the Toolbox of Validation for the 21st Century? *Altex,* 27, 253–263.

Hikasa, H., Shibata, M., Hiratani, I. & Taira, M. (2002) The *Xenopus* receptor tyrosine kinase Xror2 modulates morphogenetic movements of the axial mesoderm and neuroectoderm via Wnt signaling. *Development*, 129 (22), 5227–5239. [Online] Available from: http://www.xenbase.org/literature/article.do?method=display&articleId=6319.    [Accessed 20th October 2013].

ICCVAM (2000) *FETAX for human developmental hazard identification.* NICEATM FETAX background review document: Section 1.0, 10 Mar. [Online] Available from: http://iccvam.niehs.nih.gov/docs/fetax2000/brd/10FETAX.pdf. [Accessed 20th October 2013].

ICCVAM (2006) *In vitro test methods for detecting ocular corrosives and severe irritants.* [Online] Available from: http://iccvam.niehs.nih.gov/methods/ocutox/ivocutox.htm. [Accessed 20th October 2013].

ICCVAM (2013) *Validation study of in vitro cytotoxicity test methods.* [Online] Available from: http://iccvam.niehs.nih.gov/methods/acutetox/inv_nru_announce.htm. [Accessed 20th October 2013].

ICCVAM (2013a) *Malformations of Xenopus embryos* – photo. [Online] Available from: http://iccvam.niehs.nih.gov/docs/fetax2000/fetaxbrd2000.htm and http://iccvam.niehs.nih.gov/methods/development/dev.htm. [Accessed 20th October 2013].

ICCVAM (2014) *Interagency Coordinating Committee on the Validation of Alternative Methods*. Committee of the National Institute of Environmental Health Sciences. [Online] Available from: http://ntp.niehs.nih.gov/?objectid=62AE6E86-BB31-0E02-DD06457DC0 AB4AD7. [Accessed 20th January 2014].

ICCVAM-NICEATM meeting (2000) A Proposed Screening Method for Identifying the Developmental Toxicity Potential of Chemicals and Environmental Samples. *Minutes of the Expert Panel Meeting on the Frog Embryo Teratogenesis Assay—Xenopus (FETAX), May 16–18, Durham, North Carolina, US.*

JaCVAM (2013) Japanese Center for the Validation of Alternative Methods. Available from: http://jacvam.jp. [Accessed 20th October 2013].

Kavlock, R. J., Austin, C. P. & Tice, R. R. (2009) Commentary – toxicity testing in the 21st century: Implications for Human Health Risk Assessment. *Risk Analysis*, 29, 485–487.

Kenpei, I. (2013) *Xenopus laevis var. albino* – photo. [Online] Available from: http://it.wikipedia.org/wiki/File:Xenopus_laevis_var_albino.jpg. [Accessed 20th October 2013].

Linnenbach, M. (2013) *Xenopus laevis* – photo. [Online] Available from: http://en.wikipedia.org/wiki/File:Xenopus_laevis_1.jpg. [Accessed 20th October 2013].

MatTek Corporation (2014) [Online] Available from: http://www.mattek.com. [Accessed 20th May 2014].

Maurici, D., Aardema, M., Corvi, R. *et al.* (2005a) Alternative (non-animal) methods for cosmetics testing: current status and future prospects. Genotoxicity and mutagenicity. *Alternatives to Laboratory Animals*, 33(Suppl 1), 117–130.

Maurici, D., Aardema, M., Corvi, R. *et al.* (2005b) Carcinogenicity. *Alternatives to Laboratory Animals*, 33(1), 177–182.

National Toxicology Program (1999) *Corrositex: An in vitro test method for assessing dermal corrosivity potential of chemicals.* An independent peer review evaluation coordinated by the Interagency Coordinating Committee on the Validation of Alternative Methods (ICCVAM) and the NTP Interagency Center for the Evaluation of Alternative Toxicological Methods (NICEATM), NIH Publication Number 99-4495. [Online] Available from: http://iccvam.niehs.nih.gov/docs/dermal_docs/corprrep.pdf. [Accessed 20th October 2013].

NICEATM (2014) *National Toxicology Program Interagency Center for the Evaluation of Alternative Toxicological Methods*. Division of the National Toxicology Program of the National Institute of Environmental Health Sciences. [Online] Available from: http://ntp.niehs.nih.gov/?objectid=720160EB-BDB7-CEBA-F4A14A3D4AFF4B28. [Accessed 20th January 2014].

NHK NRU (2010) *Normal Human Keratinocyte (NHK) Neutral Red Uptake Assay*. Draft Guidance Document On Using Cytotoxicity Tests To Estimate Starting Doses For Acute Oral Systemic Toxicity Tests. OECD Series on Testing and Assessment No. 130. ENV/JM/MONO (2010) 20. [Online] Available from: http://ntp.niehs.nih.gov/iccvam/SuppDocs/FedDocs/OECD/OECD-GD129.pdf. [Accessed 20th January 2014].

NRC (2000) *Scientific frontiers in developmental toxicology and risk assessment*. Washington DC, National Academy Press (NRC).

NRC (2007) *Toxicity testing in the 21st century: a vision and a strategy*. Committee on toxicity testing and assessment of environmental agents. National Research Council. Washington DC, National Academy Press (NRC).

OCSPP (2009) *Harmonized test guidelines*, series 890. Office of Chemical Safety and Pollution Prevention, US EPA. [Online] Available from: http://www.epa.gov/ocspp/pubs/frs/publications/test_guidelines/series890.htm. [Accessed 20th January 2014].

OCSPP (2014) *Office of Chemical Safety and Pollution Prevention*. US EPA. Available from: http://www.epa.gov/aboutepa/ocspp.html. [Accessed 20th January 2014].

OECD (2005) *Guidance document on the validation and international acceptance of new or updated test methods for hazard assessment.* ENV/JM/MONO(2005)14. OECD Series on Testing and Assessment 34, 96 pp. Paris, France, OECD. [Online] Available from: http://www.oecd.org/officialdocuments/displaydocumentpdf?cote=env/jm/mono(2005)14&doclanguage=en. [Accessed 20th January 2014].

OECD (2006a) *Detailed review paper on cell transformation assays for detection of chemical carcinogens.* DRP No. 31, Fourth draft version. [Online] Available from: http://www.oecd.org/dataoecd/59/29/36069919.pdf. [Accessed 20th October 2013].

OECD (2006b) *Summary of considerations in the report from the OECD Expert Groups on Short Term and Long Term Toxicology.* ISBN 9789264035447 (PDF). [Online] DOI: 10.1787/9789264035447. Available from: http://www.oecd-ilibrary.org/environment/summary-of-considerations-in-the-report-from-the-oecd-expert-groups-on-short-term-and-long-term-toxicology_9789264035447-en;jsessionid=521qau75i131.delta. [Accessed 20th October 2013].

OECD (2008) *Draft OECD guideline for the testing of chemicals, Draft consultant's proposal.* V. 8. OECD TG 452, November 2008. [Online] Available from: http://www.oecd.org/dataoecd/30/44/41753317.pdf. [Accessed 20th October 2013].

OECD (2009) *OECD guidance document on the design and conduct of chronic toxicity and carcinogenicity studies, supporting TG 451, 452 and 453.* ENV/JM/TG(2009). [Online] Available from: http://www.oecd.org/dataoecd/62/33/42564522.pdf. [Accessed 20th October 2013].

OECD (2012) *QSAR Toolbox: User manual – Strategies for grouping chemicals to fill data gaps to assess genetic toxicity and genotoxic carcinogenicity.* [Online] Available from: http://www.oecd.org/chemicalsafety/risk-assessment/genetic%20toxicity.pdf. [Accessed 20th January 2014].

OECD (2014) *Guidelines for the Testing of Chemicals, Section 4. Health Effects.* [Online] Available from: http://www.oecd-ilibrary.org/environment/oecd-guidelines-for-the-testing-of-chemicals-section-4-health-effects_20745788. [Accessed 20th January 2014].

REACH (2006) Regulation (EC) No 1907/2006 of the European Parliament and of the Council of 18 December 2006 concerning the Registration, Evaluation, Authorisation and Restriction of Chemicals (REACH), establishing a European Chemicals Agency. *Official Journal of the European Union* L 396/1, 1–849. [Online] Available from: http://eur-lex.europa.eu/LexUriServ/LexUriServ.do?uri=OJ:L:2006:396:0001:0849:EN:PDF. [Accessed 20th October 2013].

Routledge, E.J. & Sumpter, J.P. (1996) Estrogenic activity of surfactants and some of their degradation products assessed using a recombinant yeast screen. *Environmental Toxicology and Chemistry*, 15, 241–248.

Sandos (2014) *Two adult guinea pigs (Cavia porcellus).* Photograph. [Online] Available from: http://en.wikipedia.org/wiki/File:Two_adult_Guinea_Pigs_%28Cavia_porcellus%29.jpg. [Accessed 20th January 2014].

Schultz, T.W. (2010) Adverse outcome pathways: A way of linking chemical structure to *in vivo* toxicological hazards. In: Cronin, M.T.D. & Madden, J.C. (eds.) *In silico Toxicology: Principles and Applications.* Cambridge, UK, Royal Society of Chemistry.

Siebert, H., Balls, M., Fentem, J.H. *et al.* (1996) Acute toxicity testing *in vitro* and the classification and labelling of chemicals. *Alternatives to Laboratory Animals*, 24, 499–510.

Singh, N.P., McCoy, M.T., Tice, R.R. & Schneider, E.L. (1988) A simple technique for quantitation of low levels of DNA damage in individual cells. *Experimental Cell Research*, 175, 184–191.

SkinEthic (2014) *In vitro human skin for testing dermal corrosion.* [Online] Available from: http://www.skinethic.com/EPISKIN.asp. [Accessed 20th January 2014].

Stephens, J. (2014) *Wistar rat*. Photograph. [Online] Available from: http://en.wikipedia.org/wiki/File:Wistar_rat.jpg. [Accessed 20th January 2014].

Tice, R.R., Agurell, E., Anderson, D., Burlinson, B., Hartmann, A., Kobayashi, H., Miyamae, Y., Rojas, E., Ryu, J.C. & Sasaki, Y.F. (2000) Single Cell Gel/Comet Assay: Guidelines for *In Vitro* and *In Vivo* Genetic Toxicology Testing. *Environmental and Molecular Mutagenesis*, 35, 206–221.

UNECE (2003a) *Globally Harmonized System of Classification and Labeling of Chemicals (GHS)*. Part 3. Health and environmental hazards, Chapter 3.7. Reproductive toxicity. United Nations Economic Commission for Europe. [Online] Available from: http://www.unece.org/trans/danger/publi/ghs/ghs_rev00/00files_e.html and http://www.unece.org/trans/danger/publi/ghs/ghs_rev00/English/GHS-PART-3e.pdf. [Accessed 20th October 2013].

UNECE (2003b) *Globally Harmonized System of Classification and Labeling of Chemicals (GHS)*. Part 3. Health and environmental hazards. Chapter 3.3. Serious eye damage/eye irritation. United Nations Economic Commission for Europe. [Online] Available from: http://www.unece.org/trans/danger/publi/ghs/ghs_rev03/English/03e_part3.pdf. [Accessed 20th October 2013].

UNECE (2003c) *Globally Harmonized System of Classification and Labeling of Chemicals (GHS)*. Part 3. Health and environmental hazards. Chapter 3.4. Respiratory or skin sensitization. United Nations Economic Commission for Europe. [Online] Available from: http://www.unece.org/trans/danger/publi/ghs/ghs_rev03/English/03e_part3.pdf. [Accessed 20th October 2013].

UNECE (2009a): *Globally Harmonized System of Classification and Labeling of Chemicals (GHS)*. 3rd revised edition. Part 3: Health hazards. Chapter 3.7. Reproductive Toxicity. United Nations Economic Commission for Europe, pp. 173–185. [Online] Available from: http://www.unece.org/fileadmin/DAM/trans/danger/publi/ghs/ghs_rev03/English/03e_part3.pdf. [Accessed 20th October 2013].

UNECE (2009b) *Globally Harmonized System of Classification and Labeling of Chemicals (GHS)*. 3rd revised edition. Part 3: Health hazards. Chapter 3.4. Respiratory or skin sensitization. United Nations Economic Commission for Europe, pp. 145–154. [Online] Available from: http://www.unece.org/fileadmin/DAM/trans/danger/publi/ghs/ghs_rev03/English/03e_part3.pdf. [Accessed 20th October 2013].

UNECE (2009c) *Globally Harmonized System of Classification and Labeling of Chemicals (GHS)*. 3rd revised edition. Part 3: Health hazards. Chapter 3.2. Skin corrosion/irritation. United Nations Economic Commission for Europe, pp. 121–132. [Online] Available from: http://www.unece.org/trans/danger/publi/ghs/ghs_rev03/English/03e_part3.pdf. [Accessed 20th October 2013].

US EPA (2014) United States Environmental Protection Agency. Available from: http://www.epa.gov. [Accessed 20th October 2013].

Vitro-Life Skin (2014) *Reconstructed human tissue test made in Japan for testing dermal corrosion*. [Online] Available from: http://www.jacvam.jp/en_effort/effort0101.html. [Accessed 20th January 2014].

Xenbase (2014) [Online] Available from: http://www.xenbase.org. [Accessed 20th October 2013].

Zuang, V., Alonso, M.A., Botham, P.A. *et al.* (2005) Skin irritation and corrosion in alternative (non-animal) methods for cosmetics testing: current status and future prospects. *Alternatives to Laboratory Animals*, 33(1), 35–46.

# Chapter 4

# Aquatic toxicology

*K. Gruiz & M. Molnár*

*Department of Applied Biotechnology and Food Science,
Budapest University of Technology and Economics, Budapest, Hungary*

## ABSTRACT

Aquatic toxicology is the pioneer in environmental toxicology, but still very young as an independent science. The very first developments and applications are dated back to 1960–70 when surface waters were endangered by the then applied industrial and agricultural practice.

This chapter will introduce the most important test organisms and test methods used in aquatic toxicology for the measurement of adverse effects of chemical substances to the members of the aquatic ecosystem.

Similar trends to those in human toxicology can also be identified in the field of aquatic toxicology: the use of *in silico* methods such as QSAR (Quantitative Structure Activity Relationship) is spreading and the tests on higher animals are less frequent. Instead, molecular and cell-level biomarkers for assessment and monitoring are being increasingly used.

## I INTRODUCTION TO AQUATIC TOXICOLOGY

The impacts of pollution on water quality and the conditions of aquatic habitats (surface water systems of rivers, lakes, seas and oceans, coastal zones and estuaries) have endangered the ecosystem and humanity over the last quarter of the twentieth century. The deterioration of aquatic ecosystems has been studied in detail and a wide range of engineering tools have been developed for assessing, monitoring and reducing the risk of water pollution by chemicals.

The very first model applied for aquatic toxicity assessment was the chemical model which tried to extrapolate from chemical analysis data to adverse effects. The chemical approximation failed not only because insufficient information was available on the effect of chemicals to aquatic organisms, but also because chemical models cannot handle accessibility, bioavailability and multiple interactions between contaminants, water and sediments as well as members of biota. Modern environmental management uses an integrated approach in environmental assessment and monitoring: chemical, biological and ecotoxicological methodologies are used jointly as a tool battery tailored to the specific problem of contaminants, waters or sediments.

*Aquatic toxicology* aims to understand the stresses chemical substances exert on aquatic ecosystems and to find the quantitative relationship between contaminant

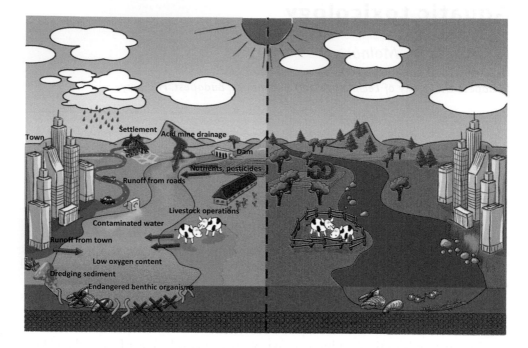

*Figure 4.1* Contaminant sources endangering the aquatic ecosystem.

concentration and adverse effects. This knowledge supports forecasting the deterioration of the ecosystem and enables sustainable management of our waters by preventing the contamination and deterioration and by the remediation of endangered ecosystems and habitats.

Two different strategies may lead to achieving these aims:

– Monitoring the environment, following the trends of ecosystem changes and comparing them to the acceptable rate of changes. Monitoring data can be obtained from the time series of aquatic ecosystem characterization. The "acceptable" rate of changes is based on the knowledge of the ecosystem's adaptive potential and actual ecosystem quality and integrity. For example, an acceptable change of 5% in the diversity of species means that 95% of ecosystem diversity remains unchanged. It is an acceptable rate of change in most of the cases. Of course a refined knowledge on the lost 5% may also be important, and may indicate unacceptable deterioration. It is typical that the roles particular species play in ecosystem quality and function are not transparent; in light of this the loss of e.g. minor components of an ecosystem may remain undetected. All these may cause large uncertainty in the outcome of assessment and monitoring.
– Another strategy is testing the response of selected species with the aim of representing the whole of the ecosystem. From these (generally laboratory) experiments it can be extrapolated to the whole ecosystem using factorial methods or relying on known probabilistic distributions. Applying the factorial extrapolation method in

effect assessment, representatives of aquatic ecosystems from a minimum of three different trophic levels should be used and tested in laboratory bioassays, microcosms or mesocosms. When results from 6–8 or more representative aquatic species are available, the species sensitivity distribution can be derived from the laboratory test result and the concentration belonging to the 5% of fraction affected (95% non-affected) can be considered as a limit value for acceptable ecosystem change (deterioration).

Ecotoxicity test results can be used in screening, monitoring and control and also for direct decision making in water management. Depending on the nature of the problem and the aim of the assessment, concepts and methods have a wide range of selection. Original new scientific data may be acquired applying individual concepts and innovative methods or by adapting existing methods to the scenario to be investigated. In other cases, e.g. for regulatory ecotoxicology purposes the stipulations of regulations should be routinely fulfilled using the prescribed standard methods.

Direct decision making based on the measured adverse effects is an efficient way of environmental management: it is practised in the USA and in some European and Asian countries for the biomonitoring of industrial discharges on a pass–fail basis (Thompson *et al.*, 2005). In the USA the National Pollutant Discharge Elimination System (US NPDS, 2004) has established aquatic toxicity bioassay criteria for effluents, sediments, and oil spills.

Besides ecosystem assessments and bioassays, chemical models still dominate environmental risk characterization. Advanced chemical analysis technologies have enabled the determination of contaminants in concentrations at the picogram and femtogram level, opening new perspectives in environmental toxicology. Although sensitivity, selectivity and reproducibility analysis bears great potential and advantages, the results of chemical analysis alone cannot truly model the fate, behavior, interactions and effects of chemical substances in the environment. Adverse effects of chemical substances depend to a great extent on the presence of other chemicals – there is always a mixture of chemicals in the environment – and the environmental compartment's quality and physico-chemical properties such as temperature, pH, redox potential, density, composition, exposed physical phases and their proportion. The effects are greatly influenced by the interaction between the substance investigated and other substances, matrices and members of the ecosystem over the short and long term. These interactions depend on the chemicals' mobility, biological accessibility and availability, which may greatly differ from the chemical availability i.e. extractability. Chemical analysis methods, even if they are chemical test batteries with complex interpretation (e.g. sequential extraction), are not sufficient to predict actual effects of contaminants or contaminant mixtures on a complex ecosystem and its members with different sensitivity. Finally, the most problematic point in chemical analysis is that only those chemicals can be analyzed which are included in an analysis program, assessment or monitoring plan. This may work properly when the contaminants have already been identified and the situation is uncomplicated, e.g. there is one contaminant without metabolites requiring special analysis methods. However, in the case of complex situations involving many interacting chemicals, matrices and living organisms, which are typical in the environment, an integrated approach is needed. This integrated approach should include water and sediment characterization and monitoring and may comprise ecosystem assessment, laboratory testing of adverse effects and chemical analysis in the form of a problem-specific test battery (Gruiz, 2009; Gruiz *et al.*, 2009). Chapter 2 in Volume 3 deals with this topic in greater detail.

## 2  HUMAN AND ECOSYSTEM EXPOSURE TO AQUATIC HAZARDS

People are not independent of aquatic ecosystems; there is a direct interconnection between them through the food chain: bioconcentration and biomagnification of contaminants in aquatic species which are consumed by people, pose high risk to human health. Man as a top-predator is exposed to the highest accumulated concentrations of chemical substances such as mercury and other toxic metals, chlorinated pesticides and POPs (Persistent Organic Pollutants). Deterioration of aquatic ecosystems occurs as an indirect effect in the form of losing the ecosystem's services, water quantity and quality and water-dependent renewable resources.

The most important ecosystem services provided by fresh-water biodiversity are the following:

–   Delivery of clean water for people;
–   Providing human food (rice, fish, crayfish);
–   Providing goods for human use such as reeds, building materials;
–   Flood attenuation in flood-prone areas;
–   Attenuation of adverse anthropogenic impacts such as micropollutants, nutrients, tensides, drugs;
–   Suppression of waterborne diseases;
–   Recreation;
–   Inspiration, religion, cultural purposes.

Any of these functions and services can deteriorate when the intact ecosystem is disturbed, when biodiversity is reduced or adversely changed and as a consequence, the community is not able to fulfill its normal function.

Urbanization, compaction and sealing of surfaces, the lack of vegetation greatly influence water cycling and transport routes and as a consequence aquatic ecosystem services and aquatic habitats. Figure 4.2 shows the percentage distribution of precipitation between groundwater, surface runoff and evapotranspiration in a pre-urban and an urban situation.

*Aquatic biodiversity* can be assessed by generally applicable methods such as the US EPA Rapid Bioassessment Procedure (RBP) (Barbour *et al.*, 1999) or by problem- and site-specific ecological tools to characterize watersheds, streams or lakes. Based on the physical, functional and biological results, the watersheds can be classified which makes management easier and more uniform. The RBP method is recommended by US EPA for aquatic habitat assessment and characterization. The most important parts of the protocol are:

–   Habitat assessment by physico-chemical parameters;
–   Periphyton assessment using laboratory or field assessment protocols;

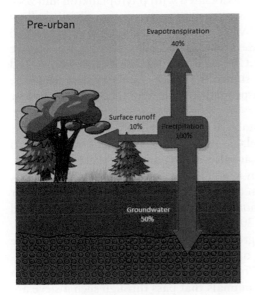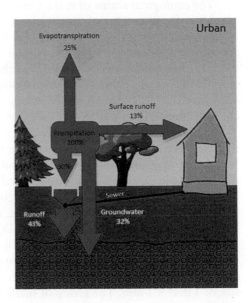

*Figure 4.2* Pre-urban and urban distribution of precipitation between groundwater, surface runoff and evapotranspiration.

– Benthic macroinvertebrate sampling and assessment protocols;
– Fish protocols for collection, identification and metrics;
– Biological data analysis.

Site- and problem-specific ecological methods can be developed for any arising problem in the habitats' suitability, biological integrity or vulnerability. Assessment protocols for some ecosystem-types already exist and are accessible on-line at the Ecological Assessment Methods Database of the USA (EAMD, 2013).

– Wetland habitats;
– Riparian community;
– Coastal region habitat;
– Streams and small stream habitats;
– Tidal rivers;
– Special habitats such as peat land, reefs, glaciers;
– Index of Biological Integrity (IBI) for plants;
– IBI for invertebrates;
– IBI for fish;
– IBI for birds;
– Aquatic vertebrates;
– Benthic organisms: (meio-, macro-, phyto- and zoobenthos);
– Floristic assessment;
– Marsh bird community, etc.

*The ecological status* of surface waters is associated with phytoplankton and zooplankton diversity. One of the main problems of surface waters, primarily of shallow lakes, is eutrophication world-wide. The EU Water Framework Directive (WFD, 2000) specifies phytoplankton to be used in the assessment of the ecological status of surface waters. Phytoplankton indices such as the *Carlson Index* (the most commonly used trophic index) can characterize the trophic status of surface waters by measuring the algal biomass. It is only applicable when suspended solid and rooted plants are in low concentration in the water.

Zooplankton communities are also suitable for ecological assessment of lakes as they are an important component of the pelagic food web. Caroni and Irvine (2010) found that zooplankton can be a good indicator for acidity and overload by nutrients. Acidic and oligotrophic lakes were typified by a high relative abundance of cladocerans, but with some key taxa groups absent from the most acidified lakes. In spite of their ability to indicate anthropogenic impact, monitoring of zooplanktons is not required by the Water Framework Directive (WFD, 2000).

Macrozoobenthos is a characteristic group of animals in sediment. They are defined as invertebrate bottom fauna living on or in the bottom. The size of these animals vary within a wide range: the "macro" fraction is retained on a sieve with a mesh size of 1 mm × 1 mm. The group of smaller animals that pass through such a sieve are called meiozoobenthos. Macrobenthic animals live in the sediment environment that is affected by eutrophication, pollution, fisheries, or, in general, by intensive influx of organic matter. Contaminants, mainly the persistent organic and inorganic ones, accumulate in the sediment and are in close interaction with the sediment-dwelling organisms. The aquatic habitats (lakes and rivers) have characteristic benthic fauna, which makes possible the monitoring and the comparison of the time-points of time series.

Indicators of ecological conditions other than species diversity may come from lower levels than ecosystem or community, i.e. from population, organism, organ, physiological, histological, biochemical or enzymatic as well as molecular levels including DNA, RNA, proteins or other type of molecules (sugars, lipids, etc.).

A genomic screening tool for identifying gene expression patterns may indicate the effect of contaminants on the activation of genes. For example the vitellogenin gene expression in male fathead minnow is an indicator of exposure to endocrine disrupting chemicals (EDCs, see in Chapter 2 in Volume 1) in an aquatic environment.

Many biochemical processes and biomolecules indicate pollution, stress and anthropogenic impact on the water ecosystem: DNA damage, stress proteins such as acetylcholine esterase, adrenaline, or oxidizers; detoxifying molecules such as antioxidant enzymes, bile metabolites, methallothioneins or immune supressors. Measuring acetylcholine esterase inhibition in river snail (*Sinotaia ingallsiana*) was useful in determining pesticide contamination. The high level of adrenaline in fish or clam as a result of the effect of chronic chemical or other stress can be an ecological indicator in aquatic habitats.

Bacteria, other micro-size organisms and planktonic organisms are relatively easy to sample and analyze, and additional stress due to catching, transporting and investigating does not influence the result as in the case of higher animals (fish or birds).

*Remote sensing* technologies used in fresh-water biodiversity monitoring can characterize aquatic habitats and special inhabitants and enable the identification of the size of water bodies, land uses around watersheds; they can measure the changes in water level, temperature, water color and density. The application of remote sensing and GIS together in monitoring water quality parameters such as suspended matter, phytoplankton, turbidity, and dissolved organic matter is especially useful. The combination with hyperspectral analysis enables the selective monitoring of any characteristic components of the aquatic environment.

Remote sensing will be a key tool in identifying priorities in water body management, e.g. conservation or risk reduction. Based on the remotely sensed data, predictive modeling of fresh-water systems can identify the most susceptible sites that need intervention or protection. US EPA carried out the "Development of Water Quality Indicators Using Remote Sensing" project (RS, 2013) to develop a library of water quality indicators that can be measured remotely.

*Management and legislation* utilized the results of aquatic toxicology very soon by creating effect-based quality criteria and quality objectives as well as monitoring and characterizing the ecological status of surface waters. The introduction of the European Water Framework Directive in 2000 (WFD, 2000) placed great emphasis on the quality and uniform management of surface waters.

Measuring water quality using the integrated approach i.e. applying chemical analysis, ecological assessment and environmental toxicity testing methods in parallel made it possible to draw a detailed picture on ecosystem health and water quality. Direct testing of adverse effects instead of predictions based on chemical models made possible to measure the effect of mixed contaminants and the results of their interactions in water. Sediments, the formerly underestimated compartment of surface waters, became one of the main risk components.

Developing DNA techniques to the level of being able to describe the metagenome of a community created a new possibility for characterizing and monitoring the whole ecosystem instead of single species.

Innovative technologies have been developed for testing adverse effects and monitoring ecosystem characteristics and health:

– Early warning by the most sensitive species approach;
– Models for simulating aquatic ecosystems, e.g. micro- and mesocosms;
– Dynamic testing of waters and water–sediment systems;
– Measuring relative sensitivity and sensitivity distribution of species;
– Utilization of specific biological or biochemical indicators for chemical substances;
– Indicators for biological pollutants such as invasive species or pests.

There are some fields where the trend of new developments can be clearly identified. Examples are: testing the effects of endocrine disruptors in water, using DNA techniques and other molecular tools, e.g. the -omics (genomics, proteomics, metabolomics, etc.) and other new indicators for measuring and monitoring stress and multiple stressors. The aggregation and utilization of all available information from different sources in an efficient and integrated way is becoming increasingly important. Knowledge gained from case studies enhances our understanding of nature and the mechanisms of the effect of chemicals and other relationships between pollutants and aquatic deterioration. For example, regional and watershed-scale environmental

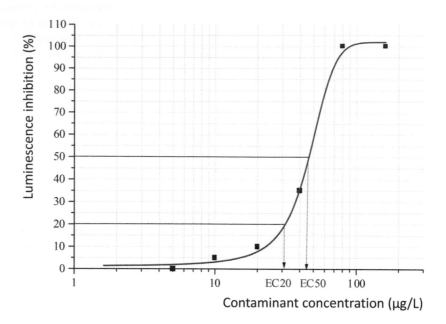

*Figure 4.3* Concentration–response curve: the main tool for bioassay evaluation.

management will benefit from mapped information on ecosystem quality based on remote sensing and evaluated in correlation with the map of chemicals.

In spite of the impressive developments outlined above, there are still many problems to be solved in the near future. One of them is the extrapolation of uncertainty from lab tests or even from micro- or mesocosm tests to the real environment. Another not fully solved problem is the integration of the ecological assessment results (e.g. diversity indices), into the risk assessment and risk management procedure.

*Ecotoxicity* – according to the common definition – involves the identification of chemical hazards to the environment. US EPA definition (US EPA, 2007) is more concrete: *ecotoxicity studies* measure the effects of chemicals on fish, wildlife, plants, and other wild organisms as well as the complex community of wild organisms, the whole of the ecosystem.

*Ecotoxicity testing* is another branch of the animal, plant or microorganism testing methods to determine whether a chemical substance or an environmental sample has an adverse effect. It uses laboratory bioassays (Figure 4.3), microcosms or mesocosms as test designs and may select the end points starting from molecular level indicators to the lethality of the test organism, or diversity of the community in the micro- or mesocosm. As an example, Figure 4.3 shows the effect of a contaminant with growing concentration on the inhibiton of *Vibrio fischeri* luminescence.

Ecological assessment and ecotoxicity testing methodologies have the same conclusion: one has to extrapolate from the test results to the real environment and this information must be used for decision making in the course of the management of the environment.

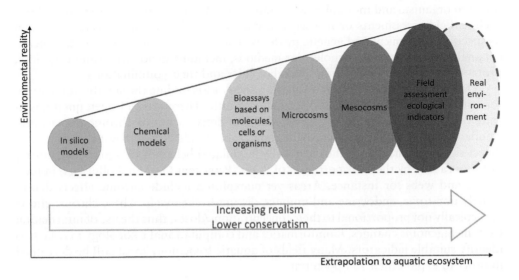

*Figure 4.4* Modeling adverse effects on the aquatic ecosystem.

One can extrapolate to the environment using chemical or mathematical models, but it is important to know that they are much farther from reality than biological or ecological models as indicated in Figure 4.4. The application of mathematical and chemical models is reasonable if their good statistics and repeatability compensate for the lack of environmental reality or if the goal of the assessment is generic, e.g. European legislation, or watershed-scale management of chemicals, and when heterogeneity, environmental parameters and interactions are handled as generic characteristics, which can be simulated by standardized test methods.

The biological and ecological methods are highly diverse. Test species can be selected to represent the average, 95% of the ecosystem members, the most sensitive species, those less sensitive than the average, or those having specific sensitivity to certain contaminants, etc. The first step is always to formulate the concept of risk assessment and to test the adverse effects. The conceptual model of the case specifies the goals, the spatial and time frames, the hot spots, transportation pathways and the water-use-specific receptors. After establishing a detailed picture of the case, the best fitting test type, set-up, exposure system, duration and test organisms to characterize aquatic risks can be selected. There is a wide variety of test organisms and abundant information on their capabilities. A complete "database" is not available today, but one can find a large number of publications on species sensitivity and the so-called Species Sensitivity Distribution (SSD), which is a relatively new area of environmental toxicology. Test organisms of different sensitivity are used for risk assessment, establishing safe contaminant concentration in the environment, screening pollution and ranking surface waters and pollutants. Appropriate aquatic indicator organisms can be applied to any environmental changes, from deterioration due to global warming, through eutrophication, to adverse effects emerging at local scale due to individual contaminants or hazardous biological agents. These indicators vary from community level down to organism and molecular levels and are used in traditional ecological methods such as field assessments or mesocosms, microcosms, laboratory bioassays, as well as specific biochemical and genetic methods. The latter ones apply traditional chemical analyses and innovative molecular methods, including genomics, transcriptomics, proteomics, lipomics or any other metabolomics and their combinations.

One of the main challenges of aquatic toxicology is finding the suitable test battery for emerging problems in aquatic ecosystem health. There are still open questions in spite of a number of model organisms that can detect acute or chronic toxicity and standardized methods that can provide comparable results. It is still not known how to detect signals of changes in the chemically modulated behavior of aquatic organisms, or those caused by the transportation of POPs (Persistent Organic Pollutants) in food chains and webs for instance. Areas yet unexplored include chronic effects due to acutely non-toxic, endocrine and immune disruptor chemicals. These chronic effects are typically not proportional to their concentrations/doses, thus the use of information from metagenome changes, bioinformatics and computational toxicology may help to identify suitable indicators. Many fields of aquatic toxicology must still be developed in order to exploit their full potential.

## 3 SOME COMMONLY USED AQUATIC TEST ORGANISMS FOR TESTING ADVERSE EFFECTS

Hundreds of aquatic species are used in standardized and non-formalized test methods. Most of the OECD test guidelines describe one methodology, but give a list of the test organism options, requiring the user to choose. The ISO tests use bacteria, algae, water plants, invertebrates living in water and sediment, macrozoobenthos (sediment-dwelling organisms) of the sediment and a great number of fish species. From among the standardized tests, many organisms are used in tests as biomarkers or indicator species, for example the members of the phyto- and zooplankton, special water plants, clams, crustaceans and fishes. The adequate test organism should be selected according to the aim of the test: i) specific effects require selective indicator organisms, for example those extremely sensitive to Zn toxicity or photosynthesis inhibition, or able to accumulate Hg, etc.; ii) the general quality of water can be characterized by the aquatic/benthic species diversity or by test organisms with a wide-spectral sensitivity, i.e. sensitive to most of the toxic metals and also to organic pollutants. In the following, we will list the most popular aquatic test organisms without attempting to be comprehensive.

### 3.1 Microorganisms: bacteria, algae and protozoa

#### Bacteria

Bacteria are easy to handle and rapidly growing organisms, which enable test methods of good reproducibility. They are considered to be able to represent their taxa in a wide context, but they are also used as toxicological "reagents" independent of their environmental role. Natural and laboratory strains, mutants, and genetically modified bacteria are also used to test toxicity both of chemicals and environmental samples.

*Figure 4.5*   Liquid culture of *Vibrio fischeri* in daylight and darkness producing luminescent light emission.

Laboratory strains, e.g. *Vibrio fischeri* or *Azomonas agilis*, are used as indicators to detect adverse effects. Natural strains are also used as endangered species or sensitive key actors of an ecosystem (responsible for nitrogen fixation or biodegradation of oil spills and xenobiotics) or as the causes of a hazard (e.g. Salmonella or other pathogens). Growth rate, metabolic activity and products can be measured as end points.

   *Vibrio fischeri* is a Gram-negative rod-shaped, heterotrophic (saprotrophic) bacterium found globally in marine environments. It moves by means of flagella and has bioluminescent properties, so it can emit bluish-green light (490 nm) due to a chemical reaction between riboflavin-5′-phosphate, luciferin and molecular oxygen. Figure 4.5 shows the luminescent light emission from a liquid culture in daylight and darkness. An enzyme called luciferase catalyzes the reaction. *Vibrio fischeri* is predominantly found in symbiosis with various marine animals. It is a key research organism for examination of microbial bioluminescence, quorum sensing, and bacterial-animal symbiosis. In environmental toxicity testing it is a generally used test organism, based on the correlation between light emission and toxic chemicals present. The test using *Vibrio fischeri* is standardized (ISO 11348, 2007) and this test organism – due to its wide-range sensitivity and easy laboratory use – is applied not only as a marine species, but also as a generic test organism for any environmental sample: water, wastewater, leachate, soil, sediment, solid waste, etc.

   *Salmonella typhimurium* is a Gram-negative, facultative anaerobic bacterium of the family of Enterobacteriaceae, genus *Salmonella*. It is used in the Ames mutagenicity assay, a short-term bacterial reverse mutation assay using histidine auxotroph *Salmonella*, which carries a mutation in the genes of histidine synthesis.

   *Azomonas agilis* is a Gram-negative bacteria, motile with peritrichous flagella, found in water and wastewater, capable of fixing atmospheric nitrogen. It belongs to the phylum Proteobacteria, family Pseudomonadaceae. It is a non-selective test organism, used in laboratory bioassays. Respiration can be followed by $O_2$ consumption, $CO_2$ production, enzyme activity or ATP production. Both chemical substances and environmental samples can be tested, applicable for polluted water and wastewater toxicity assessment and for monitoring water treatment technologies. Toxic inhibition of the dehydrogenase enzyme is caused by the deterioration of the electron transport chain on the effects of contaminants. Decreased dehydrogenase activity is indicated in the test by applying an alternative electron acceptor, which turns into red when dehydrogenase is active but remains colorless when it is not.

*Other bacteria*, first of all, *Escherichia coli* and the coliforms as well as other environmental strains from the genus *Pseudomonas, Flavobacteria, Gammaproteobacteria, Bacillus*, etc. are often used as test organisms in laboratory bioassays.

Adverse effects caused by bacteria also represent an environmental problem: in this case one can detect the hazardous bacterial strain itself or specific molecules indicating their presence. Amongst these hazardous bacteria there are human, animal and plant pathogens, e.g. human pathogenic coliforms (*Clostridium, Staphylococcus*, etc.), bacteria causing macroalgal diseases, fish pathogens (*Vibrio harveyi, Pasteurella skyesisn* or *Chryseobacterium species*), etc.

### Phototrophic bacteria and algae

Phototrophic bacteria and algae are mainly used for the indication of toxic chemical substances in water by showing reduced growth rates. Testing growth rate of bacterial and algal strains is described in the OECD test guideline No. 201.

*Synechococcus leopoliensis*, the unicellular cyanobacterium, is a member of the class Cyanophyceae and the family Synechococcaceae, used in the detection of phytotoxic pollutants such as herbicides and toxic metal cations. Found in oceans as well as in fresh-water reservoirs and lakes. The test method OECD 201 (2011) of *Fresh-water Algae and Cyanobacteria, Growth Inhibition Test* describes its use in toxicity testing (see also Table 4.1).

*Pseudokirchneriella subcapitata* (formerly known as *Selenastrum capricornutum*) belongs to the algal taxonomical groups of Raphidocelis, specifically Selenastraceae. Thalli (green shoots) unicellular or forming small colonies embedded in highly irregular and structureless mucilage. Cells shape is lunate to sigmoidal, – as shown in Figure 4.6 – pointed at both ends with evenly granulated cell walls. *Raphidocelis* species are planktonic in fresh water, and known only from central Europe. It is one of the most sensitive algal strains, which often produces the lowest toxic concentrations in environmental toxicity testing.

*Scenedesmus quadricauda* and *Desmodesmus subspicatus* (formerly known as *Scenedesmus subspicatus*) are two very similar algal species. They can be single-celled or colonial, forming 2- to 32-celled coenobia, cells arranged linearly. *Scenedesmus* species are members of the class Chlorophyceae and the family Scenedesmaceae. They are planktonic algae, mainly in eutrophic fresh-water ponds and lakes, rarely in brackish water; reported worldwide in all climates.

*Chlorella kessleri* is a member of the class Trebouxiophyceae and the family Chlorellaceae; it is a fresh-water algal species.

*Stichococcus bacillaris* is a member of the class Trebouxiophyceae and the family Prasiolaceae, it is a fresh-water algal species.

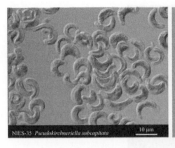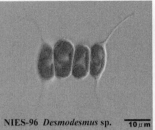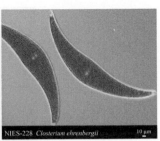

*Figure 4.6* Microscopic view of *Pseudokirchneriella subcapitata* (2014), a *Desmodesmus* species (2014) and *Closterium ehrenbergii* (2014).

*Chlamydomonas reinhardtii* is a member of the *class Chlorophyceae and the family Chlamydomonadaceae. It* is a fresh-water algal species used in the detection of phytotoxic pollutants.

*Closterium ehrenbergii* belongs to Charophyceae. Its sensitivity to nonionic and anionic surfactants enables testing agricultural chemicals and heavy metals in an aquatic environment. In laboratory bioassays morphological effects and ecotoxicity of the chemicals are measured by algal growth and in zygospore inhibition tests using *Closterium ehrenbergii.*

### Protozoa

Protozoa are a diverse group of unicellular eukaryotic animal organisms. Most of them are motile, moving by flagella or cilia or by pseudopodia (amoebas). Their size ranges between 10–50 micrometers. Some of them are easy to handle and grow in the laboratory. They may represent several trophic levels and their response to toxicants can be characterized by animal nature.

*Tetrahymena species* are non-pathogenic free-living ciliate protozoa, belonging to the order Hymenostomatida of phylum Ciliophora. Protozoans are real eukaryotic cells and ubiquitous in the aquatic and terrestrial environment. *Tetrahymena* species used as model organisms in ecotoxicology are *Tetrahymena thermophila* (PROTOXKIT F, 2013) and *Tetrahymena pyriformis.*

Among protozoa, *Tetrahymena pyriformis* is one of the most commonly used model organisms in laboratory research. The body of *Tetrahymena pyriformis* – shown in Figure 4.7 – is generally 50–60 μm long and 30 μm wide and is pear-shaped, a characteristic from which the name of the species is derived. In *Tetrahymena*, the usual cytoplasmic organelles are mitochondria, endoplasmic reticulum, Golgi complexes, ribosome, peroxisome and lysosome. Their structures reflect the physiological state of the cell. *Tetrahymena* species can grow very rapidly to high density due to their short generation time, which allows easy cultivation in suitable laboratory conditions (Sauvant *et al.*, 1999).

Current research indicates the presence of a hormonal system in *Tetrahymena* that includes receptors, hormones and signal transduction pathways as well as hormonal interactions (Csaba *et al.*, 2005).

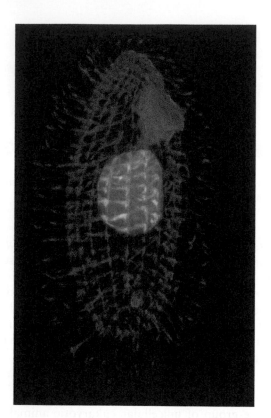

*Figure 4.7  Tetrahymena pyriformis* electronmicroscopic image (Robinson, 2006).

All these properties explain why so many studies have been performed on this organism in physiological, biochemical, pharmacological and toxicological research (Csaba *et al.*, 2005; Leitgib *et al.*, 2007; Mortimer *et al.*, 2010; Novotny *et al.*, 2006; Stefanidou *et al.*, 2008). The test protocol will be discussed in detail in this Chapter (see Section 11).

## 3.2  Fresh-water macroplants

*Lemna minor* or common duckweed (Figure 4.8) has a subcosmopolitan distribution. Its habitats are ponds, ditches, canals and slow parts of rivers and streams. It prefers water with a high nutrient level and often becomes the dominant population. It is a free-floating, monocotyledonous macrophyte with one to four fronds (leaves), each with a single root hanging in the water (OECD 221, 2006).

*Myriophyllum aquaticum* or parrot feather is an emergent, dicotyledonous macrophyte (Figure 4.9). Because of its high regeneration potential after cutting the plant into pieces, a lot of equal sprouts can easily be produced – to create good initial conditions for biotests. The Parrot feather is native to South America.

*Figure 4.8  Lemna minor or common duckweed in the cultivating vessel, under the microscope (10x) and the extracted chlorophyll content from a test series.*

*Figure 4.9  Myriophillum aquaticum (2011), Cabomba caroliniana and Ceratophyllum demersum (2009).*

*Cabomba caroliniana* is an invasive aquatic plant species, commonly used as an aquarium plant because of its delicate appearance. Its widespread occurrence and fast growth make it suitable for testing chemicals and plant inhibitory effects (Figure 4.9).

*Elodea canadensis* or Canadian waterweed is a submergent monocotyledonous macrophyte. It anchors with its roots in the sediment. *Elodea canadensis* can also develop adventitious roots at the whole sprout. It is native to North America, but as an invasive species, today it is distributed over the whole world. *Elodea canadensis* is well known for its rapid growth and its high oxygen production (Figure 4.10).

*Ceratophyllum demersum* or rigid hornwort is a cosmopolitan, submergent, dicotyledonous macrophyte. It likes a high nutrient level and summer water temperatures of 15–30°C. It anchors in the sediment without real roots and develops up to 3-m-long sprouts. It produces a lot of oxygen, so gas bubbles can be seen between its leaves (MESOCOSM, 2013).

## 3.3  Fresh-water invertebrates

### Rotifers

*Brachionus calyciflorus* is a fresh-water zooplankton belonging to the class Monogononta and the order Plioma. Rotifers are ecologically very important members of many aquatic communities. With copepod and cladoceran crustaceans they are the

*Figure 4.10   Elodea canadensis:* photos with growing magnification, the microscopic image shows a 40x magnification.

major constituents of fresh-water zooplankton with turnover rates higher than those of crustaceans. Rotifers such as *Brachionus calyciflorus* are favored test animals in aquatic toxicology because of their sensitivity to most toxicants, e.g. ROTOXKIT F (2013).

*Brachionus plicatilis* is a euryhaline (tolerate a wide range of salinity) rotifer in the family Brachionidae and is possibly the only commercially important rotifer, being raised in the aquaculture industry as food for fish larvae. It has a broad distribution in salt lakes around the world and has become a model system for studies in ecology and evolution. Marine rotifer *Brachionus plicatilis* is used in the ROTOXKIT M bioassay (2013).

### Crustacea

*Daphnia species* or water fleas are small, planktonic crustaceans between 0.2 and 5 mm in length (Figure 4.11). *Daphnia* are members of the order Cladocera, and belong to the several small aquatic crustaceans such as *Daphnia magna* (Figure 4.11), *Daphnia pulex, Daphnia pulicaria, Ceriodaphnia dubia,* water fleas are counted among the *Crustacea* and live in biotopes like big lakes and in small ponds with standing water areas. These species are used in many aquatic toxicology test systems. *Daphnia* is an indicator species of ecosystem health exhibiting consistent responses to toxins according to OECD 202 (2004) and OECD 211 (2012).

*Heterocypris incongruens* or seed shrimp is a fresh-water ostracod, living in water ponds and lakes. Ostracoda is a class of Crustacea, also known as seed shrimp because of their morphology. Ostracods are small crustaceans, typically around 1 millimeter in size; their bodies are flattened from side to side and protected by a bivalve-like, chitinous or calcareous valve or shell (Figure 4.12). A distinction is made between the valve (hard parts) and the body with its appendages (soft parts). It is a sensitive aquatic test organism.

*Thamnocephalus platyurus* is a fresh-water crustacean belonging to the class Branchiopoda and family Curculionoidea. Used in ecotoxicity tests, e.g. THAMNO-TOXKIT F (2013).

*Amphipoda* or amphipods are important components of fresh-water ecosystems, an order of malacostracan crustaceans. They have laterally compressed bodies and no

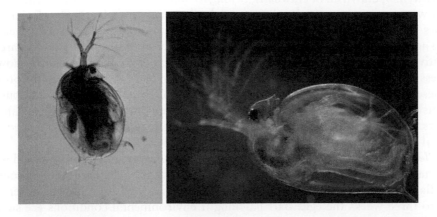

Figure 4.11  Daphnia magna or water flea, the most popular aquatic test organism.

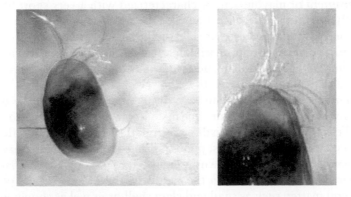

Figure 4.12  Heterocypris incongruens lives in a tiny shell.

carapace. Their name refers to the different form and size of appendages, in contrast to isopods. They are very often used in ecotoxicology, particularly the fresh-water amphipods *Gammarus pulex, Gammarus lacustris, Gammarus fasciatu* and *Hyalella azteca. Gammaridea* is the largest suborder, 5500 of the 7000 amphipods belong to it. The sensitivity to chemical substances shows a wide range, depending on the applied species, subspecies and the age of the test organism.

*Copepods* are a group of small crustaceans found in the sea and nearly every fresh-water habitat. Some of the species are planktonic (drifting in sea waters), others benthic (living in bottom sediment). Planktonic copepods are important to food chain and aquatic ecosystem health. They dominate zooplankton functioning as feed for small fish, whales, seabirds and other crustaceans in the ocean and in fresh waters. The calanoid copepod *Eurytemora herdmani* is applied as a test organism in short-term toxicity testing of effluents and toxic chemicals in waters.

*Crayfish*, crawfish, or crawdads, belong to orconetes, family Astacoidea. They are fresh-water crustaceans, feed on living and dead animals and plants. They live in water bodies that do not freeze to the bottom – mainly in brooks and streams – sheltering them against predators. They breathe through feather-like gills. Most crayfish are very sensitive to chemical contaminants in polluted water. Some invasive species however are less sensitive to toxic chemical substances and crayfish diseases. Due to the presence of these invasive species such as American crayfish (*Orconetes limosus*) and Louisiana crayfish (*Procambarus clarkii*), native species declined dramatically in European fresh waters. *Pacifastacus* species such as *Pacifastacus leniusculus* and *Cambarus* species are typical crayfish in European fresh waters.

*Plecoptera* or stoneflies: their name comes from Greek and means braided wings. This name properly characterizes their two pairs of wings which are membranous and fold flat over the back. They deposit hundreds or thousands of eggs into water. After two to three weeks the eggs start hatching if environmental conditions are suitable. Stoneflies are generally not good fliers. Most of them are seasonally aquatic, but some wingless species are exclusively aquatic from birth to death. Some species in Europe such as *Nemoura cinerea* and *Nemurella pictetii* are specific indicators of ecosystem health; some others can be used as laboratory test organisms, e.g. *Pteronarcys* species.

*Ephemeroptera* or mayflies can be characterized with a very short life span of the adults. Mayfly is a fresh-water insect whose immature stage (called naiad or nymph) usually lasts one year in water. Eggs are laid on the surface of lakes or streams, and sink to the bottom. The naiads live primarily in streams under rocks, decaying vegetation, or in the sediment. They feed on algae or diatoms. Some common species are *Baetis* sp., *Ephemerella* sp., and *Hexagenia limbata*.

*Trichoptera* or caddisflies are small moth-like insects, having two pairs of hairy membranous wings. The Greek name means hairy wing. They are also called sedge-flies or rail-flies. The larvae of caddisflies are aquatic and they live in different types of surface waters: streams, rivers, lakes, ponds, spring seeps, and temporary waters. The species *Dicosmoecus avripes* is shown in Figure 4.13. Caddisflies prepare handsome cases for their larvae. Depending on the caddisfly, these cases can be of a variety of plant or mineral matter, such as snail and clam shell or gravel as shown in the picture.

*Chironomidae* or midges from the order Diptera, are little flies, similar to mosquitoes, distributed globally. They are also called non-biting midges. *Chironomus* midges are well known from early observed polytene (giant) chromosomes in larval salina gland. Chironomids spend their larval stages in aquatic or semiaquatic habitats. The approx. 1 cm long larvae are often red colored (Figure 4.13). They form an important fraction of the macro zoobenthos of most fresh-water ecosystems. Their presence is associated with degraded or low-biodiversity ecosystems, some very adaptive species are dominant in polluted waters. Eleven subfamilies and several hundreds of species belong to the chironomids.

The total number and relative abundance of different taxa collected from water, belonging to the orders Ephemeroptera (mayflies), Plecoptera (stoneflies), and Trichoptera (caddisflies) and Chironomidae (midges) are good indicators of aquatic ecosystem health.

*Nematoda* or roundworms are probably the most diverse of all animals. Over 28,000 species have been described, but the estimate for the total species number is one million. They are microscopic in size. *Caenorhabditis elegans* has become a

*Figure 4.13* *Limnephilidae* (Trichoptera) larvae and cases (Hodges, 2013) and *Chirnomidae* (Diptera) larvae (Penrose, 2013).

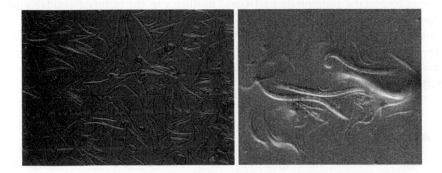

*Figure 4.14* The nematode *Panagrellus redivivus*.

laboratory model organism as it was the first multi-cellular organism to have its genome completely sequenced. Nematodes are used in laboratory assays (lethality, motility) to measure inhibitory effects of chemicals. *Panagrellus redivivus* is shown in Figure 4.14.

*Planaria* or flatworms are common to many parts of the world, living in both saltwater and fresh-water ponds and rivers. The order of Tricladida plays an important role in aquatic ecosystems and is used as a bioindicator. The 1–20 mm animals are able to regenerate their body and re-grow split or missing parts. The famous experiment in which a reflex was developed by an electric shock combined with light was carried out in 1955 on a triclad. Later on it was possible to trigger the reaction when only light was applied. Pieces of the animal exhibited the same knowledge after regeneration. Small pieces also have transferred the knowledge into animals which were fed by parts of the trained ones.

Another flatworm, *Dugesia tigrina* (Platyhelminthes, Turbellaria), is used as a model for the study of regeneration (cell re-growth) in people and animals.

*Oligochaeta* or worms, is a subclass of Annelida, which is made up of many types of aquatic and terrestrial worms. It contains fresh-water tubificids, pot worms and ice worms (Enchytraeidae), blackworms (Lumbriculidae) and several marine worms living in the pore water of sediments.

*Snails* (Mollusca, Gastropoda). Many kinds of snails can be found in fresh waters, in spite of the fact that marine species constitute the majority of snails. Snails may have lungs or gills but this fact does not constrain them to living in water with lungs or with gills as a terrestrial snail.

The purple snail or *Plicopurpura pansa*, a predator snail, naturally inhabits the high intertidal of rocky shores exposed to the open sea with high impact waves, used also as a laboratory snail. *Biomphalaria glabrata* is a species of air-breathing fresh-water snail, intermediate host for the trematode *Schistosoma mansoni*, a human-pathogen. As such, it is the subject of many laboratory research and tests. The river snail *Sinotaia ingallsiana* is used for indicating pesticide contamination of waters via acetyl-choline esterase inhibition in the snails. Some other species e.g. *Physa integra*, *Physa heterostropha*, and *Amnicola limosa* are frequently subjects of laboratory research and tests on aquatic ecosystem and environmental effects.

*Physa integra* is a small, left-handed or sinistral, air-breathing fresh-water snail, it belongs to aquatic pulmonate gastropod molluscs in the family Physidae and feeds on algae, diatoms and detritus.

*Amnicola limosa* populations are typically found in lentic environments, in slow-moving rivers and swamps of the coastal plain, often on woody debris. They belong to the class Gastropoda and genus Amnicola and appear to be sensitive to water chemistry, especially hardness and pH.

## 3.4 Aquatic vertebrates

Fish, when considering environmental toxicity testing, represent the highest aquatic trophic level. Arguments for the application of fish as priority aquatic test organism is not only their metabolic capacities, e.g. in biotransformation competence for certain chemical classes, but also the high pollution levels and frequencies of chemical spills, fish frequently being the target. There is a wide range of selection in recommended fish species which can be used for acute, prolonged and chronic toxicity and reprotoxicity tests. Fish can be used in an early developmental stage such as embryo, hatchlings (sac fry), juvenile, or as adult fish. The most popular fish species are introduced below.

*Onchorhynchus kisutch*, coho salmon or silver salmon is a species of anadromous fish in the salmon family. The traditional range of the coho salmon runs from both sides of the North Pacific Ocean, from Hokkaidô, Japan and eastern Russian, around the Bering Sea to mainland Alaska, and south all the way to Monterey Bay, California.

*Onchorhynchus mykiss*, rainbow trout, salmon trout has been introduced for food or sport to at least 45 countries, and every continent except Antarctica. In some locations such as southern Europe, Australia and South America, they have negatively impacted upland native fish species, either by eating them, outcompeting them, transmitting contagious diseases, or hybridization with closely-related species and subspecies that are native to western North America. They belong to the order Salmoniformes and genus *Oncorhynchus* (Figure 4.15).

*Salvelinus fontinalis*, brook trout, speckled trout, squaretail is a species of fish in the salmon family of order Salmoniformes. The brook trout is native to small streams,

*Figure 4.15  Onchorhynchus mykiss,* rainbow trout (2013); *Danio rerio,* zebrafish and *Lepomis macrochirus,* bluegill (2004).

creeks, lakes, and spring ponds. Some brook trout are anadromous. *Salvelinus fontinalis* prefers clear waters of high purity and a narrow pH range in lakes, rivers, and streams, being sensitive to poor oxygenation, pollution, and changes in pH caused by environmental effects such as acid rain.

*Carassius auratus,* goldfish, is a fresh-water fish in the family Cyprinidae and order Cypriniformes. It was one of the earliest fish to be domesticated, and is one of the most commonly kept aquarium fish. A relatively small member of the carp, the goldfish is a domesticated version of a less-colorful carp native to East Asia.

*Pimephales promelas,* fathead minnow is a species of temperate fresh-water fish belonging to the *Pimephales* genus of the Cyprinid family. The natural geographic range extends throughout much of North America, from central Canada south along the Rockies to Texas, and east to Virginia and the Northeastern United States. The fathead is quite tolerant to turbid, low-oxygenated water, and can be found in muddy ponds and streams that might otherwise be inhospitable to other species of fish. It can also be found in small rivers.

*Ictalurus punctatus,* channel catfish is North America's most numerous catfish species belonging to the *Ictalurus* genus of the Ictaluridae family. Channel catfish are native to the Nearctic[1], being well distributed in Lower Canada and the eastern and northern United States, as well as parts of northern Mexico. They thrive in small and large rivers, reservoirs, natural lakes, and ponds. They are cavity nesters, meaning they lay their eggs in crevices, hollows, or debris, in order to protect them from swift currents.

*Lepomis macrochirus,* bluegill, bream, brim, copper nose is a species of fresh-water fish belonging to the *Lepomis* genus of the Centrarchidae family. It is a member of the sunfish family Centrarchidae of the order Perciformes. It is native to a wide area of North America, from Québec to northern Mexico (Figure 4.15).

*Lepomis cyanellus,* green sunfish is a species of fresh-water fish in the family Centrarchidae of the order Perciformes. It is native to a wide area of North America east of the Rocky Mountains, from the Hudson Bay basin in Canada, to the Gulf Coast in the United States, and northern Mexico. The species prefers vegetated areas

---

[1] The Nearctic ecozone covers North America, Greenland and the highlands of Mexico.

in sluggish backwaters, lakes, and ponds with gravel, sand, or bedrock bottoms. Its diet may include insects, zooplankton, and other small invertebrates.

*Danio rerio*, zebrafish is a tropical fresh-water fish belonging to the minnow family (Cyprinidae) of the order Cypriniformes. It is an important vertebrate model organism in scientific research (Figure 4.15). It commonly inhabits streams, canals, ditches, ponds, and slow-moving to stagnant water bodies, including rice fields. *Danio rerio* is a common and useful model organism for studies of vertebrate development and gene function. It has a fully-sequenced genome.

The requirement of REACH regulation and of other legislations on chemicals (pesticides, biocides, etc.) to submit ecotoxicity studies to regulatory authorities to support the registration and/or approval of the products increases the number of laboratory test organisms for toxicity testing (Test methods for REACH, 2008). Both ethical issues and the cost of testing motivates research and development to find new *in silico* and *in vitro* molecular and cell-based techniques replacing and reducing the number of fish in testing chemicals. For waters of environmental origin, living animals are still the best models. The non-animal alternatives will be discussed in Chapter 1 in Volume 3.

## 3.5  Sediment-dwelling organisms

Species of the macrozoobenthos and their diversity are suitable to characterize the environmental health of the sediment ecosystem in surface waters.

*Macrozoobenthos* are defined as invertebrate bottom fauna living on or in the bottom, which are retained on a 1 mm × 1 mm mesh sieve. Several hundreds of species should be identified from a carefully taken sample. Their quantitative and qualitative parameters such as biomass, number of species or the relative abundance of the species are determined to get information on the ecological status of a river, lake or sea. The results of such an assessment are usually given in the form of diversity indices such as species richness, species evenness, Simpson's diversity index, Shannon index or some indices which measure the lack of diversity. All these output results of ecological assessment of macrozoobenthos can characterize spatial and seasonal changes, biotic or abiotic adverse effects of environmental origin and the long-term trends when regular biomonitoring is applied. The most important fresh-water and marine organism taxa to be evaluated when macrozoobenthos are assessed are listed below.

*Mollusca*: molluscs or mollusks are a large phylum of invertebrates, including Gastropoda and Bivalvia. Cephalopod molluscs such as squid, cuttlefish and octopus also belong here.

*Bivalvia*: bivalves have a shell consisting of two asymmetrically rounded halves called valves that are mirror images of each other, joined at one edge by a flexible ligament called the hinge (Figure 4.16). Their food is taken up by filter-feeding, differing from most of the molluscs.

*Gastropoda*: snails and slugs, the major part of *mollusca*, univalves, showing extraordinary diversification in habitats. They are also common individual test organisms in toxicity testing.

*Polychaeta*: polychaetes or bristle worms are a class of annelid worms, having a pair of fleshy protrusions called parapodia that bear many bristles, called chaetae on each body segment.

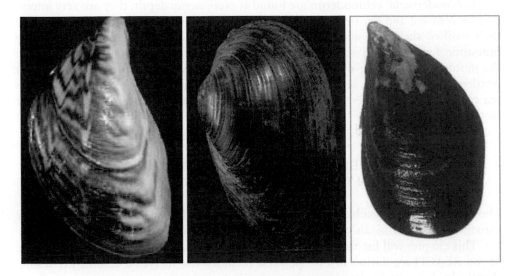

*Figure 4.16* Bivalve molluscs *Dreissenia polymorpha, Unio pictorum* and *Mytilus edulis.*

*Figure 4.17* Crab, shrimp, crayfish, and lobster.

*Crustacea*: group of arthropods, including crabs, lobsters, crayfish, shrimp, krill and barnacles. Classes of *Branchiopoda, Remipedia, Cephalocarida, Maxillopoda, Ostracoda* and *Malacostraca* belong to crustaceans. They are also used as individual test organisms in environmental toxicity tests (Figure 4.17).

*Amphipoda*: is an order of crustaceans, having appendages of different size and form. It includes the suborders of Gammaridea, Caprellidea or Corophiidea and Hyperiidea. *Corophium volutator*, one of the estuarine mud shrimps is a small (10 millimeters long) amphipod from the family Corophiidae. It inhabits the upper layers of sand on the coasts of the Netherlands, Germany, the United Kingdom and France. It has been applied as sediment test-organism in a US EPA standardized method since 1998, but its importance increased in the last years as a sediment-dwelling organism testing chronic toxicity of contaminants associated with whole sediments (Scarlett *et al.*, 2007).

*Isopoda*: isopods are an order of crustaceans, including well-known animals such as woodlice and pill bugs. Isopods lack an obvious carapace; it is reduced to a cephalic shield covering only the head. Respiration is carried out by specialized gill-like pleopods.

*Echinodermata*: echinoderms are found at every ocean depth, they are very important taxa of sea and oceanic environment as part of the community and also due to their ossified skeletons, which are major building blocks of the abiotic sedimentary limestone formations. Two main subdivisions of echinoderms are mentioned here: the motile Eleutherozoa, which encompasses Asteroidea or starfish, Ophiuroidea or brittle stars, Echinoidea, sea urchins and sand dollars and Holothuroidea, known as sea cucumbers and the sessile Pelmatazoa which consists of the crinoids, sea lilies or feather-stars.

## 4   MEASURING ADVERSE EFFECTS OF CHEMICAL SUBSTANCES ON THE AQUATIC ECOSYSTEM

The effects of chemical substances are studied to provide information for the regulatory management of chemicals or for the environmental management of water systems.

This chapter will list the OECD (2014) guidelines which primarily serve the aims of the REACH regulation in Europe (Test methods for REACH, 2008), i.e. testing chemical substances for legislative purposes. The most important requirement of regulatory toxicology is comparability: comparison of measurement results to each other, to references and to thresholds and limits; comparison between alternatives or additional options. It is possible only by using standard test organisms and standardized test methods ensuring uniform application and the comparability of the results.

The same test organisms and test methods can be used with or without modifications to test environmental water, wastewater or diluted liquid waste samples. When the test result supports the local assessment of waters and the risk-based decision on the necessary measures, the test should reflect the local situation and simulate the real environment to obtain as close an answer as possible to the response of the real ecosystem. Many of the test methods mentioned below are standardized such as the ISO Standards (2014) for environmental water samples, but some are developed for site- and problem-specific uses. Tests methods for environmental waters and wastewaters serve water monitoring and are important management tools in protecting aquatic ecosystems against toxic effluents.

Several organizations are active in developing, collecting and publishing standardized test methods to ensure harmonization world-wide. The following organizations play a key role in the field of aquatic toxicity testing:

- American Public Health Association – APHA (2014)
- American Society for Testing and Materials – ASTM International (2014)
- American Water Works Association – AWWA (2014)
- CEN Water Analyses. Published Standards – CEN Standards (2014)
- Ecological Assessment Methods Database – EAMD (2014)
- Ecotoxicity database maintained by US EPA – Ecotox (2014)
- Environment Canada – EC (2014)
- European Committee for Standardization – (CEN, 2014)
- Non-animal Methods for Toxicity Testing – AltTox (2014)
- Organization for Economic Co-operation and Development – OECD (2014)
- Society of Environmental Toxicology and Chemistry – SETAC (2014)

- Standard Methods for the Examination of Water and Wastewater, 2014
- United States Environmental Protection Agency – US EPA (2014)
- Water Bodies in Europe. Integrative Systems to Assess Ecological Status and Recovery. Methods for assessing and restoring aquatic ecosystems – WIESER (2014)
- Water Pollution Control Federation – WPCF (2014).

There are patented but not standardized tests, test kits and apparatus that have reached a certain level of statistical quality, but not used for regulatory purposes.

The trophic levels of the selected test organism are important both in the case of testing chemicals and environmental samples or when testing the environment directly, because the ecosystem members from different trophic levels together can characterize the state of the environment, the risk to food chains and to ecosystem health. Organisms from a minimum of three trophic levels should be tested when the results are intended to be used for risk assessment.

Test types, exposure scenarios and classification according to environmental relevance are important characteristics of aquatic tests (see also Chapter 1).

- Acute tests are short-term exposure tests with a duration of hours or days, generally shorter than the test organism's life span. The chemical substance to be tested is used in a relatively high concentration to be able to provoke an immediate positive response. Results are reported in terms of $EC_{50}$ or $EC_{20}$.
- Chronic tests are long-term tests lasting for days, weeks or months, including a minimum of two generations of the test organism. Continuous low concentrations are applied and NOEC (no observed effects concentration) or LOEC (lowest observed effects concentration) are the test's end points.
- Early life stage tests use exposures for less than a complete reproductive life cycle and include exposure during early, sensitive life stages of an organism. These are named as embryo-, larval, or egg-fry tests.
- Biodegradation tests measure the existence, type and rate of biodegradation of the chemical substance in water. Decrease in the concentration of the chemical substance, biodegradative enzymes/activity of the test organisms and the concentration of the metabolites and end products can be measured. Biodegradation half-life is calculated from measured data.
- Bioaccumulation tests measure the concentration of high-$K_{ow}$ chemicals in the test organism's whole body or in its fatty tissue. Test end point is the bioconcentration factor (BCF), which is the ratio of the concentration of chemical substance accumulated in the tissue to the concentration in the water.
- Effluent toxicity is characterized by so-called short-term chronic tests used to monitor the quality of municipal wastewater treatment plant effluent with the goal to ensure that the wastewater is not chronically toxic.
- Freshwater and saltwater tests apply different test organisms, test methods and standards.
- Sediment tests have priority when the contaminants have a potential to accumulate in sediment. The Sediment Quality Triad (SQT) involves an integrated sediment test using chemical and toxicological methods and field assessment.

–    Aquatic microcosm tests are multispecies tests truly simulating generic exposure scenarios or specific cases. The main differences are the size and, as a consequence, geometry and species diversity.

–    Mesocosm tests are ecosystems smaller than or a part of the real ecosystem. They may be natural or artificial systems, which contain water and sediment. Sediment can be fully natural or under fully controlled environmental conditions.

The exposure scenario in an aquatic study can be:

–    A static test where the organism is exposed to still water containing the contaminant;

–    An arrangement with recirculation where the test solution is pumped through a device to maintain water quality without reducing the concentration of the toxicant;

–    A renewal test where the test solution is periodically renewed by transferring the test organism to a fresh test medium;

–    A flow-through test exposes the organism to the toxicant by directing a flow into the test chambers where water and contaminant flux is controlled.


## 5   SOME COMMONLY USED AQUATIC TEST METHODS

The test methods standardized by OECD, ISO, US EPA and other standardization bodies are collected and listed in the following tables. These methods are established for the testing of dissolved chemical substances, surface waters, wastewaters, leachate and extracts or other waters of agricultural or industrial origin. They can be used easily or in a modified form for the testing of special liquid/solution samples. The aims of testing the adverse effect both of the chemicals dissolved in waters and of the waters for different purposes may be fulfilled by the same test design and test organism.


### 5.1   OECD guidelines for testing chemicals in aquatic environment: water, sediment, wastewater

OECD recommends validated test methods for characterizing the fate and behavior as well as the effect of chemical substances for regulatory purposes. These are listed in Table 4.1. Chemical substances are tested in a dissolved form in water. The same methods can measure the effect of environmental water samples.

As an example Figure 4.18 shows the growth curve of *Chlorella vulgaris* green alga, i.e. the result of the "Fresh-water algae growth inhibition test".

The growth curve is plotted for each dilution member of the concentration series of the toxicant to be tested. The area under the curve (mathematically known as integral) or the average specific growth rate ($\mu$) or any characteristic parameter of the growth curve can be used for the evaluation of the test. From the calculated results–alternatively read from the growth curve of each treatment and the control group, the final result of the test may be given as percentage of inhibition (I%) or as $E_rC_{50}$, $E_rC_{20}$ or $NOE_rC$, read from the concentration–I% (the response in this case) curve (Figure 4.19). $E_rC_x$ is interpreted as a concentration causing x% reduction in

*Table 4.1* OECD guidelines for testing the effects of chemical substances on aquatic organisms (OECD, 2014).

| Test guideline | Test name |
| --- | --- |
| OECD TG 201 (2013) | Fresh-water algae and cyanobacteria, growth inhibition test |
| OECD TG 221 (2006) | *Lemna sp.*, growth inhibition |
| OECD TG 202 (2004) | *Daphnia sp.*, acute immobilization test |
| OECD TG 211 (2012) | *Daphnia magna*, reproduction test |
| OECD TG 231 (2009) | *Amphibian* metamorphosis assay |
| OECD TG 219 (2004) | Sediment-water *Chironomid* toxicity using spiked water |
| OECD TG 218 (2004) | Sediment-water *Chironomid* toxicity using spiked sediment |
| OECD TG 225 (2007) | Sediment-water *Lumbriculus* toxicity test using spiked sediment |
| OECD TG 203 (1992) | Fish, acute toxicity test |
| OECD TG 204 (1984) | Fish, prolonged toxicity test: 14-day study |
| OECD TG 210 (1992) | Fish, early-life stage toxicity test |
| OECD TG 212 (1998) | Fish, short-term toxicity test on embryo and sac-fry stages |
| OECD TG 215 (2000) | Fish, juvenile growth test |
| OECD TG 229 (2009) | Fish, short term reproduction assay |
| OECD TG 230 (2009) | 21-day fish assay: short-term screening for estrogenic and androgenic activity, and aromatase inhibition |
| OECD TG 209 (2010) | Activated Sludge, Respiration Inhibition Test |
| OECD TG 224 (2007) | Determination of the inhibition of the activity of anaerobic bacteria: reduction of gas production from anaerobically digesting (sewage) sludge |

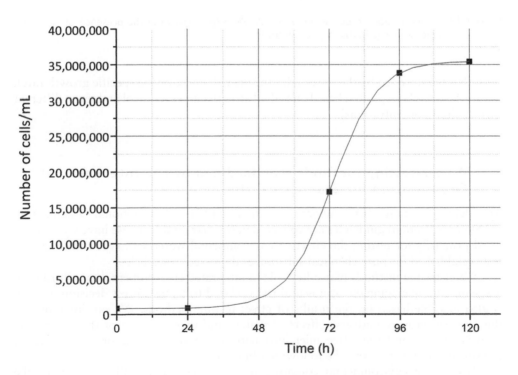

*Figure 4.18* Growth curve of *Chlorella vulgaris* (green algae).

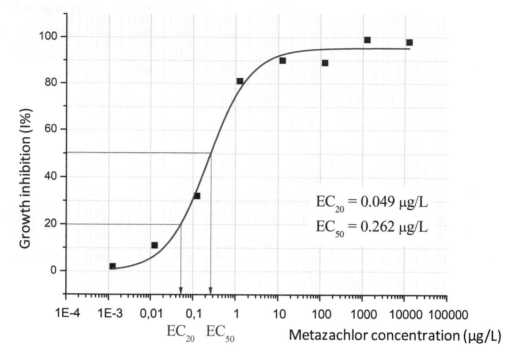

*Figure 4.19* Concentration–response curve of *Chlorella vulgaris* grown in the presence of increasing metazachlor (a herbicide) concentration.

growth (characterized by the area under the growth curve or the specific growth rate). Figure 4.19 has been plotted based on 8 algal growth curves (control and seven different metazachlor concentrations).

## 5.2 Water-testing methods standardized by the International Organization for Standardization

The identification numbers and names of the ISO (2014) (International Organization for Standardization) guidelines are collected in this paragraph. These have been developed primarily for testing the effect of environmental water samples (unlike OECD tests established for pure chemical substances dissolved in water).

Water testing methods play a key role in integrated environmental monitoring, in assessing and managing environmental risks related to water. Direct decision making is also possible based on the results of effect testing methods. We can interconnect the scale of measured adverse effects of waters and their safe use or the test results of wastewaters and their free inlet into natural surface waters, or the ecological classification of waters and the following obligations.

Toxicity of environmental contaminants adversely affects all members of the ecosystem, the food chains and food webs as well as the number and abundance

*Table 4.2*  ISO Guidelines for testing the effect of water and wastewater on bacteria (ISO, 2014).

| Test guideline | Test name |
| --- | --- |
| ISO 11348-1:2007 | Determination of the inhibitory effect of water samples on the light emission of *Vibrio fischeri* (Luminescent bacteria test) – Part 1: Method using freshly prepared bacteria |
| ISO 11348-2:2007 | Determination of the inhibitory effect of water samples on the light emission of *Vibrio fischeri* (Luminescent bacteria test) – Part 2: Method using liquid-dried bacteria |
| ISO 11348-3:2007 | Determination of the inhibitory effect of water samples on the light emission of *Vibrio fischeri* (Luminescent bacteria test) – Part 3: Method using freeze-dried bacteria |
| ISO 10712:1995 | *Pseudomonas putida* growth inhibition test (*Pseudomonas* cell multiplication inhibition test) |
| ISO/DIS 11350:2012 | Determination of the genotoxicity of water and wastewater – Salmonella/ microsome fluctuation test (Ames fluctuation test) |
| ISO 16240:2005 | Determination of the genotoxicity of water and wastewater – Salmonella/ microsome test (Ames test) |
| ISO 13829:2000 | Determination of the genotoxicity of water and wastewater using the umu-test |

of individual organisms and species. Laboratory tests on single species assume that prediction is possible from the results of single species for the whole of the ecosystem.

Prediction from single species toxicity test results is possible by extrapolation from at least three different trophic levels. In the case of aquatic ecosystems we generally use algae or plants, crustaceans and fish as suitable representatives of the aquatic ecosystem. Bacteria are also often used, mainly if biodegradation is a characteristic process in a certain environment or if a tiered assessment strategy is used.

### 5.2.1  *Standardized bacterial tests for toxicity testing of water and waste-water*

The bacterial tests standardized by ISO for water testing (in Table 4.2) are all laboratory species, they have no relevance for the aquatic environment, but they are able to give a sensitive response to chemicals with adverse effects. These tests are equally suitable for testing of new and existing chemicals as well as samples from contaminated waters, pore waters, leachates, waste extracts process-waters and wastewaters of agricultural and industrial origin.

### 5.2.2  *Standardized algal and plant tests for waters*

Algae, micro- and meso-flora of the aquatic environment are highly sensitive and widely applicable test-organisms in environmental tests. The same test methods fit to the testing of water-dissolved chemicals, waters from aquatic habitats and from other water uses, as well to assessing wastewaters. The Guidelines managed by the International Standardization Organization (ISO) and using algal and plant test organisms are gathered and listed in Table 4.3. These tests greatly overlap with the OECD tests, specified for the testing of pure chemical substances (Section 5.1).

*Table 4.3* ISO Guidelines for testing the effect of water and wastewater on algae and plant (ISO Standards, 2014).

| Test guideline | Test name |
|---|---|
| ISO/TR 11044:2008 | Scientific and technical aspects of batch algae growth inhibition tests |
| ISO 14442:2006 | Guidelines for algal growth inhibition tests with poorly soluble materials, volatile compounds, metals and wastewater |
| ISO 8692:2004 | Fresh-water algal growth inhibition test with unicellular green algae |
| ISO/DIS 8692:2010 | Freshwater algal growth inhibition test with unicellular green algae |
| ISO 10253:2006 | Marine algal growth inhibition test with *Skeletonema costatum* and *Phaeodactylum tricornutum* |
| ISO/DIS 13308: 2010–12 | Toxicity test based on reproduction inhibition of the green macroalga *Ulva pertusa* |
| ISO 10710:2010 | Growth inhibition test with the marine and brackish water macroalga *Ceramium tenuicorne* |
| ISO 20079:2005 | Determination of the toxic effect of water constituents and wastewater on duckweed (*Lemna minor*) – Duckweed growth inhibition test |

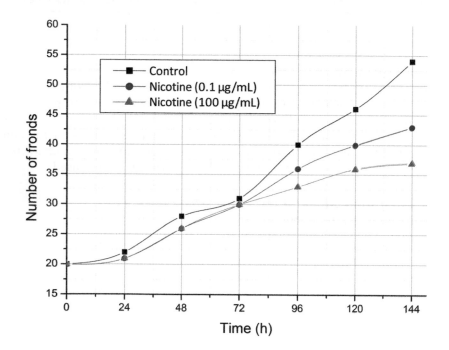

*Figure 4.20* Growth curves of *Lemna minor* (duckweed).

Figure 4.20 shows the growth curve of *Lemna minor* or common duckweed. From the number of fronds the doubling time ($T_d$) and from that, the average specific growth rate ($\mu$) can be calculated according to the formula of $T_d = \ln 2/\mu$. $E_r C_{20}$ or $E_r C_{50}$ can be determined based upon average specific growth rate values, resulted from testing the concentration series.

*Table 4.4* ISO Guidelines for testing the effect of water and wastewater on invertebrates (ISO Standards, 2014).

| Test guideline | Test name |
| --- | --- |
| ISO 6341:1996 | Determination of the inhibition of the mobility of *Daphnia magna* Straus (Cladocera, Crustacea) – Acute toxicity test |
| ISO 10706:2000 | Determination of long term toxicity of substances to *Daphnia magna* Straus (Cladocera, Crustacea) |
| ISO 20665:2008 | Determination of chronic toxicity to *Ceriodaphnia dubia* |
| ISO/DIS 14371 | Determination of fresh-water-sediment subchronic toxicity to *Heterocypris incongruens* (Crustacea, Ostracoda) |
| ISO/DIS 14380 | Determination of the acute toxicity to *Thamnocephalus platyurus* (Crustacea, Anostraca) |
| ISO 14669:1999 | Determination of acute lethal toxicity to marine copepods (Copepoda, Crustacea) |
| ISO 16303:2013 | Determination of toxicity of fresh water sediments using *Hyalella azteca* |
| ISO 16712:2005 | Determination of acute toxicity of marine or estuarine sediment to amphipods |
| ISO 19493:2007 | Guidance on marine biological surveys of hard-substrate communities |
| ISO 21427-1:2006 | Evaluation of genotoxicity by measurement of the induction of micronuclei – Part 1: Evaluation of genotoxicity using amphibian larvae |

### 5.2.3  Invertebrates using standard methods for testing water

Amongst aquatic invertebrates, crustaceans and amphipods are the most popular organisms in standardized ISO tests (Table 4.4), nevertheless all kind of arthropods, aquatic insects, planktonic animals, ciliate protozoa, nematodes, amoebas could be used for water testing. Numberless methods are existing and used for special purposes both as indicators in the assessment of natural waters, in simple laboratory assays or in more complex micro- and mesocosm tests.

### 5.2.4  Standardized fish tests for water and waste-water

Fish represents the highest-level vertebrate in water both from the taxonomic and food-chain point of view. Acute, chronic, mutagenic or teratogenic effects as well as toxicokinetics can be tested on different life stages and the test-design can also vary widely (static, flow through, prolonged, etc.). The ISO standards are shown in Table 4.5.

US EPA requires acute and chronic toxicity tests for freshwater and marine organisms for testing and biomonitoring industrial discharges in the frame of the US NPDS (2004). The water quality management system is based on the toxicity criteria shown in Table 4.6 and introduced in the relevant US EPA Guidelines (2002a, b, c).

### 5.2.5  Ecological assessment of surface waters

Both fresh-water and marine water health can be characterized by the diversity of the ecosystem and the assessed habitat. The most important element of the ecological status, the biological one, is based on five groups of aquatic flora and fauna: phytoplankton, phytobenthon, macrophytes, macroinvertebrates and fish. The most

*Table 4.5* ISO Guidelines for testing the effect of water and wastewater on fish (ISO Standards, 2014).

| Test guideline | Test name |
| --- | --- |
| ISO 7346-1:1996 | Determination of the acute lethal toxicity of substances to a fresh-water fish [*Brachydanio rerio* Hamilton-Buchanan (Teleostei, Cyprinidae)] – Part 1: Static method |
| ISO 7346-2:1996 | Determination of the acute lethal toxicity of substances to a fresh-water fish [*Brachydanio rerio* Hamilton-Buchanan (Teleostei, Cyprinidae)] – Part 2: Semi-static method |
| ISO 7346-3:1996 | Determination of the acute lethal toxicity of substances to a fresh-water fish [*Brachydanio rerio* Hamilton-Buchanan (Teleostei, Cyprinidae)] – Part 3: Flow-through method |
| ISO 10229:1994 | Determination of the prolonged toxicity of substances to fresh-water fish – Method for evaluating the effects of substances on the growth rate of rainbow trout [*Oncorhynchus mykiss* Walbaum (Teleostei, Salmonidae) ] |
| ISO 20666:2008 | Determination of the chronic toxicity to *Brachionus calyciflorus* in 48 h |
| ISO 23893-1:2007 | Biochemical and physiological measurements on fish – Part 1: Sampling of fish, handling and preservation of samples |
| ISO/TS 23893-2: 2007 | Biochemical and physiological measurements on fish – Part 2: Determination of ethoxyresorufin-O-deethylase (EROD) |
| ISO 23893-3:2013 | Biochemical and physiological measurements on fish – Part 3: Determination of vitellogenin |
| ISO 12890:1999 | Determination of toxicity to embryos and larvae of fresh-water fish – Semi-static method |
| ISO 15088:2007 | Determination of the acute toxicity of wastewater to zebrafish eggs (*Danio rerio*) |

*Table 4.6* Whole effluent toxicity (WET) methods and manuals for measuring acute toxicity to freshwater/marine organisms (US NPDS, 2004).

| US EPA Guideline | Method |
| --- | --- |
| **Fresh water acute toxicity** | |
| 2002.0 | Daphnid, *Ceriodaphnia dubia*, acute toxicity |
| 2021.0 | Daphnid, *Daphnia pulex* and *Daphnia magna*, acute toxicity |
| 2000.0 | Fathead minnow, *Pimephales promelas*/bannerfin shiner, *Cyprinella leedsi*, acute toxicity |
| 2019.0 | Rainbow trout, *Oncorhynchus mykiss* and brook trout, *Salvelinus fontinalis*, acute toxicity |
| **Marine water acute toxicity** | |
| 2007.0 | Mysid, *Americamysis bahia*, acute toxicity |
| 2004.0 | Sheepshead minnow, *Cyprinodon variegatus*, acute toxicity |
| 2006.0 | Silverside, *Menidia beryllina*, *Menidia menidia* and *Menidia peninsulae*, acute toxicity |
| **Short-term methods for estimating chronic toxicity to fresh-water organisms** | |
| 1000.0 | Fathead minnow, *Pimephales promelas*, larval survival and growth |
| 1001.0 | Fathead minnow, *Pimephales promelas*, embryo-larval survival and teratogenicity |
| 1002.0 | Daphnia, *Ceriodaphnia dubia*, survival and reproduction |
| 1003.0 | Green alga, *Selenastrum capricornutum*, growth |
| **Short-term methods for estimating chronic toxicity to marine/estuarine organisms** | |
| 1004.0 | Sheepshead minnow, *Cyprinodon variegatus*, larval survival and growth |
| 1005.0 | Sheepshead minnow, *Cyprinodon variegatus*, embryo-larval survival and teratogenicity |
| 1006.0 | Inland silverside, *Menidia beryllina*, larval survival and growth |
| 1007.0 | Mysid, *Americamysis bahia*, survival, growth and fecundity |
| 1008.0 | Sea urchin, *Arbacia punctulata*, fertilization |

*Table 4.7* European water quality standards of biological classification of the aquatic environment (CEN Standards, 2014).

| | |
|---|---|
| EN 14184:2003 | Water quality – Guidance standard for the surveying of aquatic macrophytes in running waters. |
| EN 14407:2004 | Guidance standard for the identification, enumeration and interpretation of benthic diatom samples from running waters. |
| EN 15460:2006 | Guidance standard for the surveying of aquatic macrophytes in lakes. |
| EN 15204.2007 | Guidance for phytoplankton analysis using inverse microscopy (Utermöhlmethod). |
| EN 14996:2006 | Guidance on assuring the quality of biological and ecological assessments in the aquatic environment. |

important indicator species and their abundance characterize the quality and risks in living surface waters. The European Committee for Standardization (CEN, 2014) established standards (CEN Standards, 2014) to survey key actors of aquatic ecosystems of rivers, as shown in Table 4.7. Based on these results fresh waters can be typified into classes based on EU EQS (2008) and WFD (2000).

## 6 NON-ANIMAL TESTING OF AQUATIC TOXICITY

To replace aquatic test organisms alternative testing methods such as cell-based assays, toxogenomic microarrays, and SAR or QSAR models are used to predict toxic effects to aquatic organisms. However, current *in vitro* methods for acute aquatic toxicity are neither standardized nor validated. Current *in vitro* and *in silico* approaches replacing animals in aquatic toxicity testing include (AltTox, 2007):

- Fish cell-based cytotoxicity assays;
- Fish cell-based assays with other mechanisms of toxicity as end points;
- Mammalian cell assays using cytotoxicity or other end points;
- Bacterial cell assays, usually based on luminescent reporter genes (primarily used for detection);
- Fish embryo assays based on embryo survival and pathophysiological changes;
- *In vitro* endocrine disruptor assays;
- Genomic microarrays (toxicogenomics);
- (Q)SAR and other computational programs;
- Test batteries and/or tiered testing schemes incorporating the above types of assays.

## 7 TESTING SEDIMENT

Sediments are in close relationship with surface waters; the contaminants are distributed between the solid and liquid phases according to their partition coefficient. Chemical substances with low water solubility and low mobility and high $K_{ow}$ (octanol–water partition coefficient) are sorbed or bound to the solid phase of the sediment; it is even filtrated out from the water by the suspended solid and later on settled with the organic or inorganic particulate solid matter in the form of fine particle-size bottom-sediment.

Based on the partition of chemical substances between the physical phases of sediments, the partition of toxicity or other adverse effects can also be measured, but the partition of toxicity does not accurately follow the chemical partition, because the biologically available fraction is much smaller than the total sorbed amount or the exhaustively extracted chemical content in the sediment.

The partition of toxicity between solid and water can be followed by the parallel testing of whole sediment and its pore water. Pore-water is the equilibrium state interstitial water in the sediment which can be obtained by centrifugation or by vacuum recovery. Integrated evaluation of chemically measured concentrations and toxicity provides additional information on the mobility, bioaccessibility and bioavailability and consequent risk of the contaminants bound to sediments. The most frequent cases are the following:

–   High contaminant concentration and high toxicity both in pore water and the whole sediment indicates a mobile contaminant, which is toxic and available both chemically and biologically.
–   High contaminant concentration and toxicity in pore water but lower in the whole sediment indicates highly mobile, water-soluble and bioavailable toxicants.
–   Low concentration in pore water but high concentration in the whole sediment and positive toxicity in parallel can be explained by a sorbed contaminant which is not water-soluble or otherwise chemically available but is still bioavailable due to special biological mobilization (e.g. biotensides, local pH or redox potential changes) and uptake mechanisms (chelating agents, active transport). If a highly toxic chemical substance is present in low concentration, the same combination results. It may happen that the chemical analysis does not show anything because the effective toxicant has not been included in the analysis program or cannot be analyzed due to an immeasurable, low concentration or no suitable analytical method is available.
–   Chemical analyses show high concentration in the whole sediment but it has no effect on test organisms because of very low mobility and bioavailability. This combination suggests that the contaminant's actual stability is high. In this case one has to assess toxicity and the dependence of stability and/or toxicity from environmental conditions (which refers to temperature, pH, redox potential and hydrology). To assess the risk of such cases, one has to create a worst case scenario in a dynamic test set-up to simulate the maximum mobilization potential of the sorbed (bound) chemical substance and either prove or disprove its chemical time-bomb behavior.

The ISO standardized tests methods concentrate on the field assessment of the benthic fauna as a good indicator for surface-water health. Sampling and identifying the members of the ecosystem of benthic habitat (Table 4.8) is standardized by ISO. Some species of the zoobenthos are shown in Figure 4.21 and 4.22.

Benthos can be categorized according to size, the test-methods may use any of them:

–   Macrobenthos, size greater than one mm;
–   Meiobenthos, size less than one mm but greater than 32 μm;
–   Microbenthos, size less than 32 μm.

Table 4.8 Guidelines for testing the effect of sediment (ISO Standards, 2014; CEN Standards, 2014).

| Test guideline | Test name |
| --- | --- |
| ISO/DIS 10870:2012 | Guidelines for the selection of sampling methods and devices for benthic macroinvertebrates in fresh waters |
| ISO 16665:2005 | Guidelines for quantitative sampling and sample processing of marine soft-bottom macrofauna. |
| (EN) ISO 8689-1:2000 | Biological classification of rivers – Part 1: Guidance on the interpretation of biological quality data from surveys of benthic macroinvertebrates. |
| (EN) ISO 8689-2:2000 | Biological classification of rivers – Part 2: Guidance on the presentation of biological quality data from surveys of benthic macroinvertebrates. |
| EN 13946:2004 | Guidance standard for the routine sampling and pre-treatment of benthic diatoms from rivers. |
| EN 14407:2004 | Guidance standard for the identification, enumeration and interpretation of benthic diatom samples from running waters. |
| ISO 8689-1:2000 | Guidance on the interpretation of biological quality data from surveys of benthic macroinvertebrates. |
| ISO 8689-2:2000 | Biological classification of rivers – Part 2: Guidance on the presentation of biological quality data from surveys of benthic macroinvertebrates. |

Figure 4.21 Freshwater benthic animals: larva of the mosquito Aedes notoscriptus (2014), dragonfly larva, gilled snail (2014), stonefly nymph (2014), damselfly nymph (2014), midgenymph (2014), Hirudo medicinalis, medical leech (Gjertsen, 2007).

Sediment testing includes surveying sediment-dwelling organisms and their diversity as well as laboratory bioassays on sensitive laboratory test organisms or communities. The survey can be focused on

- Suspended solid (potential sediment) to determine the partition of the toxicity;
- Bottom sediment, as habitat (the top few centimeters of the sediment);
- The impact of dredging (deeper layers should also be assessed).

*Figure 4.22* Marine benthic animals: nematode *Draconema* (2014), Christmas tree worm, fire worm, mussels on the shore, sea urchin, marine snail *Nembrotha rutilans*.

Sediment-disturbing and -reworking organisms have high impact on sediment and contaminants in sediment, which should be considered when measuring sediment toxicity in simulation tests, or microcosms.

Suspended solid, the potential sediment, should be sampled and tested to monitor the chemical pollution of surface waters and the subsequent adverse effects. The suspended solid is in close contact and interaction with water, thus the partition of the contaminants between water and solid is close to equilibrium. Therefore, there is a high probability that information obtained from adverse effects measured on suspended solids will be valid for the whole aquatic ecosystem.

Another approach is the assessment and monitoring of benthic habitats including bed sediment using an integrated chemical analytical and ecotoxicity testing methodology. When biological tools have indicated adverse effects, the next step is to identify the cause of the effect, i.e. the responsible chemical substances. For decision making on risk reduction measures (e.g. by restricting discharge or reducing the use of certain chemicals or cleaning up the sediment), risk managers should identify the contaminating chemical substance, determine its exposure level, identify the adverse effects and characterize the risk: the extent of ecosystem deterioration and predictable damage.

In the case of sediment – similar to soil – the adverse effect predictions from chemically measured concentrations often differ from actual toxicity or other measured adverse effect results. The main causes of the discrepancies are the following:

–　Presence of mixtures of chemicals which interact with each other; they may be synergistic, i.e. their joint effect is greater than their sum.

– Due to a common source, a source-specific assemblage of contaminants may occur making it impossible to assign the proper cause (one contaminant) to the adverse effect observed.
– Thousands of chemicals are released and bound to the sediment, but only a few are included into the analysis programs – generally those which have default limit values in the relevant legislation. Environmental toxicity testing can detect all of the contaminants' adverse effects.
– Biological availability and accessibility of chemical contaminants bound to sediments are small, and their chemical form, mobilization, thus their actual (active) concentration depends on a number of general and local environmental parameters from the outside temperature to the pore volume of the sediment and pH, redox potential and chemical concentrations in the pore water.

A complex and dynamic approach is needed to characterize the complex system of sediment. The Sediment Toxicity Identification Evaluation (TIE, 2007) system, developed by US EPA is based on a biologically controlled fractionation and physical/chemical manipulation of sediment samples to find, isolate or change the different groups of toxicants that are potentially present. The biological system, i.e. the test organism is used as an "indicator" to determine whether the manipulation has changed toxicity.

TIE comprises three phases: characterization, identification and confirmation:

– Characterization chiefly uses environmental toxicity tests (bioassays or other biological tools) and physical/chemical manipulations such as changes in partition (sorption–desorption), redox potential (oxidative–reductive environment), and pH (acidic–alkalic) to build a general "profile" of the causative toxicant(s). In this phase general categories of toxicants can be determined such as volatile substances, water soluble (e.g. ionic metals), biodegradable organics, highly sorbable, high-$K_{ow}$, persistent substances (e.g. PBTs), etc.
– Identification means chemical characterization and analytical measurement of the suspected toxicant.
– Confirming the presence of the identified contaminant in the water means confirming its origin, production or use which is needed for proper decision making on risk reduction or other management measures.

The TIE system separately investigates whole sediment and its pore water, also called interstitial water. Pore water is usually separated by centrifugation from the solid phase and benthic test organisms are used, shown in Table 4.9. The results of pore water tests give information on the mobile fraction of the pollutants, while tests on whole sediment clarify intensive and long term interactions between benthic organisms and the sediment sample. The tests look into the effect of the test organism on the sediment-bound contaminant, which is by several orders of magnitude higher than in the equilibrium pore water. Direct contact of the test organism with the whole sediment may result in mobilization (e.g. by secreted acids or mucilage), dermal uptake, dietary uptake and bioaccumulation over the long term. Whole sediment tests are pessimistic models because sediment-dwelling organisms can escape from contaminated matrices, but they cannot do so in the test. Another important issue is the disturbance of the sediment sample: bedded sediment in the depths of rivers and lakes is highly undisturbed, in an (or close to) equilibrium state, at stable and low redox potential. During sampling

*Table 4.9* Organisms used for pore water and whole sediment testing (TIE, 2007).

| Test organism | Pore water testing | Whole sediment testing |
|---|---|---|
| **Fresh-water benthic** | | |
| *Chironomus dilutus*, Chironomid – midge larvae | + | + |
| *Hyalella azteca*, Amphipod – scud | + | + |
| *Lumbriculus variegatus*, Oligochaete – worm | + | + |
| *Gammarus pulex*, Amphipod – crustacean | + | + |
| **Marine benthic** | | |
| *Americamysis bahia*, Mysid – shrimp | + | + |
| *Ampelisca abdita*, Amphipod (Atlantic) | + | + |
| *Eohaustorius estuarius*, Amphipod (Pacific) | + | + |
| *Leptocheirus plumulosus*, Amphipod (Atlantic) | + | + |
| *Arbacia punctulata*, Echinoderm – sea urchin | + | |
| *Strongylocentrotus purpuratus*, Echinoderm – purple urchin | + | |
| *Mytilus galloprovincialis* – mussel | | + |
| *Corophium volutator*, Amphipod – mud shrimp | + | |
| *Mercenaria mercenaria* – hard shell clam | + | |
| *Mulinia lateralis* – dwarf surf clam | + | |
| Microtox (*Vibrio fischeri*) – bacterium | + | |
| **Fresh-water pelagic** | | |
| *Ceriodaphnia dubia*, Cladoceran – water flea | + | + |
| *Daphnia magna*, Cladoceran – water flea | + | + |
| *Daphnia pulex*, Cladoceran – water flea | + | + |

and testing however the samples are disturbed and a completely new equilibrium is established, which may contribute to mobilization, oxidation or other risk-influencing changes of the contaminants. In the case of bed sediments a typical change in redox conditions may appear during sampling and transport: anaerobic or anoxic sediments from deeper layers may get in contact with air and aerobic organisms.

## 8   SEWAGE AND SEWAGE SLUDGE TESTS

Wastewater (both treated and untreated) has a higher risk, higher adverse effects on the ecosystem than receiving waters. This may be an argument to investigate wastewater toxicity and other adverse effects before its release or utilization. Sewage sludge consists of the surplus of microorganisms from wastewater treatment technology and sorbed organic and inorganic matter. The use of treated wastewater on soil as fertilizer may be an efficient solution when the toxicant content is low and the process is properly controlled. But the risks associated with sewage sludge could be high, depending on the wastewater's hazardous material content. Both wastewater and sewage sludge can be managed using environmental toxicological tools: wastewater can be discharged into receiving waters and sewage sludge onto soil if they do not show toxicity or other adverse effects in properly selected tests. Risk managers can rely on wastewater and sewage sludge toxicity results for decision making.

Another argument for testing wastewater is the risk posed by untreated wastewater to the microbiota of the sewage-treatment plants. If a toxic wastewater enters the biological treatment plant, it may kill the activated sludge and the living microorganisms which biodegrade the (dangerous, toxic) components of the wastewater. If the

Table 4.10 ISO guidelines for the testing of sewage and sewage sludge (ISO Standards, 2014).

| Test guideline | Test name |
| --- | --- |
| ISO 9888:1999 | Evaluation of ultimate aerobic biodegradability of organic compounds in aqueous medium – Static test (Zahn-Wellens method) |
| ISO 8192:2007 | Test for inhibition of oxygen consumption by activated sludge for carbonaceous and ammonium oxidation |
| ISO 9509:2006 | Toxicity test for assessing the inhibition of nitrification of activated sludge microorganisms |
| ISO 9887:1992 | Evaluation of the aerobic biodegradability of organic compounds in an aqueous medium – Semi-continuous activated sludge method (SCAS) |
| ISO 11733:2004 | Determination of the elimination and biodegradability of organic compounds in an aqueous medium – Activated sludge simulation test |
| ISO 11734:1995 | Evaluation of the 'ultimate' anaerobic biodegradability of organic compounds in digested sludge – Method by measurement of the biogas production |
| ISO 13641-1:2003 | Determination of inhibition of gas production of anaerobic bacteria – Part 1: General test |
| ISO 13641-2:2003 | Determination of inhibition of gas production of anaerobic bacteria – Part 2: Test for low biomass concentrations |
| ISO 15522:1999 | Determination of the inhibitory effect of water constituents on the growth of activated sludge microorganisms |
| ISO 18749:2004 | Adsorption of substances on activated sludge – Batch test using specific analysis methods |

microflora are killed, they cannot fulfill their task, the wastewater treatment technology is malfunctioning, and the organic contaminants in the water fall to be degraded. Those contaminants that cannot be degraded are present in the effluent or are sorbed by the flocks and other form of biofilms of the sewage sludge. The ISO tests in Table 4.10 are used to control or protect the sewage microflora. The same tests can be used for any chemical substance, contaminated environmental or waste sample in the lab, if a 'standard' or a well-known and reliable sewage sludge is available, as a complex degrading microbiota.

## 9  TESTING WASTE USING AN 'ECOTOX' TEST BATTERY

The classification of chemical substances has been harmonized with GHS by the CLP regulation that replaced the former Dangerous Substance Directive (DSD, 1967). But in the field of waste regulation – in spite of having the category H14 "Ecotoxic" containing substances and preparations which (may) present immediate or delayed risks to one or more sectors of the environment – the risk-based classification of wastes, considering the chemical composition and adverse effects has not been carried out.

The ecotoxicological characterization of waste is part of its assessment as hazardous or non-hazardous according to the European Waste List. However, as of 2007 no methodological recommendations have been provided to cover the hazard criterion H14 "ecotoxicity". Therefore, a European interlaboratory comparison (or round robin test) ("H14 EU RingTest") evaluated a biotest battery, which can be used to assess the ecotoxicity of waste material and waste eluates (Sander et al., 2008). An interlaboratory comparison is a quality assurance program for a new or existing measurement method. Usually a reference institute sends identical samples to different laboratories and the results are statistically evaluated.

In the European interlaboratory comparison 7 primary and 10 additional ecotoxicity measurement methods were evaluated and verified for solid wastes and waste eluates. The interlaboratory comparison has been executed in Europe in 2006–2007 and was managed by Umweltbundesamt or Federal Environment Agency (UBA, 2014), Germany (Moser *et al.*, 2011).

The ecotoxicological characterization of wastes is laid down in the European Standard CEN 14735 (2005), which describes the sample preparation and provides an informative collection of appropriate test procedures for the investigation of wastes. This collection of test procedures was condensed to the "basic test battery", containing 3 aquatic and 2 terrestrial methods, with two algal and two plant applications, thus the final test battery contains 4 aquatic (*Vibrio fischeri* luminescence bacterium; *Desmodesmus* and *Pseudokirchneriella* algae and *Daphnia*, the water flea) and 3

Table 4.11  Ecotoxicological characterization of waste: basic test battery.

| Standard | Test name | Test description |
|---|---|---|
| ISO 11348-1/2/3:1998 | Acute test method with the luminescent bacterium *Vibrio fischeri*. Representative of marine bacteria | This test measures the inhibition of the luminescence emitted by the marine bacterium *Vibrio fischeri* (NRRL B-11177) using freshly prepared, liquid-dried or freeze-dried bacteria. |
| ISO 8692:2004 | Fresh-water algal growth inhibition test method with *Pseudokirchneriella subcapitata*. Representative of algae | The test measures the growth inhibition of the unicellular green algae *Pseudokirchneriella subcapitata* by wastewater or waste extract. The method is applicable for substances that are easily soluble in water. With modifications to this method, as described in ISO 14442 and ISO 5667-16, the inhibitory effects of poorly soluble organic and inorganic materials, volatile compounds, heavy metals can also be tested. |
| ISO 8692:2004 | Fresh-water algal growth inhibition test method with *Desmodesmus subspicatus*. Representative of algae | The test measures the growth inhibition of the unicellular green algae *Desmodesmus subspicatus* by wastewater or waste extract. *Desmodesmus* is less sensitive than *Pseudokirchneriella*. |
| ISO 6341:1996 | Acute test method with water flea *Daphnia magna*. Representative of aquatic invertebrates | A method for the determination of acute toxicity to *Daphnia magna* Straus of chemical substances that are soluble under the conditions of test or can be maintained as a stable suspension or dispersion, treated or untreated sewage effluents, surface or groundwaters. |
| ISO 11269-2:1995 | Seedling emergence and growth method with oats, *Avena sativa*. Representative of monocotyledone plants | This method measures the toxic effects of solid or liquid form waste in soil environment, after mixing the waste to test into soil. The early stages of growth and development of plant *Avena sativa* is tested. The waste containing soil is compared to clean soil (without waste). |
| ISO 11269-2:1995 | Seedling emergence and growth method with rape *Brassica napus*. Representative of dicotyledone plants | This method measures the toxic effects of solid or liquid form waste in soil environment, after mixing the waste to test into soil. The early stages of growth and development of the plant *Brassica napus* is tested. The waste containing soil is compared to clean soil (without waste). |
| ISO 11268-1:1993 | Acute toxicity test method with earthworm *Eisenia fetida* | Earthworm lethality is measured in artificial soil, which is amended with the concentration series of the waste or waste eluate to be tested. |

terrestrial procedures, two plants (*Avena* – oats and *Brassica* – rape) and one animal test (earthworm) (Table 4.11).

The set of 10 additional ecotoxicity bioassay methods were selected in the inter-laboratory comparison (European Ring Test of Ecotoxicity of Waste, 2013). Five aquatic and five terrestrial tests were selected as being potentially appropriate for the determination of ecotoxicity of waste (Table 4.12).

*Table 4.12* Ecotoxicological characterization of waste: potentially appropriate tests.

| Standard | Test name | Test description and remarks |
| --- | --- | --- |
| ISO 13829:2000 | Genotoxicity method for water and wastewater tested by *Salmonella typhimurium* TA1535 | The umu-lux test is a genotoxicity test using the two genetically modified *Salmonella typhimurium* TA1535 strains (TL210 and TL210ctl) transformed with the luxC, D, A, B, E (luciferase gene and fatty acid reductase genes) of *Vibrio fischeri* as a reporter gene. The TL210 strain detects genotoxicants and the TL210ctl strain detects cytotoxicants. |
| ISO 10712:1995 | Bacterial growth inhibition test method with *Pseudomonas putida* | A method for determining the inhibitory effect of surface, ground and wastewater on *Pseudomonas putida* growth. Not suitable for highly colored samples or samples containing undissolved or volatile substances which react with the nutrient solution or undergo changes such as biochemical degradation. Also suitable for testing substances soluble in water. |
| DIN 38412-48:2002 | Contact test with soil bacterium *Arthrobacter sp.* Representative of soil bacteria | Bacteria and soil are in direct contact. Bacterial cell suspension is spread into the agar-medium and soil blocks or disks are put on the surface of the growth medium. The diameter of the inhibition zone (the ring without bacterial growth around the soil block) is measured. |
| ISO 20079:2005 | Duckweed growth inhibition test method with *Lemna minor* | A method for the determination of the growth-inhibiting response of duckweed to substances and mixtures contained in water, treated municipal wastewater and industrial effluents. |
| ISO/DIS 20665:2007 | Chronic toxicity test with crustacean *Ceriodaphnia dubia* | Practical but no higher sensitivity than *Daphnia magna* |
| ISO/DIS 13829:2007 | Chronic toxicity test method with rotifer *Brachionus calyciflorus* | Practical but no higher sensitivity than *Daphnia magna* |
| ISO/DIS 17512-1:2007 | Earthworm avoidance test method with *Eisenia fetida* and *Eisenia andrei* | The short test period, the exposure scenario that is similar to field conditions, high sensitivity and the use of a taxonomically higher test organism are advantages that confirm the utility of this test in waste characterization. It can be utilized as a screening tool. Further development is necessary. |
| ISO 11268-2:1998 | Earthworm reproduction test method with *Eisenia fetida*. Representative of soil-dwelling worms | A method based on placing adult earthworms in an artificial soil containing the test substance (waste or waste eluate) in different concentrations and determining the percent mortality after 7 days and 14 days. It is not applicable to volatile substances, i.e. substances for which Henry's constant or the air/water partition coefficient is greater than 1, or for which the vapor pressure exceeds 0.0133 Pa at 25°C. It does not take into account the possible degradation of the test substance. It shows high sensitivity but rather long duration. |

(continued)

*Table 4.12* Continued

| Standard | Test name | Test description and remarks |
|---|---|---|
| ISO 16387:2004 | White worm reproduction test method with *Enchytraeus albidus*. Representative of soil-dwelling annelid worms | A method for determining the effects of substances or contaminated soils on reproduction and on survival of the worm *Enchytraeus albidus* (*Enchytraeidae*). The animals are exposed to the substance to test by dermal and alimentary uptake using a defined artificial soil to which specified amounts of wastes or waste eluates are added. |
| ISO 11267:1999 | Collembolan (springtail) reproduction test with *Folsomia candida*. Representative of soil invertebrates | A long term test, in which Collembola is in contact with soil containing a dilution series of waste and waste eluates for 3 weeks. The effect on reproduction is quantified by counting the members of the first and second generation and compared to waste-free soil. |

*Figure 4.23* Closed and open mussel.

## 10  NON-STANDARDIZED BIOASSAYS AND OTHER INNOVATIVE TEST METHODS

In this chapter we introduce some non-standardized, innovative environmental toxicity measurement methods, which are not widely known or used but they are expected to spread due to their easy use and targeted application. It should be noted that in addition to the new test methods, completely new approaches are coming up e.g. non-testing approaches in ecotoxicology, molecular methods and the 3Rs of Replacement, Reduction and Refinement of animal experiments in ecotoxicology.

*Mussel monitor* is a monitoring tool for (contaminated) surface waters; it can also be used as an early warning system. It is based on the notion that when mussels meet contaminated water, they close their shells (as opposed to their normal, open state) in order to protect themselves from the contaminants and to shorten the time of exposure. Figure 4.23 shows a mussel in closed and open state. In the mussel monitor, the mussel itself is the sensor. The signals generated by the opening and closing of the shell – using an inductive electromagnet – are forwarded in the form of an electric signal to the data-processing unit, which can be several kilometers away. Analyzing the increased frequency of the shell's movement compared to the normal movement – by filtering the

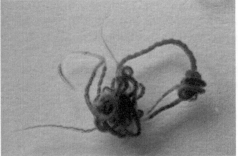

Figure 4.24 Tubifex tubifex, sludge worm or sewage worm, a common inhabitant in lake and river sediments.

signal through an adequate statistical sensor system – one can obtain a warning signal indicating that there is some problem. The cause of the problem cannot be identified from this signal, the only information is that something has triggered the protective mechanism in the mussels. The appearance of the signal initiates a series of actions which lead to a risk-decreasing intervention after toxicological testing and/or chemical analysis of the samples taken from the monitoring location, and after the source has been identified (Musselmonitor, 2013; Kramer & Foekema, 2000).

*Active biomonitoring* with fish and *Daphnia*, which can also be used as bioindicators by placing the specimen, which has been raised under supervision, into the river or lake in a flow-through cell. The water to be tested should be continuously in contact with the animals. Any of the animals' suitable measuring end points can be used as indicators: the number of living animals, their motility, behavior, proliferation, number and quality of offspring, etc. Some of the end points can be continuously observed visually, for example by cameras, and the measurements can be evaluated using an automatic evaluation system. After fitting the data to a statistical analysis, an automatic warning signal will be obtained if the system detects an anomaly that exceeds standard deviation. The frequency of the gill movement of the fish and the opening and closing of the mussel's shell are monitoring end points that are relatively easy to detect.

*Tubifex monitor* is also based on remote signal processing. The members of the *Tubificidae* family are aquatic organisms that often occur in waters contaminated by organic substances. Half of their body burrows into the bottom sediment, while the rest floats in the water. In the presence of certain contaminants, these organisms can retract a large portion of their bodies, and burrow deeper into the bottom sediment. This behavior is proportional to the concentration of the contaminating substance. The retraction can be observed both visually and with the help of a camera. By using digital image-analyzing systems, it can be quantitatively analyzed and evaluated. Based on the measurements of Leynen *et al.* (1999) the conclusion can be drawn that the movement/behavior of *Tubifex* worms shown in Figure 4.24 can be reproduced, and, by tracing and evaluating their movement, a water monitoring and early warning system can be developed.

*In situ toxicity assessment* and other *in situ* biological testing strategies play an increasingly important role in aquatic and sediment risk characterization and management. *In situ* testing has many advantages compared to laboratory testing of single samples after storage and transportation into the lab. Many contaminants are reactive, volatile and capable of being degraded and/or dissipating during transportation and storage. The change in the pH and redox potential between the time of sampling and testing may also influence the toxicity to a large extent.

*In situ* testing of adverse effects is especially important if the source of contamination is ephemeral or when the exposure varies over time and space. The mussel monitor, the tubifex monitor, and biosensors can provide continuous and integrating signals from the environment during the life span of the test organism.

Another trend of the innovations and developments in the field of *in situ* testing is the establishment of the mobile versions of laboratory bioassays to enable *in situ* decision making during site characterization, screening and mapping of hazardous contaminants, identification of hot spots, as well as in the determination of the place, time and frequency of sampling during local risk assessment.

*In situ* measured toxic effects and bioaccumulation integrates all the environmental effects of changing parameters such as temperature, water level, seasonal changes, etc. which may greatly influence bioavailability and the realization of adverse effects. On the other hand natural stressors together with the anthropogenic adverse effects may result in a confusing picture.

As a summary, we can list the advantages of *in situ* assessment of toxicity and bioaccumulation:

– Increased realism;
– Incorporation of spatial/temporal variability;
– Integration of multiple stressors;
– Reduced sample manipulation;
– Matrix-specific risk identification;
– *In situ* planning and decision making is possible.

There are some limitations too, which can be eliminated only partly by new developments:

– No control over natural exposure factors;
– Ammonia–ammonium-ion ratio is pH dependent: ammonium increases with the pH, and is more toxic, than the ionic form because the uptake of ammonia ($NH_3$) is not inhibited by any cellular mechanisms;
– Hydrogen sulfide to sulfide ion ratio is also pH dependent; the sulfide ion is more toxic and its ratio increases with pH decrease;
– Groundwater seeping into sediment influences the quality of habitat, mainly the nutrients and micro-contaminants endanger the sediment ecosystem;
– Problems associated with caging of the test organism are significant: they are not able to carry out normal behavior, e.g. they cannot escape. Cage design may influence water flux and oxygen dissolution;
– Transportation, handling, and physical stress of the test organisms influences their response;

- Predation and competition of the indigenous test organisms with the native ones may change the response;
- Need for appropriate controls and reference sites.

Some laboratory screening tests have already been modified and applied to *in situ* testing. They are useful in quick mapping of the extent of contamination at contaminated sites. These types of tests are typically rapid and can be conducted at relatively low cost and in large numbers. A quick turnaround time is useful for decision making with respect to subsequent steps in the assessment of ecological risk at a site. Some of them are introduced here in detail, based on the review of Rosen *et al.* (2009).

- Sea urchin fertilization tests: the fertilization success as end point refers to the percentage of eggs that develop fertilization membranes following 20 minutes of exposure to sperm that have been previously exposed to test samples (also for 20 minutes). These life stages are ecologically relevant because of their tendency to be negatively buoyant, and therefore, are likely to be associated with surficial sediment (Anderson *et al.*, 1996). This test needs only 5–10 mL per replicate, less than one hour exposure period, it is sensitive to a variety of contaminants, and has high ecological relevance. It needs extensive preparation time and cost.
- Microtox®, using *Vibrio fischeri* luminescent bacterium is suitable for on site use. It is moderately sensitive, its time and cost requirement is also average. Environmental relevance is low if the marine species is used for fresh water.
- QwikLite is similar to Microtox, but using a higher animal, the planktonic bioluminescent dinoflagellate. It has high ecological relevance in marine ecosystems. *Dinoflagellates* emit light only as an effect of mechanical stimulation, so that a stirrer is built into the measuring chamber. Light emission is measured generally after 24 hours exposure.
- Toxkits of ecologically relevant algal and rotifer species ensure easy-to-conduct and easy-to-use test methods, with dehydrated organisms. Rehydrated Toxkit organisms were as sensitive as laboratory species when compared to traditional standard laboratory assays and *in situ* exposures. The Toxkits have the advantage of not requiring culture facilities, they require little equipment and training, and the test can be easily and quickly conducted under a wide range of environmental conditions.
- Modification of short-term laboratory tests is a successful way to develop a portable test setup. 10 mL pore water per replicate was enough for sea urchin fertilization, mussel embryo, rotifer, and QwikLite standard tests methods after the reduction of the normal chamber size.

*In situ* tests provide more realistic exposures. The integrated application of *in situ* bioassays and laboratory testing may ensure optimal information, especially when multiple stressors are involved and improve decision making with respect to management decisions. Results from *in situ* studies will provide much greater confidence in assessing true exposures and effects occurring at a particular site. This confidence is critical when costly decisions and implications to remediate or not is at stake (Rosen *et al.*, 2009).

The time requirement of *in situ* but sometimes also laboratory tests can significantly be reduced by using conserved test organisms, e.g. lyophilized or gently dried ones. The ready-made test-kits, containing preserved test organisms able to revive instantly, can be used comfortably and quickly, without laboratory and professional preparedness for the maintenance of the test-strain. Daphnids are typical organisms suitable for long-term storage in dried form and being able to become active again when meeting the test medium. Some standardized and ready for use kits with different crustaceans are commercially available:

– DAPHTOXKIT F (2013) using *Daphnia magna*. This toxicity test adheres to OECD and ISO Guidelines.
– DAPHTOXKIT F (2013) using *Daphnia pulex*. This toxicity test adheres to the OECD Guideline.
– CERIODAPHTOXKIT F (2013) using *Ceriodaphnia dubia*. This toxicity test adheres to the US EPA Test Guideline.

Microbial communities are the most important components of the ecosystem, and therefore a number of microbiological methods are being developed which will be suitable to characterize microbiological communities and individual species. These are non-destructive, *in situ*, real-time measurement methods, which can observe details of the microbial community in action because the measurements are carried out at the sub-millimeter scale.

***Biosensors and microprobes*** are analytical instruments created by alloying electronic and biological systems: they transform biological signals measured in the water or in the soil to electrical signals. Genetic bioindicators (bioreporter genes) are genes that can create easily detectable products; thus, because they work only under certain conditions, they can enhance the selectivity and amplify the signal. The microprobes and microelectrodes can measure the characteristics of the living area and the ecosystem at the sub-millimeter scale in soils and other solid-phase samples. The measured parameters include pH, temperature, chloride ion concentration and dissolved oxygen concentration (Revsbech & Jorgensen, 1986). Microprobes are able to characterize the redox system of the soil via the measurement of ammonium (De Beer & Van den Heuvel, 1988), nitrate (Jensen *et al.*, 1993) or oxygen (Revsbech, 1989). The nitrification was monitored by means of a combined oxygen and nitrogen oxide microprobe by Christensen *et al.* (1989). Ramsing *et al.* (1993) applied a combination of oligonucleotides and microelectrodes to detect sulfate reduction.

Visible or fluorescent light provides easy-to-measure end points; thus, the most widespread bioreporter genes are those responsible for the emission of light and GFP (green fluorescent protein), which is responsible for green fluorescence. By planting these genes into selectively sensitive microorganisms, they will signal the damage in real time as part of the photosensitive biosensor, through decreased emission of light. The lux gene of *Vibrio fischeri* or *Vibrio harveyi* is only seven kilobases long. It can be built into any selectively sensitive microorganism. The GFP gained from the jellyfish species *Aequorea victoria* can be built either into a prokaryote or a eukaryote, and it will result in an easily detectable, strong green light emitted by the host.

Aquatic and terrestrial ecosystems try to adapt to the environmental circumstances, climatic conditions, seasons and contaminating substances, thus showing great flexibility. The changes in the species distribution are caused by the fact that those species that are sensitive to the contaminating substance will decline and eventually disappear, while those species that can tolerate or utilize the contaminant will gain an advantage, proliferate and their relative numbers will increase within the community. Since certain species have genes that are responsible for the coexistence with the contaminating substance, these genes will naturally proliferate in the community in a contaminated area, not just through an increase in the species' population, but also due to other mechanisms such as horizontal gene transfer between members of the community. As a global ecological trend, the diversity of the microflora in water and the soil is continuously growing, even if the contaminants exert a detrimental effect at a local level, but all in all they will trigger the evolution of further genes in the metagenome, and thus the quantity and information content in the genes will gradually increase.

*Biological and molecular methods* for characterizing the diversity and detection of species have growing application.

There are two concepts used to characterize the ecosystem of an environmental compartment, i.e. a complicated living area. The first concept suggests that we can examine all of the genes, gene products or gene activities of a community, e.g. a soil microflora, irrespective of which species' genome it belongs to. This examination can take place at the DNA level with the help of DNA chips, real-time PCRs, or through the measurement of gene products (generally enzymes) and their metabolic activities (patterns of the community's substrate utilization, for example using the BIOLOG (2013) method). The aim is to characterize the community, and statistical evaluation is needed for the correct interpretation of the result (Dobler *et al.*, 2001; Nagy *et al.*, 2010). It can be used as an early warning system if the harmful effect can be statistically separated from seasonal and climatic anomalies.

Another concept is based on the notion that only the gene that selectively appears as a direct consequence of the harmful effect, or this gene's (metabolic) product or the organism carrying the gene can be detected. In cases like these, the most selective methods are used, for example DNA hybridization, fluorescent *in situ* hybridization, or PCR. If that particular gene can be detected, the effect, too, is most likely the consequence of the harmful substance/agent.

Molecular methods are selective and sensitive by themselves, and traceability can be further enhanced using sensitivity- and detection-enhancing signals. One of the most widespread techniques, PCR itself, is based on detection after signal multiplication.

## 11  MULTISPECIES AND MICROCOSM TEST METHODS FOR AQUATIC TOXICITY

Multispecies test methods include microcosms and mesocosms. The volume of micro- and mesocosms varies from 0.1 liter to thousands of liters, there is not limit in volume or size, and it is rather their complexity than their size that distinguishes between micro- and mesocosm.

The following are the main characteristics of micro- and mesocosms:

–   They are historical, like the ecosystem itself, and irreversible in time.
–   They have a trophic structure, which is sometimes very simple, sometimes close to the real environment.
–   Evolutionary events occur in micro- and mesocosms: for example biodegrading ability or resistance to xenobiotics may arise and, as a consequence, the structure

and diversity of the community changes by adaptation to the circumstances, e.g. to the contaminant content.

– Evolution of new metabolic pathways for biodegradation (of e.g. pesticides or xenobiotics) is possible. It can also be enforced by adding a xenobiotic stepwise.
– Microcosms and mesocosms are characterized by reduced complexity compared to natural systems: the number of species is smaller.
– The organisms in a microcosm cannot move freely, they cannot leave the test vessel or escape from the test volume.
– Dynamics of the ecosystem differs from the natural one: the enforced isolation into a small scale causes changes in the dynamics of the whole system. The surface to volume ratio in a microcosm differs to a large extent from the natural systems. These changes should be distinguished from the effect of the toxicant.
– Since the solid phase (sediment) is disturbed, the natural equilibrium is not maintained, and a new equilibrium state is established in the microcosms after a transient phase. This may influence the sampling concept and the results.
– Heterogeneity: in natural ecosystems spatial and temporal heterogeneity is the key to species richness. Artificial test systems should not be heterogeneous or unique and must not lose their statistical power.
– Multispecies tests are complex systems and, unlike simple-species tests or biochemical assays, they cannot be repeated due to their dynamics and history as their past is conserved in population dynamics down to the DNA sequence.

All this information should be considered when designing and evaluating microcosm and mesocosm tests. The parallel tests indicate great differences, and certain microcosms can show a completely different behavior than others due to individual evolution.

Micro- and mesocosm tests have a number of goals: a mesocosm can be used to address ecology and management of element cycles, nutrient load due to fertilizer runoff, toxicity of natural and synthetic chemicals, and effects on or manipulation of food chains (Dibble & Pelicice, 2010). Further goals are measurement of adverse effects of biological agents and pests, biodegradation and bioaccumulation of contaminants, the mechanism of the spread of invasive species, etc. The effects of protective measures such as influencing microbial activity to enhance nutrient consumption, degrade contaminants or synthesize useful substances can be modeled. This can include simulating adverse or beneficial effects of physical, chemical and biological agents and investigating and manipulating the response of the benthic-pelagic ecosystem in micro- and mesocosms. The simulations and manipulations are aimed at conserving ecosystems and habitats as well as utilizing the ecosystem in green technologies for environmental protection, waste treatment or production technologies.

We will now introduce the type and set-up of aquatic microcosms and mesocosms based on the literature. These microcosms are mainly used for testing the effect of chemical substances on aquatic diversity (Landis & Yu, 1999).

– Mixed-flask culture mesocosm (MFC) is a small-scale microcosm with a complex community. It works in 1 liter volume for 12–14 weeks. The representative community is separately grown and then added to the test vessel. The community contains two single-cell algae or diatoms, one filamentous algal species, one nitrogen-fixing blue-green alga, one grazing pelagic microinvertebrate, one benthic macroinvertebrate and several bacterial and protozoan species. The effect of xenobiotics on the community diversity is measured.

- Standardized aquatic microcosm (SAM) is a 4 liter glass vessel with artificial sterile sediment and synthetic water. 10 algal, 4 invertebrate and 1 bacterial species are added to the water. Nutrients, pH, temperature and dissolved oxygen are measured and controlled. Test material is added on the 7th day. Parallel tests of treated and control vessels are performed and evaluated. Measured end points are pH, dissolved oxygen, nutrient level, and cell/organism counts.

- Benthic-pelagic microcosms containing both sediment and water are suitable for testing the normal functions of shallow waters or the effect of contaminated water and sediments on benthic and pelagic organisms. Contaminant partition and cycling between water and sediment, as well as sediment solid phase and pore water can be monitored. The effect of pH, redox potential and the different forms of the contaminants on benthic and pelagic organisms can be simulated. Natural or artificial sediment can be placed into the test vessel and natural or separately cultivated organisms into the microcosms.

- Eutrophication pelagic microcosm was developed for measuring chlorophyll synthesis from nutrients. Gowen et al. (1992) proposed the use of the yield of chlorophyll from dissolved nutrient to estimate the risk of marine eutrophication. The yields of chlorophyll from nitrogen were further investigated, under more controlled conditions, in microcosm experiments reported by Edwards (2001), Edwards et al. (2003), Edwards et al. (2005). The apparatus used in these microcosm experiments is a vessel containing seawater, phytoplankton, pelagic bacteria and protozoa, filtered to exclude mesozooplankton. Seawater, filtered of all organisms, and nutrient enriched, was placed in the reservoir tank and steadily pumped into the reactor vessel, giving a dilution rate of about 0.2 per day. Chlorophyll production was measured as an end point.

- Many kinds of static microcosms have been developed for authorization purposes to test aquatic biodiversity and the effect of fungicides, herbicides or insecticide on the abundance of pelagic and benthic species. After reaching a steady state, a known number of phytoplankton, zooplankton, macrozoobenthos, crustaceans and fish species are added to the benthic-pelagic microcosm and monitored over short or long term (Figure 4.25).

- The US-standardized FIFRA mesocosms are test vessels with a surface area of $5\,m^2$, a depth of at least 1.25 m, volume at least $6\,m^3$ and made of inert material. Sediment obtained from an existing pond containing natural benthic community, is placed onto the bottom of the mesocosm on trays, in a 5 cm-thick layer. Water should derive from a healthy, ecologically active pond. The water level in the microcosm should be controlled and regulated by adding or removing water. Weather should be recorded and taken into account during the evaluation of the test results. FIFRA mesocosm uses bluegill sunfish, fathead minnow, channel catfish, phytoplankton, periphyton, zooplankton, emergent insects and benthic macroinvertebrates in the test system. The test protocol prescribes the size of organisms, e.g. the biomass of fish added which should not exceed $2\,g/m^3$ of water. Smaller tanks can be used without fish.

A test substance is added after 6–8 weeks aging of the microcosm, by spraying on the water surface, pouring into the water phase if dissolved in water, or put to the bottom in the form of a soil slurry.

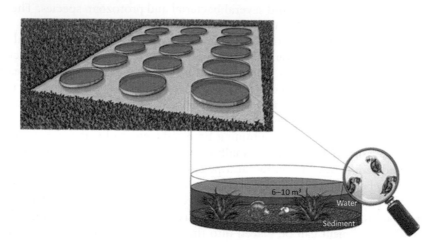

*Figure 4.25* Aquatic mesocosms (outdoor setup).

Sampling begins 2 weeks after the construction of the microcosm, it continues for 2–3 months after the last treatment with the test material. Frequency of testing depends upon characteristics of the test substance and on the treatment regime. Dosage of pesticide level, frequency and number of replicates are determined based on the objectives of the study. Measured end points are oxygen content, respiratory activity, biomass, cell and organism numbers and distribution.

–   Stream microcosms with a flow-through system can model streams.
–   Pond microcosms are models of real ponds simulating the ecosystem of the original pond.
–   Compartmentalized lake means that a natural lake is divided into compartments. A small compartment, which theoretically keeps all characteristics of the whole lake is isolated and used for testing.
–   Sediment core microcosms contain an undisturbed volume of natural sediments trying to keep the original equilibrium between the solid phase and pore-water phase of sediment to get a system as close to realty as possible.
–   Waste treatment microcosms simulate ecological technologies designed for the treatment of liquid and sludge-form wastes such as aerobic and anoxic ponds or wetlands for wastewater treatment, rhizosphere filtering or living machines. (See also Chapter 8).

## 12   DESCRIPTION OF *TETRAHYMENA PYRIFORMIS* BIOASSAY

### M. Molnár

The experimental design and test parameters of the acute (short-term) and chronic (based on reproduction) toxicity tests using *Tetrahymena pyriformis*, a ciliate protozoon, are introduced here (Figure 4.26). This test is neither standardized nor widely

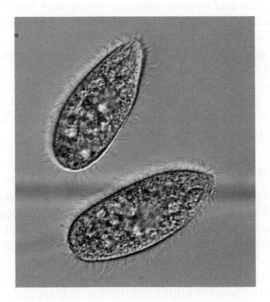

Figure 4.26  *Tetrahymena pyriformis*: a ciliate protozoon.

used, but the authors of this book have experience in applying this sensitive organism for testing toxicity of water, sediment and soil for generic and site-specific risk assessment and decision-making purposes. It is an important organism because the animal trophic level is the bottleneck and is generally represented only by *Daphnia* and fish in ecotoxicity tests. As a single-cell organism, *Tetrahymena pyriformis* is easy to handle and it grows relatively rapidly, so that we give priority to this organism to amend the test batteries for measuring aquatic ecotoxicity. Single-cell animals are good models for higher animals, inasmuch as their cells and metabolisms are identical with higher animals from many points of view.

*Tetrahymena pyriformis* – due to its wide-range sensitivity – can be applied for toxicity testing of chemical substances and environmental samples: water, wastewater, leachate, soil, sediment and solid waste. It can also be used in contact tests.

## 12.1  Experimental

A dilution series should be prepared from the chemical substances or contaminated environmental samples. At least five concentrations are to be tested simultaneously, preferably arranged in a geometric series. The lowest concentration should have no observed effect on the growth. Sterile distilled water is applied for dilution of liquid-phase samples and as a control medium.

*Tetrahymena pyriformis* A-759-b is cultivated in proteose-peptone-yeast extract medium (PPY), containing 1% proteose peptone and 0.1% yeast extract.

The Erlenmeyer flasks (100 mL) contain 600 μL of liquid-phase or 1.5 g of solid-phase environmental samples supplemented with 30 mL of PPY medium and with penicillin, streptomycin and nystatin solutions (to protect *Tetrahymena* from harmful bacteria) at final concentrations of 0.01 mg/mL, 0.1 mg/mL and 0.05 mg/mL, respectively (Leitgib *et al.*, 2007).

After vigorous mixing, 600 µL of exponentially growing culture of test organisms (initial density $1 \times 10^5$ cells/mL) are added to the flasks.

Erlenmeyer flasks are incubated with shaking (125 rpm) in a dark chamber ($20 \pm 2°C$), and the concentration of cells is determined after 0, 24, 48, 72, 96 and 120-h exposure by direct counting in a Bürker counting chamber in a light microscope or any other cell counting methods.

## 12.2  Evaluation and interpretation of the results

The mean cell-concentration values of the parallels of each dilution and the control are log-transformed and plotted versus time to produce growth curves.

Average specific growth rate can be read from the exponential phase of the growth curves and the inhibition rate (%) is calculated by comparing the treated cultures to the untreated control. Inhibition percentage is plotted as a function of concentration of the tested chemical substance or dose of the tested water, soil, sediment, etc. Dose–response analysis using a logistic model by ORIGIN 8.0 software is applied to determine effective concentration or dose values ($E_rC_x$ and $E_rD_x$) or the highest no-effect and/or the lowest effect concentrations/doses ($NOE_rC/LOE_rC$ or $NOE_rL/LOE_rL$) of the samples.

Testing pure chemical substances, the effective concentrations ($E_rC_{20}$ and $E_rC_{50}$) which cause 20% and 50% reduction in growth rate are derived statistically from a plot of percentage inhibition against the logarithm value of the concentration of the substance.

For environmental samples, $E_rD_{20}$ or $E_rD_{50}$ values were the end points of the tests, where $E_rD_{20}$ and $E_rD_{50}$ are doses (mass or volume) of the sample which caused 20% and 50% reduction in the growth rate, respectively. Instead of water dose, the rate of dilution can also be used as an end point of the tests. This means that the result can be given as a quantity of water or soil (depending on the quantity used by the test method), or as a dilution rate which causes 20%, 50% inhibition or "no effect" on the test organism. The toxicity of environmental samples can also be termed as toxicity equivalent – as a comparison to a known toxicant. The results of direct toxicity tests of environmental samples are discussed in Chapter 9 in this Volume and Chapter 2 in Volume 3.

## REFERENCES

AltTox (2007) *Ecotoxicity – Overview*. [Online] Available from: http://www.alttox.org/ttrc/toxicity-tests/ecotoxicity. [Accessed 3rd November 2013].

AltTox (2014) Non-animal Methods for Toxicity Testing. [Online] Available from: http://www.alttox.org/. [Accessed 3rd January 2014].

Anderson, B.S., Hunt, J.W., Hester, M. & Philips, B.M. (1996) Assessment of sediment toxicity at the sediment–water interface. In: Ostrander, G.K. (ed.) *Techniques in aquatic toxicology*. Ann Arbour, MI, USA, Lewis publisher.

APHA (2014) *American Public Health Association*. [Online] Available from: http://www.apha.org. [Accessed 3rd January 2014].

ASTM International (2014) *American Society for Testing and Materials*. [Online] Available from: http://www.astm.org. [Accessed 3rd January 2014].

AWWA (2014) *American Water Works Association*. [Online] Available from: http://www.awwa.org. [Accessed 3rd January 2014].

Barbour, M.T., Gerritsen, J., Snyder, B.D. & Stribling, J.B. (1999) *Rapid bioassessment protocols for use in streams and wadeable rivers: Periphyton, benthic macroinvertebrates and fish*. 2nd Edition. EPA 841-B-99-002. Washington, D.C., U.S. Environmental Protection Agency, Office of Water. [Online] Available from: http://www.epa.gov/owow/monitoring/rbp/download.html and http://water.epa.gov/scitech/monitoring/rsl/bioassessment/index.cfm. [Accessed 3rd November 2013].

BIOLOG (2013) Available from: http://www.biolog.com/company.shtml. [Accessed 3rd November 2013].

Caroni, R. & Irvine, K. (2010) The potential of zooplankton communities for ecological assessment of lakes: redundant concept or political oversight? *Biology & Environment: Proceedings of the Royal Irish Academy*, 110(1), 35–53. [Online] Available from: http://ria.metapress.com/content/1h7jh576185l71q4/. [Accessed 3rd November 2013].

CEN 14735 (2005) *Characterization of waste – Preparation of waste samples for ecotoxicity tests*. European Standard. Brussels, Belgium.

CEN (2014) European Committee for Standardization. [Online] Available from: http://www.cen.eu. [Accessed 3rd January 2014].

CEN Standards (2014) *Water Analyses. Published Standards*. [Online] Available from: http://www.cen.eu/cen/Sectors/TechnicalCommitteesWorkshops/CENTechnicalCommittees/Pages/Standards.aspx?param=6211&title=Water%20analysis. [Accessed 3rd January 2014].

CERIODAPHTOXKIT F (2013) Microbiotest with the freshwater crustacean *Ceriodaphnia dubia*. Available from: http://www.microbiotests.be/toxkits/Ceriodaphtoxkitf.pdf. [Accessed 3rd November 2013].

Chitmanat, C., Prakobsin, N., Chaibu, P. & Traichaiyaporn, S. (2008) The use of acetylcholinesterase inhibition in river snails (*Sinotaia ingallsiana*) to determine the pesticide contamination in the Upper Ping River. *International Journal of Agriculture & Biology*, 10(6), 658–660. ISSN Print: 1560-8530. ISSN Online: 1814-9596 08-014/DJZ/2008/10-6-658-660. [Online] Available from: http://www.fspublishers.org. [Accessed 3rd November 2013].

Christensen, P.D., Nielsen, L.P., Revsbech, M. & Sorensen, J. (1989) Microzonation of denitrification activity in stream sediments as studied with a combined oxygen and nitrous oxide microsensor. *Applied Environmental Microbiology*, 55, 1234–1241.

Csaba, G., Kovács, P. & Pállinger, É. (2005) Hormonal interactions in *Tetrahymena*: Effect of hormones on levels of epidermal growth factor (EGF). *Cell Biology International*, 29(4), 301–305.

DAPHTOXKIT F (2013) microbiotests with the freshwater crustaceans *Daphnia magna* or *Daphnia pulex*. Available from: http://www.microbiotests.be/toxkits/daphtoxkitf.pdf. [Accessed 3rd November 2013].

De Beer, D. & Van den Heuvel, J.C. (1988) Response of ammonium-selective microelectrodes based on the neutral carrier nonactin. *Talanta*, 35, 728–730.

Dibble, E.D. & Pelicice, F.M. (2010) Influence of aquatic plant-specific habitat on an assemblage of small neotropical floodplain fishes. *Ecology of Freshwater Fish*, 19(3), 381–389.

Dobler, R., Burri, P., Gruiz, K., Brandl, H. & Bachofen, R. (2001) Variability in microbial population in soil highly polluted with heavy metals on the basis of substrate utilization pattern analysis. *Journal of Soils and Sediments*, 1(3), 151–158.

DSD (1967) Council Directive 67/548/EEC of 27 June 1967 on the approximation of laws, regulations and administrative provisions relating to the classification, packaging and labelling of dangerous substances. [Online] Available from: http://eur-lex.europa.eu/

LexUriServ/LexUriServ.do?uri=CELEX:31967L0548:EN:HTML. [Accessed 3rd November 2013].

EAMD (2013) Ecological Assessment Methods Database. [Online] Available from: http://assessmentmethods.nbii.gov/index.jsp?page=mdetail&mid=1and http://assessmentmethods.nbii.gov/index.jsp?page=methods. [Accessed 3rd November 2013].

EC (2014) *Environment Canada*. [Online] Available from: http://www.ec.gc.ca. [Accessed 3rd January 2014].

Ecotox (2014) *Ecotoxicity database*, maintained by U.S. Environmental Protection Agency (EPA). [Online] Available from: http://cfpub.epa.gov/ecotox. [Accessed 3rd January 2014].

Edwards, V., Icely, J., Newton, A. & Webster, R. (2005) The yield of chlorophyll from nitrogen: a comparison between the shallow Ria Formosa lagoon and the deep oceanic conditions at Sagres along the southern coast of Portugal. *Estuarine Coastal and Shelf Science,* 62, 391–403.

Edwards, V.R. (2001) *The yield of marine phytoplankton chlorophyll from dissolved inorganic nitrogen under eutrophic conditions.* PhD thesis. School of Life Sciences. Napier University, Edinburgh.

Edwards, V.R., Tett, P. & Jones, K.J. (2003) Changes in the yield of chlorophyll a from dissolved available inorganic nitrogen after an enrichment event – applications for predicting eutrophication in coastal waters. *Continental Shelf Research,* 23, 1771–1785.

EU EQS (2008) *Environmental quality standards in the field of water policy* (EQS Directive). 2008/105 Directive of the European Parliament and of the Council, amending Directives 82/176/EEC, 83/513/EEC, 84/156/EEC, 84/491/EEC, 86/280/EEC and 2000/60/EC. [Online] Available from: http://eur-lex.europa.eu/LexUriServ/LexUriServ.do?uri=CELEX:32008L0105:en:NOT. [Accessed 6th November 2013].

European Ring Test of Ecotoxicity of Waste (2013) [Online] Available from: http://ecotoxwasteringtest.uba.de/h14/index.jsp?id=ecotoxtest_lu_h14rt. [Accessed 3rd November 2013].

Finney, D.J. (1978) *Statistical methods in biological assay.* 3rd edition. London, UK, Griffin, Weycombe.

Gowen, R.J., Tett, P. & Jones, K.J. (1992) Predicting marine eutrophication: the yield of chlorophyll from nitrogen in Scottish coastal phytoplankton. *Marine Ecology – Progress Series,* 85, 153–161.

Gruiz, K. (2009) Integrated and efficient assessment of polluted sites. *Land Contamination and Reclamation,* 17(3–4), 371–384.

Gruiz, K., Molnar, M. & Feigl, V. (2009) Measuring adverse effects of contaminated soil using interactive and dynamic test methods. *Land Contamination and Reclamation,* 17(3–4), 443–460.

ISO (2014) *International Organization for Standardization.* Available from: http://www.iso.org/iso/home.html. [Accessed 3rd November 2013].

ISO Standards (2014) ISO/TC 147/SC 5 *Biological methods.* Standard Catalogue. [Online] Available from: http://www.iso.org/iso/home/store/catalogue_tc/catalogue_tc_browse.htm?commid=52972. [Accessed 6th November 2013].

ISO 11348 (2007) *Water quality – Determination of the inhibitory effect of water samples on the light emission of Vibrio fischeri (Luminescent bacteria test).* [Online] Available from: http://www.iso.org/iso/iso_catalogue/catalogue_tc/catalogue_detail.htm?csnumber=40518. [Accessed 3rd November 2013].

Jensen, K., Revsbech, N.P. & Nielsen, L.P. (1993) Microscale distribution of nitrification activity in sediment determined with a shielded microsensor for nitrate. *Applied Environmental Microbiolology,* 59, 3287–3296.

Kramer, K.J.M. & Foekema, E.M. (2000) The 'Mussel monitor' as biological early warning system: the first decade. In: Butterworth, F., Gunatilaka, A. & Gonsebatt, M. (eds.) *Biomonitors and biomarkers as indicators of environmental change.* New York, Kluwer. pp. 59–87.

Landis, W.G. & Yu, M.H. (1999) *Introduction to environmental toxicology: impact of chemicals upon ecological systems.* New York, Boca Raton, Florida, CRC Press LLC.

Leitgib, L., Kálmán, J. & Gruiz, K. (2007) Comparison of bioassays by testing whole soil and their water extract from contaminated sites. *Chemosphere*, 66, 428–434.

Leynen, M., Van den Berckt, T., Aerts, J.M., Castelein, B., Berckmans, D. & Olleviera, F. (1999) The use of *Tubificidae* in a biological early warning system. *Environmental Pollution*, 105(1), 151–154.

MESOCOSM (2013) Supplier of aquatic test organisms. [Online] Available from: http://www.mesocosm.de/Supplier-of-aquatic-test-organ.220.0.html. [Accessed 6th November 2013].

Mortimer, M., Kasemets, K. & Kahru, A. (2010) Toxicity of ZnO and CuO nanoparticles to ciliated protozoa *Tetrahymena thermophila. Toxicology*, 269, 182–189.

Moser, H., Roembke, J., Donnevert, G. & Becker, R. (2011) Evaluation of biological methods for a future methodological implementation of the hazard criterion H14 'ecotoxic' in the European waste list (2000/532/EC). *Waste Management Research*, 29, 180–187.

Musselmonitor (2013) [Online] Available from: http://www.mosselmonitor.nl. [Accessed 6th November 2013].

Nagy, Zs., Gruiz, K., Molnár, M. & Fenyvesi, É. (2010) Biodegradation in 4-chlorophenol contaminated soil. In: *Proceedings CD, ConSoil 2010 Conference, Salzburg 22–24 September*, Poster No. A2-19 ISBN 978-3-00-032099-6.

Novotny, C., Dias, N., Kapanen, A., Malachova, K., Vándrovcová, M., Itavaara, M. & Lima, N. (2006) Comparative use of bacterial, algal and protozoan tests to study toxicity of azo- and anthraquinone dyes. *Chemosphere*, 63, 1436–1442.

OECD 201 (2011) OECD *Guidelines for Testing of Chemicals. Fresh-water Alga and Cyanobacteria, Growth Inhibition Test.* [Online] Available from: http://www.oecd-ilibrary.org/docserver/download/9720101e.pdf?expires=1400012008&id=id&accname=guest&checksum=54DF350B6F3390AF9427D9570AAC1009. [Accessed 4th May 2014].

OECD 202 (2004) OECD *Guidelines for the Testing of Chemicals. Daphnia sp. Acute Immobilisation Test.* [Online] Available from: http://www.oecd-ilibrary.org/docserver/download/9720201e.pdf?expires=1400011943&id=id&accname=guest&checksum=D78DBCDB42F853DEC14E00351D677F54. [Accessed 4th May 2014].

OECD 211 (2012) OECD *Guidelines for the Testing of Chemicals. Daphnia magna Reproduction Test.* [Online] Available from: http://www.oecd-ilibrary.org/docserver/download/9712171e.pdf?expires=1400011866&id=id&accname=guest&checksum=255116966BF9599B00F3E208EE661458. [Accessed 4th May 2014].

OECD 221 (2006) OECD *Guidelines for the Testing of Chemicals. Lemna sp. Growth Inhibition Test.* [Online] Available from: http://www.oecd-ilibrary.org/docserver/download/9722101e.pdf?expires=1400012047&id=id&accname=guest&checksum=0F817EAC5B6E7082FDBE1F84EF6EA4CB. [Accessed 4th May 2014].

OECD (2014) *Organization for Economic Co-operation and Development.* [Online] Available from: http://www.oecd.org. [Accessed 3rd January 2014].

OECD (2014) OECD Guidelines for the Testing of Chemicals, Section 2 Effects on Biotic Systems. [Online] Available from: http://www.oecd-ilibrary.org/environment/oecd-guidelines-for-the-testing-of-chemicals-section-2-effects-on-biotic-systems_20745761. [Accessed 6th November 2013].

PROTOXKIT F (2013) Microbiotests with the ciliate protozoan *Tetrahymena thermophila.* [Online] Available from: http://www.microbiotests.be/toxkits/protoxkitf.pdf. [Accessed 6th November 2013].

Ramsing, N.B., Kuhl, M. & Jorgensen, B.B. (1993) Distribution of sulfate-reducing bacteria, $O_2$ and $H_2S$ in photosynthetic biofilms determined by oligonucleotide probes and microelectrodes. *Applied Environmental Microbiology*, 59, 3840–3849.

Revsbech, N.P. (1989) An oxygen microsensor with a guard cathode. *Limnology & Oceanography*, 34, 474–478.

Revsbech, N.P. & Jorgensen, B.B. (1986) Microelectrodes: their use in microbial ecology. *Advances in Microbial Ecology*, 9, 293–352.

Robinson, R. (2006) Ciliate Genome Sequence Reveals Unique Features of a Model Eukaryote. *PLoS Biology*, Vol. 4/9/2006, e304. DOI:10.1371/journal.pbio.0040304.g001. [Online] Available from: http://dx.doi.org/10.1371/journal.pbio.0040304. [Accessed 6th November 2013].

Rosen, G.B., Chadwick, D., Poucher, S.L., Greenberg, M.S. & Burton, G.A. (2009) *In situ* estuarine and marine toxicity testing. A review, including recommendations for future use in ecological risk assessment. Technical Report No 1986, SSC Pacific. [Online] Available from: http://www.dtic.mil/cgi-bin/GetTRDoc?Location=U2&doc=GetTRDoc.pdf&AD=ADA51 3788. [Accessed 6th November 2013].

ROTOXKIT F (2013) Microbiotests with the freshwater rotifer *Brachionus calyciflorus*. [Online] Available from: http://www.microbiotests.be/toxkits/rotoxkitf.pdf. [Accessed 6th November 2013].

ROTOXKIT M (2013) Microbiotests with the marine rotifer *Brachionus plicatilis*. [Online] Available from: http://www.microbiotests.be/toxkits/rotoxkitm.pdf. [Accessed 6th November 2013].

RS (2013) *Development of water quality indicators using remote sensing*. [Online] Available from: http://www.epa.gov/eerd/RemoteSensing.htm. [Accessed 6th November 2013].

Sander, K., Schilling, S., Lüskow, H., Gonser, J., Schwedtje, A. & Küchen, V. (2008) *Review of the European list of waste final report executive summary*. [Online] Available from: http://ec.europa.eu/environment/waste/pdf/low_review_oekopol.pdf. [Accessed 6th November 2013].

Sauvant, M.P., Pepin, D. & Piccinni, E. (1999) *Tetrahymena pyriformis*: for toxicological studies. A review. *Chemosphere*, 38(7), 1631–1669.

Scarlett, A., Rowland, S.J., Canty, M., Smith, E.L. & Galloway, T.S. (2007) Method for assessing the chronic toxicity of marine and estuarine sediment-associated contaminants using the amphipod *Corophium volutator*. *Marine Environmental Research*, 63(5), 457–470.

SETAC (2014) Society of Environmental Toxicology and Chemistry. [Online] Available from: https://www.setac.org. [Accessed 3rd January 2014].

Standard methods (2014) Standard Methods for the Examination of Water and Wastewater. [Online] Available from: http://www.standardmethods.org. [Accessed 3rd January 2014].

Stefanidou, M., Alevizopoulos, G. & Spiliopoulou, C. (2008) DNA content of *Tetrahymena pyriformis* as a biomarker for different toxic agents. Short communication. *Chemosphere*, 74, 178–180.

Stephan, C.E. (1977) Methods for calculating an $LC_{50}$. In: Mayer, F.I. & Hamelink, J.L. (eds.) *Aquatic toxicology and hazard evaluation*. ASTM STP 634 – American Society for Testing and Materials. pp. 65–84.

Test methods for REACH (2008) Council Regulation (EC) No 440/2008 of 30 May 2008 laying down test methods pursuant to Regulation (EC) No 1907/2006 of the European Parliament and of the Council on the registration, evaluation, authorisation and restriction of chemicals (REACH). *Official Journal*, L 142(31/05/2008), 0001–0739. [Online] Available from: http://eur-lex.europa.eu/LexUriServ/LexUriServ.do?uri=CELEX:32008R0440:EN:HTML. [Accessed 3rd November 2013].

THAMNOTOXKIT F (2013) Microbiotests with the crustacean *Thamnocephalus platyurus*. [Online] Available from: http://www.microbiotests.be/toxkits/thamnotoxkitf.pdf. [Accessed 6th November 2013].

Thompson, K.C., Wadhia, K. & Loibner, A.P. (eds.) (2005) *Environmental toxicity testing*. Oxford, Blackwell Publishing.

TIE (2007) *Sediment toxicity identification evaluation (TIE). Phases I, II, and III Guidance Document*. Washington, DC, United States Environmental Protection Agency, Office of Research and Development, 20460, EPA/600/R-07/080. [Online] Available from: http://www.epa.gov/nheerl/publications/files/Sediment%20TIE%20Guidance%20Document.pdf. [Accessed 6th November 2013].

UBA (2014) Umweltbundesamt. [Online] Available from: http://www.umweltbundesamt.de/index-e.htm. [Accessed 6th November 2013].

US EPA (2002a) *Methods for measuring the acute toxicity of effluents and receiving waters to fresh-water and marine organisms*. EPA-821-R-02-012. U.S. Environmental Protection Agency Office of Water (4303T). [Online] Available from: http://water.epa.gov/scitech/methods/cwa/wet/upload/2007_07_10_methods_wet_disk2_atx.pdf. [Accessed 6th November 2013].

US EPA (2002b) *Short-term methods for estimating the chronic toxicity of effluents and receiving waters to freshwater organisms*. EPA-821-R-02-013. U.S. Environmental Protection Agency Office of Water (4303T). [Online] Available from: http://water.epa.gov/scitech/methods/cwa/wet/upload/2007_07_10_methods_wet_disk3_ctf.pdf. [Accessed 6th November 2013].

US EPA (2002c) *Short-term methods for estimating the chronic toxicity of effluents and receiving waters to marine and estuarine organisms*. EPA-821-R-02-014. U.S. Environmental Protection Agency Office of Water (4303T). [Online] Available from: http://water.epa.gov/scitech/methods/cwa/wet/upload/2007_07_10_methods_wet_disk1_ctm.pdf. [Accessed 6th November 2013].

US EPA (2007) *Ecotoxicity database*. [Online] Available from: http://www.epa.gov/opp00001/science/efed_databasesdescription.htm. [Accessed 6th November 2013].

US EPA (2014) *United States Environmental Protection Agency*. [Online] Available from: http://www.epa.gov. [Accessed 3rd January 2014].

US NPDS (2004) US Environmental Protection Agency, Office of Enforcement and Compliance Assurance NPDES Compliance inspection manual, EPA 305-X-04-001. [Online] Available from: http://www.epa.gov/compliance/resources/publications/monitoring/cwa/inspections/npdesinspect/npdesinspect.pdf. [Accessed 6th November 2013].

WFD (2000) *Directive 2000/60/EC of the European Parliament and of the Council of 23 October 2000 establishing a framework for Community action in the field of water policy*. [Online] Available from: http://eur-lex.europa.eu/LexUriServ/LexUriServ.do?uri=CELEX:32000L0060:EN:HTML. [Accessed 6th November 2013].

WIESER (2014) *Methods for assessing and restoring aquatic ecosystems*. Water bodies in Europe – Integrative Systems to assess Ecological Status and Recovery. [Online] Available from: http://www.wiser.eu. [Accessed 3rd January 2014].

WPCF (2014) Water Pollution Control Federation. [Online] Available from: http://www.wef.org. [Accessed 3rd January 2014].

## PHOTOS

*Aedes notoscriptus* (2014) *Aedes notoscriptus* mosquito larva. [Online] Available from: http://uq.edu.au/integrative-ecology/evolution-of-inducible-defences-in-mosquito-larvae. [Accessed 6th November 2013].

Bluegill (2004) Photo of Briandykes. [Online] Available from: http://en.wikipedia.org/wiki/File:Bluegill_sc.jpg. [Accessed 6th November 2013].

*Ceratophyllum demersum (2009)* Aquarium fish. [Online] Available from: http://diszhal.info/nov enyek/Ceratophyllum_demersum.php. [Accessed 6th November 2013].

*Closterium ehrenbergii* (2014) Alga Resource Database, microscopic photo. [Online] Available from: http://www.shigen.nig.ac.jp/algae/images/strainsimage/nies-0228.jpg. [Accessed 6th November 2013].

Damselfly nymph (2014) [Online] Available from: http://scioly.org/wiki/index.php/File: Damselnymph.jpeg from: Water Quality Macroorganism List. http://scioly.org/wiki/index. php/Water_Quality/Macroorganism_List. [Accessed 6th November 2013].

*Desmodesmus* species (2014) Alga Resource Database, microscopic photo. [Online] Available from: http://www.shigen.nig.ac.jp/algae/images/strainsimage/nies-0096.jpg. [Accessed 6th November 2013].

*Draconema* (2014) Darwin Project. [Online] Available from: http://www.reefed.edu.au/home/ explorer/animals/marine_invertebrates/nematode_worms. [Accessed 6th November 2013].

Gilled snail (2014) [Online] Available from: http://scioly. org/wiki/images/c/ca/Unknown.jpeg from: Water Quality Macroorganism List. http://scioly.org/wiki/index.php/Water_Quality/ Macroorganism_List. [Accessed 6th November 2013].

Gjertsen, K.R. (2007) *Hirudo medicinalis*, medical leech. [Online] Available from: http:// en.wikipedia.org/wiki/File:Sv%C3%B8mmende_blodigle.JPG. [Accessed 6th November 2013].

Hodges, J.C. Jr. (2013) Trichoptera, Limnephilidae: *Dicosmoecus avripes* larva & case. [Online] Available from: http://freshwater-science.org. http://freshwater-science.org/Education-and-Outreach/Media-Galleries/Invertebrates.aspx?imagepath=%2fEducation-and-Outreach%2f Media-Galleries%2fMedia%2fimages%2fmacro062-jpg. [Accessed 3rd November 2013].

Midgenymph (2014) [Online] Available from: http://scioly.org/wiki/index.php/File:Midgenymph. jpeg from: Water Quality Macroorganism List. http://scioly.org/wiki/index.php/Water_ Quality/Macroorganism_List. [Accessed 6th November 2013].

*Myriophillum aquaticum* (2011) Aquarium fish. [Online] Available from: http://diszhal.info/ novenyek/Myriophyllum_aquaticum.php. [Accessed 6th November 2013].

Penrose, D. (2013) Diptera, Chironomidae larvae. [Online] Available from: http://freshwater-science.org, http://freshwater-science.org/education-and-outreach/media-galleries/inverteb rates.aspx?imagepath=/Education-and-Outreach/Media-Galleries/Media/images/macro100-jpg). [Accessed 6th November 2013].

*Pseudokirchneriella subcapitata* (2014) Alga Resource Database, micoscopic photo. [Online] Available from: http://www.shigen.nig.ac.jp/algae/images/strainsimage/nies-0035. jpg.[Accessed 6th November 2013].

Rainbow trout (2013) Photo of TerminaVelocity. [Online] Available from: http://static4.wikia. nocookie.net/__cb20130730132445/acnl/images/0/01/RainbowTroutIRL.jpg. [Accessed 6th November 2013].

Stonefly nymph (2014) [Online] Available from: http://scioly.org/wiki/index.php/File:Stonefly_ nymph.jpeg from:  http://scioly.org/wiki/index.php/Water_Quality/Macroorganism_List. [Accessed 6th November 2013].

Pictures without reference are commercially available or prepared by the authors.

Chapter 5

# Terrestrial toxicology

K. Gruiz, M. Molnár, V. Feigl, Cs. Hajdu, Zs. M. Nagy,
O. Klebercz, I. Fekete-Kertész, É. Ujaczki & M. Tolner
Department of Applied Biotechnology and Food Science,
Budapest University of Technology and Economics, Budapest, Hungary

## ABSTRACT

Soil is the key compartment of the terrestrial ecosystem, the scene of element cycling, the habitat of myriads of bacteria and other microorganisms, plants, fungi and animals. The intensively interacting solid, liquid and gas phases of the soil together with the complex microbiota ensure an enormous capacity of genetic, chemical, biochemical and biological activities, storage and buffering capacity of the terrestrial habitat.

Exposure to chemical substances endangers the delicate equilibrium of the soil ecosystem and leads to soil deterioration. Science still cannot precisely identify a healthy soil structure and function, so assessment of deviations from a healthy state and identification of the unacceptable scale of changes are difficult both in theory and practice. Terrestrial toxicology may provide valuable information about chemical risk in soil. Terrestrial toxicology seeks to understand the soil's capability to buffer adverse effects by capturing and releasing toxicants. The toxicity buffering capacity together with air, water and nutrient management protects soil habitats and at the same time serves the benefit of the environment as a whole. Simplified chemical, biological and ecological models as well as complex field assessments are applied in terrestrial ecotoxicology depending on the aim of the survey. Mobility, bioavailability, partition between physical phases, degradation, biological uptake, bioaccumulation and any other outcome of the interaction of the soil matrix and its inhabitants with the contaminant play an important role in chemical substances exerting their effect.

This chapter describes the basic knowledge of test organisms and test methods used in terrestrial toxicology. Microcalorimetry, which is introduced in detail, is not a typical toxicity testing tool although heat production is directly related to all the organism's activities. The authors introduce the application of TAM (Thermo Activity Monitor) for testing the effect of chemical substances in soil.

## I  INTRODUCTION

Chemical contaminants may be present in the soil and sediments for a long time during their life cycle. They often enter the soil via direct processes such as fertilization, plant nutrient supply, pesticides application and disposal of wastes on or into the soil. Atmospheric deposition from airborne contaminants, first of all dry particulate matter with adsorbed contaminants, acid rain and pollutants dissolved or suspended in precipitation also deliver considerable amount of organic and inorganic micropollutants into the soil. Floods transport both suspended and dissolved matter onto the soil, polluting the flood area. Contaminants of subsurface waters are filtrated and concentrated by soil solids.

Contamination and soil deterioration endangers ecosystem services attributed to soil. Most of the ecosystem services are based on good health and function of the soil, and the activity of the terrestrial ecosystem. The most important ecosystem services the soil provides are:

Supporting services:

– Nutrient cycling and primary production which underlie the delivery of all other services but are not directly accessible to people.

Provisioning services:

– Groundwater for human uses;
– Food, crops, wild food, spices;
– Pharmaceuticals, biochemicals, industrial products;
– Energy (geothermal, biomass for fuel).

Regulating services:

– Carbon sequestration and climate regulation;
– Water cycling;
– Water and wastewater purification;
– Element cycling;
– Natural hazard, pest and disease control;
– Waste decomposition and detoxification;
– Contaminant elimination;
– Crop pollination.

Cultural services:

– Recreation and ecotourism service;
– Cultural, intellectual and spiritual inspiration;
– Scientific discovery.

The aim of soil toxicology is to understand how contaminants affect the soil ecosystem and, as a consequence, the ecosystem services. Soil toxicology determines the quantitative relationship between contaminant concentration and adverse effects. This knowledge enables forecasting and managing the soil ecosystem in order to prevent or stop soil deterioration. Risk levels acceptable over the long term ensure good-quality and sustainable use of soil at local, regional and global levels.

Soil health, diversity and functions can be characterized by monitoring soil on a regular basis, identifying the adverse trends and trying to reverse them. The main global soil deterioration trends are the prime targets of soil monitoring, including:

– Loss of diversity;
– Loss of organic matter/humus content;

- Acidification;
- Sodification;
- Erosion;
- Compaction;
- Contamination by chronic and acute emissions from point and diffuse sources.

Adverse effects on the terrestrial environment include:

- Effects on soil functions and well balanced nutrient cycling, ensuring nutrients for plants, microorganisms and soil invertebrates;
- Effects on plant biomass production and plant activities, such as participating in nutrient cycling and ensuring equilibrium of gases in the atmosphere;
- Effects on invertebrates, which represent food for other organisms, and covers essential roles as pollinators, detrivores, saprophages and pest controllers;
- Effects on terrestrial vertebrates, domestic and wild species;
- Accumulation of toxic compounds in food through the food chain, endangering higher animals and humans;
- Effect on humans through cultivated and wild species, by the deterioration of the landscape, and endangering recreation in the nature.

The extent of contamination is continuously growing world-wide. Its origin is emission from point sources (production units, chemical plants) and diffuse deposition from polluted air, diffuse pollution from mining and ore processing, from agriculture using persistent chemicals (nutrients and pesticides) as well as legal or illegal waste disposal on soils.

Humans are exposed to contaminated soil as shown in Figure 5.1:

- Directly (children play on the ground and sometimes eat soil; farmers and gardeners are in intensive contact with soil);
- Via the use of groundwater and surface waters (drinking, bathing, irrigation, etc.);
- Through dust inhalation;
- By consuming plant and animal products contaminated through the food chain.

Ecosystem health can be characterized by ecological assessment methods, describing species diversity and their changes in soil, or ecotoxicological assessment using laboratory bioassays, microcosms or mesocosms and their extrapolation to the real terrestrial ecosystem.

- The trends of ecosystem changes in soil can only be properly interpreted if the healthy soil characteristics and the acceptable rate of change are known. This acceptable rate of change is almost impossible to specify because the soil biota have an extremely high adaptive potential and their survival and activity do not necessarily mean soil health. Species diversity and activity of up to $10^9$ cells in one gram of soil together with higher soil-dwelling organisms and plants can characterize soil health. The limit of acceptable adverse changes is normally 5% in terms of species diversity. Soil diversity assessment faces rather high uncertainties due to spatial heterogeneity, climatic and seasonal changes and biological adaptation. Our insufficient knowledge of species diversity in healthy soil, for example

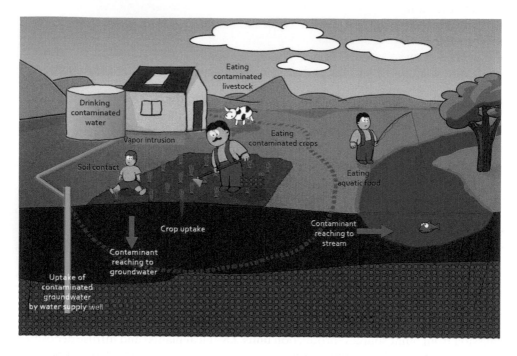

*Figure 5.1* Humans endangered by contaminated soil.

the hidden minor components of the microbial community or the rate of healthy natural adaptation and deviations may cause serious problems when assessing and interpreting species diversity.

– Testing the response of certain species which can represent the soil ecosystem as a whole, is another strategy in soil health management and risk assessment. It can be extrapolated from soil toxicity tests on representative test organisms to the soil ecosystem using safety factors or known distributions for extrapolation to the real ecosystem. In practice we use soil-living microbes, plants and animals representing three different trophic levels used to determine the limit concentrations of *effect* and *no effect* – the lowest concentration which shows a measurable effect or the highest tested concentration which fails to show any effect. These values are used for risk management and decision making. The main uncertainty of this strategy derives from the representativeness of the selected test organisms.

Soil ecotoxicity results can be used for screening, monitoring and control as well as for direct decision making in management in general or for specific contaminated sites.

Most European countries regulate contaminated site management by standard requirements and methodologies. The basic theory suggests that contaminated site management should be *risk based*, giving priority to environmental risks considering (future) land use and including economic and social impacts. All contaminated sites and soils are individual cases where non-identified mixtures of contaminants together with their derivatives and metabolites typically occur. This situation together with local environmental conditions results in a unique combination and requires site-specific management. This means that one should assess locally relevant adverse effects on local land users (ecosystem and humans) and make decisions based on the site-specific risk which arises from the results of actual adverse effects.

Toxicity and other adverse effects typically can serve as information for direct decision making in the first exploratory step of a tiered assessment and in recurrent cases under a generic or constant environment where toxic effect is the only variable of risk assessment. In the first assessment tier a *yes* or *no* answer on toxicity may decide on the exclusion from or inclusion in further assessment. Quality and further use of *ex situ* remedied soil (e.g. in a soil treatment plant) can be decided based on its toxicity. Effect-based decisions can facilitate rapid and efficient risk management in many cases.

Direct decision making *in situ*, based on measured soil toxicity or other observed adverse effects can be used to decide whether:

– The in-depth contaminated site assessment should be continued;
– Sampling and further analysis are needed;
– Contaminated soil should be removed from the environment, the fraction to be removed should be delineated;
– Treated, remedied soils should be utilized;
– Plants should be applied to deteriorated, contaminated or remedied soils.

Figure 5.2 shows the different forms of contaminants and their binding to and transport by the different components of the soil matrix:

– Volatilization;
– Dissolution in pore water, groundwater or runoff;
– Chemical reactions with the environmental constituents (air, water, components of the matrix) and contaminants with each other;
– Biological transformation by microbial, plant and animal impact as well as by free enzymes bound to soil particles;
– Binding to humus and soil organic matter;
– Binding to oxides and clay minerals;
– Sorption to roots;
– Uptake by microorganisms, plants or animals.

The combination of several contaminants and all the above-mentioned transport and fate characteristics varied with the environmental conditions (temperature, pH, redox potential, osmotic characteristics, etc.) are endless and hardly predictable.

Chemical models often fail in contaminated soil assessment due to the strong interaction between contaminants and the complex solid matrix of the soil. The consequence of the various interactions is that partition, chemical availability (by water, by solvents or by other contaminants) and biological accessibility/availability of contaminants may significantly differ from each other. The interactions between contaminants, the three phases of the soil matrix and soil-dwelling organisms greatly influence the effect of contaminants on soil community. The same interactions influence the response

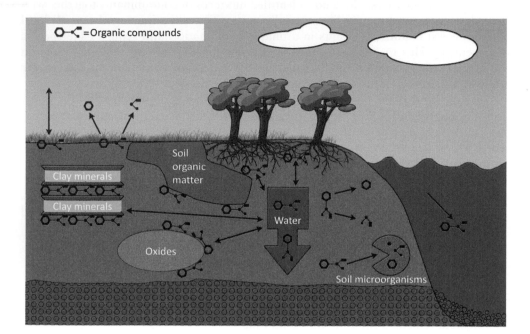

*Figure 5.2*   Fate and behavior of a contaminant in soil.

of the test organisms in a bioassay or other tests for measuring ecotoxicity. Extrapolation from chemical data to the risk of contaminants on soil ecosystem or soil-using humans is a multi-step procedure handicapped by multiple uncertainties. There have been many trials to refine the chemical models, e.g. by incorporating biomimetic chemical reagents or sequential contaminant extraction, or parallel extraction by different solvents (see Chapter 2.6). However, these complicated and rather expensive chemical analytical methods still cannot integrate all the biological and genetic factors of the ecosystem and simulate the current rate adverse effects of contaminants. In addition, these complex chemical models based results are often difficult to interpret.

Complex environmental situations such as the soil itself with the three physical phases, complicated matrix materials, milliards of living cells and mixtures of contaminants require an integrated approach for characterization and monitoring (Gruiz *et al.*, 2001; Gruiz, 2009; Gruiz *et al.*, 2009). Integrated approach means the simultaneous application of physico-chemical and biological, toxicological and ecological methods, as well as the integrated evaluation and interpretation of their results.

The strong binding of pollutants to the solid matrix may lead to high contaminant concentrations in soil. At the same time, high sorption capacity of soil reduces contaminant availability to water and biological systems. The buffering and filtering behavior of the soil is beneficial for the waters and the ecosystem, but this "filter" is not separated, it is in close contact and dynamic interaction with the environment thus the bound contaminants can be released again, depending on actual environmental conditions. Organic contaminants are strongly bound to the soil's organic fraction (humus) and inorganic contaminants both to clay minerals and humus, thus the transport of contaminants by soil moisture and groundwater and their uptake by the biota is restricted by their retention in soil-solids. The high concentration of contaminants may turn the soil into a chemical time bomb, which may explode when environmental conditions reduce the sorption and toxicity buffering capacity of the contaminated soil.

This is why direct toxicity testing of soils and dynamic test methods are so important: they can measure actual toxicity and changes in contaminant mobility and availability. The mobility and availability of a contaminant in soil is the result of diffusion, dissolution, emulsification, suspension, solubilization in water, or in water-based ionic (diluted acids or alkali) and organic solvents (chelating and the complexing agents, biotensides). These are key processes in seeking material balances in soil and are in continuous variation as the response of soil to external conditions, to which different equilibria belong. Soil analysis and testing require a strict and problem-specific concept for the measurement of the effects under conditions that are capable of truly modeling the concrete problem. When the conceptual model does not fit to the problem, the assessor can easily under- or overestimate the risk posed by soil contaminants.

Toxicity end points of terrestrial ecosystems can be selected within a wide range according to the type of chemical substance, source, transport routes, land uses and receptors. Any change in an organism, which can clearly be attributed to the adverse effect of a contaminant, can be applied as an end point. Options are:

- Mineralization by soil microorganisms: any enzyme or metabolic pathway taking part in biodegradation, such as
  o hydrolyzing enzymes: cellulose, hemicellulose, carbohydrate, lipid and protein hydrolyzing enzymes;
  o oxidation and respiration: ATP production, methane oxidation, [14]C labeled substrate oxidation, dehydrogenase activity, oxygen uptake, $CO_2$ generation or $O_2$ consumption using endogenous or added substrates;
  o specific xenobiotics-degrading enzymes and mechanisms;
  o damage to microbial activities in hydrolyzing and denitrifying enzymes, in nitrification or nitrogen fixation.
- Plant germination, growth and metabolism:
  o decrease in photosynthesis rate;
  o changes in the composition of chlorophyll or other photosynthetic pigments;
  o damage to evapotranspiration;
  o enzymatic alterations;
  o effects on biochemical plant protection mechanisms, increased susceptibility to pathogens (viruses, bacteria, fungi) or pests;
  o ineffective pollination, reduction of gain of biomass, loss of leaves, mortality.
- Interactions between soil microbes and plants:
  o loss of mycorrhiza, rhizoplane and root microflora;
  o diminished nodulation of leguminous plants.
- Soil animals and their molecular, cellular, population or community level responses:
  o changes in enzyme synthesis
  o endocrine and immune system disruption;

- o stress response by producing stress proteins such as metallothioneins or heat-shock proteins;
  - o adverse effects on reproduction rates, natural sex ratio, increased susceptibility to infectious diseases and mortality.
- – Ecosystem structure, population densities and diversities:
  - o adverse changes in population dynamics, species diversity, yield, energy transfer and food chain structure.
- – Plant–animal interactions:
  - o influence of adversely affected plants on animals;
  - o influence of adversely affected animals on plants.

*Toxicity bioassays* may substantially diminish uncertainty regarding biological availability, uptake, and biological toxicant effects, by ensuring completely controlled, standard circumstances. Besides the strengths of single-species tests such as simplified and interpretable systems, there are also weaknesses, first of all the lack of environmental realism.

*Multispecies test systems* such as terrestrial model systems (microcosms) offer a potentially useful compromise between the simplicity of laboratory tests and the complexity of field assessments. Inexpensive test systems using field microbiota communities and soil microinvertebrates (nematodes, mites, springtails, and other microarthropods) have been used to investigate the effects of toxicants in terrestrial systems since the 1980s. Community-level adverse effects on soil microbiota and the disruption of food web structures can be measured in microcosm systems. A shift in bacterial community composition and, as a consequence, changes in soil activities can be detected by biochemical tests, measuring enzyme activities in multiwell systems such as BIOLOG (2013).

**In situ** *testing and biomonitoring* is the most realistic type of toxicity testing. In this test organisms are exposed to contaminated soil in the field, wherein plants and animals are subject to both realistic conditions and natural variations in exposure. This testing protocol is best applicable to sessile organisms, for example plants, or caged animals, but other disadvantages and unrealistic conditions, e.g. the lack of avoidance must be taken into account. Biological monitoring may integrate and evaluate cumulative impacts from a variety of natural and anthropogenic stressors.

Terrestrial toxicity is an adverse environmental condition for soil-living microbes, soil-dwelling organisms, and all kind of plants, terrestrial wildlife and of course humans, using terrestrial ecosystem services. Soil plays a role in the life cycles of water, chemical elements and organic matter, including contaminants. When the biodegradability or the bioaccumulative potential of toxic substances is the question, biotesting provides information on what happens when contaminated soil meets the organism. Biodegradation of chemical substances in soil can be easily measured both in static and dynamic test systems, and this is introduced in Chapters 3 and 8. Nevertheless, predicting the environmental fate and behavior of chemicals in soil and their effect on living organisms over the long term from measured results is burdened with uncertainties due to heterogeneity and individual character of the soils of different areas. For long-term predictions the analyst has to investigate and compare the uncertainties of the methodological options in every individual case and determine which type of methodology should be used for extrapolation: relatively well-reproducible and standardized chemical analytical methods, laboratory bioassays, microcosms, mesocosms or the combination of any of these.

## 2  TERRESTRIAL TEST ORGANISMS

Individual members of the soil-living community, a well-defined group of the ecosystem (e.g. plants), or soil biota as a whole can be used to test fate, behavior and adverse effects of chemical substances in soil or to explore and observe the impacts of contaminants on the soil ecosystem. The organisms to be tested can belong to either micro-, meso- or macro-fauna and -flora.

The proper selection of test species is a key issue in terrestrial ecotoxicity testing. We have to choose species that are relevant to the aim of the assessment (screening, monitoring, early warning, etc.), are representative for the functional roles played by the community in question (e.g. mineralization, nitrogen-fixing, decontamination, bioaccumulation), and are sensitive to the contaminants or other exposures to be characterized. They have to fulfill some general requirements (ASTM, 1998; Laskowski *et al.*, 1998):

–   Rapid life cycles;
–   Uniform reproduction and growth;
–   Ease of culturing and maintenance in the laboratory;
–   Uniformity of population-wide phenotypic characteristics;
–   Routes of exposure similar to that encountered in the field.

Cell density in soil can be measured directly as a cell concentration, or through the activity of the living microbiota. The restrictions of traditional, cultivation-based methods can be eliminated by DNA and omic techniques.

Species diversity in soil means the abundance and the relative distribution of plant or soil-dwelling mesofauna members. Diversity assessment requires identifying and counting relevant species or members after proper sampling. When diversity of microorganisms of the soil is the issue, the task is more complicated. Billions of species and cells live in one gram soil while only a few can be grown in the laboratory and isolated on special growth media to identify them by using e.g. colony descriptors. However, it is now known from the comparison of direct microscopic observation and new DNA fingerprinting techniques that the organisms that can be grown in artificial laboratory conditions represent a tiny percentage of the total number. It is possible to extract DNA directly from soil and estimate the density of different species by analyzing the complexity of this whole community DNA, the metagenome (see also Chapter 1). Traditional and innovative methods are available for the complex characterization of soil microbiota from the assessment of the totality of the genetic material (genome) and proteins (proteome) to biochemical, physiological and behavioral characterization of the whole microbiota of the soil.

Traditionally, in practice, ecotoxicology is based on two main concepts:

– Use of specific indicator organisms to extrapolate to the whole soil biota and function;
– Direct measurement of the soil's biological, biochemical and enzymatic functions, based on the aggregated activity of the microorganism community.

Species diversity cannot be determined because of the large number of viable but nonculturable microorganisms in soil.

All microbiological characteristics of the soil can be studied using the modern molecular biology techniques: determining presence, density and activity of individual organisms, species diversity and the aggregated density and activities of the community.

Contaminated soil may show unusual characteristics, for example good biochemical activities and enzyme functions, including contaminant biodegradation, accompanied by completely unusual species distribution and missing indicator species. It may be the result of long-term adaptation of the microbiota to the contaminant. Another extreme situation is when the dominating species are not sensitive to the contaminant, but the whole community gradually becomes inactive. The reason for that may be the disappearance of some sensitive minor components with an essential role within the community. These kinds of changes cannot be handled as regular trends due to too high variability in soil microbial diversity, only a case-by-case interpretation is possible.

In this chapter some selected soil biota members are introduced, mainly those which are used in practice as indicator organisms and/or laboratory species for soil toxicity testing. The discussed species include microorganisms, terrestrial plants, and some members of the micro-, meso- and macrofauna as well as higher members of the terrestrial ecosystem.

## 2.1   Soil-living bacteria and fungi as test organisms

Using bacteria for soil testing has many advantages: it is representative for soil microflora and active soils contain $10^9$ bacteria/gram soil. Bacteria and fungi are directly responsible for decomposition of organic matter in terrestrial ecosystems. They fill a critical position in the soil food web and all nutrient cycling pathways and processes. They are easy to maintain and grow, the time and cost requirement of the tests is generally low, the number of replicates is arbitrary, and the statistics of the tests are good. Microbial activities, population density and species distribution are extremely sensitive to soil conditions, continuously changing parameters such as moisture content and temperature, as well as to seasonal alterations. This means in the practice that the test strategy should be based on a comparison with a negative/untreated control or reference instead of using the absolute values of microbial activities for measuring adverse effects.

*Pseudomonas* species are Gram-negative rod-shaped bacteria, mainly living in soil. *Pseudomonas* is a genus of gammaproteobacteria, belonging to the family *Pseudomonadaceae*. There are 191 *Pseudomonas* species described in the literature. Most of them are very active in biodegradation and mineralization; and they are responsible for the degradation of a number of organic soil contaminants of petroleum origin. The best-known species are *Pseudomonas fluorescens*, *Pseudomonas putida* and *Pseudomonas syringae* as biodegrading strains of xenobiotics. *Pseudomonas aeruginosa* is increasingly recognized as an emerging opportunistic pathogen, frequently occurring in petroleum-contaminated soils.

*Pseudomonas putida* is a saprotrophic species. It demonstrates a very diverse metabolism, including the ability to degrade organic solvents such as toluene. It was the ever first bacterium pioneering patenting of living organisms. It may play a key role in contaminated soil bioremediation. ISO 10712:1995 standard method uses *Pseudomonas putida* growth inhibition for measuring the toxic effects of chemicals or contaminated environmental samples.

*Azotobacter* species are Gram-negative, aerobic soil-dwelling bacteria, belonging to the *Pseudomonadaceae* family. There are around six species in the genus, some of which are motile by means of peritrichous flagella while others are not. They are typically polymorphic (different sizes and shapes depending on age and life conditions). Old cells tend to form cysts; they have the ability to fix atmospheric nitrogen by converting it to ammonia. Many species are used in soil toxicology tests: *Azotobacter chroococcum*, *Azotobacter beijerinckii*, *Azotobacter nigricans* and *Azotobacter vinelandii*. *Azotobacter croococcum* is applied in the direct-contact soil block test.

*Azomonas agilis* found in soil and water is closely related to other species of the *Pseudomonas* and *Azotobacter* genus in the same Pseudomonadaceae family. It is a motile, Gram-negative bacterium, capable of fixing atmospheric nitrogen. The main difference, compared to *Azotobacter*, is that *Azomonas* does not form cysts. It is a widespread test organism for contaminated soil, eluates and waters. Its former name is *Azotobaceter agile*.

*Rhizobia* fixing nitrogen in symbiosis with plants are sensitive to a wide range of contaminants. *Rhizobia* form nodules on the plants' roots, this is their habitat as shown in Figure 5.3. Their nodule formation and count can be utilized as end points.

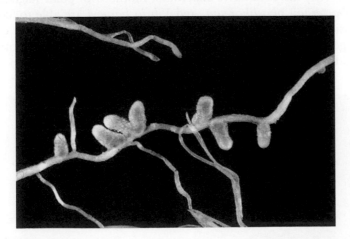

*Figure 5.3* Root nodules of a 4-week-old *Medicago italica* inoculated with *Sinorhizobium meliloti* (Ninjatacoshell, 2009).

*Arthrobacter species* from the Gram-positive bacterial genus *Arthrobacter* are widely distributed in soil. They are able to metabolize and, as a consequence, degrade a variety of toxic soil contaminants. *Arthrobacter* can grow in the presence of hexavalent chromium, and is able to reduce it to less toxic trivalent chromium. They are able to biodegrade persistent pesticides including organophosphate insecticides. *Arthrobacter chlorophenolicus* A6 can survive at extremely high concentrations of the toxic pollutant 4-chlorophenol.

*Arthrobacter globiformis* forms small colonies on nutrient agar, ranging in color from yellow to white and measuring 2 mm in diameter on average. It is used for soil toxicity testing in standardized contact tests.

*Bacillus* genus of Gram-positive rod-shaped bacteria and a member of the division Firmicutes is one of the best known soil-living microorganisms and the best understood prokaryotes, in terms of molecular biology and cell biology. Bacillus species can be obligate aerobes or facultative anaerobes. Under stressful environmental conditions, the cells produce oval endospores that can stay dormant for extended periods. Bacillus genus used in soil toxicity testing in many forms: e.g. *Bacillus subtilis* and *Bacillus pumilus* in contact test, many different *Bacillus* species and strains in growth tests, and enzyme-activity tests.

*Bacillus subtilis*, also called hay bacillus or grass bacillus is very commonly found in soil. It can convert (decompose) natural organic compounds and contaminants into simple organic or mineral products. A high number of exoenzymes produced enable them to attack the big-size molecules of dead organic matter taking part actively in mineralization and element cycling. *B. subtilis* is used in growth inhibition tests for screening toxic metal cations in soil both in form of contact tests in solid soil and testing soil extracts and eluates. The macroscopic view in Figure 5.4 shows the chain of cells in a laboratory grown culture and the staining with 4′,6-diamidino-2-phenylindole (DAPI, a fluorescent stain) of the smear under fluorescence microscope.

*Vibrios* are Gram-negative bacteria with curved rod-shaped cells. Some of them are pathogen s (e.g. *Vibrio cholerae*) which can cause foodborne infection, usually from seafood. They are facultative anaerobes, motile with polar flagella. *Vibrio fischeri, Photobacterium phosphoreum,* and *Vibrio harveyi* are able to 'communicate' by light emission, produced through bioluminescence and regulated by quorum sensing. The lux gene from these natural strains has been transformed into other indicator bacteria used for environmental analyses and testing.

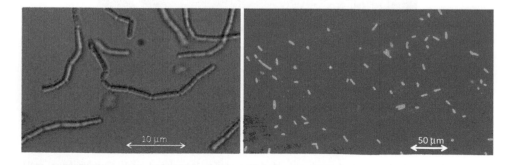

*Figure 5.4  Bacillus subtilis under the microscope: native and DAPI-stained cells.*

*Vibrio fischeri* is a popular bacterial strain also used for soil testing, nonetheless it is not a soil-living, but a marine species, but its wide range of sensitivity, the strain availability and the standardized test method makes possible its use for the testing of contaminated soil and sediment in tests using soil-extracts or leachates and whole soil.

*Cellulose-degrading bacteria and fungi* including actinomycetes degrade cellulose when it is associated with pentosans such as xylans and mannans, and it undergoes rapid decomposition. The best known cellulose-degrading bacteria are: *Fibrobacter succinogenes, Cellulomonas sp., Clostridium thermocellum.*

*Aerobic bacteria* can be considered as a whole, characterized by density, activities and diversity. The activity and response of soil-living bacteria as a whole can also be utilized for the characterization of soil health, its dynamic adaptability and loadability. Using the site-specific soil in the tests (sampled from the contaminated site) one can directly measure the response of the targeted soil ecosystem, without extrapolation. But to use whole soil for generic risk assessment of chemical substances is always problematic, due to the unachievable requirement of an everlasting, unabated reference soil with a standard response.

The growth and total activities of the soil microbiota as toxicity end points, (e.g. respiration rate, consumption and conversion rate of substrates, production of metabolites) can provide a quantitative picture on soil function. The rise of the response, the time of adaptation, etc. can be investigated in a dynamic way but the qualitative changes in diversity cannot be detected in growth tests. Cell numbers and activities should be amended with information on the microbial diversity in soil to obtain complete information.

Zak *et al.* (1994) suggest that a sound interpretation of the pollution-induced effects on the structure and function of soil microbial communities must be based on measuring the integrating parameters such as litter decomposition or nutrient mineralization (e.g. using the BIOLOG plate).

*Pollution-induced Community Tolerance* (PICT) is a toxicological tool used for characterizing soil community tolerance to the contaminants present over a long time. Community tolerance is formed by adaptation, using various mechanisms: switching on existing, but mute, genes responsible for tolerance, increasing the relative abundance of the tolerant species, creating new gens by random mutation–selection mechanism and spreading useful genes by horizontal gene transfer. Any of these processes are good indicators to prove PICT, what is an indirect evidence for the presence of contaminants.

*Anaerobic bacteria* dominate the habitat in deep and water-saturated soils. They can be handled as a whole, characterized by density, activities and diversity. Anaerobic soil has a microbiota highly different from aerobic soil, in spite of the fact that some microbes are the same as aerobic ones; these are the facultative anaerobes, able to switch their metabolism from aerobic to anaerobic, using alternative forms of respiration, such as nitrate- or sulfate-respiration. Atmospheric oxygen is not available for soil organisms under low redox potentials, so that they can utilize alternative electron acceptors, such as nitrate, iron, manganese, sulfate, carbonate, and, in contaminated soils, exotic pollutants such as chloride. Anaerobic metabolism of soil-microorganisms occurs in normal soil, as shown by the redox potential gradient in every soil. Anaerobic microbes and processes play a key role in the biodegradation of certain soil contaminants such as chlorinated contaminants degraded by reductive dechlorination. In spite of their reduced activity compared to aerobic ones, anaerobic bacteria play an important role, mainly in the two-phase soil and groundwater and in the elimination of its contaminants. The density and diversity of anaerobic microbes in deeper soils and groundwater suitably characterize anaerobic soil habitats.

*Microfungi* as a separate group of microbes may characterize soil by their density and activities. They are of special importance in agricultural soils where their roles and species distribution is well known and utilized.

*Soil enzymes of microbial origin* and their activities are good indicators of microbial activity, heterotrophy and mineralization in soil ecosystems. When evaluating the absolute values of enzyme activity, results should be compared to the enzyme activity of the healthy soil community. This requires monitoring data or time-series information. The other concept to utilize enzyme activities of soils is the use of a standard-quality test soil for generic testing of chemical substances or environmental samples.

Any of the enzymes of the respiration chain, i.e. energy production (biodegradation, mineralization) as well as of biosynthesis can be used to test toxicity. The commonly measured enzymes are beta-glucosidase, carboxymethylcellulase, n-acetylglucosaminidase and acid/alkaline phosphatases.

Proteomics – referring to the study of proteome, the totality of proteins – has not yet been adapted to soil because of the many problems of analysis which arise. However, it can be applied in a complementary way to find a relationship between composition and activity of soil microflora. Proteogenomics, the science of the metagenome responsible for the diversity of proteins in soil microflora, is a promising tool to overcome the problems of soil proteome analysis (Armengaud *et al.*, 2013).

## 2.2  Terrestrial plants for soil toxicity testing

As primary producers, plants capture the solar energy in their structure, thus their function influences the whole terrestrial ecosystem and modifies the landscape and land uses. Measuring and/or predicting toxicant impacts on plants enables environmental and agricultural managers to identify potential thresholds of contamination above which floral establishment, growth, and reproduction may be compromised, and direct mortality is a concern.

The general advantage of plants as test organisms is their ready availability, for example seeds, which can be purchased in bulk, their easy activation by germination, their relatively long life and usability of up to several months, and in all life stages. Low costs and high reliability in site- and problem-specific applications makes their application feasible. When using plants for soil toxicity testing, one must be aware of the great variability in sensitivity between species (Fletcher *et al.*, 1990).

*The long list of terrestrial plants* reflects the trials to find the best fitting plant to the topic of testing. Available test plants should represent both dicotyledonous and monocotyledonous species, the two major groups of plants. Traditionally, plants have been used for

–   testing toxicity of soil by inhibition of germination and plant growth;
–   plant injury by destruction of cellular structures;
–   effects of contaminants on plant metabolism, respiration and photosynthesis and
–   for testing contaminant uptake.

*Intact plants or plant tissue* can be used for soil testing. ASTM methodology for terrestrial plant toxicity testing lists nearly 100 plant taxa that have been used in phytotoxicity testing (Davy *et al.*, 2001). The most commonly used test plants for soil testing are:

– *Agrostis gigantea*, redtop, a perennial grass from the monocotyledonous family Poaceae;
– *Allium cepa*, onion, family Liliaceae, sensitive to herbicides;
– *Amaranthus retroflexus*, redroot or pigweed, Amaranthaceae family, a flowering plant, sensitive for herbicides;
– *Arabidopsis thaliana*, thale cress, Brassicaceae family, sensitive response by reproduction;
– *Artemisia filifolia* silver sagebrush or sand sage Asteraceae family, sensitive for herbicides;
– *Avena sativa*, oat, monocot cereal from the family Poaceae;
– *Beta vulgaris*, sugar beet, a core eudicot for the family Chenopodiaceae, a monoculture plant, sensitive to herbicides;
– *Brassica napus*, rape, close relative of cabbage and mustard, family Brassicaceae. It is also used in standardized tests for measuring plant growth inhibition.
– *Brassica rapa*, turnip field mustard or turnip mustard is a plant widely cultivated as a leaf vegetable, a root vegetable (see turnip), and an oilseed (rapeseed oil). It is used for the study of whole life-cycle assessment and other biochemical research.
– *Cucumis sativus*, cucumber, an eudicot from family Cucurbitaceae;
– *Glycine max*, soybean, a eudicot from family Fabaceae, representative of nitrogen-fixing plants in symbiosis with rhizobia;
– *Helianthus annuus*, sunflower, Asteraceae family, sensitive response by growth and reproduction;
– *Ipomoea purpurea*, morning glory, flowering plant from the family Convolvulaceae;
– *Lactuca sativa*, lettuce, a temperate annual or biennial plant of the daisy family Asteraceae. It is very sensitive to some herbicides.
– *Lolium perenne*, perennial ryegrass, family Poacea, sensitive to herbicides and plays role in the wild food chain;
– *Medicago sativa*, alfalfa, family Fabacea;
– *Phaseolus vulgaris*, garden bean, family Fabaceae, sensitive plant;
– *Pisum sativum*, pea, an eudicot from family Fabaceae, sensitive to herbicides as a non-target plant;
– *Raphanus sativus*, radish, an edible root vegetable of the Brassicaceae family
– *Sinapis alba*, white mustard, an eudicot from family Brassicaceae, the most widespread test plant used for germination, root and shoot-growth testing as well as for the testing of bioaccumulation (Figure 5.5).
– *Solanum lycopersicum*, tomato, family Solanaceae, sensitive to herbicides;
– *Sorghum species*, many of them, such as energy grasses, broom sorghum, or broomcorn and sorghum for biofuels are products of industrial agriculture (Figure 5.6). These may be cultivated on deteriorated or contaminated land under controlled conditions.
– *Trifolium repens*, white clover, family Fabaceae;

*Figure 5.5  Sinapis alba* seedlings, just germinated from seeds on soil and different plants grown on a soil containing agar plate in a laboratory soil test.

*Figure 5.6  Sorghum sudanense* (energy grass), *Sorghum vulgare* (broomcorn) and *Zea mays* (maize) in a contaminated soil remedial field experiment.

- *Triticum* spp., wheat, a monocot cereal from family Poaceae;
- *Zea mays*, maize or corn, a monocot from Poaceae family (Figure 5.6).

Markwiese (2001) gives four pieces of advice regarding the use of plants, their seeds, standardized and non-standardized species:

- When purchasing seeds of non-standard test species from landscaping companies, it is important to gather as much information as possible about the seed lot and to buy enough seeds from the same lot for running the experiment.
- While collecting seeds or other plant material from the field, it is critical that only seeds, seedlings or cuttings from the species of interest be collected. Details regarding location, field conditions time in growing season, and potential confounding factors (e.g., previous pesticide and/or herbicide use in the area) must be recorded for the collection event.
- Selection of species for testing needs pretesting seed germination and early seedling growth. The following recommendations should be considered:

○   as a reference toxicant: an effective and nonselective biocide, e.g. boric acid,
○   as control soils: artificial and soils collected from the field,
○   as end points: mortality, shoot length, root length, shoot wet mass, root wet mass, and shoot and root dry mass.

–   Data evaluation and interpretation of phytotoxicity tests on relatively unstudied species must be supported by detailed ecological characterization (ASTM, 1998) to be able to distinguish ecotoxicological effects specific to the species. Tests should be accomplished by running rigorous control tests with enough replicates to achieve the desired statistical test power.

In addition to toxicity assessment, plant growth tests are applied to see if the planned plant is able to grow and function on a certain deteriorated or contaminated soil.

The following can be used as end points of plant tests:

–   Germination rate;
–   Root or shoot growth, root length and plant height;
–   Biomass quantity: wet and dry mass;
–   Photosynthetic activity, simplest way via chlorophyll content;
–   Respiration rate;
–   Enzyme activities;
–   Metabolites specific for selected metabolic pathways;
–   The concentration of the contaminating and accumulated chemical substances in different parts or in the whole plant;
–   Plant reproduction: both quantitative and qualitative indicators are useful end points;
–   Chromosome aberration;
–   Biomarkers such as antioxidant enzyme activity can be sensitive indicators of toxicant-related stress;
–   Abnormal plant growth, morphology, flowering.

Plants may show a very specific response to environmental stress; this is why their use as laboratory test organisms often fails: they do not represent the average of the plant community, rather the opposite, they give unique and specific responses, not suitable for generalization.

With the increasing information on plant functional genomics, the association between ecological functions and molecular stress responses, genes or gene products can be mapped and utilized as a basis for molecular ecotoxicology. The identification of signaling substances, enzymes and genes in plants involved in any plant response to contaminants or reaction against pathogens, xenobiotics or any other environmental stressors increases the number of bioindicators of plant origin to be used in environmental toxicology. Some of these are listed below:

–   Detoxification: plant glutathione S-transferase enzyme family play a role in the response on oxidative stress and detoxification. Glutathione and glutathione S-transferase participate not only in antioxidative and detoxification reactions but also in redox regulation of the expression of protective genes in infected cells (Edwards *et al.*, 2000; Sandermann, 2004).

– Defense response: cross-talk between ethylene and jasmonate signaling pathways determines the activation of a set of defense responses against pathogens and herbivores in plants (Lorenzo *et al.*, 2003). Ethylene and jasmonate synthesis and release from plants can be used as ecotoxicological end point.

– Phytosensing pathogens: plants possess defense mechanisms to protect against pathogen attack. Inducible plant defense is controlled by signal transduction pathways, inducible promoters and cis-regulatory elements corresponding to key genes involved in defense, and pathogen-specific responses. Identified inducible promoters and cis-acting elements could be utilized in plant sentinels, or phytosensors, by fusing these to reporter genes to produce plants with altered phenotypes in response to the presence of pathogens. Pathogen-inducible genes as well as those responsive to the plant defense signal molecules salicylic acid, jasmonic acid, and ethylene are suitable stress indicators (Mazarei *et al.*, 2008).

– Salicylic acid and nitric oxide play a role in programmed cell death and induced resistance (Sandermann, 2004).

– Cis-regulatory elements (cis-elements are present on the same DNA molecule as the gene they regulate) and transcription factors regulating gene promoters play a role in response to environmental stress, e.g.
  o Ozone-induced gene regulation (ethylene);
  o Xenobiotic-responsive cis-elements;
  o Activation of DNA sequences by heavy metal ions, etc.

## 2.3   Soil fauna members as test organisms

Soil fauna together with soil microorganisms are responsible for the harmonic operation of mineralization and element cycling in the soil.

*Invertebrates of the soil fauna* regulate the decomposition process by preparatory mincing of dead organic matter – also called detritus – before mineralization by soil microorganisms. They play an important role in soil, in spite of the fact that soil invertebrates count for less than 10% considering respiration and energy production in terrestrial environments. In addition to initialing decomposition, they are essential in predating and controlling the decomposer microbiota. Soil-dwelling invertebrates together with the biodegrading microbiota are responsible for mineralization in soil participating in the short turnover of nutrients in terrestrial systems.

The use of invertebrates for soil testing is dated for the beginning of the 1990s, as it is indicated by the book of Løkke and Van Gestel (1998) on soil vertebrates toxicity. Laskowski *et al.* (1998) selected the test organisms for testing in the same book.

Soil animals, suitable for adverse-effect testing may belong to the soil micro-, meso- or macrofauna. The most important groups and some individuals are introduced in the following.

*Soil microfauna* covers single-cell animals such as amoeba and ciliophora. They are better known as inhabitants of waters but they also show high diversity in soils. They contribute to mineralization and, as key members of the food web, to feeding mineralizing soil microflora. Foissner *et al.* (2002, 2003) have found huge ciliate diversity in Austrian forest soils as well in in arid areas of Namibia.

*Tetrahymena*, the best-known unicellular animal test organisms, have been used for biochemical studies for many years. They fulfill all the requirements to be met by test organisms: they are easy to maintain and grow, genetically stable and their genome is well-known. They represent animal cells regarding biochemistry, physiology and metabolism. The authors of this chapter have developed a soil toxicity test method using the ciliate *Tetrahymena pyriformis* for soil extracts and for soil suspension with direct contact between solid soil and *Tetrahymena* cells. They are described in Chapter 4 as the aquatic test organisms, but they can also be applied for soils and soil leachates.

*Nematoda* are the most frequent organisms on the earth, four of every five multicellular animals on the planet are nematodes. These organisms occupy any niche in terrestrial environments that provides an available source of organic carbon (Bongers & Ferris, 1999).

Nematodes are of high ecological relevance at contaminated sites because their permeable cuticle provides direct contact with soil contaminants in their microenvironment (Bongers & Ferris, 1999). They occupy key positions in the soil food web; they can be placed into four trophic groups: bacterivores, fungivores, omnivore predators, and plant feeders (Zak & Freckman, 1991).

Acute nematode tests were developed by Vainio (1992) and Donkin and Dusenberry (1993) with the end point of survival and parasitism. Kammenga and van Koert (1992) worked out a nematode reproduction test and calculated both $EC_{50}$ and NOEC values from the measured data. The nematode chronic toxicity test of Niemann and Debus (1996) is based on the measured end point of abundance.

In addition to mortality, the ratio of adult to juvenile nematodes, termed the maturity index (MI), is also useful for characterizing soil stressors (Bongers & Bongers, 1998). The MI has been shown to be a sensitive metric for monitoring natural and anthropogenic stressors for certain nematode populations (Yeates & Bongers, 1999).

*Caenorhabditis elegans* is a free-living, transparent nematode, about 1 mm in length. It lives in temperate soil environments, used extensively as a model organism. *C. elegans* was the first multicellular organism to have its genome completely sequenced.

*Panagrellus redivivus*, sour paste nematode, a free-living species, is a tiny roundworm about 50 μm in diameter and 1 mm in length as shown by the microscopic photo in Figure 5.7. It is widely used in aquacultures as food for a variety of fish and crustacean species. This microworm has also been used in genetic studies.

*Pristionchus pacificus* is similar to *Caenorhabditis elegans* in many points of view, but differs in one important feature; similar to all species of the family Diplogastridae: it has an embryonic molt.

*Soil mesofauna* consist of bigger size animal organisms than the microscopic organisms of the microfauna. Four groups are discussed in detail: Tardigrada, Acari, Collembola and Enchytraeidae.

*Tardigrada* is a phylum of 1–1.5 mm dwelling in water and soil and occurring over the entire world. Tardigrades have a cylindrical body and eight legs, their movement is typically slow, similar to that of the bear, this is why they are also called water bear or moss piglets (Figure 5.8). They live on lichens and mosses, hunting nematodes, rotifers and protozoans, but also consuming plants. They can pierce both animal and plant cells by their sharp-edged stylets and suck out the cell fluid. They are highly adaptive to extreme environmental conditions.

*Figure 5.7* Microscopic view of *Panagrellus redivivus*, a free-living nematode.

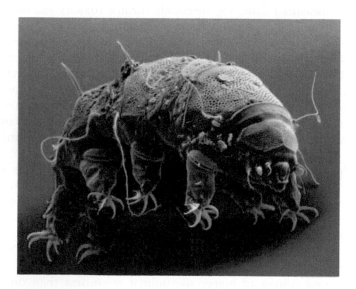

*Figure 5.8* 3D Tardigrade *Echniscus granulatus*, also called water bear or moss piglet (European Atlas of Soil Biodiversity, 2010).

*Acari*, mites and ticks are microarthropods, a diverse group of arthropods which are generally omnivorous and reduce the size of dead organic matter fragments. There are fungal feeders like oribatids (moss mites or beetle mites) and fungiphagous prostigmatid mites amongst them. Mites meet several requirements of good laboratory

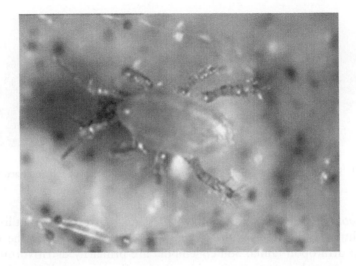

*Figure 5.9  Hypoaspis aculeifer*, a new test organism in soil toxicity testing (CEH, 2013).

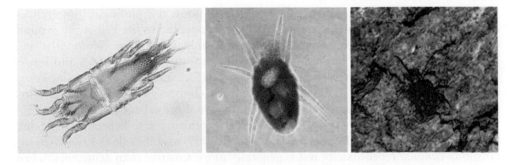

*Figure 5.10*  Microscopic mite from a bird's nest, plant pathogen microscopic mite and velvet mite.

test species such as ease of handling, measurable reproductive output, and monotypic genomic structure. Testing requires low cost, average laboratory and small space. Acute toxicity and inhibition of reproduction can be measured on the soil mite, *Hypoaspis (Geolaelaps) aculeifer* (Figure 5.9), and the predator mites of *Typhlodromus pyri* and *Phytoseiulus persimilis*. *Hypoaspis aculeifer* is used in the standardized tests of OECD TG 226 (2008). *Oppia nitens*, an herbivorous and fungivorous oribatid mite, lives in Canadian soils and is used in soil tests on survival and reproduction. Figures 5.10 and 5.11 show some mite and tick species.

Van Gestel and Doornekamp (1998) used the oribatid mite of *Platynothrus peltifer* in the test developed for soil testing. Survival, reproduction and feeding rate was measured and the results given as $LC_{50}$, $EC_{50}$ and NOEC. Krogh (1995) developed a sublethal test using a predatory mite, and measured its survival and reproduction.

*Figure 5.11*   Sheep tick (Bartz, 2009), common blood sucking tick well fed and hard deer tick.

*Collembola* or springtails are small, wingless microarthropods that live primarily in the soil litter layer. Along with the mites, collembola make up the majority of the soil litter arthropod fauna. Soil collembola are primarily detritivorous, living on partially decomposed vegetation or on the soil microflora, especially fungi. Collembola contribute to soil organic matter breakdown and nutrient cycling by shredding organic matter into fragments, feeding on humus and feces in the litter, and feeding on decaying roots (Bardgett & Chan, 1999).

Collembola shown in Figure 5.12 exhibit significant morphological differences. *Folsomia candida,* the little white springtail, highly adapted to life below the ground: no visual apparatus, no pigmentation, short appendages. *Isotoma violacea* is at a lower level of adaptation: has developed visual apparatus, pigmentation and longer appendages. Collembola have become popular test organisms in toxicity bioassays because they are easy to maintain in laboratory cultures, have uniform genomic structure (many species are parthenogenic), and have short generation times.

*Folsomia candida* has been selected as a test organism for soil chronic (reproduction) toxicity testing by OECD (OECD TG 232, 2009) and ISO (ISO 11267, 2014). *Folsomia candida* reprotoxicity test is generally more sensitive than acute earthworm mortality tests (Crouau *et al.,* 1999). The OECD test guideline characterizes *Folsomia candida* and *Folsomia fimetaria* as soil-dwelling. Collembola is an ecologically relevant species for ecotoxicological testing. Their thin exoskeleton is greatly permeable to air and water, and they represent an arthropod species with a different route and rate of exposure compared to earthworms and enchytraeids (OECD TG 232, 2009).

*Enchytraeidae,* potwurm, iceworm, or whiteworm belongs to a group of tiny worms popular as live food cultured by aquarists, also used as laboratory test organisms in ecotoxicity testing.

*Enchytraeus albidus* or whiteworms are one of the most popular cultured worms. *Enchytraeus albidus* Henle 1873 is the test organism of the OECD test guideline (OECD TG 220, 2004). In addition to testing hardly dissolvable chemicals in artificial soil, ISO 16387 (2004) gives recommendations on the testing of soil quality by the modification of the original method. *Enchytraeus buchholzi* is another species applied to laboratory tests.

*Soil macrofauna* is the group of soil-living animals which are visually observable such as earthworms, insects, spiders, etc.

*Figure 5.12* Springtails *Entomobrya nivalis* (Duine, 2014), *Isotoma viridis (Duine, 2011)* and *Isotomurus* unifasciatus (Duijne, 2013).

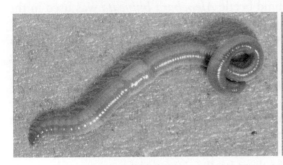 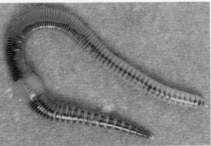

*Figure 5.13* *Allolobophora chlorotica* (soil-living earthworm) and *Eisenia veneta* (compost earthworm) (Earthworm, 2013).

*Lumbricidae*, earthworms are the priority animal test organisms for soil and compost. In many countries soil ecotoxicity data are limited to earthworms. In the beginning it was used for toxicity screening of hazardous waste sites and US EPA endorsed this test with *Eisenia foetida*. It has been studied extensively, producing a large data set on the toxicity and bioaccumulation of a number of compounds. Other species used as test organisms in laboratory and field tests are *Eisenia andrei* and *Lumbricus terrestris*. Figure 5.13 shows *Allolobophora chlorotica*, an endogeic soil-living earthworm and *Eisenia veneta*, a compost earthworm (Earthworm, 2013).

*Arthropods* are widely applied in soil toxicity testing both in the laboratory and field tests. Different phyla and classes belong to soil-living *Arthropods* such as

- Class Arachnida (Chelicerata) with spiders, mites, ticks, scorpions, allies
- Class Malacostraca (Crustacea), order: Isopoda
- Class Diplopoda, millipedes, (Myriapoda)
- Class Chilopoda or centipedes (Myriapoda)
- Class Insecta, insects (Hexapoda), with beetles, crickets, flies, bees and wasps, ants and termites, etc.

*Spiders* play an important role in biotic interactions in terrestrial systems so that they have been the focus of several toxicant impact assessments in recent years. In a review of invertebrate ecotoxicological test systems, Van Gestel and Van Straalen (1994) note that spiders seem to be a particularly sensitive group among arthropods.

*Pardosa*, a large number of species of the wolf spider genus are used in standardized tests by the International Organisation for Biological Control (IOBC, 2000) as non-target organism affected by pesticides.

*Beetles* and their larvae play an essential role in soil by aerating the soil, pollinating blossoms, controling insect and plant pests and decompose dead materials. The larva is usually the principal feeding stage of the beetle life cycle. Larvae tend to feed voraciously once they emerge from their eggs. Some feed externally on plants, such as those of certain leaf beetles, while others feed within their food sources.

*Poecilus cupreus* and other carabid beetles have been the focus of several toxicological investigations (Heimbach *et al.*, 1994; Heimbach & Baloch, 1994). Heimbach and coworkers used carabid beetles to evaluate pesticide-contaminated soils by caging the beetles in field enclosures. Staphylinid beetle *Aleochara bilineata* was used in the survival and reproduction tests developed by Samsøe-Petersen (1992).

*Dung beetles* are the most abundant species among organisms forming a complex food web in livestock dung (Hanski & Cambefort, 1991). Other members of the dung community include mites, nematodes and annelids.

*Aphodius constans* and *Onthophagus taurus* dung beetles (*Scarabaeidae*) are applied in a new OECD laboratory test, for veterinary pharmaceuticals, especially parasiticides. The test organism is planned to be applied not only in laboratory but also in field toxicity assessments and monitoring (OECD, 2008)

*Diptera*, flies and its larvae are used for toxicity and developmental toxicity testing of chemicals substances and also for the dung itself originating from livestock treated with the chemical.

*Scathophaga stercoraria* and *Musca autumnalis* dung flies are considered to be suitable indicator species for estimating the developmental toxicity of parasiticides on dung-dependent Diptera because the species cover a wide geographic range. Both species are dung-dependent, multi-voltine, do not undergo obligate diapause and are easy to culture and have a short life cycle which makes it possible to determine effects on development and survival in the laboratory (OECD TG 228, 2008).

*Cockroaches* belong to Neoptera (Insecta) and are applied in non-standardized tests in terrestrial toxicology. *Blattella germanica* is one of the best known household species. It is relatively small, about 1.3 cm to 1.6 cm long; although it has wings, it is unable to sustain flight. German cockroach is omnivorous and a scavenger.

*Crickets* are representatives of macroarthropod detrivores (harmonize with p. 231) in the order Orthoptera and class Insecta. In fact, crickets are known to exhibit a high degree of cannibalism (Crawford, 1991). Crickets are widespread in most terrestrial systems. These organisms serve a valuable purpose by consuming and processing plant litter and are prey for other animals. *Acheta domesticus*, house cricket and *Gryllus pennsylvanicus*, field cricket belong to Gryllinae.

*Social insects* within the Hymenoptera order such as wasps, bees and ants are popular test organisms both in ecological and contaminated site assessment and biocide and pesticide hazard assessment.

*Wasps* play an important role in natural control of pest insects and are therefore important in food chains. Parasitic wasps are applied in biological plant protection mainly in foliage. Their sensitivity to pesticides is an important characteristic, so that they are applied to testing the effect of pesticides as non-target insects. *Aphidius rhopalosiphi*, *Trichogramma cacoeciae* and *Encarsia formosa* are popular parasitic wasps, available in commerce.

*Bees*, for example the best-known *Apis mellifera* grown by beekeepers and apiaries, are good sources of well-kept, homogenous populations of bees. For test purposes, young adult worker bees of the same race should be used, from adequately fed, healthy, as far as possible disease-free and queen-right colonies with known history and physiological status. They could be collected in the morning of use or in the evening before test and kept under test conditions until the next day. Bees treated with chemical substances, such as antibiotics, anti-varroa, etc., should not be used for toxicity test for four weeks from the time of the end of the last treatment (OECD TG 213, 1998).

*Ants and termites* from the Formicidae family are valuable bioindicators of environmental quality. Relevant attributes of ants include:

- Active throughout most of the year;
- Relative abundance;
- High species richness;
- Many specialists;
- Occupation of higher trophic levels;
- Easily sampled;
- Easily identified (Lobry de Bruyn, 1999; Whitford *et al.*, 1999).

Termites are generally the primary detritivores in warm desert systems, actively consuming standing dead vegetation, plant litter and feces. In desert grasslands, termites consume 50 percent or more of all photosynthetically fixed carbon (Whitford, 1991). Termites also serve as an important food source for higher trophic levels. In less humid soils the role of earthworms is taken over by termites, ants, and tenebrionids (Van Gestel & Van Straalen, 1994).

*Isopoda*, crustaceans, include familiar animals such as woodlice and pill bugs. They can consume in excess of 35% of their body weight in a single 24-hour period in laboratory cultures. They are able to survive for up to 180 days without food (Drobne & Hopkin, 1994). Isopods are considered to be good candidates as standard test species because they are common, easy to handle, and generally respond quickly to environmental contamination (Paoletti & Hassall, 1999). Several common species are readily available from commercial marketers of biological educational material, and gives ease of culturing with better-understood species e.g., *Porcello scaber* and *Oniscus asellus*. Drobne *et al.* (2002) found the microflora of *Porcello scaber* as suitable end point for toxicology testing. Living species such as *Trachelipus rathkii* and *Armadillidium nasatum* can be studied in gardens and greenhouses as non-target organisms of pesticides and biocides. Hornung *et al.* (1998) developed both a growth test with the end points of survival, growth and feeding rate and a reproduction test with the end points of survival, reproduction and oorsoption.

*Diplopoda* (millipede) exposed to high metal contamination accumulate toxic metal in their intestine tissues. They have been studied as indicators of high risks due to metals (Köhler *et al.*, 1995).

*Chilopoda* (centipede) are able to accumulate not only metals but also persistent pesticides and other persistent organic substances. The assimilation of zinc, cadmium, lead and copper by the centipede *Lithobius variegatus* was studied (Hopkin & Martin, 1984). *Lithobius mutabilis* lives in Central Europe; it plays an important role as a dominant epigeic predator especially in woodlands.

*Gastropoda* (molluscs and slugs) are the biggest animals in size being in intensive contact with soil through feeding and skin contact. They are used not only in toxicity testing but also in behavior tests, such as avoidance and herbivory. Slug *Arion rufus* L. proved itself to be sensitive in contaminated soil showing differences in herbivory (form of predation in which plants are consumed) (Mench & Bes, 2009).

*Vertebrates,* the higher members of the terrestrial macrofauna, primarily birds and mammals, have been used both for environmental and human toxicity assessments. The results of the tests developed for the evaluation of human health effects can also be used for the estimation of the environmental/ecological effects of industrial chemicals and pesticides already tested. Properly selected mammalian end points can serve for both human health and ecological risk assessment, and thus fulfill the requirement for minimizing animal testing. Animal welfare consideration fosters non-destructive techniques in vertebrate testing such as biochemical markers.

*Birds* represent both terrestrial and water ecosystems and food chains. Avian tests are required for pesticide registration and play a role in food chain assessments and bird species protection. They are easy to cultivate with the possibility to test reproduction via the eggs. OECD TG 205 (1984) recommended more bird species for testing of chemicals such as:

– *Anas platyrhynchos*, mallard duck;
– *Colinus virginianus*, bobwhite quail;
– *Columba livia*, pigeon;
– *Coturnix coturnix japonica*, Japanese quail;
– *Phasianus colchicus*, ring-necked pheasant;
– *Alectoris rufa*, redlegged partridge.

*Mammals* typically living in soil are mole and rodents, which are exposed via respiration, skin contact and nutrition, like all the other smaller organisms; nevertheless they are able to escape from deteriorated or contaminated soil to the neighborhood or to a farther distance. In some cases higher terrestrial wild animals are also targeted receptors of pollution.

*Mole, rodents and higher wild animals* (Figure 5.14), for example small mammals, which sensitively react to environmental stress (Elliott & Root, 2006; Torre *et al.*, 2004), are suitable organisms for field assessments and diversity and abundance evaluation of ecosystems. Ecological assessment methods (EAMD, 2013) also recommend and assess mammals, including endangered species in areas of ecological value.

## 3  MEASURING TERRESTRIAL TOXICITY: END POINTS AND METHODS

For testing fate, behavior and adverse effects of chemical substances, or to measure the adverse effect of contaminated soil, many of the soil-living micro-, meso- and macro-fauna and -flora members can be used.

*Figure 5.14* Rat, mole (Hill, 2005) and prairie dog (Weiss, 2011).

Soil biodiversity and the changes in the density and distribution of species indicate and characterize truly soils' health or deterioration. On the other hand the complete assessment of all species, their density and distribution in the soil is not possible. The solution is that only some indicator species are assessed for their density and share and by comparing these results to the diversity of the healthy soil, the scale of deterioration can be estimated. Further methodological simplification is the use of well-known, healthy and uniform test organisms instead of naturally occurring endemic ecosystem members, and controlled conditions instead of the non-controllable and heterogenic natural circumstances. The use of bioassay may exclude many 'natural' uncertainties which may overrule the loss in environmental reality of the bioassays compared to the natural environment. The concept of characterizing soil species diversity in its complexity can be fulfilled by studies on the metagenome and related molecules extracted from soil. Through metagenome, metatrascriptome and other ome classes (proteome, metabolome) arising from whole soil, the soil community can be considered as one entire 'organism', a dynamic and adaptive biological system. With the development and perfection of the molecular techniques the interpretation of the results will also make a step forward, creating an efficient ecotoxicological tool.

## 3.1  Soil biodiversity

Soil biodiversity assessment is a practical tool that can characterize soil as a natural habitat. Biodiversity assessment is based on mapping either the total ecosystem, or part of it or only some characteristic bioindicator species.

Soil species diversity and their changes would be the most appropriate end point for the characterization of environmental health, but our knowledge and our financial and labor potential is not enough for regular and comprehensive ecosystem monitoring. The correlation between spatial and temporal ecological diversity changes and the production and use of chemicals could be the ideal tool for the risk assessment, identification and management of chemical substances, but only highly reduced surrogates are available today.

Characterization of the total ecosystem – including soil-living microorganisms, plants and animals – can hardly be achieved using traditional tools such as counting the individual organisms, and calculating the species distribution after their identification. Investigation of the metagenome (total DNA) of soil biota is an innovative molecular tool for the ecosystem as a whole and it is suitable for the characterization of following changes at the DNA and RNA levels.

Characterization of soil biota and microbiota requires high expertise and experience. Reduction of the ecosystem to a part or even to some indicator organisms is a major simplification, which may deform the picture as a whole. Another conceptual problem is that a target is difficult to specify because it is hard to define what normal is, what can be used to characterize the diversity of a healthy ecosystem. Of course major changes, large-scale deteriorations can be easily observed, but not the early signs which are important for efficient risk prevention.

Soil community has a strong impact on soil processes and on the changes of these processes in time and space. The foodweb – the structure and interactions between the groups of soil-living organisms – reflects both the dynamics of populations and the dynamics in material, energy, and nutrient cycles but sustainability (stability and biodiversity) cannot be characterized based on food-web models alone. The model that truly characterizes soil sustainability must describe the interaction between the living community and the soil's function. It also must include functional redundancy i.e. the soil's capability to continue to function even if the species composition changes e.g. a species disappears and another takes over its place in the food web.

Another problem is that the *diversity index*, the result of the complex ecological assessment, cannot be integrated into the contaminant-specific quantitative risk assessment scheme. The reason is that the results of biodiversity assessments are not quantitative values, but indexes on a relative scale not in direct relation with contaminants and their concentrations but with the aggregate of unidentified impacts.

Classification of soil biodiversity measurement methods:

– Direct observation and counting visually or by microscopy;
– Functional tests:
  o Enzyme analysis, proteome and activities;
  o Metabolic activity profile, metabolome analysis;
  o Carbon-source utilization profiles;
  o Protein analysis, proteome analysis using proteomics;
  o Functional gene arrays.
– Taxonomic tests:
  o Nucleic acid fingerprinting techniques, genomics and transcriptomics;
  o Cloning;
  o Gene arrays, hybridization;
  o Massively parallel sequence analysis;
  o Phospholipid fatty acid analysis (PLFA).

The biodiversity assessment based on counting and identification can be applied to macroplants and the members of meso- and macrofauna but not to soil microorganisms because most of the soil-living microorganisms cannot be isolated and cultivated under laboratory circumstances. Molecular techniques such as the analysis of the genome (DNA), the transcribed RNA, proteins, the active gene-products and any metabolite of the biochemical processes of the organism can be applied without isolating the organism. These methods can be used for individuals, for populations or for the whole community without isolating any of the microbes from the soil, but analyzing the whole genome, the whole RNA (transcriptome), the whole of proteins (proteome) or

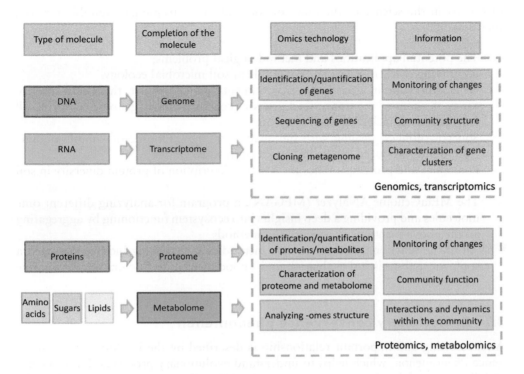

*Figure 5.15*  Enviromics: the study of envirome, the total complement of environmental characteristics, conditions, and processes required for life.

metabolites (metabolome) isolated directly from soil. Another possibility is the cloning of the metagenome and the analysis of the gene products expressed by a host cell, e.g. *E. coli* or yeast. Screening phylogenetic and functional marker genes certainly fulfills the requirement of ecotoxicology. The "only" information still needed is the list and combination of genes which corresponds to the functional or taxonomic indicator characteristics. "Enviromics" is the special study of environmentally critical genes and chemicals. It tries to collect all the environmental components and their genetic pairs (Omics, 2014; Genomic Glossaries, 2014) and to reflect on the interaction between the environment and the inhabiting community: how the genome dwells within the environment, and how the genomic expression shapes and is shaped by the environment (Anthony, 2001). Figure 5.15 shows an overview of the soil "-omes" and the omics techniques.

The soil microbiota is extremely complex and variable, and the detection of genes and characterization of the genome and other "omes" such as transcriptome, proteome, and metabolome in soil has increased our knowledge of soil microbiota. It became possible to study unculturable microorganisms and their possible functions, metabolic pathways, and the relationship between composition and activity of soil microbiota. Therefore, omics application in soil sciences is advancing and spreading rapidly in spite of the difficulties of DNA and other ome extraction from soils. The new book by Nanniperi *et al.* (2014) gives a comprehensive overview of the state of the art in the science of the omes of soil and omics techniques suitable for soil, focusing on:

- The potential applications and methodological problems;
- Identifying marker genes that play a role in soil microbial ecology,
- Gene expression in a particular environment and its product, the metatranscriptome, the sum of the transcribed genes in contrast to metagenome (covering all existing genes, including inactive ones);
- Soil volatilomics, dealing with the analysis of the multitude of volatile organic compounds produced, stored or degraded in soils;
- Poteomics and proteogenomics for a better description of protein diversity in soil microbiota;
- The MEtaGenome ANalyzer (MEGAN), a program for analyzing different ome molecules and providing a deep insights into ecosystem functioning by aggregating all the above-mentioned concepts and methods.
- All of the functional and taxonomic methods characterizing biodiversity need a sophisticated statistical evaluation and a good concept to find or create a suitable reference.

## 3.2   Evolutionary convergence phenomenon

An interesting and important relationship is described by the Evolutionary Convergence Phenomenon, which helps to understand evolutionary processes that influence soil diversity and the phenotypes of species. Traditionally, convergence means the independent evolution of similar phenotypes by distant genealogical species. Convergence can be explained by adaptation to similar environments by adjusting phenotype and morphology. For example, if a species is divided or separated and then the separated part adapts to aquatic or terrestrial conditions, its morphology will differ greatly from the original after a period of time. The process of adaptation to living conditions in soil may result in a high level of convergence. For example, the visual apparatus of soil-dwelling organisms is reduced, pigmentation is lost and the size of appendages is reduced, while special structures essential for life below the ground such as chemo- and hydroreceptors are distributed diffusely on the body, not only in the oral region.

The scale of adaptation to the soil by losing and acquiring special morphological and phenotypical markers indicates the extent of stability of the habitat. In a stable soil environment diverse factors such as water, temperature and organic matter vary only slightly over the short and medium term, as compared to large variations in above-ground environments. There is obviously no light in soil at depths greater than a few millimeters. As a result of all of these factors combined, edaphic (living in soil) microarthropods are sensitive and unable to survive abrupt variations in environmental factors. They are particularly sensitive to soil degradation and to the disturbances caused, for example, by agricultural cultivation and trampling.

Because of the great variation and the difficulties in interpreting the biodiversity assessment results of healthy or contaminated and deteriorated soils, the simplified model of bioassays is a good alternative. Instead of a complete biodiversity assessment, the results obtained on test organisms of a minimum of three trophic levels allow a prediction for the whole soil ecosystem, although with high uncertainty. Bioassay results are suitable for risk assessment and decision making about the necessary intervention for a deteriorated or contaminated soil. In some cases, direct decision on risk reduction or soil use can also be made.

Field assessment and bioassay application involve microcosms and mesocosms which can simulate the most important characteristics of the real environment, but are still easy to control and monitor. The microflora of the soil can be observed both by traditional microbiological or by new molecular techniques such as metagenomics and transcriptomics.

## 3.3  Terrestrial bioassays for testing chemical substances and contaminated soil

Laboratory bioassay measures chemically induced adverse effects on soil organisms, including microorganisms, soil-living micro-, meso- and macrofauna, birds or mammals. The study designs include acute systemic, dietary, and reproductive (also known as 'life-cycle') toxicity testing. Figure 5.16 shows the lethal (% death) effect by a soil artificially contaminated with diesel oil on Collembola. In this test, diesel oil is added to the soil, the test is run in a temporarily ventilated, closed system, where the actual concentration of diesel oil can be controlled. But in the case of real contaminated soil, neither the components of the complex pollutant, nor the selective changes of these components are known. The end point is a certain amount (mass) of soil which caused 50% decrease in the number of surviving animals. In this case the result is given as gram soil, – which has no information for the decision makers or managers, only for special experts. The amount of soil which causes 50% lethality is different in every single test type, according to the test setup. To be able to interpret the toxicity expressed in gram

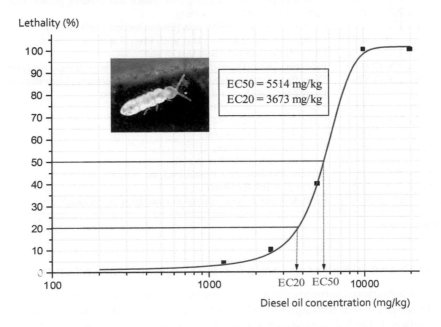

*Figure 5.16*  Collembola's response as a function of diesel oil concentration in soil.

soil, toxicity limit values must be created in soil gram. Subsequently it must be decided if the measured value is above or under this limit, and the exceedance rate should be calculated. This value is proportional to the risk characterization ratio (RCR). Another concept is the application of uniform equivalents, as described in Chapter 9.

## 4   STANDARDIZED AND NON-STANDARDIZED TEST METHODS

In this chapter we give an overview on the guidelines of soil toxicology test methods, recommended and standardized by OECD, ISO and other organizations for standardization.

### 4.1   OECD standards for testing chemical substances in soil and dung with terrestrial organisms

OECD guidelines were prepared to support the legislation of chemical substances. Formerly the Dangerous Substance Directive (1967), nowadays REACH Regulation (2006) requires the use of the OECD guidelines (Table 5.1).

*Table 5.1* OECD guidelines for testing of chemicals by terrestrial tests.

| Guideline number | Test name |
| --- | --- |
| OECD No. 216 (2000) | Soil microorganisms: nitrogen transformation test |
| OECD No. 217 (2000) | Soil microorganisms: carbon transformation test |
| OECD No. 227 (2006) | Terrestrial plant test: vegetative vigor test |
| OECD No. 208 (2003) | Terrestrial plant test: seedling emergence and seedling growth test |
| OECD No. 207 (1984) | Earthworm, acute toxicity tests |
| OECD No. 222 (2004) | Earthworm reproduction test (*Eisenia fetida/Eisenia andrei*) |
| OECD No. 220 (2004) | Enchytraeid reproduction test with several species |
| OECD No. 232 (2009) | Collembolan reproduction test in soil |
| OECD No. 226 (2008) | Predatory mite (*Hypoaspis (Geolaelaps) aculeifer*) reproduction test in soil |
| OECD No. 228 (2008) | Dung flies (*Scathophaga stercoraria, Musca autumnalis*) laboratory tests |
| OECD (2010, draft) | Dung beetle (*Aphodius constans*) laboratory test |
| OECD No. 213 (1998) | Honeybees, acute oral toxicity test |
| OECD No. 214 (1998) | Honeybees, acute contact toxicity test |
| OECD No. 205 (1984) | Avian dietary toxicity test |
| OECD No. 206 (1984) | Avian reproduction test |
| OECD No. 304A (1981) | Inherent biodegradability in soil |
| OECD No. 307 (2002) | Aerobic and anaerobic transformation in soil |
| OECD No. 312 (2004) | Leaching in Soil Columns |
| OECD GD 54 (2006) | Breakdown of organic matter in litter bags in the field** |

**Field microcosms.

### 4.2   ISO and other standards for testing soil and sediment

Various international standardization organizations ensure uniform methods and comparable results in soil testing, similarly to other areas of environmental management and practice. Table 5.2 summarizes the most well-known soil test standards, including

Table 5.2  ISO and other standardized test methods for testing soil quality and toxicity (ISO, 2014).

| Standard | Title of the standard method |
|---|---|
| ISO 10381-6:2009 | Sampling – Part 6: Guidance on the collection, handling and storage of soil under aerobic conditions for the assessment of microbiological processes, biomass and diversity in the laboratory |
| ISO 23611-1-6:2006-12 | Sampling of different groups of soil invertebrates |
| ISO 11063:2012 | Method to directly extract DNA from soil samples |
| ISO/DIS 17601:2013 | Estimation of abundance of selected microbial gene sequences by quantitative real-time PCR from DNA directly extracted from soil |
| ISO/TS 29843-1:2010 | Determination of soil microbial diversity – Part 1: Method by phospholipid fatty acid analysis (PLFA) and phospholipid ether lipids (PLEL) analysis |
| ISO/TS 29843-2:2011 | Determination of soil microbial diversity – Part 2: Method by phospholipid fatty acid analysis (PLFA) using the simple PLFA extraction method |
| ISO 14240-1:1997 | Determination of soil microbial biomass – Part 1: Substrate-induced respiration method |
| ISO 14240-2:1997 | Determination of soil microbial biomass – Part 2: Fumigation-extraction method |
| ISO 23753-1:2005 | Determination of dehydrogenase activity in soils – Part 1: Method using triphenyltetrazolium chloride (TTC) |
| ISO 23753-2:2005 | Determination of dehydrogenase activity in soils – Part 2: Method using iodotetrazolium chloride (INT) |
| ISO/DIS 10871:2009 | Determination of the inhibition of dehydrogenase activity of Arthrobacter globiformis – Solid contact test using the redox dye resazurine |
| ISO 16072:2002 | Laboratory methods for determination of microbial soil respiration |
| ISO 17155:2012 | Determination of abundance and activity of soil microflora using respiration curves |
| ISO 11266:1994 | Guidance on laboratory testing for biodegradation of organic chemicals in soil under aerobic conditions |
| ISO 14239:1997 | Laboratory incubation systems for measuring the mineralization of organic chemicals in soil under aerobic conditions |
| ISO 15473:2002 | Guidance on laboratory testing for biodegradation of organic chemicals in soil under anaerobic conditions |
| ISO 15685:2012 | Determination of potential nitrification and inhibition of nitrification – Rapid test by ammonium oxidation |
| ISO 14238:2012 | Determination of nitrogen mineralization and nitrification in soils and the influence of chemicals on these processes |
| ISO/TS 22939:2010 | Measurement of enzyme activity patterns in soil samples using fluorogenic substrates in micro-well plates |
| ISO 11269-1:2012 | Determination of the effects of pollutants on soil flora. Part 1: Method for the measurement of the inhibition on root growth |
| ISO 11269-2:2012 | Determination of the effects of pollutants on soil flora. Part 2: Effects of chemicals on the emergence and growth of higher plants |
| ISO 22030:2005 | Chronic toxicity in higher plants |

(continued)

*Table 5.2* Continued

| Standard | Title of the standard method |
| --- | --- |
| ISO 17126:2005 | Determination of the effects of pollutants on soil flora – Screening test for emergence of lettuce seedlings (*Lactuca sativa* L.) |
| ISO 29200:2013 | Assessment of genotoxic effects on higher plants – *Vicia faba* micronucleus test |
| ISO/TS 10832:2009 | Effects of pollutants on mycorrhizal fungi – *Glomus mosseae* spore germination test |
| ISO 11268-1:2012 | Effects of pollutants on earthworms (*Eisenia fetida*) – Part 1: Determination of acute toxicity using artificial soil substrate |
| ISO 11268-2:2012 | Effects of pollutants on earthworms (*Eisenia fetida*) – Part 2: Determination of effects on reproduction |
| ISO 11268-3:2014 | Soil quality – Effects of pollutants on earthworms – Part 3: Guidance on the determination of effects in field situations |
| ISO 16387:2014 | Effects of pollutants on Enchytraeidae (*Enchytraeus sp.*). Determination of effects on reproduction |
| ISO 17512-1:2008 | Avoidance test for determining the quality of soils and effects of chemicals on behavior – Part 1: Test with earthworms (*Eisenia fetida* and *Eisenia andrei*) |
| ISO 11267:1999 | Inhibition of reproduction of Collembola (*Folsomia candida*) by soil pollutants |
| ISO 17512-2:2011 | Avoidance test for determining the quality of soils and effects of chemicals on behaviour – Part 2: Test with collembolans (*Folsomia candida*) |
| ISO 20963:2005 | Effects of pollutants on insect larvae (*Oxythyrea funesta*) – Determination of acute toxicity |
| IOBC (2000) (Candolfi *et al.*, 2000) | Non-target arthropod acute and chronic laboratory tests (surface dwellers like *Aleochara bilineata, Poecilus cupreus, Pardosa spec.*) |
| ISO/DIS 18311:2013 | Method for testing effects of soil contaminants on the feeding activity of soil dwelling organisms — Bait-lamina test |
| ISO 15952:2006 | Effects of pollutants on juvenile land snails (Helicidae) – Determination of the effects on growth by soil contamination |
| ISO 10872:2010 | Determination of the toxic effect of sediment and soil samples on growth, fertility and reproduction of *Caenorhabditis elegans* (Nematoda) |
| ISO 16191:2013 | Determination of the toxic effect of sediment and soil on the growth behavior of *Myriophyllum aquaticum* – Growth test |
| ISO 21338:2010 | Kinetic determination of the inhibitory effects of sediment, other solids and colored samples on the light emission of *Vibrio fischeri* (kinetic luminescent bacteria test) |

ISO = International Organization for Standardization.
IOBC = International Organization for Biological Control.

new tests on DNA extraction from whole soil, the rapid nitrification test, the new earthworm reproduction test, the *Vicia faba* micronucleus test and the simple PLFA method. The standards cover sampling, chemical methods, including DNA and other omics technologies, biological methods, laboratory bioassays, simulation methods and field assessment. A special trend is testing soil as a whole using any of the chemical, DNA or biological and ecological methods.

US EPA (2014) Chemical Safety and Pollution Prevention Harmonized Test Guidelines (OCSPP, 2014) has developed a series of harmonized test guidelines for use in the testing of pesticides and toxic substances in soil and to provide uniform data for the regulatory and management purposes. Some of them are presented in Table 5.3.

*Table 5.3* Some selected soil tests from the US EPA harmonized test-guidelines.

| *Toxicity in soil* | |
| --- | --- |
| 850.3200 (2012) | Soil microbial community toxicity test |
| 850.3040 (2012) | Field testing for pollinators |
| 850.4000 (2012) | Background and special considerations – Tests with terrestrial and aquatic plants, cyanobacteria, and terrestrial soil-core microcosms |
| 850.4100 (2012) | Seedling emergence and seedling growth |
| 850.4150 (2012) | Vegetative vigor |
| 850.4230 (2012) | Early seedling growth toxicity test |
| 850.4300 (2012) | Terrestrial plants field study |
| 850.4600 (2012) | Rhizobium-legume toxicity |
| 850.4800 (2012) | Plant uptake and translocation test |
| 850.4900 (2012) | Terrestrial soil-core microcosm test |
| 850.2500 (2012) | Field testing for terrestrial wildlife |
| *Fate and behavior in soil* | |
| 835.1210 (1998) | Soil thin layer chromatography |
| 835.1220 (1998) | Sediment and soil adsorption/desorption isotherm |
| 835.1230 (2008) | Adsorption/desorption (batch equilibrium) |
| 835.1240 (2008) | Leaching studies |
| 835.2410 (2008) | Photodegradation in soil |
| 835.3300 (1998) | Soil biodegradation |
| 835.4100 (2008) | Aerobic soil metabolism |
| 835.4200 (2008) | Anaerobic soil metabolism |
| 835.6100 (2008) | Terrestrial field dissipation |
| 835.6300 (2008) | Forestry dissipation |
| 835.8100 (2008) | Field volatility |

## 4.3  Testing waste: a terrestrial test battery for solid waste

Environmental toxicity and other adverse effects of wastes can be tested using three terrestrial, three aquatic and one genotoxicity test according to the recommendation of the European Ring Test (correctly: European Interlaboratory Comparison) (Römbke *et al.*, 2009). The terrestrial tests for wastes are shown in Table 5.4.

## 5  NON-STANDARD TERRESTRIAL TOXICITY TEST METHODS

Terrestrial habitats and soil differ from aquatic habitats mainly in their intrinsically heterogenic character. The origin of this heterogeneity is based on the geochemical

*Table 5.4* Standardized terrestrial tests suitable for solid wastes.

| Standard or guideline | Title/name of the method |
| --- | --- |
| ISO 10871:2008 | Solid-contact test using *Arthrobacter globiformis* ISO NWI – Dehydrogenase activity 20% |
| ISO 11269-2:2005 | Determination of the effects of pollutants on soil flora. Part II: Effects of chemicals on the emergence and growth of higher plants (*Brassica napus*) – Biomass 30% |
| ISO 11268-1:1993 | Effects of pollutants on earthworms (*Eisenia fetida*) Part I: Determination of acute toxicity using artificial soil substrate – Mortality 20% |
| ISO 17512-1:2007a | Avoidance test for determining the quality of soils and effects of chemicals on behavior – Part I: Test with earthworms (*Eisenia fetida* and *Eisenia andrei*) |

heterogeneity of the bed/host rock and differences in elevation, positions and consequent exposure to sun, water and wind, etc. Heterogeneity of soil matrix is combined with the heterogeneity of the communities and individuals using the soil as habitat. All these mean that site- and problem-specific tests may play a prime role in terrestrial habitat and soil assessment. Non-standard species may be used to more accurately represent the functional roles played by local flora or fauna. Non-standard methods, mainly site-specific microcosms and mesocosms can simulate certain soil situation or terrestrial ecosystem.

The cost of tests on non-standard species can be higher than standardized ones due to the additional tasks of defining optimal conditions for the organism's growth, establishing the organism's sensitivity to site-related contaminants, and establishing control condition responses. These additional costs may be justified by obtaining surplus information about adverse effects on resident species.

## 5.1   Some aspects of problem-oriented and site-specific soil testing

Most of the environmental problems are unique and have an individual risk profile, which is reflected by the conceptual model. Standardized test methods often fail to meet the needs of site-specific risk management requiring site-specific test organisms, or models and test setups that suite to the problem. The problems may include an area or a certain species to be monitored, contaminated sites to be assessed, or a contaminant to be detected.

Before planning the assessment or monitoring of an environmental problem, the aim of testing should be clarified, the main risks identified, the concept of testing, the selection of field or laboratory methods decided. It should also be decided how to deal with the soil matrix effects and the soil's physical phases, whether to apply single- or multispecies tests, etc. These aspects will be discussed below. Assignment of sampling and monitoring points is also an important issue and it will be discussed in Chapters 3 and 4 in Volume 3.

### 5.1.1  Soil community response

Any parameter of the whole soil that characterizes soil in general may be suitable for assessing and monitoring soil activity and health from the environmental point of view. These can be:

– The concentrations of microbial cells or higher organisms, abundance and diversity of

   o   aerobic heterotrophic cells;
   o   facultative and obligate anaerobes;
   o   cyanocteria;
   o   plants;
   o   microfungi and macrofungi;
   o   single-cell animals;
   o   multi-cell animals, such as micro- and meso- and macrofauna members: nematodes, worms, collembolans, insects, snails, vertebrates.

– Characteristics of the metagenome, the chemical, biochemical, metabolic and physiological parameters of the whole biota of the soil. These indicators use innovative DNA and other molecular techniques such as specific microscopes, genomics, proteomics, enzyme analysis, carbon-source utilization profiles, functional gene arrays, phospholipid fatty acid analysis (PLFA), nucleic acid finger printing techniques, cloning the metagenome, and identifying transcripts, gene arrays, massively parallel sequence analysis, etc. The following chemical species can be used as indicators:

   o   DNA, its quantity and quality, the guanine-cytosine (GC) content of DNA and RNAs;
   o   Structural building chemicals such as structural proteins, muramic acid, ergosterol, technoic acid, phospholipids, lipopolysaccharides, glucoseamine, diaminopimelic acid;
   o   Functional proteins, mainly enzymes such as oxidases, glucosidase, saccharase, xilanase, proteases, lipases, urease, amidases, esterases (phosphor mono-, di- and triesterases), cellulases, kitinase, arilsulfatase, etc.

– Functions of living soil, its energy production (ATP), respiration, denitrification, biodegradation, biosynthesis.
– Dynamic functions: increase or decrease in any of the products or soil functions further to the effect of environmental parameters or stresses such as habitat deterioration or the presence of contaminants.

There are plenty of chemical analytical, enzymological and biological techniques for extracting as well as selective methods for direct testing the relevant parameter in the soil.

In addition to single tests, innovative test batteries and multi-well test systems do exist, integrating the experience of experts and helping the design, evaluation and interpretation of measured data.

The dynamic changes in soil activity and function as a response to the presence of adversely affecting physical, chemical or biological agents can be followed on site or can be simulated or even provoked in microcosms, mesocosms or other type of dynamic and interactive tests.

The above-listed potential end points can be used in all strategies of soil toxicity testing.

### 5.1.2   Concepts for characterizing soil functioning and health

There are two main concepts for the characterization of soil health and its normal functioning: monitoring absolute values of characteristic indicators, or testing on representatives and extrapolate to the whole.

–   Measuring the **absolute values** of the selected soil characteristics, or a set of characteristics, meaning the absolute value of cell concentrations, contents and activities. These results can be compared to the healthy average or to the formerly recorded results of the same soil. In the case of existing monitoring data the trends can be characterized by these absolute values. When one gathers data from one single time-point, the changes due to seasonal, climatic and incidental events are hardly to be distinguished from the adverse effects of chemicals, therefore the statistical evaluation of testing will be very weak.
–   Testing **selected members** of the soil community and extrapolation from their response to the entire soil. In this case organisms from a minimum of three different trophic levels are selected and tested to exclude soil toxicity or to predict no-effect values for risk assessment. In such type of soil testing one can select standard test organisms, which are – with high probability – not relevant to the soil to be tested, or can use non-standard test organisms well representing the biota of the soil in question. Both site-specific and generally used or standard test organisms can be included in test batteries according to the aim of the test. Based on the results of the laboratory test batteries or of the tested remedied soil, one can decide whether the soil or waste is hazardous or not.

### 5.1.3   Aims of testing whole soil response

The two main objectives of whole soil tests are: testing the fate and effect of chemical substances in soil and testing the adverse effect of contaminated soil, sediment or solid waste on representative organisms.

There are two main concepts for testing adverse effects of pure chemical substances, as well as waste and soil contaminated with chemicals:

1   To use the formerly listed soil indicators for measuring **the adverse effects of chemical substances** in soil environment, the chemical substance is added in different concentrations into a well-defined and stable natural soil, containing the original microbiota (the 'test organism'). From the response of the spiked soils, one can identify the highest no-effect concentration of the chemical substance in the investigated soil ($NOEC_{in\ soil}$), the lowest-effect concentration ($LOEC_{in\ soil}$) or the concentration resulting in 50% or any other percentage inhibition in the measured soil characteristic.

Using this concept the huge adaptive and *toxicity buffering capacity of the soil* can be experienced i.e. by comparing test results with chemical substances dissolved in water.

2   Standard healthy soil may be suitable for *testing solid waste or contaminated soil*, mainly in such cases when mixing waste or contaminated soil into healthy soil (e.g. into agricultural soil) is a true model of a real scenario, e.g. waste disposal on soil or soil amendment using compost, etc. In this concept contaminated soil/solid waste is mixed into healthy soil in incremental concentrations and the effect of the mixture on the relevant soil activity is measured as an end point. In this concept soil is used as a 'test organism' and the test is evaluated similarly to single organism bioassays, but the result is obtained as contaminated soil mass NOEL, LOEL or $ED_{50}$, instead of contaminant concentration. These results cannot be integrated into a chemical-substance-centered risk assessment concept, but can be used for classification of wastes/contaminated soils and for direct decision making on their use or disposal. Applying this concept, another problem arises: the dose–response curve *may be deformed due to possible positive interactions* between test soil and waste/contaminated soil. For example a waste with high organic matter content may activate soil mineralization, or the microflora and other components of the otherwise contaminated soil could substitute the missing components, the bottleneck of the test soils.

### 5.1.4   *Consequences of the effect of soil matrix on the test methodology*

Testing soils is more difficult than testing water due to the three equivalent phases (gas, liquid and solid), the partition of the contaminants between the phases and the limited availability of test organisms (Sijm *et al.*, 2000; Horvath *et al.*, 2000; Fenyvesi *et al.*, 2002).

There are some important aspects, which have to be considered when deciding on the concept of soil testing. Availability and accessibility should be considered according to the differences between soil matrix and the test system's matrix. The following scenarios should be distinguished:

– Liquids, containing dissolved contaminants infiltrate the soil, and solid soil particles compete with living organisms for the dissolved chemicals. If the sorption capacity of soil particles is very strong, biological access will be reduced.
– Contaminated soil (to be tested) is added to a water-based test system: desorption and dissolution are the limiting factors of biological availability for aquatic organisms. Terrestrial organisms can interact and enhance bioavailability.
– Contaminated solid (contaminated soil, sediment or waste) is mixed into soil: the contaminants have to redistribute between four phases of the mixture: soil solid and solid waste, pore water of soil and moisture content of solid waste until establishing a new equilibrium.
– Testing soil may involve nutrients together with contaminants (e.g. plant nutrients in a plant test) in the test system modifying the response of the test organism.

Figure 5.17 shows the changes in soil toxicity during a remediation process based on biodegradation. Toxicity increases in the beginning due to the biological mobilization of the sorbed contaminants prior to biodegradation. Soil microorganisms have

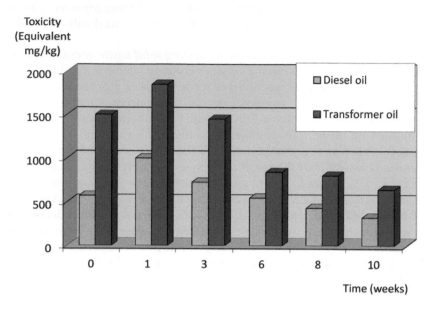

*Figure 5.17* Time dependence of soil toxicity in a soil remediation process based on biodegradation.

their own set of tools (biotensides, complexing agents) to increase access to sorbed or otherwise bound substances. The run of time-dependent toxicity shown by Figure 5.17 is typical in the case of aged, sorbable, but still biodegradable contaminants, e.g. most of the petroleum-based hydrocarbon mixtures. After making the sorbed molecules biologically available, the microbial community degrades the contaminants and lowers toxicity of the soil.

Table 5.5 shows a matrix of the potential test scenarios including the most important interactions between the test systems containing test organism(s) and the samples to be tested. As Table 5.5 indicates, the following two cases represent different situations:

– Testing pure chemicals mixed into groundwater or soil representing a generic scenario to simulate their fate and behavior plus their adverse effects on soil ecosystem members (representing generic or specific ecosystems), and
– Contaminated soils or interacting liquid phases (runoff, leachate, flood water) containing various chemical substances (representing the contaminated site).

Nevertheless, the same tests are still often applied to both pure chemicals and contaminated soil testing. In ISO 11268-3 standardized earthworm test – a guidance for the determination of the effects of chemicals on earthworms in field situations – Kula *et al.* (2006) recommended splitting the current guideline into two fields of application: one for testing chemicals and pesticides (i.e. within the scope of OECD) and another one for testing soil quality (i.e. within the scope of ISO). They indicate a major need for guidance concerning the interpretation of effects determined in such complex field tests.

Table 5.5 Laboratory bioassays for testing chemicals in real contaminated soil.

| Sample to be tested<br>Test organism<br>Test system | Effect of a chemical substance dissolved in water<br>Simulation test for chemicals dissolved in water | Effect of contaminated environmental runoff, leachate, pore water, wastewater, sludge (with unidentified mixture of contaminants and LB)<br>Liquid phase environmental sample testing | Effect of chemical substance mixed into standard model soil or site-specific reference soil (with SB)<br>Simulation tests for contaminants mixed into the soil | Effect of contaminated soil or solid waste (with un-identified mixture of contaminants and SB)<br>Solid phase environmental sample testing |
|---|---|---|---|---|
| Aquatic test organism(s) (ATO) in liquid medium (generally sterilized*). *The response of ATO is measured.* | Water-based system, the model of groundwater contaminated by dissolved chemicals. Bioassay is based on the interaction between ATO and dissolved (diluted) pure chemicals. | Water-based system, the liquid phase environmental sample is tested by ATO. Bioassay is based on the interaction between ATO and the mixture of dissolved or otherwise solubilized contaminants and other dissolved chemical substances (with inhibitory or stimulatory effects). Interaction of ATO with natural biota (LB) maybe significant in non-sterilized samples. | Water-based slurry system modeling water erosion, leaching, infiltration, resuspension and the fate of the chemicals during these. Bioassay is based on the interaction between ATO and desorbed, mobilized, solubilized chemicals. Soil texture, composition and pre-incubation with the chemical (availability) are influential. | Water-based system modeling water erosion, leaching, infiltration, resuspension and the fate of soil contaminant mixture when interacting with water. Interaction between ATO and mobilized, solubilized contaminants or other chemical constituents of soil. Interaction between ATO and SB may be significant. Soil texture and composition is integrated into the response. |
| Terrestrial test organism(s) (TTO) in standard/model soil (may or may not be sterilized*). *The response of TTO is measured.* | 3-phase wet soil or 2-phase slurry. Modelling irrigation, flood, liquid nutrients or manure application on soil and the infiltration of dissolved contaminants. Bioassay is based on the interaction of TTO with partitioned (from dissolved to sorbed) chemical. Interaction of TTO with SB may be influential. The influence of TTO on contaminants may be weak (due to dilution of its biologically active products, e.g. exoenzymes or exudates). | 3-phase wet soil or 2-phase slurry. Modelling effect of contaminated liquids on soil ecosystem. Bioassay is based on the interaction of TTO with partitioned (from dissolved to sorbed) contaminants or other chemicals resulted from liquid sample. Interaction between indigenous biota (LB+SB) and TTO may be significant in non-sterilized soil, but also occurs in sterilized soil. The influence of the test organism on the contaminant may be weak. | A) Chemical substance mixed into standard soil or B) site-specific reference soil. 3- or 2-phase soil. Modeling direct contamination of soil with chemicals. Direct contact and close interaction between TTO and the partitioned (mainly sorbed, partly dissolved) chemical. Interaction of TTO with living or dead SB may be significant and is unpredictable. Influence of TTO on contaminants may be significant. Avoidance is possible. | 3- or 2 phase soil. Direct contact and interaction of TTO with solid phase bound and partly desorbed mixture of chemicals (contaminants, nutrients, etc.). Nutrients, stimulants from soil are influential. Soil texture and composition as well as contaminant bounding is influential. Influence of TTO on contaminants may be (locally) significant. Avoidance is possible. |

(continued)

# Table 5.5 Continued

| Sample to be tested Test organism Test system | *Effect of a chemical substance dissolved in water* Simulation test for chemicals dissolved in water | *Effect of contaminated environmental runoff, leachate, pore water, wastewater, sludge (with unidentified mixture of contaminants and LB)* Liquid phase environmental sample testing | *Effect of chemical substance mixed into standard model soil or site-specific reference soil (with SB)* Simulation tests for contaminants mixed into the soil | *Effect of contaminated soil or solid waste (with un-identified mixture of contaminants and SB)* Solid phase environmental sample testing |
|---|---|---|---|---|
| Real (site-specific or reference) soil with indigenous microbiota (SB). **The response of SB is measured.** | 3-phase wet soil or 2-phase slurry. Simulating real soil. The interaction of SB with the partitioned (partly dissolved, partly sorbed) chemical is intensive. Strong interaction of SB with soil matrix. The influence of the contaminant to be tested may be significant. | 3-phase wet soil or 2-phase slurry. Simulating the effect of contaminated liquid phase compartments on real soil. Bioassay is based on the interaction of SB with the partitioned (dissolved, sorbed) mixture of contaminants and other chemicals. Interaction of SB with soil matrix is strong. SB interacts with LB. The influence of SB on the contaminants may be significant. | 3- or 2-phase soil. Modeling direct contamination of soil with chemicals. Bioassay is based on the interaction between SB and the chemical mixed in soil and partitioned between soil phases. Interaction of SB with soil matrix is strong. The reaction of SB and its influence on contaminant mobilization and degradation may be significant. Soil conditions and nutrient content may be synergic (strengthening) or antagonistic (canceling) the effect of the chemical substance to be tested. | 3- or 2-phase soil. Based on the interaction of SB with the sorbed mixture of contaminants and other chemicals. Interaction of SB with soil matrix is very strong. The reaction of SB and its influence may be significant. Soil used for receiving and diluting the contaminated soil or waste causes further interactions between two soils or soil and waste: contaminants and nutrients, inhibitors and stimulants in changing ratios along the dilution series, with different equilibrium states. |
| Terrestrial test organism(s) in real contaminated soil. **TTO and SB response together is measured** in real contaminated soil. This case is typical in soil testing, because SB cannot be excluded, neither in the case of sterilized samples. | 3-phase wet soil or 2-phase slurry. Modeling the fate and effect of a dissolved chemical in contaminated soil containing adapted SB. Bioassay is based on the interaction of TTO and SB with partitioned (partly dissolved, partly sorbed) contaminant. The influence of the adapted SB on the contaminant to be tested may be significant. | 3-phase wet soil or 2-phase slurry. Simulating the effect of contaminated liquid phase compartments on contaminated soil. Bioassay is based on the interaction of TTO, SB and LB with partitioned (partly dissolved, partly sorbed) mixture of contaminants and other chemicals. Interaction of TTO with LB and SB. The influence of adapted SB on the contaminants of the tested liquid may be significant. | 3-phase wet soil or 2-phase slurry. Simulating the fate of the chemical and its direct effect on contaminated soil containing adapted SB. Bioassay is based on the interaction of TTO and SB with sorbed contaminants. Interaction of TTO with adapted SB. The influence of adapted SB on the tested contaminant may be significant. Avoidance is possible. | 3- or 2-phase real soil. Simulating the addition of solid phase soil/waste into contaminated soil: the fate and effect of contaminants can be followed. Bioassay is based on the interaction of TTO and adapted SB with the solid form (mainly sorbed) mixture of contaminants and other chemicals. Interaction of TTO with adapted SB. The influence of TTO and SB on the contaminants in the tested waste may be significant. Close interaction of TTO with soil matrix. Avoidance is possible. Soil used for receiving the contaminated soil or waste and diluting the sample causes further interactions between two soils or soil and waste: contaminants and nutrients, inhibitors and stimulants in changing ratios along the dilution series with different equilibrium states. |

ATO: aquatic test organism TTO: terrestrial test organism LB: biota of the liquid phase of environmental sample SB: soil biota.
*Effectivity of sterilization of environmental samples is questionable and may cause further problems due to chemical changes and dead biomass.

Table 5.5 summarizes the test scenarios, modeling or simulating the fate and effect of pure chemicals and the tests for soil or solid waste, soil leachates and other liquid phase environmental samples. The table contains information on the possible interactions during testing (between soil matrix, contaminant(s), soil biota and the test organism). Some other key factors are also highlighted. The table intends to cover bacterial, plant and animal test organisms as well as test setups of bioassays, leaching tests or microcosms. A dilution series for studying the dose–response function of contaminated soil is usually difficult to produce. Dilution series of chemicals in soil can be best prepared by adding exactly weighed masses of the chemical substance (e.g. 1.0 mg, 2.0 mg, 5 mg, 10 mg, etc.) into soil samples of the same mass (e.g. 50 g). When mixing liquid samples to soil, the same volumes of a water-based dilution series should be added to the soil. When contaminated solid matter (waste or contaminated soil) is mixed into the receiving soil, the result of the mixing should be calculated (e.g. in percentage), and the same amounts of the solid mixture should be tested.

### 5.1.5   Field assessment or laboratory testing?

This question can be answered after having defined the assessment target and the conceptual risk model associated with the problem, the potential contaminants and land uses.

The field assessment's results may be representative for the site if the assessment (monitoring) concept truly reflects the real distribution of risks and the key risk components, for example the most sensitive species and those suffering the greatest exposure.

On the other hand, the results of field assessment can only be quantified on a relative scale, and the use of these relative results is restricted to the site in question, cannot be generalized and used for other sites or different regions. The statistical power of field assessment is generally weak due to non-repetitions and interactions with unidentified co-existing strengthening or canceling effects.

Field microcosms may avoid unexpected and disturbing external effects as they are partly isolated and better controlled than real nature. Of course if a small part of the environment is isolated for test purposes, a great part environmental reality is lost, but replicates can increase validity of the results.

Moving towards simpler models of the soil environment, the loss of environmental relevance can be compensated for by using a good concept, for example, by selecting the priority risk components and the most important participants.

When long-term processes such as biodegradation, bioaccumulation and food-chain effects play a role in the quality and health of the soil, a test system should be created which simulates the complex natural soil, where these long-term processes can proceed and their effects can be monitored. Soil microcosms, even laboratory microcosms can meet these requirements.

A test concept suited to soil contaminants which are easy to mobilize must ensure the contact between the contaminated soil and the test organism. Direct contact tests – also called solid contact or contact tests – are such tests. The role of contact tests is to ensure the interaction between contaminated soil and the test organism, and let the strongly bound contaminants be mobilized or otherwise influenced by the organism. Bioassays on soil extracts or leachates can model runoffs, infiltrates, drainage and leachates from the soil to monitor contaminant transport from solid soil to water. Contaminated precipitation, flood or liquid waste disposal can be modeled by irrigating dissolved contaminants onto the ground. If the soil should be assessed as a habitat or an object used by humans or by an ecosystem, soil as whole should be tested. When the partition of the contaminant between solid and liquid phases plays a role in the formation of risk, soil as a whole and pore water should be simultaneously tested.

Chemical models fulfill the requirement of being exact, reproducible and capable of characterizing the risk of a potential soil contaminant; e.g. estimating the bioavailable fraction from the extracted amount by biomimetic agents. *In silico* toxicology, i.e. the use of mathematical models has growing importance in soil toxicology, but the number of reliable mathematical models is yet small due to the relatively small amount of data (e.g. compared to aquatic toxicology).

## 5.2  Ecological assessment: field testing of habitat quality, diversity of species and abundance of indicator organisms

Site- and problem-specific ecological assessment methods can be created for characterizing habitat quality and suitability as well as biological integrity or vulnerability of habitats. Some protocols for terrestrial ecosystems have been standardized by US EPA. These are accessible online from the Ecological Assessment Methods Database of the USA.

–  Floristic quality assessment index – FQAI;
–  Habitat assessment model;
–  Habitat evaluation procedure – HEP;
–  Index of biological integrity (IBI) for birds;
–  Index of biological integrity (IBI) for invertebrates;
–  Index of biological integrity (IBI) for plants;
–  GIS based wetland assessment (Montana Natural Heritage);
–  Multi-scale assessment of watershed integrity – MAWI;
–  Remotely-sensed indicators for monitoring condition of natural habitat in watersheds;
–  Soil management assessment framework;
–  Uniform mitigation assessment method – UMAM;
–  Variables for assessing reasonable mitigation in new transportation – VARMINT;
–  Wetland value assessment methodology – WVA;
–  Wildlife habitat appraisal procedure – WHAP.

A Guidance Document has been established for terrestrial field studies by Fite *et al.* (1988) for US EPA, ensuring a standardized protocol for the various assessments. This area has been developed for terrestrial field dissipation studies for pesticides and a guidance document has been established by several interested organizations under the auspices of NAFTA (Corbin *et al.*, 2006).

### 5.2.1  Abundance and diversity of soil microbiota

Measuring pollution-induced changes in soil microbiota diversity, metabolic activity (for example ability to biodegrade contaminants) or community tolerance

(PICT = Pollution-Induced Community Tolerance) is based on shifts in diversity toward new metabolic pathways or tolerant microbial communities in the presence of contaminants (Rutgers & Breure, 1999). In this case, microbial community shifts may be characterized by carbon substrate utilization.

### 5.2.2 The use of carbon substrate utilization patterns for ecotoxicity testing

In a normal carbon substrate utilization study the carbon substrates are contained in a multi-well plate consisting of a freeze-dried medium, a standard set of carbon substrates (up to 96 wells) and a redox dye for monitoring microbial activity. The soil suspension that contains the soil microbial community is uniformly inoculated in all wells of the plate and incubated under controlled laboratory conditions. When the community in any of the wells is able to utilize the carbon source present in a certain well, the redox dye (tetrazolium violet) turns purple, showing the oxidation of the carbon substrate. Each substrate in the Biolog$^{TM}$ plate represents a different test system and provides information on the microbial oxidation reaction (binary 0 or 1, rate constants, etc.) that is specific to the inoculated microbial community (Rutgers & Breure, 1999). The wells together give a fingerprint on the metabolic activity of the community.

The same system can be used for ecotoxicity testing purposes. In this case a healthy microbial community is inoculated into the wells and single chemicals, extracts, leachates or pore water from contaminated soil or contaminated soil itself is added to the healthy microbial community of the wells.

The relative sensitivity of the microbial community to toxicant exposure is calculated as the toxicant concentration that causes a 50% decrease in activity (substrate utilization) in comparison to the untreated control. When comparing an adapted soil community to a non-adapted one, higher $EC_{50}$ ($ED_{50}$) will be measured in the case of adaptation. This assessment concept can be applied for both tolerance assessment (Rutgers et al., 1998) and the assessment of biodegrading potential of the soil community (Nagy et al., 2010). Both are associated with the presence of contaminant in soil over a long term.

Tolerance is manifested as an enhanced ability to use a substrate (higher $EC_{50}$) relative to an unexposed population. The concept of PICT encompasses the phenomenon that, given variation in species' sensitivities, toxic effects will reduce the survival and growth rate of the most sensitive organisms within a population, thus increasing the average tolerance of the community.

### 5.2.3 Dung-dwelling organisms, a not yet standardized field study

To register veterinary medicinal products (VMPs) as parasiticides on pastured animals, legislation in the European Union requires an environmental risk assessment to test the potential non-target effects of fecal residues on dung-dwelling organisms. Products showing adverse effects in single-species laboratory tests require further, higher-tier studies to assess the extent of the adverse effects on entire communities of dung-dwelling organisms under more realistic field or semi-field conditions. Currently, there are no documents specifically written to assist researchers in conducting higher-tier studies or to assist regulators in interpreting the results of such tests in an appropriate context. The members of the SETAC (2013) Advisory Group DOTTS (Dung Organism Toxicity Testing Standardization) provide a description on dung fauna in Central and Southern Europe, Canada, Australia, and South Africa. This document briefly reviews the organisms that make up the dung community and their role in dung degradation, identifies key considerations in the design and interpretation of experimental studies, and makes recommendations on how to proceed (Jochmann et al., 2011).

The veterinary parasiticide *ivermectin* was selected as a case study compound within the project ERAPharm (2013) (Environmental Risk Assessment of Pharmaceuticals). Current ERA clearly demonstrates unacceptable risks for all investigated environmental compartments, and several gaps in the existing guidelines for ERA of pharmaceuticals have been indicated and improvements suggested. In addition, guidance is lacking for the assessment of effects at higher tiers of the ERA, e.g. for field studies or a tiered effects assessment in the dung compartment (Liebig et al., 2010).

### 5.2.4   Effects of pollutants on earthworms in field situations: avoidance

The ability of organisms to avoid contaminated soils can act as an indicator of toxic potential in a particular soil. Based on the escape response of earthworms and collembolan, avoidance tests with these soil organisms have great potential as early screening tools in site-specific assessment. These tests are becoming more common in soil ecotoxicology because they are ecologically relevant and have a shorter duration time compared to standardized soil toxicity tests. The avoidance of soil-dwelling invertebrates, however, can be influenced by the soil properties (e.g. organic matter content and texture) that affect the behavior of the test species in the exposure matrix.

Earthworms are suitable organisms for measuring avoidance and for the evaluation of hazardous wastes sites. Avoidance tests show higher sensitivity compared to acute toxicity tests based on the deterioration because organisms generally exhibit behavioral responses at lower levels of stress than those that acute toxicity tests can detect. Avoidance is an ecologically relevant end point that can potentially indicate sub-lethal stress within a short period of time, the test is easy to perform in a soil matrix, and an avoidance test has the potential of specialized applications for soil testing (Natal-da-Luz et al., 2008).

'Dual-control' test data have shown that, in the absence of a toxicant, worms did not congregate, instead distributed themselves fairly randomly with respect to the two sides of the test chambers, that is, they did not display behavior that might be mistaken for avoidance. In tests on artificial soil spiked with reference toxicants and hazardous site soils, worms avoided soils that contained toxic chemicals. Avoidance behavior proved in most cases a more sensitive indicator of chemical contamination than acute tests. Determination of avoidance was possible in 1 to 2 days, much less than the current duration of acute and sub-lethal earthworm tests (Yardley et al., 1996).

This means that the earthworm avoidance test for soil assessments is an efficient alternative to acute and reproduction tests (Hund-Rinke & Wiechering, 2001) and is a sensitive screening method. Currently, two test designs, a two-chamber system and a six-chamber system, are in the standardization process.

Other animals than earthworms have also been used in avoidance tests: the springtail *Folsomia candida*, other springtails and the white-worm *Enchytraeus albidus* (Enchytraeidae). All of them are ecologically relevant soil species and are commonly used in standardized toxicity tests. Their rapid reaction to a chemical exposure can be used as a toxicological measurement end point that assesses the avoidance behavior (Amorim *et al.*, 2008).

## 5.3  Non-standardized contact bioassays: description of some  tests

Tests of contaminated soil as well as potential contaminants in the soil environment should be adjusted to the risk scenario to be modeled by the test. Table 5.5 explains that extracts and leachate made from soils mainly represent the risk of soil posed to subsurface or surface waters. If we assess soil as a habitat, direct contact between soil and test organism is an important condition, which ensures realistic and dynamic interactions between contaminants, soil matrix components and the organisms present. In order for the test results to show this integrated effect, representing actual risk, we have to make the actors to interact with each other in a more or less realistic way.

In the case of generic testing, this kind of soil-specific realism is a disturbing condition, which should be excluded and 'abstract models' with very low environmental relevance but high reproducibility and precision should be preferred.

In the following, we shall introduce some of the useful contact tests the authors have had successful experience with.

### 5.3.1  Single species bacterial contact tests

*Vibrio fischeri* is generally used as a test organism for testing liquid samples, but it can also be used in soil suspensions. The problem of detection is that the soil particles absorb a part of the luminescent light the bacterium emits. When the control soil differs in texture and composition from the tested soil, this is another problem because the light absorption of the two differs. As a solution, soil sample is compared to itself by measuring light emission immediately after soil addition and 30 minutes later. This experimental setup assumes that the measurable light intensity decreases immediately after soil addition only due to the presence of light-absorbing solid particles of the soil. Within 30 minutes, the effect of the toxic chemical is effectuated by further decreasing light emission. The difference in light intensity of these two time points represents the real inhibitory effect of the contaminated soil on bacterial luminescence.

The tests need 2 grams of soil, which is suspended in 2 mL of 2% NaCl solution. A minimum of five-step (seven is better) dilution series is prepared from the contaminated soil. After measurement of the reference luminescence intensity, $50 \mu L$ of each member of the dilution series is added to the test medium that contains the luminescent test organism. The light production of the test bacterium in these tests is measured by a luminometer. Figure 5.18 introduces the luminescent bacterial culture, the sample preparation and the portable equipment.

*Azomonas agilis* bioassay is based on the dehydrogenase activity inhibition caused by the toxic effect of the contaminated soil to be tested. 100 mL sterile medium is supplemented with 1 mL 1% 2,3,5-triphenyl tetrazolium chloride (TTC) as an artificial electron acceptor and with the test bacteria previously incubated on a rotary shaker at 28°C for 72 h. The stock solution is injected into the tubes that contain the dilution series (dilution factor 2) of the contaminated soil. After incubation in dark,

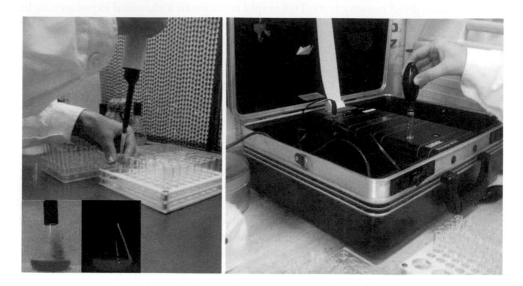

*Figure 5.18* Luminescence inhibition of contaminated soil measured by a portable luminometer.

the red-colored formasan (produced by the reduction of TTC) is determined visually (semi-quantitative method) or after extraction by organic solvents (quantitative determination).

The inhibited growth and metabolic activities of several sensitive bacterial or microfungal strains can indicate toxicity. Therefore these activities can be used as end points for testing contaminated soil. When growing in suspension, density or color as a result of microbial growth cannot be measured directly from the suspension due to the disturbance of soil particles. Plating and counting of colonies grown from living cells is feasible to measure growth. The metabolic activity of the microorganisms can be monitored by the microplate technique (Figure 5.19) from the soil/sediment suspension. Substrate and indicator is placed into the microwells (ready-made or home-prepared) and the suspended soil sample with the cells is poured on that. The developed color of the indicator after incubation proves the metabolic activity. The type and number of substrates and parallels is optional. The properly compiled metabolic fingerprint can show activity and adaptation compared to reference or to baseline when testing time series. It is important to note that soil suspension differs from three-phase soil significantly, it can be considered as a model of water eroded soil, flooded soil or re-suspended sediment (see also Table 5.5).

Bacterial species of *E. coli, Pseudomonas, Bacillus, Azomonas, Azotobacter, Arthrobacter,* yeasts such as *Saccharomyces cerevisiae, Rhodotorula, Torula, Candida* species and filamentous fungi such as *Penicillium, Aspergillus, Trichoderma,* etc. have been successfully used by the authors for soil suspension and sediment slurry contact testing.

***Bacillus subtilis*** and ***Azotobacter croococcum*** growth test using an agar-diffusion method and soil discs is an efficient innovative test method developed by Gruiz

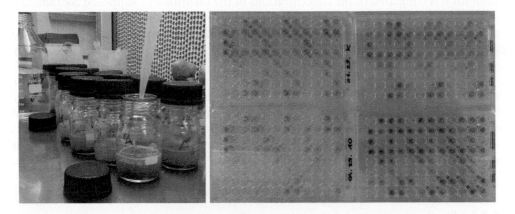

*Figure 5.19* Soil suspension and positive metabolic activity shown by coloration in the microplate.

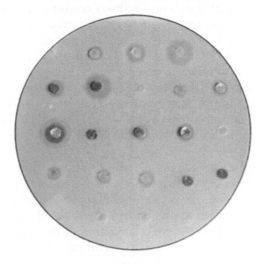

*Figure 5.20* Agar-diffusion test on metal sensitive *Bacillus subtilis*.

*et al.* (2001) and applied to screening of metal-contaminated soil at watershed-scale (Figure 5.20).

Soil is fixed with plain agarose gel and diskettes cut out from this gelled soil are placed on the surface of a relatively high-density bacterial culture containing agar medium. The agar-based soil disks and the nutrient agar-medium are prepared from the same agar-agar, ensuring similar transport character and capacity. The transport between the two media is free, the interaction between bacterial exudates and soil-contaminants can take place. The agar-medium is poured into 20-cm-diameter round or 30 × 30 cm rectangular trays, where 20–40 samples can be tested in parallel. The inhibition zone without bacterial growth around the soil diskette is measured in mm.

Inhibition is compared to the concentrations of the identified or supposed contaminating substances (mixed into soil) or to any well-known contaminant, when the chemicals responsible for soil toxicity have not been identified. This fast and simple test method is suitable for screening a great number of soil and sediment samples and for toxicity mapping of large sites.

The *Bacillus* species which has medium sensitivity, and as such is suitable for screening and finding the hot spots at an extended contaminated area, were selected by the authors specifically for the purpose of screening a small watershed with unidentified contaminants (Gruiz, 2005). Hot spots were identified by 500 soil samples using this rapid bioassay. In a second step, contaminants have been identified from the hot spot samples using chemical analysis.

### 5.3.2　Single species animal contact tests

The protozoon ***Tetrahymena pyriformis*** reproduction inhibition test characterizes the toxic effect of the contaminated soils on a primary consumer in the food chain. *T. pyriformis* stock is grown in a proteose peptone yeast extract medium (PPY), containing 1% proteose peptone and 0.1% yeast extract. The test goes on in a soil suspension of 0.25 g of soil in 5 mL of PPY medium supplemented with penicillin-, streptomycin- and nystatin (antibiotics) solutions to prevent the infectious damage of the test organism when contacting soil. 100 µL of six-day-old test organisms (about 1000 cells/mL) are used for the inoculation of the test tubes. Incubation period is 3–4 days, in the course of which sampling and counting of the cells is carried out 6–7 times. A growth curve can be drawn from the cell counts and $E_rC_{50}$, NOEC or LOEC is read from the curve.

*Tetrahymena* tests on real contaminated soils showed great differences between the results of contact tests in soil suspension and on extracts from the same soils. A soil historically contaminated with organic pollutants and another one with extremely high toxic metal content failed to exhibit toxicity when the extract was tested, but showed high toxicity after being in direct contact with the test organism for a period of time. A third soil with low contaminant concentration but high water-soluble metal content showed toxic effect measured both from extract and in direct contact (Leitgib *et al.*, 2007).

***Folsomia candida***, a collembolan, is a popular test organism as it is easy to maintain and handle in tests. An OECD guideline exists for testing reproduction, including mortality as a range-finding test. The test requires a synchronized culture and its preparation needs some experience. Otherwise it is very easy to perform: a two-fold dilution series is prepared from the contaminated soil (OECD soil or healthy reference soil) at final concentrations of 6.25%. Ten to twenty pieces of twenty-day-old springtails are gently delivered from the gypsum block (1) into the trap by weak air flow. A synchronized culture is used for acute testing. The animals are transferred from the trap into 250-mL test flasks (2) containing 20 g of the soil mixtures with equilibrium moisture (3). Test flasks are incubated at 20°C in the dark for 14 days. At the end of the incubation period, each soil in the test flask is flooded with distilled water and the survivor animals, floating on the surface (4) are counted. Figure 5.21 shows the test steps described above.

Figure 5.21   Folsomia candida test in jar: placing 10–20 animals into the jar and counting the survivors.

The avoidance end point is more sensitive than lethality in acute toxicity tests. In avoidance tests, two containers have to be used with a path between the two, easily available for the little animals.

### 5.3.3   Plant tests

Plant tests were only used for testing waste extracts over a long period of time. ISO has standardized chronic plant tests with two species: a rapid-cycling variant of turnip rape (*Brassica rapa* CrGC syn. Rbr) and oat (*Avena sativa*). The standardized tests are recommended for both contaminated soil and pure chemicals in soil (ISO 22030, 2005).

White mustard, *Sinapis alba* also enjoys a widespread use for acute and chronic toxicity assessment in laboratory bioassays. Based on a Hungarian Standard (MSZ 21976-17, 1993) for waste leachate, germination and root and shoot growth test was developed by Gruiz *et al.* (2001) for soil toxicity. A two-fold dilution series was prepared by mixing the contaminated soil with OECD standard or healthy reference soil. 5 g wet mass of the soil mixtures is put into 10-cm diameter Petri dishes and brought to equal moisture content. 20 seeds of high viability (germination ability >90%) are arranged on the soil surface. The test dishes are kept in the dark at 20°C and plant growth is evaluated after 72 h (Figure 5.22). The number of germinated seeds and the length of the grown root and shoot of the seedlings are measured as it is shown in Figure 5.23.

*Sinapis alba* growth test is a useful screening and assessment test, but root growth may give false negative results due to the roots avoiding contaminated spots in heterogeneous soils. Therefore the inhibition of shoot length as an end point has higher relevance for toxicity. Another problematic point of this test is that artificial soil used for dilution may have a too low nutrient content, while the contaminated soil has a high content. The adverse effect (growth inhibition) of the mixture (of the contaminated and diluting artificial soils) may be lower due to high nutrient content in spite of the presence of toxicants, compared to the artificial reference soil.

*Sinapis alba*, similar to any other plants, can be used for chronic testing by evaluating the biomass produced in long-term experiments.

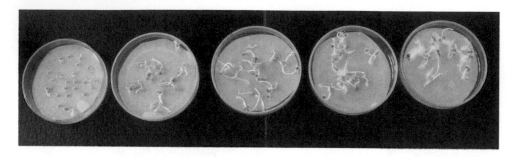

*Figure 5.22*  Plant growth test in Petri dishes: a dilution series from soil extracts.

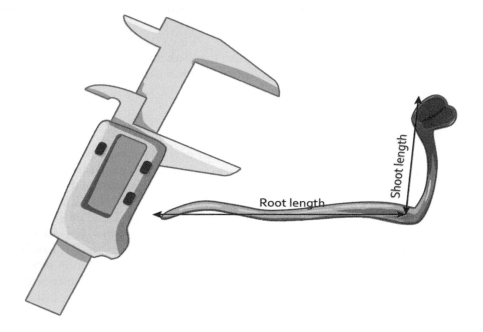

*Figure 5.23*  Root and shoot length of a seedling.

A third application of *Sinapis alba* is its utilization in ***rapid bioaccumulation*** tests. Feigl *et al*. (2009) developed a 5-day test for measuring the metal bioaccumulation in the seedlings grown in soils contaminated with metals. The rapid bioaccumulation test proved to be suitable for monitoring of remediation of metal-contaminated soil and waste rock by *in situ* soils remedied by chemical stabilization (Feigl *et al*., 2009).

### 5.3.4  *Soil as a test organism*

Soil itself can be used as 'a test organism' in ecotoxicity tests. In this concept healthy soil – with the active microbiota – is the 'test organism' and contaminated soil or pure chemicals are added to it in increasing concentrations. Any relevant soil characteristics, microbiological indicators or soil activities, including respiration, nitrification, cellulase enzyme activity, etc. can be measured as end points.

It should be noted that the chemical substance or contaminated soil can inhibit or stimulate the microbiota. Toxic chemicals will decrease soil activities such as respiration and enzyme activities while biodegradable contaminants may increase metabolic activities. On the other hand, toxic chemicals may inhibit respiration immediately after their addition to the soil, but may stimulate it later on, after a period of adaptation. Changes in diversity, e.g. in the community's species diversity or in the genome diversity of the metagenome may explain the changes in the activities. The background is that suppressed genes or minor species may come to the fore, or completely new genes can emerge.

In these types of toxicity tests, the reduced respiration of the test soil ($CO_2$ production or $O_2$ consumption) is the response to the contaminant's presence. The respiration rates of the concentration series prepared from the contaminant (in the healthy test-soil) is evaluated one by one and the contaminant-concentration–response (respiration) curve an $EC_{50}$ or a NOEC value is read. If a contaminated soil (with unidentified contaminant content) is tested, it should be added to the healthy test soil in increasing ratio and the amount of the contaminated soil which caused 50% (90, 20 or 10%) inhibition in the respiration of the test soil can be read from the soil dose–response curve. Another feasible test end point is the highest amount which has not affected the respiration (No Observed Effect Dose).

*Soil respiration* can be tested under aerobic or anaerobic conditions. Open and aerated test systems with three-phase soil create a realistic model of aerobic conditions for a soil community. Two-phase soil (solid and pore water) in a closed system can ensure anaerobic conditions. Closed-bottle soil tests are between the two conditions that is why the results are often difficult to interpret. It is similar to wastewater BOI determination: continuously decreasing atmospheric oxygen is available for the soil microbes. Soil microbes in the closed bottle have to switch from aerobic to anoxic or anaerobic metabolism after a certain decrease of available oxygen. These changes make the evaluation rather complicated, given that the pressure drop in the closed bottle is generally the end point measured, based on the assumption that pressure drop and $O_2$ consumption are in close relationship as $CO_2$ is absorbed by alkali. In addition to the appearance of alternative respiration forms, soil microbes can produce different volatile organic compounds and gases, for example, $H_2$ and $N_2$, which greatly influence pressure in the bottle.

*Soil nitrification* is also a suitable indicator for soil health and toxicity. Nitrification activity inhibition tests measure the potential ammonium oxidation activity of the soil. The method has been developed by Leitgib *et al.* (2007) based on the original method of Berg and Rosswall (1985).

An uncontaminated garden soil ensures a 100% nitrification activity and this high activity is then decreased by the added contaminant or contaminated soil. A minimum five-step dilution series is prepared from contaminated soil. 5 g of soil mixture is suspended in 0.1 mL $NaClO_3$ (1.5 M) and 20 mL $(NH_4)_2SO_4$ (1 mM) solutions. Test flasks are incubated on a rotary shaker (220 rpm) at 28°C for 5 hours. To determine the initial concentration of nitrite, the soil is suspended in distilled water – instead of $(NH_4)_2SO_4$ – at 20°C for 5 hours. At the end of the incubation period, the nitrification process is terminated by the addition of 5 mL of 2 M KCl. After centrifuging (4500 rpm, 2 s) samples are sieved on an N-free filter. The nitrite content is measured by photometry at 538 nm.

Similar to respiration and nitrification inhibition in healthy whole soil due to chemical substances or contaminated environmental samples, toxicity measuring methods based on other enzymes or enzyme systems can be created. The end point can be the residual substrate, the enzyme and the metabolites, all measuring substrate utilization. Depending on the test setup, it can be measured one by one or in a multiwell system such as in the BIOLOG system described earlier (Figure 5.19).

The application of these tests in the practice of contaminated soil characterization, risk assessment and risk reduction is described in Chapter 3 in Volume 3.

## 6  MULTISPECIES TERRESTRIAL TESTS

Multispecies tests are most widespread in soil testing due to the extremely high microbial activity and the complexity of the microbiotic system in soils, giving its intrinsic entity and making an artificial simulation impossible. Another argument for using multispecies systems in soil tests is the complexity of interactions between the soil's physical phases, strong matrix effects, multispecies microbiota, higher organisms and single or complex contaminants. The best solution in many cases is the application of real soil with a complex matrix and microbiota. Some multispecies test methods apply not only soil microorganisms, but also plants and soil-dwelling animals. Several authors published their results in the 1990s on successful applications of terrestrial microcosms for ecotoxicity testing, for example Mothes-Wagner *et al.* (1992), Morgan and Knacker (1994), Edwards *et al.* (1997), Sheppard (1997) and Olesen and Weeks (1998).

### 6.1  Classification of multispecies soil tests

Soil micro- or mesocosms are typical test setups for multispecies studies. Their design and structure are determined by the aim of the study, which can be the fate and behavior of the contaminants, their effects on the diversity of the soil and on the food web, or long-term multiple effects. In multispecies tests the community of soil-living organisms integrates all primary and secondary effects of a contaminant and can aggregate the adverse effects of several contaminants. The interaction between the members of the microbial or animal community can also be observed and the result utilized for risk assessment (Förster *et al.*, 1996). A few published test setups are listed below:

- Terrestrial microcosm system for measuring respiration in closed bottles;
- Terrestrial microcosm system for measuring respiration in flow-through system;
- Terrestrial microcosm for substrate-induced respiration technique (SIR);
- Terrestrial model ecosystem (TME) (Knacker 1998; Förster *et al.*, 1999);
- Soil litter bags;
- Cotton-strip method (Kratz, 1996);
- Bait-lamina test (ISO, 2014);
- Root microcosm system for the investigation of root growth, root exudates and the mycorrhizosphere;
- Soil core microcosms (EPA, 1987; ASTM, 1993);

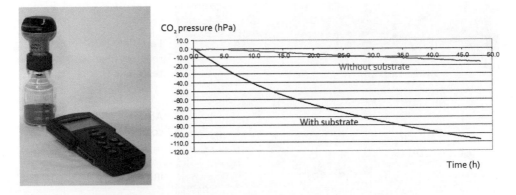

*Figure 5.24*  Closed-bottle system and respiration curve with and without substrate addition.

- Soil in jar;
- Terrestrial microcosm chamber (TMC): a model ecosystem with real or synthetic soil medium with invertebrates, rodents and agricultural crops. Sometimes also voles (e.g. gray-tailed *Microtus canicaudus*) are called into the testing;
- Versacore;
- Soil lysimeters.

Many of the published studies confirmed that microcosm tests have many advantages, are more sensitive and realistic than laboratory bioassays (Teuben & Verhoef, 1992; Vink & Van Straalen, 1999). Some of these microcosms are introduced in detail in the following paragraphs.

### 6.1.1  Terrestrial microcosm system for measuring respiration

Respiration is generally measured in soil-filled columns or other type of closed containers with or without controlled air flow. The pressure continuously decreases in the closed bottle without air flow when $CO_2$ is eliminated by an alkaline absorber. Palmborg and Nordgren published respiration test setups, results and evaluation for soil toxicity assessment in the book of Torstensson (1993).

Figure 5.24 shows the respiration curve measured in the closed-bottle system with and without substrate induction. When using a flow through a system, alkali absorbers treat the air before entering the soil system to eliminate $CO_2$ (Figure 5.25). The air leaving the column is generally analyzed for $CO_2$ or $O_2$. The produced $CO_2$ and the consumed $O_2$ is proportional to soil respiration. In such microcosms, we can measure the respiration rate of soil contaminated with a series of growing concentration of chemical substances or contaminated soil. The results can be used to draw the concentration-response curve and read NOEC, LOEC or $EC_{50}$ values.

### 6.1.2  Terrestrial microcosm for substrate-induced respiration technique (SIR)

The terrestrial microcosm introduced above can be supplemented with injection. After reaching steady state, the substrate is injected as a short impulse. The type and size of

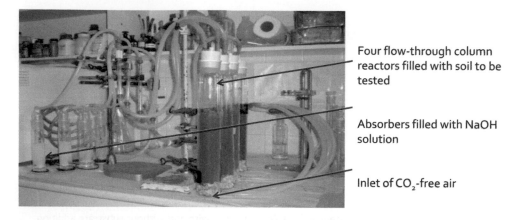

Four flow-through column reactors filled with soil to be tested

Absorbers filled with NaOH solution

Inlet of $CO_2$-free air

*Figure 5.25* Soil respiration measurement in flow-through columns.

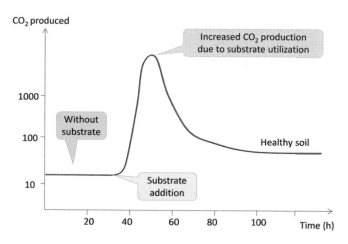

*Figure 5.26* Soil respiration curve with substrate addition.

the response on the suddenly introduced energy source may be characteristic for the soil, depending on its activity and adaptability. Figure 5.26 shows the respiration curve of a high-quality forest soil: $CO_2$ production increases suddenly after the injection of the energy substrate. On the other hand, having a high-quality, active and adaptable soil enables the effect of chemical substances on soil respiration to be measured. Its time-dependence graph can be drawn based on the changes in the rate of respiration, and time, intensity and size of the response can be read from this curve.

### 6.1.3 Terrestrial model ecosystems (TME)

TMEs are semi-field tests for studying effects and the fate of chemicals in soil. There are both indoor and outdoor protocols for these tests. TME uses the litter layer of the soil for testing soil function. The litter layer is the uppermost soil layer consisting of plant residues in a relatively non-decomposed form. This layer is the source of energy for mineralizing soil biota and also for humus formation from the residue of mineralized organic matter.

TME is also recommended to be used for testing pesticides in soil because specific exposure to plant protection products takes place in the soil litter layer, including direct exposure, uptake via food and food web transfer (biomagnification). It is an important energy resource in soil and thus of vital importance for maintaining organism communities and their great biodiversity. Not protecting the natural processes and organisms in litter will fail the EU's objectives regarding biodiversity, soil erosion, organic matter decline, and the integrity of soil and soil biota. Therefore, the litter layer should be taken into consideration in the environmental risk assessment of plant protection products. It was the summary of the European Food Safety Authority (EFSA, 2014) in an overview on the composition of litter in an agricultural context (EFSA Opinion, 2010), the underlying processes which play a role in litter decomposition and an outline of how to consider the litter layer in the environmental risk assessment of plant protection products (Shäffer *et al.*, 2011).

### 6.1.4  *The cotton strip assay*

The cotton strip assay uses the loss of tensile strength in a standardized cotton material as a parameter. It needs an equipment to measure tensile strength. A major shortcoming of this method is that the degradation of pure cotton compared to natural litter is an unacceptable simplification (Kratz, 1996). This is the main reason why this test has been substituted by the soil litter bag test.

### 6.1.5  *Soil litter bag*

Toxicant effects on carbon mineralization can be quantified in several ways. One of the simplest techniques encloses pre-weighed plant litter in a mesh bag, bury it, and collect and weigh the bag's contents after a period of time, and compare the mass loss relative to similarly bagged litter in reference soil (Figure 5.27).

OECD has issued the Guidance Document (OECD No 56, 2006) to identify possible and suitable approaches to the evaluation of chemical impacts, and in particular the impact of plant protection products on soil organic matter breakdown. Methods identified in this document, and specifically the litter bag test method, add important tools to a battery of existing standardized protocols for assessing chemical impacts on the soil biota communities. Procedures outlined in this Guidance Document are primarily intended for the evaluation of agricultural chemicals but they can also be applied to ecological risk assessment at contaminated sites as well as to laboratory chemical toxicity testing (Kula & Römbke, 1998; Römbke *et al.*, 2003; Schäffer *et al.*, 2011).

Heath *et al.* (1964) already described in 1964 that the litter-bag method offers a simple and efficient tool for measuring decomposition in terrestrial systems. This method is of high ecological relevance because the actual rates of decomposition with (presumably) native material can be measured *in situ*, providing real-time data for toxicant impacts on decomposition at a study site. In addition, various functional and taxonomic groups can be chemically restricted (e.g., with fungicides, bactericides, and pesticides) or physically restricted (e.g. the mesh size of the bag can exclude certain

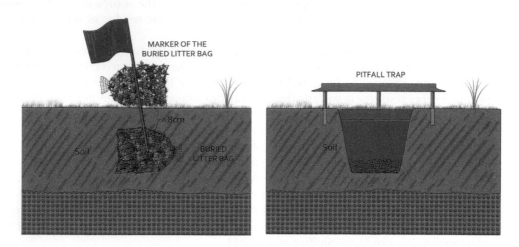

*Figure 5.27* Soil litter bag (left) and pitfall trap (right).

size groups of microarthropods) in order to evaluate their contribution to the process. An aspect of the litter-bag method that detracts from its ecological relevance is simply that the litter is not in direct contact with contaminated soil, and thus it presents a microenvironment different from buried litter in native surroundings (De Jong, 1998).

### 6.1.6  Pitfall traps

The pitfall trap – shown in Figure 5.27 – is a relatively simple field assessment tool for collecting arthropods, mainly insects such as ground beetles, crickets, etc. from contaminated land. Pitfall traps are placed at the contaminated site or the landfill and next to a reference site. Abundance, diversity and bioaccumulation of the animals fallen into the pit can be determined in comparison with the reference.

### 6.1.7  Bait lamina

Bait lamina is an alternative to the litter-bag method, also concentrating on the litter layer of the soil and its biological activity. Small bait portions (standard mixture of cellulose, bran flakes and active coal, in a ratio of 70:27:3) are fixed in holes pierced in PVC strips (Figure 5.28) that are then exposed to the soil's biological decomposition activity. Soil invertebrates and soil microorganisms progressively degrade the bait placed in the soil substrate. It is assumed that the disappearance of the bait material is directly associated with the feeding activity of soil invertebrates, even if microbial processes and microbiogenic metabolism may play a minor role (von Törne, 1990; Kratz, 1998).

The general requirement for the bait material is that a mixture of natural (powdered) materials are used that can be eaten by soil-dwelling animals and have enough consistency, elasticity and stability to be placed easily in moist, fine and coarse soil without any damage. The bait material mixture, proposed by the *terra protecta* GmbH company, consists of cellulose, bran flakes and active coal. Materials from field sites

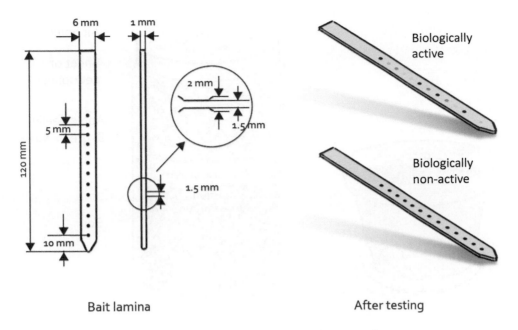

Bait lamina

After testing

*Figure 5.28*  Bait sticks or lamina™ are suitable for laboratory and field testing of metabolic activity in soil.

such as powdered leaves of birch (*Betula pendula* L.) or different grass species (e.g. *Calamagrostig epigaeios* L.), have already been used.

The bait-lamina strips are left in the soil/substrate until about 10–40% of the baits are perforated. Since the necessary exposure time depends on the site and on the moisture content of the soil, feeding activity assessment can require an exposure between 7 (in soils with good moisture conditions) and 20 days (dryer soils). The most favorable conditions are easily obtained by a short pre-test.

The exposed baits are evaluated after washing the strips carefully under flowing tap water and examining the strips on a lighted bench. Differentiation is made only between 'bait eaten' ("1", light falls through the bait) and 'bait not eaten' ("0", light does not fall through the bait). Utilized bait-lamina strips can be reused if refilled after soaking and cleaning in water: the company selling it offers a cleaning and refilling service of used strips (terra-protecta, 2013).

### 6.1.8  Soil in jar

Soil in jar is a simple test method for soil monitoring under controlled conditions in open flowerpots or locker jars with a volume of a few mL to 10–50 liters. Both 3-phase or 2-phase soils can be modelled in jars; layers of different texture and drainage can be established on the bottom. Soil or contaminated environmental samples of soil, sediment or water can be treated or mixed with chemical substances. In addition to soil microbes, soil invertebrates and plants grown on the soil surface can be added.

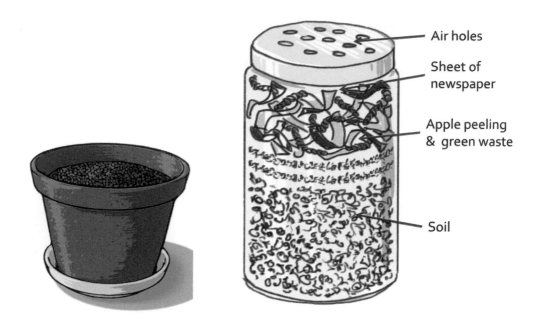

*Figure 5.29* Soil in pot (left) and soil in jar (right).

The measured end point can be abundance, diversity, or any physico-chemical change in the soil. Soil can be analyzed after careful sampling.

Soil-in-jar type microcosms (Figures 5.29) were applied to model soil contamination by flood that was transporting mine waste as sediment and mixing sediment to soil or placing a sediment layer on the top of the soil. Flood microcosms clearly showed an enhanced weathering process that occurs when putting sediment on soil surface and interacting with soil (Gruiz, 2005).

Soil-in-jar type microcosms were efficiently applied to mobilization and immobilization laboratory tests dealing with toxic metals. The tests included different chemical treatments and incubation (followed by physico-chemical analyses), analysis of the emerged toxic effects on soil microbiota, plants and animals (ecotoxicological tests), and the plant uptake of toxic metals (bioaccumulation test) (Feigl *et al.*, 2008).

### 6.1.9 Soil lysimeters

Lysimeters can model the infiltration of precipitation into the soil and the results of this complex transport process over the short and long term. Lysimeter sizes range from laboratory microlysimeters of 50 mL through 200, 500, and 1000 mL to field lysimeters of a few cubic meters.

Regardless of the size, the concept shown in Figures 5.30 and 5.31 is always the same: the infiltrating precipitation flows through the system and dissolves, extracts, leaches substances from the soil and washes them into deeper layers or removes them from the soil-filled column (Vaszita *et al.*, 2009). *In situ* placed lysimeters need a sampling installation at the outflow level of the experimental soil column.

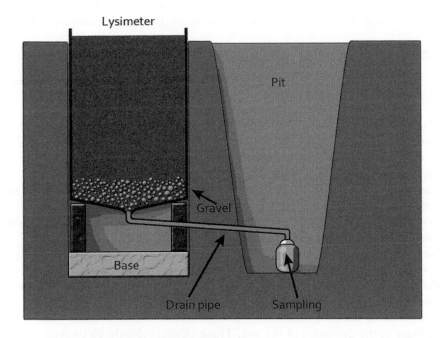

*Figure 5.30*  Field lysimeter filled with soil: buried below the surface.

*Figure 5.31*  Lysimeters filled with soil above the surface with built-in sensors.

Figure 5.30 shows a simple underground lysimeter and an adjacent pit for sampling the outflow. Figure 5.31 illustrates a more sophisticated setup: open-air lysimeters installed on the ground. They can operate as either open or closed columns. Observing and sampling the outflow is easier than with the underground lysimeters. The pictured large-size lysimeters are equipped with special sensors integrating temperature, pH and conductivity measurement devices to monitor the transport of ions in soils which have been deteriorated or endangered by acidification, sodification or contamination.

## 6.2   Characteristics of multispecies toxicity tests

The most important characteristics of multispecies toxicity tests are summarized in the next paragraph.

- Micro- and mesocosms are multispecies test methods: depending on the aim of the testing, 'only' soil microbiota or also invertebrates or vertebrates are involved.
- Size varies from 0.1 liter to a few cubic meters.
- They are 'historical': like the ecosystem itself they are irreversible in time.
- They have a trophic structure, with trophic levels, sometimes very simple, sometimes close to the real environment.
- Evolutionary events occur in the micro- and mesocosms: strains, able to use contaminants as energy sources or resistant to xenobiotics, arise.
- New metabolic pathways for biodegradation (of pesticides, xenobiotics) can develop spontaneously or under an external effect.
- Reduced complexity compared to the field: the number of species is generally smaller than in natural systems.
- Dynamics of the ecosystem: the enforced isolation into a small scale results in changes in the dynamics of the soil. These changes should be distinguished from the toxicant's effect.
- Heterogeneity: in natural ecosystems spatial and temporal heterogeneity is the key to species richness and adaptability. Artificial test systems are less heterogeneous, less unique; this ensures their statistical power as opposed to field cases.
- Multispecies tests are complex systems, with dynamics and history, so they are not repeatable unlike simple species tests or biochemical assays. The past is conserved in population dynamics down to the DNA sequence.

All this information should be considered in designing and using micro- and mesocosms for environmental toxicity testing. These specific characteristics make the evaluation rather complex; therefore the usual statistical methods do not work for micro- and mesocosms.

## 6.3   Evaluation and monitoring of microcosms

In order to monitor micro- or mesocosms, these separate ecosystems, biotic and abiotic components need to be observed.

- Biotic factors refer to all living organisms present in an ecosystem such as microbes, plants and animals, etc. They can be monitored by biological methods and tools.
- Abiotic factors refer to all non-living or physical factors present in an ecosystem: soil texture, geochemistry, oxygen, elements and salts, temperature, moisture, water-forms, pH, redox potential, etc. These parameters can be monitored using physico-chemical analysis methods and tools.
- Micro- and mesocosms are (should be) stable and self-sustaining systems, meaning inside cycling of energy, water and elements as well as their balance. The most

important cycles in these individual systems are the water cycle, energy cycle, organic matter mineralization and humus formation, carbon and nitrogen cycles, sulfur, phosphor, iron and microelement cycles.

– Limiting factors may change the survival, metabolism and adaptation of the micro-biota in micro- and mesocosms. Such limiting factors are oxygen or alternative electron donors, nutrient source, and water.

– Any component of these self-sustaining mini-ecosystems can be monitored: biotic and abiotic components and their interactions and relationships.

Micro- and mesocosms can be monitored by both built-in measurement systems (electrodes to continuously measure pH, redox potential, conductivity, $CO_2$, etc.) and by sampling the soil to analyze its biotic and abiotic components in the laboratory. The time series of the results will provide information about the behavior of the micro- or mesocosm, the changes in the qualitative and quantitative characteristics and the physico-chemical properties of the soil.

Similar to the real environment, an integrated methodology comprising the combination of physico-chemical analyses, biological-ecological assessment and tox-icological testing is the most appropriate tool to monitor micro- and mesocosms. Simultaneously acquired physico-chemical and biological-ecotoxicological informa-tion should be evaluated and compared to obtain a true picture of the complex environment and a reliable prognosis on its characteristics and responses.

Statistical evaluation of micro- and mesocosms should be adjusted to the acquired data. The evaluation of these data requires multivariate techniques, see Chapter 8.

# 7 MICROCALORIMETRY – A SENSITIVE METHOD FOR SOIL TOXICITY TESTING

## K. Gruiz, V. Feigl, Cs. Hajdu & M. Tolner

Microcalorimetry is not a typical method for measuring ecotoxicity. It is chiefly used to characterize physico-chemical reactions and for biological studies in pharmacology. In this chapter the heat production by soil living microorganisms, plants and animals is used as end point for soil toxicity characterization.

## 7.1 Background of microcalorimetric heat production by living organisms

All chemical, physical and biological processes trigger a net flow of heat. The response of an organism to adverse effects is accompanied by increased or decreased heat pro-duction. Microcalorimeters can detect very small heat flows, for example $\pm 50\,\mathrm{nW}$ ($0.5 \times 10^{-6\circ}\mathrm{C}$) with the help of TAM (Thermo Activity Monitor) (Figure 5.32). This means that heat production and its change can be a sensitive end point of a bioassay using a microcalorimeter. The selective response to a chemical substance or to a con-taminated environmental sample can be measured and compared to the response of an untreated control soil.

The reaction of bacteria, fungi, whole microbial communities, small animals and plants to certain environmental parameters and conditions such as temperature (Querioz *et al.*, 2000; Meissner & Schaarschmidt, 2000; Burger & Qian, 2008) or $O_2$ and $CO_2$ concentration in air (Zhou *et al.*, 2000) has already been investigated using TAM.

Sparling (1983) and Critter *et al.* (2002a,b, 2004a,b) compared respirometry, microorganism counting and calorimetric heat production of microbial biomass and found significant correlations. Barros *et al.* (1995, 1997) found that the total heat production of a soil microbial community significantly correlated with soil moisture content and soil organic matter content.

An overview on calorimetry of soil (Rong *et al.*, 2007) summarizes the applications on soil, mainly measuring microbial activity (Barros, 2007; Wadsö, 2009). However, ecotoxicological applications similar to those we introduce here are not yet available.

The first application of microcalorimetry in soil toxicity testing published by Gruiz *et al.* in 2010 drew the attention to a possible new tool to increase sensitivity and selectivity in toxicity testing as well as to study the effect mechanism. The concept and test design of toxicity measurement by TAM was similar to other toxicity tests based on measuring the relation between concentration/dose and response of soil-living organisms or their community.

A TAM was used to measure the heat production of the whole soil (including its microbiota) or that of the applied test organism in a direct contact scenario. This means that the organisms are placed into the soil thus ensuring a close interaction with soil through respiration of soil air, ingestion of dissolved and solid-bond matter as well as having dermal contact with all three soil phases and the contaminants. Well-known bacterial (*Azomonas agilis*), plant (*Sinapis alba*) and animal (*Folsomia candida*) organisms were tested for heat production and compared with traditional ecotoxicological end points. The heat response to toxic metals (Cu and Zn) and to organic pollutant exposure in contaminated soils was measured. PCP (pentachlorophenol), a pesticide and disinfectant (its sodium-salt dissolves easily in water); DBNPA (2,2-dibromo-3-nitril-propionamide) a quick-kill biocide (easily hydrolyzes under both acidic and alkaline conditions) and diesel oil (a mixture of hydrocarbons) were used as model organic contaminants (Gruiz *et al.*, 2010). Some results are introduced here.

## 7.2  Experimental setup

Experiments in the microcalorimeter are carried out in a perfectly closed sterile glass ampoule of 5 mL or 20 mL volume.

0.5–2.5 g air-dried, grained, sieved (2 mm sieve), sterilized soil is placed into the ampoule and is wetted with distilled water or growth media one day before the measurement and incubated at 25°C. After the incubation period the test organisms are placed into the ampoules:

– *Azomonas agilis* (soil bacterium) 48 h culture, $2 \times 10^8$ cells/200 μL);
– *Sinapis alba* (white mustard) 10 seeds from a controlled order (10 seeds);
– *Panagrellus redivivus* (nematode) 500 μL from the 5-fold dilution of a standardized culture;
– *Folsomia candida* (collembolan) 50 insects from a 14-day-old synchronized culture.

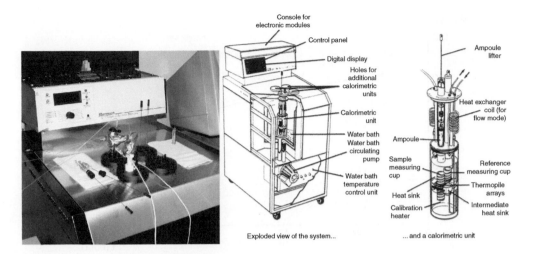

*Figure 5.32* TAM Microcalorimeter with four test vessels (red caps), its inner design and the calorimetric unit (LKB Bromma, 2013).

Samples from contaminated soil or artificially contaminated solid samples can be tested. As an alternative no test organism is added to the soil, but the heat production of the soil's own microbiota is detected. In this case the heat production of the soil's own microbiota is measured.

Heat production is measured by a TAM 2277 (Thermal Activity Monitor by LKB Bromma) device (Figure 5.32). In the TAM microcalorimeter the thermal energy difference between the reference and the test-ampoule is converted into electric energy and is measured in $\mu W$ units by a pair of Peltier elements. Recent/new TAM models are able to measure simultaneously 24 or 48 samples. The equipment shown in Figure 5.32 contains only 4 microcalorimetric test vessels.

Microcalorimetric results are evaluated using specific software, Digitam for Windows v. 4.0. The power-time curves obtained during the real-time measurements were used for reading the characteristic parameters of the curve: the initial slope ($m$) of the curve (after the equilibration between the ampoule and the thermal bath of the microcalorimeter), maximum power ($P_{max}$) and the time of the appearance of the maximum ($t_{max}$). The authors created an $S \times P_{max}/t_{max}$ factor to magnify the trends, as $S$ and $P_{max}$ usually decrease and $t_{max}$ increases as functions of contaminant concentration. $EC_{20}$ and $EC_{50}$ values were calculated from the data obtained from dilution series of the contaminants in soil. The results were compared to the results of traditional toxicity measuring methods, evaluated in the conventional way (see also Chapter 9).

## 7.3  Heat response of *Folsomia candida* to the effect of diesel oil

*Folsomia candida* is a soil-dwelling springtail, extremely sensitive to volatile soil contaminants because its main exposure route is the cutaneous respiration, thus it is highly exposed to soil air and solid sorbed volatile contaminants. Collembola placed onto the

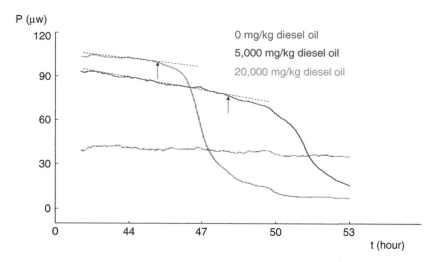

P (µw)

0 mg/kg diesel oil
5,000 mg/kg diesel oil
20,000 mg/kg diesel oil

*Figure 5.33* The effect of diesel oil on heat production by *Folsomia candida*.

contaminated soil in the calorimetric ampoule increased its heat production proportionally to the diesel oil concentration. Maximum heat production was observed at the beginning – $108\,\mu W$ for the 20,000 mg/kg and $95\,\mu W$ for the 5000 mg/kg diesel contaminated soil. This enhanced heat production gradually decreased during the test. Figure 5.33 shows the decreasing heat production as a function of time due to the effect of the diesel oil in contaminated and untreated control soil. At a certain time point the decrease of heat production declines steeply (shown by the arrows) until it reached zero. This moment was: 45.2 h and $102\,\mu W$ for 20,000 mg/kg and 48.2 h and $75\,\mu W$ for 5000 mg/kg diesel oil concentration. The uncontaminated control showed no change in heat production over 53 hours.

The number of animals survived in the closed ampoule was counted at the end of the measurement. The results were: 0 animals in soil contaminated with 20,000 and 5000 mg/kg diesel oil and 48 (from 50) animals in the unpolluted control soil. The traditional test method in an open jar provided similar results: 0 animals alive in soil contaminated with 20,000 and 5000 mg/kg diesel oil and 9 (from 10) animals alive (after 1 week) in the non-polluted control soil.

Heat production used as an end point can help differentiate between 5,000 and 20,000 mg/kg diesel oil contamination, since the time of decline in heat production (probable time of death) is different. Heat production instead of lethality provides the benefit of a much shorter time required: results are available in 2–3 days instead of the one week in the lethality test. Measuring several concentrations between 0 and 5000 mg/kg, the highest no-effect concentration can be precisely identified.

## 7.4   Heat response of *Panagrellus redivivus* on contaminated soil

The traditional test on the nematode *Panagrellus redivivus* is a 2-week-long reproduction test. It is moderately sensitive to most of the organic soil contaminants,

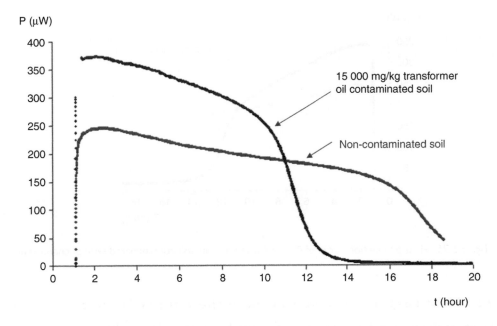

P (µW)

15 000 mg/kg transformer oil contaminated soil

Non-contaminated soil

t (hour)

*Figure 5.34* Heat production curve of *Panagrellus redivivus* in soil contaminated with transformer oil.

15,000 mg/kg transformer oil (not containing polychlorinated biphenyls) or 250 mg/kg phenanthrene (a polycyclic aromatic hydrocarbon) do not show any inhibition in the reproduction of this little animal.

The heat production by the nematode in a soil contaminated with 15,000 mg/kg transformer oil (Figure 5.34) increased by 50% compared to the non-contaminated soil indicating an increased metabolic activity. As a consequence, increased energy production (increased respiration rate), i.e. the decline in the curve occurs 50% earlier than in the non-contaminated soil. Decline of heat production is caused by the death of the animals, due to poisoning and/or oxygen depletion in the closed vessel. Similar trend has been measured by the nematode in soil contaminated with 250 mg/kg phenanthrene (Figure 5.35).

The response of elevated heat production can be measured within three hours, which is a very quick response compared to the two-week time requirement of the reproduction test. In the microcalorimeter, *P. redivivus* does not propagate, just lives, feeds, and tries to survive during the 16–18 hours of the test. Increased heat production is characteristic of those concentrations that are not strongly inhibitory or lethal, against which an increased metabolic activity may act successfully for some time. At higher concentrations where increased activity does not help survival, heat production decreases, similarly to conventional end points. The produced heat is less in 500 mg/kg phenanthrene-contaminated soil than in the non-contaminated control soil.

Sensitivity of the heat-response-based measurement is much greater than the sensitivity of lethality or the reproduction end points, meaning that much smaller contaminant concentration is sufficient for the heat response.

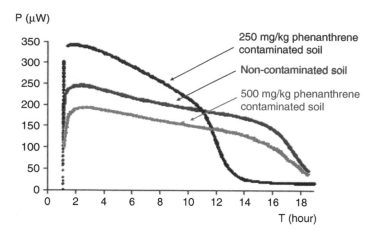

*Figure 5.35* Heat production curve of *Panagrellus redivivus* in soil contaminated with phenanthrene.

## 7.5 Heat response of *Sinapis alba* to the effect of toxicants in soil

Plants in the microcalorimeter are continuously growing during the test. Germinated seeds are placed into the relatively large test vessel where they grow and metabolize in the closed system.

The effect of Cu- and Zn-contaminated soils on *Sinapis alba* (white mustard) is shown in Figure 5.36.

The higher the Cu and Zn concentration, the smaller the maximum of the power–time curve ($P_{max}$) and the greater its delay in time ($t_{max}$). The slope of the curve ($\mu W/h$) is inversely proportional to the concentration.

The lowest concentrations in the experiments were 40 mg/kg Cu and 50 mg/kg Zn, which are much less than the soil screening values ($Cu_{EUcountries}$: 100–500 mg/kg, $Zn_{EUcountries}$: 200–1000 mg/kg). 40 mg/kg Cu concentration caused 26% inhibition, 50 mg/kg Zn caused 55% inhibition compared to non-contaminated soil. $EC_{50}$ of 156 mg/kg Cu soil was calculated for copper based on heat production, whose order of magnitude is similar to $EC_{50}$ based on shoot growth. $EC_{50}$ of Zn based on heat production is about 46 mg/kg Zn soil. The same Zn concentration did not cause any inhibition in traditional growth-based tests and is 4–20 times less than the environmental threshold in European countries.

## 7.6 Heat production response of *Azomonas agilis* to toxicants

*Azomonas agilis* is a typical soil-dwelling bacterium, a priority test organism used for soil toxicity tests. The bacterium placed in the microcalorimeter ampoule shows metabolism and growth (propagation) during the 24-hour test. The result of these metabolic activities is heat production.

Figure 5.37 shows some of the power-time curves characterizing heat production of the bacterium in soil contaminated with DBNPA (2,2-dibromo-3-nitrilopropionamide,

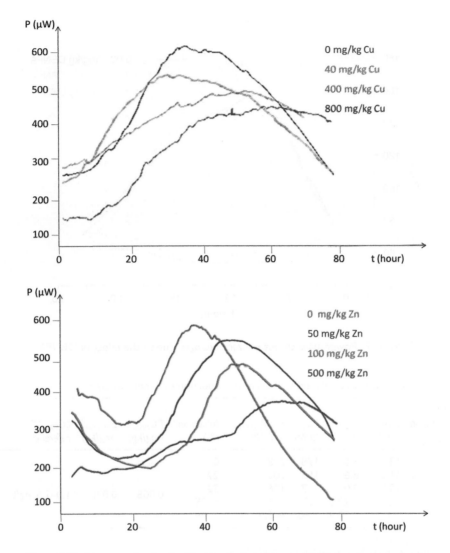

*Figure 5.36*  Heat production by *Sinapis alba* in copper- and zinc-contaminated soils.

a biocide). The concentration series cover only 0–2 mg/kg, however a very large drop can be seen between 0.02 and 0.2 mg/kg where 22% inhibition (relative to non-contaminated soil) drops to 98% (see also Table 5.5).

Table 5.6 contains the characteristic parameters of the power–time curves of *Azomonas agilis* affected by DBNPA contaminated soil. Inhibition in %, $EC_{20}$ and $EC_{50}$ values can be read from the concentration–response curve. The higher the concentration of DBNPA in the soil, the smaller the slope (S = slope in $\mu W/h$) and the maximum ($P_{max}$) of the curve, indicating reduced heat production. There is a shift in the power maximum in time ($t_{max}$) too. The values of $EC_{20} = 0.008$ and

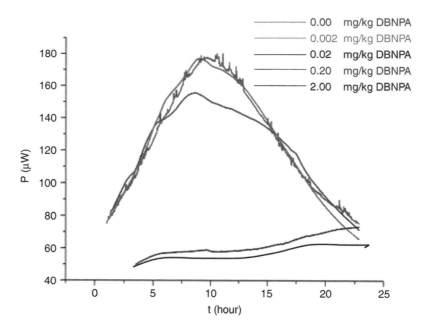

*Figure 5.37* Power–time curves of *Azomonas agilis* due to the effect of DBNPA.

*Table 5.6* Effect of DBNPA-polluted soils on heat production by *Azomonas agilis*.

| DBNPA concentration (mg/kg) | Slope ($\mu W/h$) | $t_{max}$ (h) | $P_{max}$ ($\mu W$) | $S \times P_{max}/t_{max}$ ($\mu W^2/h^2$) | Inhibition* (%) | $EC_{20soil}$* (mg/kg) | $EC_{50soil}$* (mg/kg) | $LC_{50aq}$** reference |
|---|---|---|---|---|---|---|---|---|
| 0.000 | 13.9 | 9.6 | 178 | 258 | 0 | | | |
| 0.002 | 11.2 | 8.6 | 155 | 203 | 22 | | | |
| 0.02 | 13.5 | 9.0 | 117 | 176 | 32 | 0.008 | 0.038 | 1.0–1.5 mg/L |
| 0.2 | 1.3 | >22.8 | >73 | ~4.1 | ~98 | | | |
| 2.0 | 1.1 | >22.8 | >60 | ~2.9 | ~99 | | | |

*Calculated from $S \times P_{max}/t_{max}$.
**$LC_{50}$ aquatic (fish, Daphnia) from MSDS DBNPA, 2013.

$EC_{50} = 0.038$ mg DBNPA/kg soil have been determined from $S \times P_{max}/t_{max}$, calculated from the slope, while $t_{max}$ and $P_{max}$ have been read from the curve.

Traditional *Azomonas agilis* growth inhibition parallel tests on the same soil and at the same DBNPA concentrations did not show any toxicity.

No soil screening value exists for DBNPA, only aquatic $LC_{50}$ values are available (MSDS, 2013), which are 1.5 orders of magnitude higher than the $EC_{50}$ measured in soil by microcalorimetry (generally, toxicity in soil is 2–4 orders of magnitude smaller than in water). This means that the sensitivity of the microcalorimetric test using *Azotobacter agilis* to DBNPA is much greater than that of the aquatic test organisms of *Daphnia* and fish.

## 7.7 Evaluation and interpretation of the microcalorimetric heat production results

The first exploratory results shown by the graphs demonstrate that the heat response of a test organism to contaminated soil can be a new and sensitive end point in environmental toxicology and risk assessment of chemicals and contaminated environmental samples.

The response to low concentrations of contaminants in soil is due to the increased activity of the test organisms. Higher concentrations of the contaminants cause inhibition in heat production. A concentration/dilution series has to be tested to be able to distinguish between the responses to low and high concentrations of the toxicant and to plot the concentration–response curve. Microcalorimetric testing of serial concentrations (dilutions) from chemical substances or contaminated environmental samples will identify the borderline between the initial activation and the following inhibition.

At low concentrations of Zn (20 and 200 mg/kg) and Cu (4 and 40 mg/kg), increased heat production was observed both in animal (nematode and collembolan) and bacterial (*Azomonas*) test organisms. This may have been caused by real stimulation or fight for survival and switching on the test organism's defense mechanism. The initial period of increased heat production may yield important information for "ome" research (genome, proteome or metabolome) and the development of new techniques in the field of genomics, proteomics, metabolomics, since heat production is part of the complex transcriptional and translational processes associated with the response to toxicants or other biologically active substances.

Later in time, heat production ($P$) becomes inversely proportional to the contaminant concentration even in those cases when an increase occurred at the beginning. The time ($t_{max}$) and the maximal intensity ($P_{max}$) together with the slope of the curve ($S$) can be a good and easy-to-calculate inhibition indicator. The best indicator would be the integrated amount of the produced heat (the surface area under the power–time curve), but this has not been solved yet. The usual test end points were applied: inhibition in percentage ($I\%$) was calculated compared to a control and $EC_x$ was read from the concentration–response curve.

A comparison of traditional and microcalorimetric test methods shows that microcalorimetry is more sensitive and much faster. With the traditional test methods no change has been found in some cases, however microcalorimetry enabled a response to be measured. For example based on the traditional method one would conclude that 500 mg/kg Zn has no adverse effect on *Sinapis alba*, but the microcalorimetry indicated 92% inhibition in heat production at this concentration. Also the $EC_{20}$ and $EC_{50}$ and mainly the NOEC (No Observed Effect Concentration) value is much lower in the case of the microcalorimetric test compared to the traditional growth inhibition test. This means that the heat response can be utilized as an early indicator of adverse effects.

## 7.8 Summary of microcalorimetric toxicity testing: experiences and outlook

Microcalorimetry in soil ecotoxicity is a potential useful new tool both for quantifying the hazard of chemicals and direct toxicity testing of environmental samples. It was concluded that heat production of the test organisms is inversely proportional to the concentration of toxicants in soil. Compared to traditional test methods (growth and lethality), microcalorimetrical detection of heat production provides an extremely sensitive and selective end point in soil ecotoxicity testing, which can indicate minor changes in the test organisms' energy production and general metabolism. Heat production can be selectively detected and it does not encounter the difficulties of direct contact tests where microbes or little animals are counted in the soil or the length of roots grown in soil is measured.

Metabolic activities that follow the exposure of an organism to toxic agents are always accompanied by heat flow, usually by additional heat production at the beginning. Heat production decreases later on simultaneously with the upcoming traditional responses. This means that by measuring heat production a primary metabolic signal is used as a toxicity test end point, which can be detected in the early phase of the long metabolic pathway leading to the traditionally measured dramatic end points (growth inhibition, death rate or reprotoxic effect). Innovative toxicity end points such as genome, proteome or metabolome may also act as early indicators, but the metabolic pathways and mechanisms should be clarified and the key point identified. Heat production studies can be of great help in such developments.

Heat production is an integrative signal, it includes all metabolic changes accompanied by energy production. This may be an advantage compared to single receptors, which cannot represent the organism or the culture as a whole.

Three major areas of application can be envisaged for microcalorimetry:

–   Research into the mechanism of the organisms' response to toxic agents and supplying information for other test developments, for example, testing genome and other "omes";
–   Identifying the real NOEC (No Observable Adverse Effect Concentration) and LOEC (Lowest Observable Adverse Effect Concentration) values;
–   Spreading direct contact soil and sediment tests by creating an end point which is easy to measure and is exempt from matrix effects and other disturbing influences.

The main advantages are:

–   Broad-spectral, generic;
–   Highly sensitive;
–   Rapid;
–   Dynamic.

Applicability:

–   For testing chemical substances in water or in soil;
–   For testing environmental samples;
–   The response of soils' own microbiota can also be tested;
–   Any kind of small organism or mixture of organisms can be used in the test;
–   Static or growing organisms or cultures can be tested.

Disadvantages:

–   No widespread or routine methodology;
–   Expensive equipment.

Proposed developments:

- Use of multichannel equipment;
- Evaluation by integration of the produced heat;
- Identifying metabolites parallel to increased heat production;
- Preparation of a standardized protocol.

It can be concluded from the results that heat production is an early indicator showing what cannot be measured and seen using conventional methods, in other words, it shows the 'no effect'. That is to say it measures dynamic changes within the concentration range of contaminants considered non-harmful. It may be true that the risk at these concentrations is low enough to be accepted, but it must be borne in mind that the calm on the surface conceals sensitive responses against contaminants and increased energy demand to implement early responses.

## 7.9 Acknowledgement to microcalorimetry research

The authors thank Prof. Wolfgang Sand and Dr. Thore Rohwerder for the opportunity to carry out the measurements at the Biofilm Center, University of Duisburg-Essen. The measurements were financed by the Hungarian-German TéT project (Hungarian National Office for Research and Technology), the LOKKOCK project (GVOP-3.0-0257-04, National Competitiveness Program of Hungary) and the MOKKA project (NKFP-3-020/05, Hungarian National R&D Fund).

## REFERENCES

Amorim, M.J.B., Novais, S., Römbke, J. & Soares, A.M.V.M. (2008) Avoidance test with *Enchytraeus albidus* (Enchytraeidae): Effects of different exposure time and soil properties. *Environmental Pollution*, 155(1), 112–116.

Anthony, J.C. (2001) The promise of psychiatric enviromics. *The British Journal of Psychiatry.* Supplement 40, 8–11. [Online] Available from: DOI:10.1192/bjp.178.40.s8. [Accessed 10th December 2013].

Armengaud, J., Hartmann, E.M. & Bland, C. (2013) Proteogenomics for environmental microbiology. *Proteomics*, 13, 2731–2742. [Online] Available from: DOI:10.1002/pmic.201200576. [Accessed 10th December 2013].

ASTM (1993) Standard guide for conducting a terrestrial soil-core microcosm test. American Society for Testing and Materials. *Annual Book of Standards*, 1197, 546–557.

ASTM (1998) *Standard guide for conducting terrestrial plant toxicity tests.* E-1963-98. Philadelphia, PA, American Society for Testing and Materials.

Bardgett, R.D. & Chan, K.F. (1999) Experimental evidence that soil fauna enhance nutrient mineralization and plant nutrient uptake in montane grassland ecosystems. *Soil Biology & Biochemistry*, 31(7), 1007–1014.

Barros, N., Fetjoo, S. & Balsa, R. (1997) Comparative study of the microbial activity in different soils by the microcalorimetric method. *Thermochimica Acta*, 296, 53–58.

Barros, N., Gomez-Orellana, I., Feijóo, S. & Balsa, R. (1995) The effect of soil moisture on soil microbial activity studied by microcalorimetry. *Thermochimica Acta*, 249, 161–168.

Barros, N., Salgado, J. & Feijoó, S. (2007) Calorimetry and soil. *Thermochimica Acta*, 458, 11–17.

Bartz, R. (2009) Mite *Ixodus ricinus*. Photo. [Online] Available from: http://en.wikipedia.org/wiki/File:Ixodus_ricinus_5x.jpg. [Accessed 10th January 2014].

Berg, P. & Rosswall, T. (1985) Assay of nitrification (short-term estimations). In: Alef, K. & Nannipieri, P. (eds.) *Methods in Applied Soil Microbiology and Biochemistry.* London, Great Britain, Academic Press. pp. 241–242.

BIOLOG (2013) *Carbon sources in EcoPlate, Microbial community analysis.* [Online] Available from: http://www.biolog.com/pdf/eco_microplate_sell_sheet.pdf. [Accessed 10th December 2013].

Bongers, T. & Bongers, M. (1998) Functional diversity of nematodes. *Applied Soil Ecology*, 10, 239–251.

Bongers, T. & Ferris, H. (1999) Nematode community structure as a bioindicator in environmental monitoring. *Trends in Ecology & Evolution*, 14, 224–228.

Candolfi, M.P., Blümel, S., Forster, R. *et al.* (2000) *Guidelines to evaluate side-effects of plant protection products to non-target arthropods.* IOBC, BART and EPPO Joint Initiative. IX, Gent, IOBC/wprs. pp. 158. ISBN: 92-9067-129-7.

CEH (2013) *Soil Biodiversity and Ecosystem Function, Biological Indicators of Soil Quality.* Centre for Ecology & Hydrology. [Online] Available from: http://www.ceh.ac.uk/sci_programmes/SoilsAnimals-Hypoaspisaculeifer.html. [Accessed 10th December 2013].

Corbin, M., Eckel, W., Ruhman, M., Spatz, D. & Thurman, N. (2006) *NAFTA Guidance Document* for *Conducting Terrestrial Field Dissipation Studies,* NAFTA. [Online] Available from: http://ebookbrowsee.net/terrestrial-field-dissipation-guidance-pdf-d75151390. [Accessed 10th January 2014].

Crawford, C.S. (1991) Macroarthropode detrivores. In: Polis, G.A. (ed.) *Ecology of desert communities.* Tuscon, University of Arizona Press. pp. 89–112.

Critter, S.A.M., Freitas, S.S. & Airoldi, C. (2002a) Comparison between microorganism counting and a calorimetric method applied to tropical soils. *Thermochimica Acta*, 394, 133–144.

Critter, S.A.M., Freitas, S.S. & Airoldi, C. (2002b) Microbial biomass and microcalorimetric methods in tropical soils. *Thermochimica Acta*, 394, 145–154.

Critter, S.A.M., Freitas, S.S. & Airoldi, C. (2004a) Microcalorimetric measurements of the metabolic activity by bacteria and fungi in some Brazilian soils amended with different organic matter. *Thermochimica Acta*, 406, 161–170.

Critter, S.A.M., Freitas, S.S. & Airoldi, C. (2004b) Comparison of microbial activity in some Brazilian soils by microcalorimetric and respirometric methods. *Thermochimica Acta*, 410, 35–46.

Crouau, Y. & Pinelli, E. (2008) Comparative ecotoxicity of three polluted industrial soils for the Collembola *Folsomia candida*. *Ecotoxicology and Environmental Safety*, 7(3), 643–649.

Crouau, Y., Chenon, P. & Gisclard, C. (1999) The use of *Folsomia candida* (*Collembola, Isotomidae*) for the bioassay of xenobiotic substances and soil pollutants. *Applied Soil Ecology*, 12, 103–111.

Dangerous Substance Directive (1967) [Online] Available from: http://ec.europa.eu/environment/chemicals/dansub/home_en.htm. [Accessed 10th December 2013].

Davy, M., Petrie, R., Smrchek, J., Kuchnicki, T. & Francois, D. (2001) *Proposal to update non-target plant toxicity testing under NAFTA.* Joint Presentation by: Health Canada – Pest Management Regulatory Agency and US EPA – Office of Prevention, Pesticides, and Toxic Substances. [Online] Available from: http://www.epa.gov/scipoly/sap/meetings/2001/june/sap14.pdf. [Accessed 10th January 2014].

De Jong, F.M.W. (1998) Development of a field bioassay for the side effects of pesticides on decomposition. *Ecotoxicology and Environmental Safety*, 40, 103–114.

Donkin, S.G. & Dusenberry, D.B. (1993) A Soil Toxicity Test Using the Nematode Caenorhabditis elegans and an Effective Method of Recovery. *Archives of Environmental Contamination and Toxicology*, 25, 145–151.

Drobne, D. & Hopkin, S.P. (1994) Ecotoxicological laboratory test for assessing the effects of chemicals on terrestrial isopods. *Bulletin of Environmental Contamination and Toxicology*, 53, 390–397.

Drobne, D., Rupnik, M., Lapanje, A., Strus, J. & Janc, M. (2002) Isopod gut microflora parameters as end points in toxicity studies. *Environmental Toxicology and Chemistry*, 21(3), 604–609.

EAMD (2013) *Ecological Assessment Methods Database*. [Online] Available from: http://assessmentmethods.nbii.gov/index.jsp?page=mdetail&mid=1and http://assessmentmethods.nbii.gov/index.jsp?page=methods. [Accessed 10th December 2013].

Earthworm (2013) Earthworm Society of Britain, Earthworm ecology. Available from: http://www.earthwormsoc.org.uk/earthworm-information/earthworm-information-page-2. [Accessed 10th December 2013].

Edwards, R., Dixon, D.P. & Walbot, V. (2000) Plant glutathione S-transferases: enzymes with multiple functions in sickness and in health. *Trends in Plant Science*, 5(5), 193–198. [Online] Available from: DOI:10.1016/S1360-1385(00)01601-0. [Accessed 10th December 2013].

Edwards, C.A., Knacker, T., Pokarshevskii, A.A., Subler, S. & Parmelee, R. (1997) Use of soil microcosms in assessing the effect of pesticides on soil ecosystems. In: *Environmental behaviour of crop protection chemicals*. Vienna, International Atomic Energy Agency. pp. 435–451.

EFSA (2014) European Food Safety Authority. [Online] Available from: http://www.efsa.europa.eu/. [Accessed 10th May 2014].

EFSA Opinion (2010) Scientific Opinion on the importance of the soil litter layer in agricultural areas. EFSA Panel on Plant Protection Products and their Residues (PPR). Parma, Italy, European Food Safety Authority (EFSA). *EFSA Journal*, 8(6), 1625. [Online] Available from: http://www.efsa.europa.eu/en/efsajournal/doc/1625.pdf. [Accessed 10th May 2014].

Elliott, A.G. & Root, B.G. (2006) Small Mammal Responses to Silvicultural and Precipitation-Related Disturbance in Northeastern Missouri Riparian Forests. *Wildlife Society Bulletin*, 34(2), 485–501. [Online] Available from: http://www.bioone.org/doi/abs/10.2193/0091-7648%282006%2934%5B485%3ASMRTSA%5D2.0.CO%3B2?journalCode=wbul. [Accessed 10th December 2013].

EPA (Environmental Protection Agency) (1987) Soil Core Microcosm Test 797.3995. *Federal Register (USA)*, 52 (187), 36363–36371. Washington, D.C., USA.

ERAPharm (2013) A Project on Environmental Risk Assessment of Pharmaceuticals. [Online] Available from: http://www.erapharm.org/. [Accessed 10th December 2013].

European Atlas of Soil Biodiversity (2010) Jeffery, S., Gardi, C., Jones, A., Montanarella, L., Marmo, L., Miko, L., Ritz, K., Peres, G., Römbke, J. & van der Putten, W.H. (eds.) Luxembourg, European Commission, Publications Office of the European Union.

Feigl, V., Anton, A., Fekete, F. & Gruiz, K. (2008) Combined chemical and phytostabilisation of metal polluted soil – From microcosms to field experiments. In: *Proceedings of the 10th International UFZ-Deltares/TNO Conference on Soil-Water Systems in cooperation with Provincia di Milano, ConSoil 2008, 3–6 June, 2008, Milano*. ISBN 978-3-00-024598-5. 'CD' Theme E. pp. 823–830.

Feigl, V., Anton, A., Fekete, F. & Gruiz, K. (2009) Combined chemical and phytostabilization: field application. *Land Contamination & Reclamation*, 17(3–4), 579–586.

Feigl, V., Uzinger, N. & Gruiz, K. (2009) Chemical stabilisation of toxic metals in soil microcosms. *Land Contamination & Reclamation*, 17(3–4), 485–496.

Fenyvesi, É., Gruiz, K., Molnár, M., Murányi, A., Szaniszló, N. & Szejtli, J. (2002) Cyclodextrin-Bioavailability Enhancing Agent in Biodegradation of Organic Pollutants in Soil – In: *Book of Abstracts of European Conference on Natural Attenuation, October 2002, Heidelberg, Germany*. pp. 190–191.

Fite, E.C., Turner, L.W., Cook, N.J. & Stunkard, C. (1988) Guidance Document for Conducting Terrestrial Field Studies. Washington, US Environmental Protection Agency. p. 67.

Fletcher, J.S., Johnson, F.L. & McFarlane, J.C. (1990) Influence of greenhouse *versus* field testing and taxonomic differences on plant sensitivity to chemical treatment. *Environmental Toxicology and Chemistry*, 9, 769–776.

Foissner, W., Agatha, S. & Berger, H. (2002) Soil ciliates (Protozoa, Ciliophora) from Namibia (Southwest Africa), with emphasis on two contrasting environments, the Etosha Region and the Namib Desert. *Denisia*, 5, 1–1459.

Foissner, W., Berger, H., Xu, K. & Zechmeister-Boltenstern, S. (2003) A huge, undescribed soil ciliate (Protozoa: Ciliophora) diversity in Austrian natural forest stands. – *Abstract book of 7th Central European Workshop on Soil Zoology, April 14–16, 2003, Ceske Budejovice, Czech Republic*. ISBN 80-86525-01-5. [Online] Available from: http://bfw.ac.at/300/pdf/diana_soil_ciliate.pdf. [Accessed 10th December 2013].

Förster, B., Eder, M., Morgan, E. & Knacker, T. (1996) A microcosm study of the effects of chemical stress, earthworms and microorganisms and their interactions upon litter decomposition. *European Journal of Soil Biology*, 32, 25–33.

Förster, B., Jones, S.E., Knacker, T., Ufer, A., Sousa, J.P. & Van Gestel, C.A.M. (1999) Standardisation of a Terrestrial Model Ecosystem (TME) for the Assessment of the Fate and of the Effects of Chemicals on the Soil Biocenosis. *Mitteilungen der Deutschen Bodenkundlichen Gesellschaft*, 89, 229–232.

Genomic Glossaries (2014) *Biopharmaceutical Glossary homepage*. Cambridge Healthtech Institute. [Online] Available from: http://www.genomicglossaries.com/content/omes.asp. [Accessed 10th May 2014].

Gruiz, K. (2005) Biological tools for soil ecotoxicity evaluation: Soil testing triad and the interactive ecotoxicity tests for contaminated soil. In: Fava, F. & Canepa, P. (eds.) *Soil Remediation Series, NO 6*, pp. 45–70. Italy, INCA.

Gruiz, K. (2009) Risk assessment and environmental data interpretation. *Land Contamination & Reclamation*, 17(3–4), 511–513.

Gruiz, K., Feigl, V., Hajdu, Cs. & Tolner, M. (2010) Environmental toxicity testing of contaminated soil based on microcalorimetry. *Environmental Toxicology*, 25(5), 479–486.

Gruiz, K., Horváth, B. & Molnár, M. (2001) Környezettoxikológia. Environmental toxicology (In Hungarian). Budapest, Hungary, Műegyetem Publishing Company.

Gruiz, K., Molnár, M. & Feigl, V. (2009) Measuring adverse effects of contaminated soil using interactive and dynamic test methods. *Land Contamination & Reclamation*, 17(3–4), 445–462.

Hanski, I. & Cambefort, Y. (1991) *Dung beetle ecology*. Princeton, NJ, USA, Princeton University Press.

Heath, G., Edwards, C. & Arnold, M. (1964) Some methods for assessing the activity of soil animals in the breakdown of leaves. *Pedobiologia*, 4, 80–87.

Heimbach, U. & Baloch, A.A. (1994) Effects of three pesticides on Poecilus cupreus (Coleoptera: Carabidae) at different post-treatment temperatures. *Environmental Toxicology and Chemistry*, 13, 317–324.

Heimbach, U., Leonard, P., Khoshab, A., Miyakawa, R. & Abel, C. (1994) Assessment of pesticide safety to the carabid beetle, *Poecilus cupreus*, using two different semifield enclosures. In: Donker, M.H., Eijsackers, H. & Heimbach, F. (eds.) *Ecotoxicology of Soil Organisms*. FL, Lewis. pp. 273–285.

Hill, M.D. (2005) *Mole*. Photo. [Online] Available from: http://en.wikipedia.org/wiki/File:Closeup_of_mole.jpg. [Accessed 10th January 2014].

Hopkin, S.P. & Martin, M.H. (1984) Assimilation of zinc, cadmium, lead and copper by the centipede *Lithobius variegatus* (Chilopoda). *Journal of Applied Ecology*, 21(2), 535–546.

Horváth, B., Gruiz, K., Molnár, M. & Kovács, Sz. (2000) Assessment of long term ecological risk of toxic metals by examining the alteration of their bioavailability. In: *Proceedings of ConSoil 2000, 7th International Conference on Contaminated Soil*. Leipzig, Thomas Telford. pp. 896–901.

Hornung, E., Farkas, S. & Fischer, E. (1998) Test on the isopod *Porcellio scaber*. In: Løkke, H. & Van Gestel, C.A.M. (eds.) *Handbook of soil invertebrate toxicity*. Chichester, John Wiley & Sons. pp. 207–226.

Hund-Rinke, K. & Wiechering, H. (2001) The potential of an earthworm avoidance test for evaluation of hazardous waste sites. *Journal of Soils and Sediments*, 1(1), 15–20.

IOBC (2000) International Organisation for Biological Control. [Online] Available from: http://www.iobc-wprs.org/. [Accessed 10th December 2013].

ISO (2014) *Biological Methods. Standard Catalogue.* [Online] Available from: http://www.iso.org/iso/home/store/catalogue_tc/catalogue_tc_browse.htm?commid=54366. [Accessed 10th January 2014].

ISO 11267 (2014) *Soil quality – Inhibition of reproduction of Collembola (Folsomia candida) by soil contaminants.* [Online] Available from: http://www.iso.org/iso/home/store/catalogue_tc/catalogue_detail.htm?csnumber=57582. [Accessed 10th May 2014].

ISO 16387 (2004) *Soil quality – Effects of pollutants on Enchytraeidae (Enchytraeus sp.) – Determination of effects on reproduction and survival.* [Online] Available from: http://www. iso.org/iso/iso_catalogue/catalogue_tc/catalogue_detail.htm?csnumber=30946. [Accessed 10th December 2013].

ISO 22030 (2005) *Soil quality – Biological methods – Chronic toxicity in higher plants.* [Online] Available from: http://www.iso.org/iso/iso_catalogue/catalogue_tc/catalogue_detail.htm?csnumber=36065. [Accessed 10th December 2013].

ISO/TC 190/SC 4N (2012) *Soil quality – Method for testing effects of soil contaminants on the feeding activity of soil dwelling organisms – Bait-lamina test.* [Online] Available from: http://standardsproposals.bsigroup.com/Home/Proposal/1562. [Accessed 10th December 2013].

Jeffery, S., Gardi, C., Jones, A., Montanarella, L., Marmo, L., Miko, L., Ritz, K., Peres, G., Römbke, J. & van der Putten, W.H. (eds.) (2010) *European Atlas of Soil Biodiversity*. Luxembourg, European Commission, Publications Office of the European Union.

Jochmann, R., Blanckenhorn, W.U., Bussière, L., Eirkson, C.E., Jensen, J., Kryger, U., Lahr, J., Lumaret, J.P., Römbke, J., Wardhaugh, K.G. & Floate, K.D. (2011) How to test nontarget effects of veterinary pharmaceutical residues in livestock dung in the field. *Integrated Environmental Assessment and Management. Special Issue: Traits-Based Ecological Risk Assessment (TERA): Realizing the Potential of Ecoinformatics Approaches in Ecotoxicology*, 7(2), 287–296. [Online] Available from: http://onlinelibrary.wiley.com/doi/10.1002/ieam.v7.2/issuetoc. [Accessed 10th December 2013].

Kammenga, J.E. & Van Koert, P.H.G. (1992) *Sublethal soil toxicity test with the nematode Plectus acuminatus. Draft test protocol for the Office of the Soil Research Programme.* Wageningen, Department of Nematology. Agricultural University.

Knacker, T. (ed.) (1998) *The use of Terrestrial Model Ecosystems (TME) to assess environmental risks in ecosystems.* TME Project First Summary Progress Report.

Köhler, H.R., Körtje, K.H. & Alberti, G. (1995) Content, absorption quantities and intracellular storage sites of heavy metals in Diplopoda (Arthropoda). *BioMetals*, 8(1), 37–46.

Kratz, W. (1996) Die Cotton-Strip Methode im Vergleich zu herkömmlichen bodenbiologischen Untersuchungsmethoden im Freiland. *Mitteilungen der Deutschen Bodenkundlichen Gesellschaft*, 81, 17–20.

Kratz, W. (1998) The bait-lamina test. General aspects, applications and perspectives. *Environmental Science and Pollution Research*, 5(2), 94–96. [Online] Available from: dx.doi.org/10.1007/BF02986394. [Accessed 10th December 2013].

Krogh, P.H. (1995) Laboratory toxicity testing with a predacous mite. In: Løkke, H. (ed.) Effects of Pesticides on Meso- and Microfauna in Soil. *Danish Environmental Protection Agency*, Nr. 8, pp. 185.

Kula, C. & Römbke, J. (1998) Testing organic matter decomposition within risk assessment of plant protection products. *ESPR – Environmental Science & Pollution Research*, 5, 55–60.

Kula, C., Heimbach, F., Riepert, F. & Römbke, J. (2006) Technical recommendations for the update of the ISO Earthworm Field test Guideline (ISO 11268-3). *Journal of Soil Science*, 6, 182–186.

Laskowski, R., Kramarz, P. & Jepson, P. (1998) Selection of species for soil ecotoxicity testing. In: Løkke, H. & Van Gestel, C.A.M. (eds.) *Handbook of Soil Invertebrate Toxicity Tests*. Chichester, Wiley. pp. 21–31.

Leitgib, L., Kálmán, J. & Gruiz, K. (2007) Comparison of bioassays by testing whole soil and their water extract from contaminated sites. *Chemosphere*, 66, 428–434.

Liebig, M., Fernandez, A.A., Blübaum-Gronau, E., Boxall, A., Brinke, M., Carbonell, G., Egeler, P., Fenner, K., Fernandez, C., Fink, G., Garric, J., Halling-Sørensen, B., Knacker, T., Krogh, K.A., Küster, A., Löffler, D., Cots, M.A.P., Pope, L., Prasse, C., Römbke, J., Rönnefahrt, I., Schneider, M.K., Schweitzer, N., Tarazona, J.V., Ternes, T.A., Traunspurger, W., Wehrhan, A. & Duis, K. (2010) Environmental risk assessment of ivermectin: A case study. *Integrated Environmental Assessment and Management Special Issue: Environmental Risk Assessment of Pharmaceuticals (ERAPharm)*, 6(S1), 567–587. [Online] Available from: http://onlinelibrary.wiley.com/doi/10.1002/ieam.96/full. [Accessed 10th December 2013].

LKB Bromma (2013) BioActivity Monitoring Seminar Notes. Bromma, Sweden, LKB-Producter AB.

Lobry de Bruyn, L.A. (1999) Ants as bioindicators of soil function in rural environments. *Agriculture, Ecosystem, Environment*, 74, 425–441.

Løkke, H. & Van Gestel, C.A.M. (eds.) (1998) *Handbook of soil invertebrate toxicity tests*. Chichester, UK, John Wiley & Sons.

Lorenzo, O., Piqueras, R., Sánchez-Serrano, J.J. & Solano, R. (2003) Ethylene response factor integrates signals from ethylene and jasmonate pathways in plant defense. *The Plant Cell*, 15, 165–178.

Markwiese, J.T., Ryti, R.T., Hooten, M.M., Michael, D.I. & Hlohowskyj, I. (2001) Toxicity bioassays for ecological risk assessment in arid and semiarid ecosystems. *Reviews of Environmental Contamination and Toxicology*, 168, 43–98.

Mazarei, M., Teplova, I., Hajimorad, R.M. & Stewart, N.C. (2008) Pathogen phytosensing: Plants to report plant pathogens. *Sensors*, 8, 2628–2641.

Meissner, K. & Schaarschmidt, T. (2000) Ecophysiological studies of *Corophium volutator* (Amphipoda) infested by microphallid trematodes. *Marine Ecology Progress Series*, 2000(202), 143–151.

Mench, M. & Bes, B. (2009) Assessment of the ecotoxicity of topsoils from a wood treatment site. *Pedosphere*, 19(2), 143–155. ISSN 1002-0160/CN 32-1315/P.

Morgan, E. & Knacker, T. (1994) The Role of Laboratory Terrestrial Model Ecosystems in the Testing of Potentially Harmful Substances. *Ecotoxicology*, 3, 213–233.

Mothes-Wagner, U., Reitze, K., Seitz, K.A. (1992) Terrestrial Multispecies Toxicity Testing. 1. Description of the multispecies assemblage. *Chemosphere*, 24, 1653–1667.

MSDS of DBNPA (2013) *Material safety data sheet of DBNPA*. [Online] Available from: http://www.accepta.com/prod_docs/2028.pdf. [Accessed 10th December 2013].

MSZ 21976-17 (1993) *Testing municipal waste – Seedling test*. Hungarian Standard.

Mvuijlst (2009) *Orchesella cincta* a common springtail. Photo. [Online] Available from: http://en.wikipedia.org/wiki/File:Orchesella_cincta.jpg. [Accessed 10th January 2014].

Nagy, Zs., Gruiz, K., Molnár, M. & Fenyvesi, É. (2010) Biodegradation in 4-chlorophenol contaminated soil. In: *Proceedings CD, ConSoil 2010 Conference, Salzburg 22–24 September, 2010*, ISBN 978-3-00-032099-6.

Nannipieri, P., Pietramellara, G. & Renella, G. (2014) *Omics in Soil Science*. Norfolk, Caister Academic Press.

Natal-da-Luz, T., Römbke, J. & Sousa, J.P. (2008) Avoidance tests in site-specific risk assessment–influence of soil properties on the avoidance response of Collembola and earthworms. *Environmental Toxicology and Chemistry*, 27(5), 1112–7.

Niemann, R. & Debus, R. (1996) Nematodentest zur Abschätzung der chronischen Toxizität von Bodenkontaminationen. *Zeitschrift für Umweltchemie und Ökotoxikologie*, 8, 255–260.

Ninjatacoshell (2009) *The root nodules of a 4-week-old* Medicago italica *inoculated with* Sinorhizobium meliloti. Photo. [Online] Available from: http://commons.wikimedia. org/wiki/File:Medicago_italica_root_nodules_2.JPG. [Accessed 10th December 2013].

OCSPP (2014) [Online] US EPA *Office of Chemical Safety and Pollution Prevention* – Harmonized test guidelines. Available from: http://www.epa.gov/ocspp/pubs/frs/home/guidelin.htm. [Accessed 10th January 2014].

OECD (2008) Guidance document on the determination of the toxicity of a test chemical to dung beetles. *OECD environmental health and safety publications series on testing and assessment. No. XX*. [Online] Available from: www.oecd.org/dataoecd/42/48/44052428.pdf. [Accessed 10th January 2014].

OECD No 56 (2006) *OECD Environmental health and safety publications series on testing and assessment*. No. 56 Guidance document on the breakdown of organic matter in litter bags. Paris, Environment Directorate. [Online] Available from: http://www.oecd.org/official documents/displaydocumentpdf/?cote=env/jm/mono%282006%292923&doclanguage=en. [Accessed 10th December 2013].

OECD TG 205 (1984) *OECD guideline for testing of chemicals: Avian dietary toxicity test*. OECD Publishing. [Online] Available from: http://www.oecd-ilibrary.org/environment/ test-no-205-avian-dietary-toxicity-test_9789264070004-en;jsessionid=9mleptjh4pqnl.x-oecd-live-02. [Accessed 10th December 2013].

OECD TG 213 (1998) *OECD guidelines for the testing of chemicals: Honeybees, acute oral toxicity test*. OECD Publishing. [Online] Available from: http://www.oecd-ilibrary.org/ environment/test-no-213-honeybees-acute-oral-toxicity-test_9789264070165-en;jsessionid= 9mleptjh4pqnl.x-oecd-live-02. [Accessed 10th December 2013].

OECD TG 220 (2004) *OECD guidelines for the testing of chemicals: Enchytraeid Reproduction Test*. OECD Publishing. [Online] Available from: http://www. oecd-ilibrary.org/environment/test-no-220-enchytraeid-reproduction-test_9789264070301- en;jsessionid=9mleptjh4pqnl.x-oecd-live-02. [Accessed 10th December 2013].

OECD TG 226 (2008) *OECD guidelines for the testing of chemicals: Predatory mite (Hypoaspis (Geolaelaps) aculeifer) reproduction test in soil*. OECD Publishing. [Online] Available from: http://www.oecd-ilibrary.org/environment/test-no-226-predatory-mite-hypoaspis- geolaelaps-aculeifer-reproduction-test-in-soil_9789264067455-en;jsessionid=9mleptjh4pqnl .x-oecd-live-02. [Accessed 10th December 2013].

OECD TG 228 (2008) *Determination of developmental toxicity of a test chemical to Dipteran dung flies (Scathophaga stercoraria L. (Scathophagidae), Musca autumnalis De Geer (Muscidae)*. [Online] Available from: http://www.oecd-ilibrary.org/environment/ test-no-228-determination-of- developmental-toxicity-of-a-test-chemical-to-dipteran-dung- flies- scathophaga-stercoraria-l-scathophagidae-musca-autumnalis-de-geer-muscidae_97892 64067479-en.[Accessed 10th December 2013].

OECD TG 232 (2009) *OECD guidelines for testing chemicals: Collembolan reproduction test in soil.* OECD Publishing. [Online] Available from: http://www.oecd-ilibrary. org/environment/test-no-232-collembolan-reproduction-test-in-soil_9789264076273-en; jsessionid=9mleptjh4pqnl.x-oecd-live-02. [Accessed 10th December 2013].

Olesen, T.M.E. & Weeks, J.M. (1998) Use of a model ecosystem for ecotoxicological studies with earthworms. In: Sheppard, S., Bembridge, J., Holmstrup, M. & Posthuma, L. (eds.) *Advances in Earthworm Ecotoxicology.* Pensacola, SETAC Press. pp. 381–386.

Omics (2014) *Omics wiki site.* [Online] Available from: http://omics.org/index.php/Main_Page. [Accessed 10th May 2014].

Palmborg, C. & Nordgren, A. (1993) Soil respiration curves, a method to test the influence of chemicals and heavy metals on the abundance, activity and vitality of the microflora in soils. Guideline 17. In: Torstensson, L. (ed.) *Guidelines. Soil biological variables in environmental hazard assessment.* Report No 4262. Solna, Sweden. Swedish Environmental Protection Agency. pp. 157–166.

Paoletti, M.G. & Hassall, M. (1999) Woodlice (Isopoda: Oniscidea): their potential for assessing sustainability and use as bioindicators. *Agriculture, Ecosystems and Environment,* 74, 157–165.

Queiroz, C.G.S., Mares-Guia, M.L. & Magalhaes, A.C. (2000) Microcalorimetric evaluation of metabolic heat rates in coffee (*Coffea arabica* L.) roots of seedlings subjected to chilling stress. *Thermochimica Acta,* 351, 33–37.

REACH Regulation (2006) 1907/2006 of the European Parliament and of the Council. *Official Journal of the European Union.* [Online] Available from: http://eur-lex. europa.eu/LexUriServ/LexUriServ.do?uri=OJ:L:2007:136:0003:0280:en:PDF. [Accessed 10th December 2013].

Römbke, J., Moser, T.H. & Moser, H. (2009) Ecotoxicological characterisation of 12 incineration ashes (MWI) using six terrestrial and aquatic tests. *Waste Management,* 29(9), 2475–2482.

Römbke, J., Heimbach, F., Hoy, S., Kula, C., Scott-Fordsmand, J., Sousa, P., Stephenson, G. & Weeks, J. (eds.) (2003) *Effects of plant protection products on functional end points in soil* (EPFES Lisboa 2002). Pensacola, USA, SETAC Publication. pp. 92.

Rong, X.M., Hunag, Q.Y., Jiang, D.H., Cai, P. & Liang, W. (2007) Isothermal Microcalorimetry: A review of applications in soil and environmental sciences. *Pedosphere,* 17, 137–145.

Rutgers, M. & Breure, A.M. (1999) Risk assessment, microbial communities and pollution-induced community tolerance. *Human and Ecological Risk Assessment,* 5, 661–670.

Rutgers, M., Van't Verlaat, I.M., Wind, B., Posthuma, L. & Breure, A.M. (1998) Rapid method for assessing pollution-induced community tolerance in contaminated soil. *Environmental Toxicology and Chemistry,* 17, 2210–2213.

Samsøe-Petersen, L. (1992) Laboratory method to test side-effects of pesticides on the rove beetle *Aleochara bilineata* – adults. In: Hassan, S.A. (ed.) Guidelines for testing the effects of pesticides on beneficial organisms: description of test methods. *IOBC/WPRS Bulletin,* XV(3), 82–88.

Sandermann, H. (ed.) (2004) *Molecular ecotoxicology of plants. Series: Ecological studies.* Vol. 170. ISBN 978-3-540-00952-8.

Schäffer, A., van den Brink, P., Heimbach, F., Hoy, S., De Jong, F.M.W., Römbke, J., Ross-Nickoll, M. & Sousa, J.P. (2011) *Semi-field methods for the environmental risk assessment of pesticides in soil* (PERAS). Boca Raton, FL, CRC Press. [Online] Available from: DOI:10.1080/03601234.2011.549806. [Accessed 10th December 2013].

SETAC (2013) Society of Environmental Toxicology and Chemistry. [Online] Available from: http://www.setac.org/. [Accessed 10th December 2013].

Sheppard, S.C. (1997) Toxicity Testing using Microcosms. In: Bitton, G. *et al.* (eds.) *Soil Ecotoxicology.* Boca Raton, Lewis Publ. pp 345–373.

Sijm, D., Kraaij, R. & Belfroid, A. (2000) Bioavailability in soil or sediment: exposure of different organisms and approaches to study it. *Environmental Pollution*, 108, 113–119.

Sparling, G.P. (1983) Estimation of microbial biomass and activity in soil using microcalorimetry. *Journal of Soil Science*, 34, 381–390.

Teuben, A. & Verhoef, H.A. (1992) Relevance of micro- and mesocosm experiments for studying soil ecosystem processes. *Soil Biology and Biochemistry*, 24, 1179–1183.

terra-protecta (2013) [Online] Available from: http://www.terra-protecta.de/englisch/ks-info-en.htm#01. [Accessed 10th December 2013].

Torre, I., Arrizabalaga, A. & Flaquer, C. (2004) Three methods for assessing richness and composition of small mammal communities. *Journal of Mammalogy*, 85, 524–530.

Torstensson, L. (ed.) (1993a) *Guidelines. Soil biological variables in environmental hazard assessment.* Report No 4262. Solna, Sweden, Swedish Environmental Protection Agency.

US EPA (2014) *US Environmental Protection Agency.* [Online] Available from: http://www.epa.gov. [Accessed 10th January 2014].

Van Gestel, C.A.M. & Van Straalen, N.M. (1994) Ecotoxicological test systems for terrestrial invertebrates. In: Donker, M.H., Eijsackers, H. & Heimbach, F. (eds.) *Ecotoxicology of Soil Organisms.* FL, Lewis. pp. 205–228.

Van Gestel, C.A.M. & Doornekamp, A. (1998) Tests on Oribatid Mite Platynothrus peltifer. In: Løkke, H. & Van Gestel, C.A.M. (eds.) *Handbook of soil invertebrate toxicity.* Chichester, John Wiley & Sons. pp. 113–130.

Vainio, A. (1992) Guideline for laboratory testing of the side-effects of pesticides onentomophagous nematodes Steinernema spp. In: Hassan, S.A. (ed.) Guidelines for testing the effects of pesticides on beneficial organisms: description of test methods. *IOBC/WPRS Bulletin*, XV(3), 145–147.

van Duinen, J. (2011) *Isotoma viridis.* Photo. [Online] Available from: https://www.flickr.com/photos/fotos-janvanduinen/5339898788. [Accessed 10th June 2010].

van Duinen, J. (2013) *Isotomurus unifasciatus.* Photo. [Online] Available from: https://www.flickr.com/photos/fotos-janvanduinen/11179011905/in/set-72157623058753453. [Accessed 10th June 2010].

van Duinen, J. (2014) *Entomobrya nivalis.* Photo. [Online] Available from: https://www.flickr.com/photos/fotos-janvanduinen/sets/72157623058753453. [Accessed 10th June 2010].

Vaszita, E., Gruiz, K. & Szabó, J. (2009) Complex leaching of metal sulfide containing mine waste and soil in microcosms. *Land Contamination & Reclamation*, 17(3–4), 465–474.

Vink, K. & Van Straalen, N. (1999) Effects of benomyl and diazinon on isopodmediated leaf litter decomposition in microcosms. *Pedobiologia*, 43, 345–59.

von Törne, E. (1990) Assessing feeding activities of soil living animals. I. Bait-lamina-tests. *Pedobiologia*, 34, 89–101.

Wadsö, I. (2009) Characterization of microbial activity in soil by use of isothermal microcalorimetry. *Journal of Thermal Analysis and Calorimetry*, 95, 843–850.

Weiss, L. (2011) *Prairie Dog.* Photo. [Online] Available from: http://en.wikipedia.org/wiki/File:Black-tailed_Prairie_Dog-Wichita_Mountain_Wildlife_Refuge-1.jpg. [Accessed 10th January 2014].

Whitford, W.G. (1991) Subterranean termites and long-term productivity of desert grasslands. *Sociobiology*, 19, 235–243.

Whitford, W.G., van Zee, J., Nash, M.H., Smith, W.E. & Herrick, T.E. (1999). Ants as indicators of exposure to environmental stressors in North American desert grasslands. *Environmental Monitoring and Assessment*, 54, 143–171.

Yeardley, Jr. R.B., Lazorchak, J. & Gast, L.C. (1996) The potential of an earthworm avoidance test for evaluation of hazardous waste sites. *Environmental Toxicology and Chemistry*, 15(9), 1532–1537. [Online] Available from: http://cfpub.epa.gov/si/si_public_record_Report.cfm?dirEntryID=13719. [Accessed 10th December 2013].

Yeates, G.W. & Bongers, T. (1999) Nematode diversity in agroecosystems. *Agriculture, Ecosystem & Environment*, 74, 113–135.

Zak, J. & Freckman, D. (1991) Soil communities in deserts: microarthropods and nematodes. In: Polis, G. (ed.) *The Ecology of Desert Communities*. Tucson, AZ, University of Arizona Press. pp. 55–58.

Zak, J.C., Willig, M.R., Moorhead, D.L. & Wildman, H.G. (1994) Functional diversity of microbial communities: A quantitative approach. *Soil Biology and Biochemistry*, 26, 1101–1108.

Zhou, S., Criddle, R.S. & Mitcham, E.J. (2000) Metabolic response of Platynota stultana pupae to controlled atmospheres and its relation to insect mortality response. *Journal of Insect Physiology*, 46, 1375–1385.

Chapter 6

# Advanced methods for chemical characterization of soil pollutants

*Gy. Záray & I. Varga*
*Cooperative Research Centre of Environmental Sciences,*
*Eötvös Loránd University, Budapest, Hungary*

## ABSTRACT

The chemical model and chemical analyses are dominant in current environmental risk management of chemical substances and contaminated land. Management based on the chemical model characterizes the state of the environment based on the prognosed or measured environmental concentrations, and manage the risk mechanically with the aim to reach a chemically based threshold.

Advanced chemical analytical methods used in various stages of environmental management starting from site assessment, through technology monitoring to evaluation of the risk and the outcome of risk reduction measures are briefly introduced. The methods aiming to determine inorganic and organic pollutants in soil and sediment are the focus of this chapter.

Inductively coupled plasma optical emission and mass spectrometry, and X-ray fluorescence spectrometry are used for the analysis of inorganic compounds, while organic compounds are investigated using the entire scope of analytical procedure (sample pretreatment, extraction, cleanup, preconcentration, separation and detection techniques). Analytical capabilities of these methods (e.g. detection limits, dynamic range, time demand of analysis) and some typical examples, literature on the analysis of pesticides, veterinary pharmaceuticals and petroleum hydrocarbons are reviewed.

The selection of the adequate chemical analytical method that suits the management task in terms of sensitivity, speed and precision is equally important. This chapter concentrates on the analytical techniques and the quality of the results obtained. Advice on their adequate applications, possible combinations and integration into targeted tool batteries for different environmental management activities can be found in Volume 3.

## 1  INTRODUCTION

Chemical analysis was believed to be the main tool for environmental toxicology and the management of chemicals and contaminated land. The chemical approach is still considered a major tool in environmental management, mainly in regulatory risk management. The chemical model may ease the handling of the environment in a uniform and standardized way, but chemical analyses alone cannot provide a true picture of the environment. They must be part of an integrated methodology in combination with biological and (eco)toxicological methods.

Uncertainties in environmental fate and behavior of chemicals and their effective proportions are accounted for by using additional transport and fate models, bioavailability models and refined target values. Environmental variables in sensitivity and adaptation are usually handled by extrapolations, applying assessment or uncertainty factors (often in the range of 10 to 100). If this is the case, the benefits of chemical analysis precision are lost. On the other hand, if ecological or human damage is prognosed, observed or demonstrated, precise chemical analysis is required to detect and quantitatively determine the contaminants. High-sensitivity analytical methods are also needed for chemicals that are hazardous in very small concentrations over the long term. Effect mechanism and toxicokinetic studies require advanced analytical methods to search for target molecules that play a role in the resulting responses and molecular mechanisms. Environmental risk management requires an integrated approach, balanced and joint application of chemical analyses and a study into the adverse effects. The adequate chemical analytical methods and their suitability to the environmental management tool battery are important for efficiency of environmental monitoring and risk assessment.

The determination of the chemical composition of soils is a sophisticated analytical task. Depending on the chemical information required, the analysis is focused on the investigation of inorganic or organic compounds of natural origin or those being pollutants due to different anthropogenic activities. Inductively coupled plasma optical emission spectrometry (ICP-OES) (Nölte, 2003), inductively coupled plasma mass spectrometry (ICP-MS) (Thomas, 2008) and X-ray fluorescence (XRF) spectrometry (Beckhoff et al., 2006) are currently the most widely used analytical techniques for elemental analysis due to their multi-elemental capabilities and large dynamic ranges for characterizing different soils. These methods can also be applied to measure uptake and distribution of various toxic elements such as arsenic and lead by the vegetables grown on the soil (McBride, 2013) or copper zinc and zinc oxide nanoparticles by free living nematodes (Sávoly 2014).

The determination of organic compounds requires a proper extraction procedure to separate organic pollutants from the inorganic matrix (Raynie, 2006). Efficient extraction methods result in exhaustive extraction of the soil samples which makes it possible to measure the total contaminant concentration, while less efficient extraction methods might give an estimate of the easily available (bioavailable) fraction of the contaminants. For details see Chapter 7. After the cleanup and preconcentration steps, the extracts are investigated by a powerful gas chromatograph-mass spectrometer (GC-MS) (Hubschmann, 2009) or a high-performance liquid chromatograph-mass spectrometer (HPLC-MS) (Barcelo, 1996) system to separate, identify and quantify different organic analytes. In this chapter, the determination of some typical organic pollutants (pesticides and pharmaceutical residues, petroleum hydrocarbons) is demonstrated.

# 2   ANALYTICAL METHODS FOR THE DETERMINATION OF INORGANIC COMPOUNDS

## 2.1   ICP-based analytical methods

### 2.1.1   Sample preparation

Since nebulization of sample solutions is the traditional method of introduction into the ICP (Nölte, 2003; Thomas, 2008), the soil samples must be dissolved. Dried and homogenized or size-fractionated soil samples can be fully dissolved in a mixture of $HF+HCl+HNO_3$, or $HF+HNO_3$ using a pressure and temperature controlled microwave assisted digestion system (Kingston & Jassie, 1988). However, the dissolution of silicates can often be avoided. Microwave (MW)-assisted acidic soil extraction using *aqua regia* or nitric acid is usually sufficient to receive reliable analytical information about the inorganic pollutants, e.g. toxic heavy metals. It should be noted that the total salt concentration of sample solutions to be nebulized is limited. The recommended concentrations for ICP-OES and ICP-MS generally are 1.0 and 0.1%, respectively.

### 2.1.2   Inductively coupled plasma as photon and ion source

Inductively coupled plasma is an electrodeless discharge in a gas (mostly argon) at atmospheric pressure, generated by an electromagnetic field at a frequency of 27.12 or 40.7 MHz. A water-cooled coupling coil arranged around the quartz plasma torch (Figure 6.1) is connected to a radio frequency generator. The torch consists of three coaxial tubes with gas entering tangentially by a side tube and creating a vertical flow. The outer gas flow, termed the coolant flow, protects the tube walls and acts as the main plasma support gas, usually of 10–15 L min$^{-1}$. The small diameter ($\sim$1.5 mm) of the injector tube is instrumental in producing a high velocity gas jet by which the sample aerosol can be introduced into the plasma to form a cooler axial channel.

To initiate the discharge, free electrons are generated by spark from a Tesla coil, and the electrical energy applied to the coil is converted into kinetic energy of electrons. The free electron path prior to the collision with an argon or analyte atom, to which its energy is transferred, is only about 1 μm thus, the plasma is heated forming a bright discharge. Due to the skin effect, the plasma has a toroidal form; therefore, wet or dry aerosol particles can be introduced into the plasma with a good efficiency resulting in a self-absorption-free source for the atomic emission spectrometry. The temperature in the induction region amounts to 10,000 K. However, the gas kinetic temperature varies between 5,000 and 7,000 K in the central channel above the mouth of the torch. This temperature is high enough to atomize the aerosol particles and to excite and ionize the analyte atoms.

For the introduction of liquid samples into the ICP, the solutions are nebulized by pneumatic, ultrasonic or hydraulic high pressure nebulizers (HHPN) combined with spray chambers. The cut-off diameter of spray chambers generally is 5 μm. The nebulization efficiency (volume of solution transported to the ICP/volume of solution transported to the nebulizer) is relatively poor (1–3%) for pneumatic nebulizers

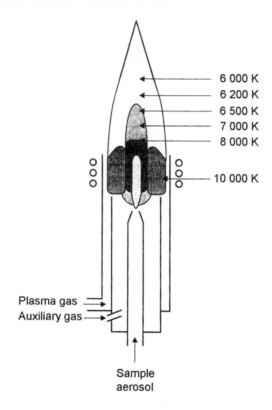

*Figure 6.1* Schematic structure of ICP torch.

(Meinhard, cross flow, glass frit), better for ultrasonic nebulizers (~15%) and the highest for the HHPN (~50%). In the last two cases, a desolvation unit is necessary to eliminate the loading of the plasma with a significant amount of solvents.

Instead of dissolving soil samples, the soil particles can be transported into the ICP by nebulization of soil suspension with a concentration of 0.5–1.0% using Babington-type nebulizers (Ebdon & Goodall, 1992). This method is applicable if the soil sample is finely ground (d < 5 μm) and homogeneous. The uncertainty of analytical data can be reduced by addition of surfactants (e.g. Triton-X) to the slurry or by selecting the optimal pH range (Varga *et al.*, 1996). The transport efficiency can be increased by desolvation of slurry aerosols (Hartley *et al.*, 1993).

Another way to transport the analytes of soil samples to the plasma is generating dry aerosols by electrothermal vaporization (ETV) of highly volatile elements (As, Cd, Pb, Zn) (Záray & Kántor, 1995). Figure 6.2 demonstrates the set-up of this unit. 1–10 mg dried and uniformly ground soil sample in a graphite boat can be introduced into the graphite furnace. Following the drying (105°C) and the ashing steps (to remove organic compounds) at 500–600°C, the temperature of the graphite boat is rapidly increased to 1,600°C resulting in hot vapors of volatile elements which are mixed with cool Ar gas forming a well-transportable dry aerosol. In order to eliminate the

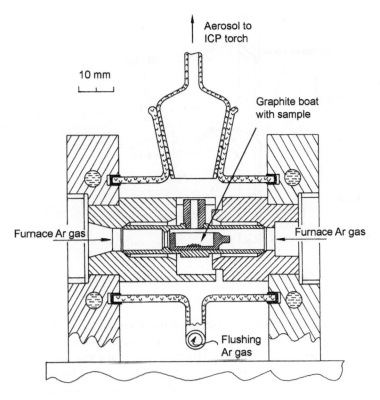

*Figure 6.2* Setup of electrothermal vaporization unit for ICP-OES or ICP-MS investigations.

transport losses between the ETV unit and the ICP, it is recommended to apply chemical modifiers, e.g. sodium selenite. In this case, the formation of dry aerosols is also supported by the chemical condensation of vapors. During the last decade, the ETV-ICP-MS ultrasonic slurry sampling method was mostly applied to determine highly volatile elements in soils and sediments (Dias *et al.*, 2005; Tseng *et al.*, 2007). For the determination of mercury, the cold vapor generation technique coupled with ICP-MS offers a powerful way. The detection limit is 6 ng/g and 300 mg samples are used for analysis (Picoloto *et al.*, 2012).

Collisions of analyte atoms with high energy electrons or metastable argon atoms are primarily responsible for the excitation and ionization processes in the analytical channel of ICP. Analytical data about the elements can be collected by measuring the emitted photons at the elements' characteristic wavelengths or by detecting the ions of stable isotopes through application of quadrupole, time-of-flight or high-resolution mass spectrometers. The ions are sampled from the plasma through a water-cooled interface consisting of a sampler and a skimmer cone with holes of 1.0 and 0.8 mm diameter. These cones are generally made of nickel; however, platinum-iridium cones are preferred for the investigation of reactive samples. Figure 6.3 shows the schematic structures of the interface unit located between the plasma and the mass spectrometer.

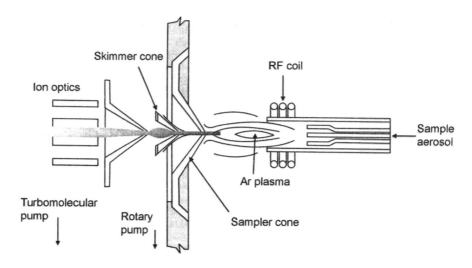

*Figure 6.3*  Schematic structure of ICP-MS interface.

Independently of the detection system (optical emission or mass spectrometer), there are different spectral interferences which cause errors in the analytical data. In the case of ICP-OES, overlapping of spectral lines of different elements is the most critical phenomenon. Therefore, it is recommended to apply a monochromator to check the spectral interferences at the wavelengths of the analytical lines in the multichannel optical emission spectrometers or a multiple line detection for all elements investigated.

Applying mass spectrometers, the ionized stable isotopes of analyte elements are detected at different m/z values. Unfortunately, this technique also suffers from spectral interferences originating from different sources. The first one is the isobaric overlap. This phenomenon occurs if two elements have stable isotopes of essentially the same mass. Fortunately, all elements have a minimum of one interference-free isotope. However, their abundance is not the highest in many cases. Doubly charged ions provide the second source of spectral interferences. They generate a number of isotopic overlaps at half of the mass of the parent elements. The third group includes polyatomic ions formed of the atoms of the plasma gas, the solvents (mostly water and acids) and the analyte elements. A few characteristic polyatomic ions and the disturbed elements are listed in Table 6.1.

There are various methods to eliminate these spectral interferences: application of high-resolution mass spectrometers ($R = 10,000$); use of collision cells to destroy the polyatomic ions; mathematical correction; modification of plasma parameters (e.g. application of helium instead of argon); or removal of disturbing matrix elements prior to ICP-MS measurements.

### 2.1.3  Analytical figures of merit

The ICP-OES or ICP-MS techniques can determine about 70–75 elements simultaneously and the quantification limits for solutions are in the ppb and ppt range,

*Table 6.1* Spectral interferences caused by polyatomic ions.

| Mass | Polyatomic ion | Disturbed stable isotope |
|------|----------------|--------------------------|
| 24 | $^{12}C_2^+$ | Mg |
| 25 | $^{12}C_2H$, $^{12}C^{13}C^+$ | Mg |
| 26 | $^{12}C^{14}N^+$ | Mg |
| 28 | $^{14}N_2^+$, $^{12}C^{16}O^+$ | Si |
| 29 | $^{14}N_2H^+$ | Si |
| 30 | $^{14}N^{16}O^+$ | Si |
| 31 | $^{14}N^{16}OH^+$ | P |
| 32 | $^{16}O_2^+$ | S |
| 33 | $^{16}O_2H^+$ | S |
| 39 | $^{38}ArH^+$ | K |
| 41 | $^{40}ArH^+$ | Ca |
| 44 | $^{12}C^{16}O_2^+$ | Ca |
| 45 | $^{12}C^{16}O_2H^+$, $^{13}C^{16}O_2^+$ | Sc |
| 46 | $^{32}S^{14}N^+$ | Ti |
| 47 | $^{31}P^{16}O^+$ | Ti |
| 48 | $^{31}P^{16}OH^+$, $^{32}S^{16}O^+$ | Ti |
| 49 | $^{32}S^{16}OH^+$ | Ti |
| 50 | $^{34}S^{16}O^+$ | Ti,V |
| 51 | $^{35}Cl^{16}O$, $^{34}S^{16}OH^+$ | V |
| 52 | $^{40}Ar^{12}C^+$, $^{36}Ar^{16}O^+$, $^{35}Cl^{16}OH^+$, $^{40}Ar^{16}C^+$ | Cr |
| 53 | $^{37}Cl^{16}O^+$ | Cr |
| 54 | $^{40}Ar^{14}N^+$, $^{37}Cl^{16}OH^+$ | Fe, Cr |
| 55 | $^{40}Ar^{14}NH^+$ | Mn |
| 56 | $^{40}Ar^{16}O^+$ | Fe |
| 57 | $^{40}Ar^{16}OH^+$ | Fe |
| 63 | $^{31}P^{16}O_2^+$ | Cu |
| 64 | $^{32}S_2^+$, $^{32}S^{16}O_2^+$ | Zn |
| 66 | $^{34}S^{16}O_2^+$ | Zn |
| 67 | $^{35}Cl^{16}O_2^+$ | Zn |
| 69 | $^{37}Cl^{16}O_2^+$ | Ga |
| 70 | $^{38}Ar^{32}S^+$ | Ge |
| 71 | $^{40}Ar^{31}P^+$ | Ge |
| 72 | $^{40}Ar^{32}S^+$, $^{38}Ar^{34}S^+$ | Ge |
| 74 | $^{40}Ar^{34}S^+$ | Ge |
| 75 | $^{40}Ar^{35}Cl^+$ | As |
| 76 | $^{40}Ar^{36}Ar^+$ | Se |
| 77 | $^{40}Ar^{37}Cl^+$ | Se |
| 78 | $^{40}Ar^{38}Ar^+$ | Se |
| 80 | $^{40}Ar_2^+$ | Se |

respectively. Considering the dilution factors, the quantification limits of inorganic compounds of soil samples are in the ppm and ppb range. It should be emphasized that the dynamic ranges of ICP-OES and ICP-MS methods amount to 4–5 and 7–8 orders of magnitudes, respectively. This means that ICP-MS makes it possible to determine major, minor and trace elements simultaneously.

## 2.2 X-ray fluorescence spectrometry

XRF spectrometry is a well-established method of elemental analysis. The determination of chemical composition through the measurement of XRF spectra has become a routine analytical method, employed in a wide variety of research areas ranging from material research to biomedical and environmental sciences. XRF is a non-destructive method, for which only simple sample preparation is necessary. The elements which can be analyzed by commercial laboratory-scale equipment range between $Z = 11$ (Na) and $Z = 92$ (U). The analysis of elements below Na requires specific conditions: a detector with a thin entrance window, measurements in vacuum, special sources for excitation (e.g. synchrotron radiation or special X-ray tubes with low Z anodes).

The basic form of XRF spectrometry consists of two steps: the first is the ionization process in the K shell caused by high-energy photons (photoelectric effect) and the second is the neutralization of the K-shell by an electron removed from another atomic shell having lower binding energy such as L, M, N, etc. The difference of the two binding energy values (K and L, M, N, etc.) will be carried by a fluorescence photon that is emitted during the relaxation process.

Due to the quantified binding energies of the different electron shells in the atoms, the emitted-fluorescence photons have a well-defined energy at a given wavelength. By measuring the energy of the photons, information can be gained on the two atomic shells involved in this process. The emitted photons are called characteristic radiation because they exactly identify the two shells between which the process takes place. The energy spectrum of these photons is a line spectrum, characteristic of the atomic number of the atom. These characteristic lines in the X-ray spectra provide information on the elements contained in the analyzed material.

In each X-ray emission spectrum, along with the characteristic lines, there is also a continuous component of X-ray radiation. This part of the spectrum originates from the electrons decelerated by the Coulomb field of the target material's nucleus. The electrons do not excite the X-ray energy levels of the atoms during the deceleration process, but they lose their kinetic energy on their way between the nuclei of the target material, and emit electromagnetic radiation at different wavelengths and different angles from the original direction. One portion of this radiation is in an X-ray region called *Bremsstrahlung*. Figure 6.4 shows a typical X-ray spectrum demonstrating the intensity distribution of the characteristic X-ray lines emitted by the major, minor and trace elements of a calcareous loam soil certified reference material (BCR No. 141).

### 2.2.1 Sample preparation

Quick and easy sample preparation is one of the reasons why XRF-spectrometry has become a popular analytical method for soil analysis. However, the very sample preparation is a major source of analytical error. First of all, matrix effects including particle size, uniformity and homogeneity are responsible for errors. To eliminate or reduce matrix effects, two groups of sample preparation procedures have been developed:

–   Fusion-based methods: dry and ground ($d < 2$ mm) soil samples are mixed with $Li_2B_4O_7$ in a ratio of 1:2–1:6 and heated to 900–1,100°C in graphite or platinum crucibles applying muffle furnace or RF-coil (Sweileh & Vanpeteghem, 1995; Johnson *et al.*, 1999);

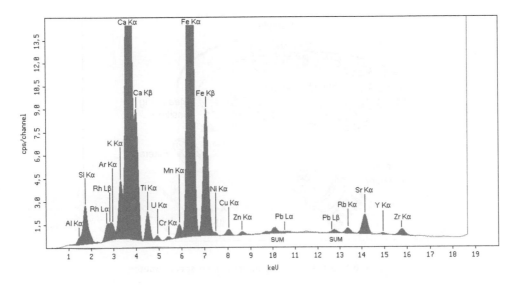

*Figure 6.4* XRF spectrum of calcareous loam soil certified reference material (BCR141).

– Pelleting techniques: dry and finely ground (d < 60 μm) soil samples are mixed and homogenized with different binding materials e.g. boric acid (dos Anjos *et al.*, 2000), cellulose (Hayumbu *et al.*, 1995), polyvinyl alcohol (Anda *et al.*, 2009), phenolic resin (Longerich, 1995), and pressed to pellets applying a pressure of 200–300 MPa.

### 2.2.2 Basic equipment and set-up for XRF analysis

Several XRF technical set-ups and experimental approaches have been developed and used routinely for the quantitative characterization of a great variety of materials. In general, any type of XRF spectrometer consists of an X-ray source producing the excitation beam, focusing devices for controlling shape and size of the X-ray beam, and the detection system. Figure 6.5 shows the simple set-up of the PANalytical MiniPal 2 equipment.

### 2.2.3 X-ray sources

There are various possible approaches, e.g. X-ray tubes, radioactive sources, synchrotron facilities, ion accelerators, and electron sources used to generate X-rays. The operation principle of each type of X-ray tubes is based on the interactions between electrons and atoms in the anode material. In an X-ray tube a cathode emits electrons that are accelerated by an electric field. After reaching the surface of the anode, they interact with the anode atoms (van Grieken & Markowitz, 2004) resulting in the deceleration of free electrons and excitation of atomic shell electrons. Finally, this procedure yields characteristic X-ray photons and *Bremsstrahlung* radiation, as mentioned above. The energy of both types of radiations originates from the energy loss

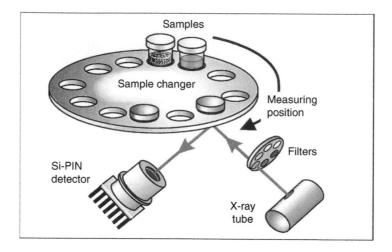

*Figure 6.5* Set-up of X-ray fluorescence spectrometer.

of the electrons. The high voltage used for accelerating the electrons is set between 20–300 kV depending on the type of X-ray tube, spectrometer, and atomic number of the element being tested.

For special XRF analysis such as *in-situ* or in-field type measurements, sometimes radioactive isotopes ([57]Co, [109]Cd, [241]Am) are used as X-ray sources. The advantage of these types of X-ray sources is that the application is very simple, and the analysis is often less expensive than that using X-ray tubes. The use of radioisotopes as X-ray sources is based on the emission of characteristic X-ray lines from a selected radioactive element. In this decay process the nucleus of the atom captures one electron on the K-shell. After the decay process, the atomic number of the daughter atom is decreased by one unit. Since one K-electron is missing, the excited atom relaxes, and emits characteristic photons which transport the excitation energy of the atom.

### 2.2.4  Detectors

The detection of X-rays is one of the most important steps in an X-ray-based study which gathers the information coded in the emitted characteristic and scattered X-rays on the atomic and molecular structure of the irradiated material. X-ray detectors can be divided into two main groups, (Van Grieken & Markowicz, 2004):

i)   Detectors used in Wavelength Dispersive X-Ray Fluorescence (WD-XRF) analysis, which are based on the Bragg reflection of X-rays on a crystal surface;
ii)  The Energy Dispersive X-Ray Fluorescence (ED-XRF) detectors.

Wavelength dispersive X-ray detectors measure the X-ray at a specific wavelength diffracted by a perfect single crystal. The wavelength of the impinging X-ray and the crystal lattice spacing are related by Bragg's law. The basic difference between the wavelength- and the energy-dispersive detection mode is that the WD-XRF detector only reads counts at a selected wavelength and it does not produce a broad spectrum.

The operation principle of the second group of detectors is the interaction between X-ray photons and semiconductor materials. The photons that enter a semiconductor single crystal (Si, Ge) excite some electrons of the crystal atoms, and they pass their energy on to the crystal during this process. Due to the additional energy received from the absorbed photons, the excited electrons will not bind to the crystal atoms; therefore, they will be able to move freely in the whole semiconductor crystal. When applying voltage to this crystal, the free electrons are forced to move out of the semiconductor detector and the sum of their electric charge can be measured. The electric charge is to the absorbed energy in the crystal. The X-ray detectors act similarly to a calorimeter. The deposited energy delivered by the absorbed photons is transformed into free electric charge.

The most frequently used semiconductor detector material uses single-crystal Si for detection of soft (E < 25–30 keV) X-rays and high-purity Ge (HPGe), GaAs or CdZnTe for hard (E > 25–30 keV) X-rays, and $\gamma$ rays. Conventional Si detectors such as Si(Li), essentially have a semiconductor diode structure, which is made up of three semiconductor layers that have different conductive properties. In the Si(Li) detectors a high resistivity intrinsic layer is sandwiched between n- and p-type layers. The thickness of the central layer doped by Li amounts to 2–3 mm. Due to random electron generation, the amount of collected charges greatly depends on the temperature of the detector crystal.

Since the X-ray radiation does not hit the middle part of the detector (the so-called depleted region), it is free of electric charge. If an X-ray photon is absorbed in the depleted region, free charges will be generated, i.e. electrons and holes of electrons with positive charge. Because of the relatively strong electric field in the depleted region, the free charges are collected separately by two electrode layers located on the surface of the semiconductor detector. With increasing temperature, the number of the free electrons becomes greater, resulting in a temperature-dependent constant electric current (leakage current) in the detector output. In order to reduce the thermally generated leakage current and to keep its value as low as possible, the detector must be cooled down to the temperature of liquid nitrogen ($\approx$77 K). Without cryogen cooling, this type of X-ray sensor would only continuously detect intensive noise. The necessity of intensive cryogen cooling mode limits the possible application of semiconductor detectors due to the complicated technical conditions for producing and storing liquid nitrogen. In order to avoid this limitation, new types of Si detectors were developed and introduced for use in X-ray spectroscopy in the last two decades. These devices need a higher operating temperature than those mentioned above (220–240 K) whereas some of them, e.g. the silicon drift detector (SDD), can operate at room temperature. Due to the absence of the cryogenic cooling unit, the size of the high-temperature semiconductor detector decreased dramatically, which allowed portable X-ray spectroscopic devices to be manufactured.

The SDD detectors have remarkable, advantageous spectroscopic properties, such as: *i*) very fast signal detection (about $10^5$ cps) whereas the detection capability of conventional Si(Li) detectors is no more than $10^4$ cps; *ii*) low noise, with energy resolution similar to that of Si(Li) detectors (120–140 eV at 5.89 keV). However, it should be noted that their detection efficiency is much lower than that of the Si(Li) detectors.

### 2.2.5 Quantification

The quantification methods and algorithms can be divided into two major groups (Jenkins *et al.*, 1995): *i*) empirical methods, when the quantification procedure is based on the use of standard samples that have a composition similar to that of the material analyzed; *ii*) fundamental parameter methods based on physical and mathematical models which describe the physical processes occurring between the sample atoms and the excitation X-ray beam. A mathematical relationship is obtained between the elemental compositions of the sample material and the characteristic X-ray intensities as detected by the X-ray technique applied (Lachance & Claisse, 1994).

The key problem is that the attenuation of the sample at the X-ray excitation energy and at the energy of the characteristic lines is unknown; therefore, the experimental determination of these values is rather complicated. In order to circumvent this problem, the most effective and simplest solution is offered by an empirical method known as the standard addition procedure. The method fits X-ray analysis of environmental samples well because this procedure is primarily suitable for the determination of trace and minor elements. The principle of the method is to add a known amount of the element under test to a given mass amount of the sample material. The procedure is repeated 5–8 times using different masses. A set of samples is eventually available, all having the same matrix composition, but with known different concentration of the investigated element. This allows a calibration curve to be constructed for the element being tested. If the sample contains only a trace amount of the analyte, the calibration curve is linear. For most environmental samples this assumption is acceptable. On the other hand, if the sample contains high concentrations of the element to be analyzed, the first step is to dilute the matrix so that the diluted sample can be used as the working matrix for the standard addition method.

The Fundamental Parameter Method (FPM) is based on the physical description of fundamental atomic processes between the excitation X-ray beam and the atoms of the sample. The theory gives an exact mathematical relationship between the concentration of the sample elements and their characteristic X-ray fluorescence intensities (K, L, and M lines).

### 2.2.6 Analytical figures of merit

Laboratory-scale XRF-spectrometers working with X-ray tubes offer quantification limits for different elements from Na to U in the ppm concentration range. Portable XRF-devices have higher quantification limits, and the analytical results are only qualitative or semi-quantitative due to the matrix effect. Since the calibration curves are linear in a concentration range of 3–4 orders of magnitude, the XRF spectrometry is a powerful and rapid method to provide multi-elemental information about major and minor inorganic compounds in soils and sediments.

### 2.2.7 Comparison of XRF and ICP-based analytical techniques

Comparing the analytical figures of merit, time demand and cost of analysis, as well as the "green character" of these multi-elemental analytical methods, the following conclusions can be drawn:

- ICP-based techniques offer lower detection limits and larger dynamic range than XRF spectrometry, therefore the ICP-based techniques can be recommended for trace analysis.
- Calibration can be performed more reliably for the ICP-based techniques compared to XRF spectrometry where the matrix effect has a strong influence on the analytical signal. Therefore, this analytical method needs standard samples with the same or similar matrix composition to that of the sample to be analyzed.
- Due to the time-consuming sample preparation, ICP-based techniques need longer time for producing analytical data than XRF-spectrometry.
- The cost of analysis for the ICP-based techniques is considerably higher than that for XRF spectrometry due to the consumption of high-purity argon and high-purity chemicals used for sample digestion.
- Due to the chemical-free sample preparation procedure, the XRF-spectrometry is a "green" analytical technique.

In addition to this theoretical comparison of ICP-based techniques and XRF spectrometry, it should be mentioned that it is useful to experimentally compare the efficiency of acidic digestion or extraction of different analytes from soil samples. Some authors demonstrated that special attention should be paid to chromium which had a low recovery in MW assisted digestion with *aqua regia* (Amorosi & Sammartino, 2011) or $HNO_3$-HF-$H_2O_2$ (Congiu *et al.*, 2013).

## 3 ANALYTICAL METHODS FOR ANALYSIS OF ORGANIC POLLUTANTS

The presence and bioavailability of organic pollutants and their metabolites in soil can adversely affect animal and human health, plants and soil microorganisms. The most important organic pollutants in soil can be classified into the following groups (a more detailed grouping of the soil contaminants can be found in Volume 1, Chapter 2):

- Pesticides, including herbicides, fungicides, bactericides, rodenticides, repellents, insecticides applied in agriculture;
- Veterinary and human pharmaceutical residues (antibiotics, anti-inflammatories, steroids, hormones, analgesics) and their metabolites originating chiefly from veterinary medicine;
- Polychlorinated biphenyls, terphenyls, naphtalenes, paraffins;
- Polybrominated biphenyls and diphenyl ethers; hexabromocyclododecane applied as flame retardants;
- Petroleum hydrocarbons.

Their identification and quantitative determination in soils and sediments need a multi-step analytical procedure:

- Sample pretreatment (drying and grounding);
- Extraction of analytes from the soil matrix;
- Cleanup of extracts in order to remove the interfering matrix compounds;
- Preconcentration/enrichment of analytes;
- Separation and detection of analytes using GC-MS or LC-MS techniques.

## 3.1 Sample pretreatment

The ideal soil sample for extraction is dry, and the particle size is smaller than 0.5 mm. Air-drying, oven-drying or freeze-drying (lyophilization) can be applied to remove water. The latter two methods provide a lower volatile compound recovery due to the possible losses. Another problem in soil analysis might be that the soil particles can form aggregates preventing an efficient extraction. Therefore, it is recommended to mix the soil sample with inert material such as sand (Crescenzi *et al.*, 2000) or diatomaceous earth (Eskilsson & Bjorklund, 2000).

## 3.2 Extraction of analytes from soil samples

The aim of the extraction step is the quantitative transfer of a small quantity of organic pollutants from a relatively large amount of soil sample into a liquid organic phase. Polar (e.g. methanol, acetonitrile) or nonpolar (e.g. benzene, n-hexane) organic solvents can be applied for extraction depending on the chemical properties of the organic pollutants and the nature of the soil sample.

The relatively old, but simple and economic Soxhlet extraction technique (Figure 6.6) is widely used in laboratory practice. 50–100 g soil samples are usually placed into the thimble chamber. The solvent vapor travels up through a distillation branch, and floods the chamber housing the thimble of soil. A condenser cools the solvent vapor, which drips back onto the soil sample. The desired compounds will then dissolve in the warm solvent. Due to the repetition of vaporization, condensation and extraction steps the desired compounds are concentrated in the distillation flask. After extraction the solvent is removed, typically by means of a rotary evaporator, yielding the extracted compounds. The non-soluble portion of the extracted solid remains in the thimble, and is usually discarded. The efficiency of this extraction technique is high, but its time demand is 8–36 hours.

Modern extraction techniques (Raynie, 2006) were developed to facilitate sample preparation and to reduce analysis time, exposure of laboratory personal to toxic chemicals and the cost of sample preparation. The improved extraction methods need instrumental systems, and the enhanced efficiency of these methods is the result of elevated solvent temperature. The higher temperature leads to favorable kinetic, diffusion and solubility parameters.

### 3.2.1 Supercritical fluid extraction (SFE)

Gases can be converted to a supercritical fluid state above a critical temperature and pressure. For example, $CO_2$ exists in a supercritical fluid state above $T = 304.2$ K and

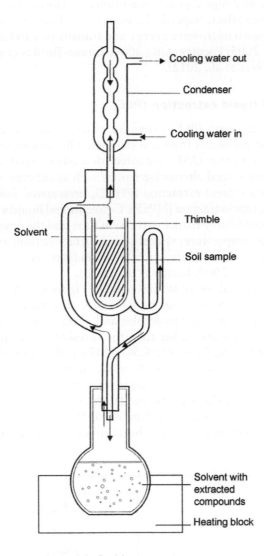

*Figure 6.6* Soxhlet extraction system.

72.8 bar where its density is closer to that of the liquid state, while its viscosity and diffusion capability are near to those of the gaseous state. Therefore, this material is an excellent solvent. The efficiency of SFE is similar to the Soxhlet technique, however, extraction takes only 10–30 minutes (Smith, 1999; Rissato *et al.*, 2005).

### 3.2.2  Microwave assisted extraction (MAE)

MW devices applied in the laboratory practice operate at the frequency of 2.45 GHz and in the power range of 100–1000 W. Microwave energy can be transferred to kinetic energy of molecules with high dipole momentum or to ions, based on dipole-rotation and ionic-conductance effect, respectively. In most applications the solvent is selected as the medium to absorb microwave energy and transform it to heat in the soil sample (Padron-Sanz *et al.*, 2005; Fuentes *et al.*, 2006; Esteve-Turillas *et al.*, 2006; Azzouz & Ballesteros, 2012; Perez *et al.*, 2012).

### 3.2.3   Pressurized liquid extraction (PLE)

Pressurized liquid extraction (PLE) offers an efficient way to achieve quantitative extraction of organic pollutants from soil samples. This technique was initially called accelerated solvent extraction (ASE), patented for a commercial device by Dionex. In the literature there are several alternative names such as subcritical solvent extraction (SSE), pressurized hot solvent extraction (PHSE), pressurized fluid extraction (PFE), and high pressure solvent extraction (HPSE). Conventional liquid solvents and elevated temperatures can considerably reduce extraction time and the amounts of organic solvent applied. At high temperature, viscosity and surface tension are less than those at room temperature. In addition, diffusion rate and mass transfer are higher resulting in higher solubility (Richter, 2000; Konda *et al.*, 2002).

The basic experimental set-up is illustrated in Figure 6.7. Most PLE applications operate within the temperature range of 50–150°C in which there is a compromise between the degradation of target molecules and extraction efficiency. The pressures applied in the PLE process are higher than the threshold value required to keep the solvents in their liquid state. Generally 6.5–10 MPa is the optimal range. PLE can be carried out in static or dynamic mode. In static mode, a fixed volume is used and the efficiency of PLE depends on the analyte mass-transfer equilibrium between the solvent and the matrix. In the dynamic mode, the solvent is passed through the soil sample at a fixed flow rate. The time demand of PLE depends on the experimental parameters of the extractor as well as the chemical properties of analytes and the

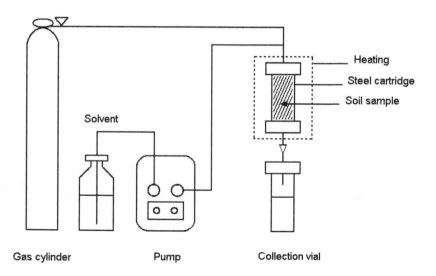

*Figure 6.7*  Pressurized liquid extraction equipment.

selected solvents. Approximately 5–15 minutes are in most cases sufficient to achieve extraction efficiency above 90%.

### 3.2.4   Ultrasonic assisted extraction (UAE)

Sound waves are mechanical vibrations in gases or solids. The frequencies of ultrasonic waves applied in practice vary from 20 kHz to 1–2 GHz. If sound waves are traveling through a medium, they generate expansion and compression cycles. In a liquid the expansion cycle results in negative pressure creating bubbles or cavities. The process by which vapor bubbles form, grow and undergo implosive collapse known as "cavitation". Rapid adiabatic compression of gases and vapors within the bubbles or cavities produces extremely high temperatures in these hot spots (about 5,000°C), and the pressure amounts to roughly 1,000 bar (Suslick, 1994). Since the size of bubbles is very small compared to the total liquid volume, the heat is rapidly dissipated resulting in no remarkable change in the environmental conditions. Due to these processes, the ultrasound assisted leaching for extraction of different compounds from soil matrices is an efficient method (Luque-Garcia & Luque de Castro, 2003; Ho et al., 2012). The efficiency of extraction depends on the polarity of the solvent, the homogeneity of the matrix and the ultrasonic treatment time (Babic et al., 1998; Goncalves & Alpendurada, 2005).

The advantages and disadvantages of the extraction methods discussed above are summarized in Table 6.2.

Table 6.2  Comparison of advantages and disadvantages of different extraction techniques.

|         | Advantages | Disadvantages |
|---------|------------|---------------|
| Soxhlet | Simple, inexpensive equipment<br>Matrix neutral<br>Large size of sample<br>Filtration not required | Large amount of solvent<br>High time demand (up to 8–36 h) |
| SFE | Small amount of solvent<br>Time demand 30–60 min<br>Filtration not required<br>$CO_2$ is environmentally friendly | Expensive equipment<br>Matrix dependent<br>Small size of sample |
| PLE | Matrix independent<br>Small amount of solvent<br>Large amount of sample<br>Time demand 15–20 min<br>Filtration not required | Expensive equipment<br>Cleanup necessary |
| MAE | Matrix independent<br>Small amount of solvent<br>Large size of sample<br>Time demand 15–20 min | Expensive equipment<br>Filtration required<br>Cleanup necessary<br>Polar solvent needed<br>Degradation possible |
| UAE | Simple, inexpensive equipment<br>Any solvent can be used<br>Small amount of solvent<br>Time demand 5–20 min | Filtration required<br>Cleanup necessary |

## 3.3 Cleanup process

Since the extract generally contains co-extracted compounds causing interferences during quantification, they must be removed. For example, the co-extractives can change the characteristic of the chromatography column, or halogen impurities may result in interference in the case of electron capture detection. Adsorption methods, solvent partition, gel or ion-exchange chromatography and distillation are used as cleanup techniques. The selection of adsorbent depends on the chemical properties of the co-extracted compounds. A short column of alumina, charcoal, silica gel or Florisil is sufficient to remove disturbing compounds (Dobor et al., 2010; Varga et al., 2010). The adsorbents can be activated by pretreatment with bases, acids, organic solvents and/or strong heating depending on their physico-chemical properties. It is recommended to check regularly the activity of the adsorbents by recovery studies.

## 3.4 Preconcentration/enrichment of analytes

Enrichment of the analyte is required in some cases before the separation and detection processes. The choice of enrichment techniques depends on the volatility and solubility of the contaminant, e.g. pesticide residues, to be measured, the degree of concentration required and the nature of the analytical technique to be applied. There are two options that meet these requirements. One is the solvent removal by lyophilization, freeze concentration, distillation, ultrafiltration and reverse osmosis. The second possibility is the isolation of the analyte, e.g. pesticide residues mostly by solid-liquid extraction (SPE) using polymeric adsorbents, activated carbon, polyurethane foam plugs, etc.

## 3.5 Separation and detection techniques

Prior to the separation step, the derivatization of analytes is recommended in most cases in order to improve specificity, selectivity and sensitivity resulting in lower quantification limits (Toys'oka, 1999; Görög, 2004). In the case of gas chromatography, the aim of derivatization is to increase the volatility and thermal stability of analytes and to eliminate the overlapping of analytical signals. For these purposes, silylation, acylation and alkylation/esterification reactions are generally used. Table 6.3 contains the list of the most common reagents and the target functional groups. The derivatization should be made prior to the injection of the sample into the GC unit.

*Table 6.3* Commonly used derivatization reagents and target functional groups.

| Reaction type | Target functional group | Reagent |
| --- | --- | --- |
| Silylation | –COOH, –OH, acidic and basic –NH groups | N,O-bis-trimethylsilyl-trifluoroacetamide, N-methyl-N-trimethylsilyl-trifluoroacetamide |
| Alkylation | –COOH, acidic –OH, –NH groups | Diazomethane, (Perfluoro)alkyl halides, Quaternary ammonium/sulfonium salts |
| Esterification | –COOH | Alcohols, (perfluoro)alkyl chloroformates |
| Acylation | –OH, basic –NH groups | (Perfluoro)acyl halides, (perfluoro)acyl anhydrides |

In the case of liquid chromatography, sensitivity of UV-VIS and fluorescence detections can be increased by formation of chromophore groups on the target molecules or by a reaction with derivatization reagents (e.g. o-phthalaldehyde, benzoxadiazole, 5-dimethylaminonaftil-1-sulfonil, 9-fluorenil-methyloxycarbonylchlorid) resulting in molecules with high fluorescence yield. It is advantageous in mass spectrometric detection if the fragmentation only leads to a few fragments with high intensity. Trifluoroacetyl ($CF_3CO^-$) or $t$-butyl-dimethylsilyl derivatives are usually produced for this purpose before the mobile phases enter the detector unit. Due to this sequence, the separation and detection can be optimized separately.

For identification and quantitation of different organic pollutants, the capillary GC and HPLC coupled with mass spectrometric detection are the most powerful analytical techniques. Their application makes the multiresidue analysis of pesticides (Paya et al., 2007; Sanchez Brunete et al., 2004) or pharmaceutical pollutants (Jacobsen, 2004; Sebők, 2008) possible. However, it should be noted that reliable analytical information can also be achieved by the application of simpler detection systems, e.g. flame ionization and photoionization detectors for measurements of hydrocarbon compounds, or electron capture detector for chlorinated pesticides using GC or fluorescence detector for HPLC.

## 3.6  Applications

### 3.6.1  Pesticide analysis

Nowadays there is a need to regularly undertake systematic surveillance and to monitor pesticide residues in soil, surface and groundwater, and food commodities, etc. in order to keep pesticidal pollution within a safe level. Amongst these environmental compartments, soil is the largest reservoir of pesticides. Therefore the investigation of agricultural soils is a key issue in the environmental chemistry studying the fate and impact of pesticides residues on the living systems. A handbook gives an excellent overview about the analytical methods applied for the determination of pesticide residues in the environment (Nollet & Rathore, 2010).

A review of the literature suggests that acetone is the most preferred extracting solvent (Babic et al., 1998; Vigh et al., 2001) for the isolation of pesticides from soil. A mixture of acetone with hexane (Suri & Joia, 1996; Sharma et al., 2006) and dichloromethane (Babic et al., 1998; George et al., 2007) is also favored. A variety of other solvents e.g. hexane (Fuentes et al., 2006) or ethyl acetate (Castro et al., 2001; Sanchez Brunete et al., 2004) are also used. GC (George et al., 2007), HPLC (Kang et al., 2007), ultra performance liquid chromatography (Kovalczuk et al., 2008), and capillary electrophoresis (Cooper et al., 2000) are well adaptable for separation. The EPA method for pesticide determination in water, soil, sediment, biosolids, and tissue applies high resolution GC-MS (EPA, 2007). Single or tandem MS equipment offer the best capability for detecting pesticides. Due to the high number of chemical compounds which belong to this group of organic pollutants, several authors developed analytical methods for multiresidue analysis of pesticides (Bao et al., 1996; Castro et al., 2001; Sanchez Brunete et al., 2004; Rissato et al., 2005). The detection limits are in the range of 0.1–10 µg/kg depending on the target molecules and simultaneously 200–300 pesticides can be identified and quantified.

### 3.6.2  Veterinary pharmaceuticals

Veterinary pharmaceuticals have been in use for many years as feed additives, and for prophylactic, metaphylactic and therapeutic purposes. Because of the persistence of different veterinary pharmaceuticals in liquid manure, these compounds reach the soil after the fertilization process and have already been shown to persist in this environmental compartment. The typical organic pollutants from this group are the antibiotics, antiparasitic agents and the coccidiostats. The most frequently used substance classes are listed below (Boxall *et al.*, 2004):

–  Aminoglycosides (gentamicin, neomycin B, streptomycin);
–  β-lactams (amoxicillin, ampicillin, benzylpenicillin, cefazolin);
–  Macrolides (oleandomycin, tylosin);
–  Sulfonamides (sulfachloropyridazine, sulfadicizine, sulfadimethoxine, sulfamethazine, sulfathiazole);
–  Tetracyclines (chlorotetracycline, doxycycline, oxytetracycline).

Several authors developed analytical methods for their identification and quantitative determination. For example, persistent tetracycline residues in soil fertilized with liquid manure were determined by a LC-MS/MS system applying electrospray ionization following an extraction with ethyl acetate and citrate buffer. The detection limits for oxytetracycline and chlorotetracycline were 1 and 2 µg/kg, respectively (Hamscher *et al.*, 2002). Simultaneous extraction of tetracycline, macrolide and sulfonamide antibiotics from agricultural soil was carried out using pressurized liquid extraction, followed by solid-phase extraction and LC-MS/MS investigation. Detection limits were in the concentration range of 1–5 µg/kg (Jacobsen *et al.*, 2004). A rapid technique based on dynamic microwave–assisted extraction coupled on-line with solid-phase extraction was developed for the determination of sulfonamides including sulfadiazine, sulfameter, sulfamonomethoxine and sulfaquinoxaline in soil by applying LC-MS/MS equipment. The detection limits amounted to 1.4–4.8 µg/kg (Chen *et al.*, 2009). A rapid analytical method was developed based on ultrasonic extraction, solid-phase extraction and LC-MS/MS measurement for simultaneous determination of veterinary antibiotics and hormones in broiler manure, soil and manure compost (Ho *et al.*, 2012).

### 3.6.3  Petroleum hydrocarbons

For total petroleum hydrocarbon (TPH) analysis three methods – based on gas chromatography, infrared spectrometry and gravimetry – have become popular in the practice. Currently the GC-based methods are preferred since they detect a broad range of hydrocarbons with appropriate sensitivity and selectivity, and they can be used for identification and quantification of TPH. Chromatographic columns are commonly used to determine TPH compounds, approximately in the order of their boiling points. The analytes between $C_6$ and $C_{25}$ or $C_{36}$ can be detected with a flame ionization detector (FID). The highly volatile compounds (below $C_6$) usually cannot be quantitatively detected due to the interference caused by the solvent peak. The detection limits are in the range of 5–10 µg/g soil. By changing the FID to a photoionization detector (PID), the aromatics can be selectively detected. The typical detection limit for light aromatics (benzene, ethylbenzene, toluene and xylene, BTEX) is 5 µg/g.

GC-MS systems are used to measure concentrations of target volatile and semivolatile petroleum constituents. The advantage of this technique is the high selectivity and the ability to confirm compounds by identifying them through retention time and unique spectral pattern. The detection limits for volatile and semivolatile analytes in soil are 20 ng/g and 50 ng/g, respectively. It should be noted that the total ion chromatogram of petroleum hydrocarbons is similar to the GC-FID signals.

In the field of petroleum hydrocarbon analysis focused on soil or sediment matrices, numerous papers have been published in the past years. In order to demonstrate the typical analytical procedures, the following four publications are discussed: soils contaminated with hydrocarbons were investigated by GC-FID following an accelerated solvent extraction with dichloromethane-acetone (1:1) at the temperature of 175°C for 5 minutes (Richter, 2000). Aliphatic hydrocarbons ($C_9$-$C_{27}$ including pristane and phytane) were determined applying continuous microwave-assisted extraction coupled on-line with liquid-liquid extraction for cleanup purposes and GC-MS technique. The limits of detection values were 0.1–0.2 µg/g (Serrano & Gallego, 2006). For the determination of PAH compounds in sediments, a microwave-assisted extraction was applied using 1M KOH/methanol in combination with solid-phase extraction and GC-MS detection (Itoh et al., 2008). A novel microwave-assisted extraction procedure using ionic liquid (1-hexadecyl-3-methyldimidazolium bromide) extractant and HPLC with fluorescence detector offers an environmentally friendly way for the analysis of PAH compounds (Pino et al., 2008).

## 3.7 Recent developments and future trends

In order to improve the isolation and identification of volatile and semivolatile organic compounds present in complex mixtures, the multidimensional gas chromatography (MDGC) has been developed. The fundamentals and new applications are summarized in a valuable review paper (Seeley & Seeley, 2013). This analytical technique employs two or more gas chromatographic separations in a sequential fashion. Since the majority of MDGC separations use two columns, they are called two-dimensional gas chromatography (2-D GC). The material transport between the two columns that have substantially different selectivity can be regulated by special interfaces. In the case of the heart-cutting 2-D GC equipment only the selected analytes and disturbing compounds that coeluate on the primary column are subjected to a secondary separation step. The comprehensive 2-D GC (GCxGC) technique subjects each sample component to both separation stages. The primary separation is similar to a temperature-ramped single-column GC-separation, whereas the secondary separations are essentially a series of fast isothermal analyses conducted with an increasing temperature. Since the peaks that enter the detector after the GCxGC separation have widths in the order of magnitude of 100 ms, the detector must have a fast response. For this purpose the time-of-flight mass spectrometers (TOFMS) offer the best analytical capabilities (Silva et al., 2012). The usefulness of the GCxGC-TOFMS analytical system in soil analysis was recently demonstrated for dioxins and polychlorinated dibenzofurans (de Vos et al., 2011), and for PAHs (Pena-Abaurrea et al., 2012; Manzano et al., 2012).

Another important research area is the development of a rapid, simple and portable micro-GC (µGC) system for separation and detection of analytes with various volatilities and polarities within a few minutes. The multi-point on column detection applying optofluidic ring resonators (Sun *et al.*, 2010) or Fabry-Pérot cavity sensors (Liu *et al.*, 2010) makes it possible to perform field measurements to localize polluted areas or monitor the migration of organic pollutants.

## REFERENCES

Amorosi, A. & Sammartino, I. (2011) Assessing natural contents of hazardous metals in soils by different analytical methods and its impact on environmental legislative measures. *International Journal of Environment and Pollution*, 46, 164–177.

Anda, M., Chittleborough, D.J. & Fitzpatrick, R.W. (2009) Assessing parent material uniformity of a red and black soil complex in the landscapes. *CATENA*, 78, 142–153.

Azzouz, A. & Ballesteros, E. (2012) Combined microwave-assisted extraction and continuous solid-phase extraction prior to gas chromatography-mass spectrometry determination of pharmaceuticals, personal care products and hormones in soils, sediments and sludge. *Science of Total Environment*, 419, 208–215.

Babic, S., Petrovic, M. & Telan Macan, M.K. (1998) Ultrasonic solvent extraction of pesticides from soil. *Journal of Chromatography A*, 823, 3–9.

Bao, M.L., Pantani, F., Barbieri, K. *et al.* (1996) Multi-residue pesticide analysis in soil by solid-phase disk extraction and gas chromatography/ion-trap mass spectrometry. *International Journal of Environmental Analytical Chemistry*, 64, 223–245.

Barcelo, D. (1996) *Application of LC-MS in environmental chemistry*. Elsevier.

Beckhoff, B., Kanngiesser, B., Langhoff, N. *et al.* (2006) *Handbook of Practical X-ray Fluorescence Analysis*. Berlin, Springer.

Boxall, A.B.A., Fogg, L.A., Kay, P. *et al.* (2004) Veterinary medicines in the environment. *Reviews of Environmental Contamination and Toxicology*, 182, 1–91.

Castro, J.C., Sanchez Brunete, C. & Tadeo, J.L. (2001) Multiresidue analysis of insecticides in soil by gas chromatography with electron-capture detection and confirmation by gas chromatography-mass spectrometry. *Journal of Chromatography*, 918, 371–380.

Chen, L., Zeng, Q., Wang, H. *et al.* (2009) On-line coupling of dynamic microwave assisted extraction to solid-phase extraction for the determination of sulfonamide antibiotics in soil. *Analytica Chimica Acta*, 648, 200–206.

Congiu, A., Perucchini S. & Cesti, P. (2013) Trace metal contaminants in sediments and soils: comparison between ICP and XRF quantitative determination. E3S Web of Conferences, *Proceedings of the 16th International Conference on Heavy Metals in the Environment*, 1, 09004. [Online] Available from: http://dx.doi.org/10.1051/e3sconf/20130109004. [Accessed 6th August 2014].

Cooper, P.A., Jessop, K.M. & Moffatt, F. (2000) Capillary electrochromatography for pesticides analysis: Effects of environmental matrices. *Electrophoresis*, 21, 1574–1579.

Crescenzi, C.D., Corcia, A., Nazzari, M. *et al.* (2000) Hot phosphate buffered water extraction coupled on-line with liquid chromatography/mass spectrometry for analyzing contaminants in soil. *Analytical Chemistry*, 72, 9050–3055.

Dias, L.F., Miranda, G.R., Saint'Pierre, T.D. *et al.* (2005) Method development for the determination of cadmium, copper, lead, selenium and thallium in sediments by slurry sampling electrothermal vaporization inductively coupled plasma mass spectrometry and isotopic dilution calibration. *Spectrochimica Acta B*, 60, 117–124.

Dobor, J., Varga, M., Yao, J. *et al.* (2010) A new sample preparation method for determination of acidic drugs in sewage sludge applying microwave assisted solvent extraction followed by gas chromatography – mass spectrometry. *Microchemical Journal*, 94, 36–41.

dos Anjos, M.J., Lopes, R.T., de Jesus, E.F.O. *et al.* (2000) Quantitative analysis of metals in soil using X-ray fluorescence. *Spectrochimica Acta B*, 55, 1189–1194.

Ebdon, L. & Goodall, P. (1992) Thermochemical effects in hexafluoroethane (freon-116) modified argon inductively coupled plasmas. *Spectrochimica Acta B*, 47, 1247–1257.

EPA (2007) *Method 1699: Pesticides in Water, Soil, Sediment, Biosolids, and Tissue by HRGC/HRMS*. [Online]. Available from: http://water.epa.gov/scitech/methods/cwa/ bioindicators/upload/2008_01_03_methods_method_1699.pdf. [Accessed 6th January 2014].

Eskilsson, C.S. & Bjorklund, E. (2000) Analytical scale microwave-assisted extraction. *Journal of Chromatography A*, 902, 227–250.

Esteve-Turrillas, F.A., Pastor, A., de la Guardia, M. (2006) Comparison of different mass spectrometric determination techniques in the gas chromatographic analysis of pyrethroid insecticide residues in soil after microwave-assisted extraction. *Analytical and Bioanalytical Chemistry*, 3848, 801–809.

Fuentes, E., Báez, M.E. & Reyes, D. (2006) Microwave assisted extraction through an aqueous medium and simultaneous cleanup by partition on hexane for determining particles in agricultural soils by gas chromatography: A critical study. *Analytica Chimica Acta*, 578, 122–130.

George, T., Beevi, S.N., Priya, G. (2007) Persistence of chlorpyrifos in acidic soils. *Pesticide Research Journal*, 19, 113–115.

Goncalves, C. & Allpendurada, M.F. (2005) Assessment of pesticide contamination in soil samples from an intensive horticulture area using ultrasonic extraction and GC-MS. *Talanta*, 65, 1179–1189.

Görög, S. (2004) Derivatization of Analytes. In: Townshend, A. (ed.) *Encyclopedia of Analytical Sciences*. Elsevier.

Hamscher, G., Sczesny, S., Hoper, H. *et al.* (2002) Determination of persistent tetracycline residues in soil fertilized with liquid manure by LC-ESI-tandem MS. *Analytical Chemistry*, 74, 1509–1518.

Hartley, J.H.D., Hill, S.J. & Ebdon, L. (1993) Analysis of slurries by inductively-coupled plasma-mass spectrometry using desolvation to improve transport efficiency and atomization efficiency. *Spectrochimica Acta B*, 48, 1421–1433.

Hayumbu, P., Haselberger, N., Markowicz, A. *et al.* (1995) Analysis of rock phosphates by X-ray fluorescence spectrometry. *Applied Radiation and Isotopes*, 46, 1003–1005.

Ho, Y.B., Zakaria, M.P., Latif, P.A. *et al.* (2012) Simultaneous determination of veterinary antibiotics and hormone in broiler manure, soil and manure compost by liquid chromatography-tandem mass spectrometry. *Journal of Chromatography A*, 1262, 160–168.

Hubschmann, H.J. (2009) *Handbook of GC-MS: fundamentals and applications*. Wiley-VCH.

Itoh, N., Numata, M. & Yarita, T. (2008) Alkaline extraction in combination with microwave-assisted extraction followed by solid-phase extraction treatment for polycyclic aromatic hydrocarbons in a sediment sample. *Analytica Chimica Acta*, 615, 47–53.

Jacobsen, A.M., Halling-Sorensen, B., Ingerslev, F. *et al.* (2004) Simultaneous extraction of tetracycline, macrolide and sulfonamide antibiotics from agricultural soils using pressurized liquid extraction followed by solid phase extraction and liquid chromatography tandem mass spectrometry. *Journal of Chromatography A*, 1038, 157–170.

Jenkins, R., Gould, R.W. & Gedcke, D. (eds.) (1995) *Quantitative X-ray spectrometry. In: Practical spectroscopy* series 20, New York, Marcel Decker, Inc.

Johnson, D.M., Hooper, P.R., Conrey, R.M. (1999) XRF analysis of rocks and minerals for major and trace elements on a single low dilution Li-tetraborate fused bead. *Advances in X-Ray Analysis*, 41, 843–867.

Kang, B.K., Gagan, J., Singh, B. *et al.* (2007) Persistence of imidacloprid in paddy and soil. *Pesticide Research Journal*, 19, 237–238.

Kingston, H.M. & Jassie, L.B. (1988) Introduction to microwave sample preparation: theory and practice. Washington DC, American Chemical Society.

Konda, L.N., Füleky, G. & Morovjan, G. (2002) Subcritical water extraction to evaluate desorption of organic particles in soil. *Journal of Agricultural and Food Chemistry*, 58, 2338–2343.

Kovalczuk, T., Lacina, O., Jech, M. *et al.* (2008) Novel approach to fast determination of multiple pesticide residues using ultraperformance liquid chromatography-tandem mass spectrometry (UPLC-MS/MS). *Food Additives & Contaminants*, 25, 444–457.

Lachance, G.R. & Claisse, F. (1994) *Quantitative X-ray fluorescence analysis.* West Sussex, UK, Wiley.

Liu, J., Sun, Y., Howard, D.J. *et al.* (2010) Fabry-Perot cavity sensors for multi-point on-column micro-gas chromatography detection. *Analytical Chemistry,* 82, 4370–4375.

Longerich, H.P. (1995) Analysis of pressed pellets of geological samples using wavelength-dispersive X-ray fluorescence spectrometry. *X-Ray Spectrometry*, 24, 123–136.

Luque-Garcia, J.L. & Luque de Castro, M.D. (2003) Ultrasound: a powerful tool for leaching. *Trends in Analytical Chemistry*, 33, 41–47.

Manzano, C., Hoh, E. & Simonich, S.L.M. (2012) Improved separation of complex polycyclic aromatic hydrocarbon mixtures using novel column combinations in GCxGC/TOF-MS. *Environmental Science and Technology*, 46, 7677–7684.

McBride, M. B. (2013) Arsenic and Lead Uptake by Vegetable Crops Grown on Historically Contaminated Orchard Soils. *Applied and Environmental Soil Science*, Vol. 2013, Article ID 283472, 8 pages.

Nollet, L.M.L. & Rathore, H.S. (2010) *Handbook of Pesticides: method of pesticide residues analysis.* Boca Raton, CRC Press.

Nölte, J. (2003) *ICP Emission Spectrometry.* New York, Wiley-VCH.

Padron-Sanz, C.R., Halko, R., Soza Ferrara, Z. *et al.* (2005) Combination of microwave assisted micellar extraction and liquid chromatography for the determination of organophosphorous pesticides in soil samples. *Journal of Chromatography A*, 1078, 13–21.

Paya, P., Anastassiades, M., Mack, D. *et al.* (2007) Analysis of pesticide residues using the Quick Easy Cheap Effective Rugged and Safe (QuEchERS) pesticide multiresidue method in combination with gas and liquid chromatography and tandem mass spectrometric detection. *Analytical and Bioanalytical Chemistry,* 389, 1697–1714.

Pena-Abaurrea, M., Ye, F., Blasko, J. *et al.* (2012) Evaluation of comprehensive two-dimensional gas chromatography time-of-flight-mass spectrometry for the analysis of polycyclic aromatic hydrocarbons in sediment. Journal of *ChromatographyA*, 1256, 222–231.

Pérez, R.A., Albero, B., Miguel, C. *et al.* (2012) Determination of parabens and endocrine-disrupting alkylphenols in soil by gas chromatography-mass spectrometry following matrix solid-phase dispersion or in-column microwave-assisted extraction: a comparative study. *Analytical and Bioanalytical Chemistry*, 402, 2347–2357.

Picoloto, R.S., Wiltsche, H., Knapp, G. *et al.* (2012) Mercury determination in soil by CVG-ICP-MS after volatilization using microwave-induced combustion. *Analytical Methods*, 4(3), 630–636.

Pino, V., Anderson, J.L., Ayala, J.H. *et al.* (2008) The ionic liquid 1-hexadecyl-3-methylimidazolium bromide as novel extracting system for polycyclic aromatic hydro-carbons contained in sediments using focused microwave-assisted extraction. *Journal of Chromatography A*, 1182, 145–152.

Raynie, D.E. (2006) Modern extraction techniques. *Analytical Chemistry*, 78, 3997–4003.

Richter, B.E. (2000) Extraction of hydrocarbon contamination from soils using accelerated solvent extraction. *Journal of Chromatography A*, 874, 217–224.

Rissato, S.R., Galhiane, S.M., Apon, B.M. *et al.* (2005) Multiresidue analysis of particles in soil by supercritical fluid extraction / gas chromatography with electron-capture detection and confirmation by GC-MS. *Journal of Agricultural and Food Chemistry*, 53, 62–69.

Sanchez Brunete, C., Albero, B. & Tadeo, J.L. (2004) Multiresidue determination of pesticides in soil by gas chromatography-mass spectrometry detection. *Journal of Agricultural and Food Chemistry*, 52, 1445–1451.

Sávoly, Z., Nagy, I.P. & Záray, Gy. (2014) Analytical methods for chemical characterization of nematodes. In: Davis, L.M. (ed.) *Nematodes: Comparative Genomics, Disease Management and Ecological Importance*. Nova Publishers. ISBN: 978-1-62948-764-9.

Sebők, Á., Vasanits-Zsigrai, A., Helenkár, A., *et al.* (2008) Multiresidue analysis of pollutants as their trimethylsilyl derivatives by GC-MS. *Journal of Chromatography A*, 1216, 2288–2301.

Seeley, J.V. & Seeley, S.K. (2013) Multidimensional gas chromatography: Fundamental advances and new applications. *Analytical Chemistry*, 85, 557–578.

Serrano, A. & Gallego, M. (2006) Continuous microwave-assisted extraction coupled on-line with liquid-liquid extraction: Determination of aliphatic hydrocarbons in soil and sediments. *Journal of Chromatography A*, 1104, 323–330.

Sharma, I.D., Dubey, J.K. & Patyal, S.K. (2006) Persistence of fenazaquin in apple fruits and soil. *Pesticide Research Journal*, 18, 79–81.

Silva, B.J.G., Tranchida, P.Q., Purcaro, G. *et al.* (2012) Evaluation of comprehensive two-dimensional gas chromatography coupled to rapid scanning quadrupole mass spectrometry for quantitative analysis. *Journal of Chromatography A.*, 1255, 177–183.

Smith, R.M. (1999) Supercritical fluids in separation science – The dreams the reality at the future. *Journal of Chromatography A*, 856, 83–115.

Sun, Y., Liu, J., Howard, D.J. *et al.* (2010) Rapid tandem-column micro-gas chromatography based on optofluidic ring resonators with multi-point on-column detection. *Analyst*, 135, 165–171.

Suslick, K.S. (1994) *The year book of science and future*. Chicago, Encyclopedia Britannica.

Suri, K.S. & Joia, B.S. (1996) Persistence of chlorpyrifos in soil and its terminal residues in wheat. *Pesticide Research Journal*, 8, 186–190.

Sweileh, J.A. & Vanpeteghem, J.K. (1995) The analysis of furnace slag and roaster feed by X-ray-fluorescence spectroscopy using a fusion preparation. *Canadian Journal of Applied Spectroscopy*, 40, 8–15.

Thomas, R. (2008) *Practical Guide to ICP-MS: A Tutorial for Beginners*. CRC Press.

Toys'oka, T. (1999) *Modern derivatization methods for separation sciences*. New York, Wiley.

Tseng, Y.J., Liu, C.C. & Jiang, S.J. (2007) Slurry sampling electrothermal vaporization inductively coupled plasma mass spectrometry for the determination of As and Se in soil and sludge. *Analytica Chimica Acta*, 588, 173–178.

van Grieken, R. & Markowicz, A.A. (eds.) (2004) *Handbook of X-Ray Spectrometry*. London, Wiley.

Varga, I., Csempesz, F., Záray, G. (1996) Effect of pH of aqueous ceramic suspensions on colloidal stability and precision of analytical measurements using slurry nebulization inductively coupled plasma atomic emission spectrometry. *Spectrochimica Acta B*, 51, 253–259.

Varga, M., Dobor, J., Helenkár, A. *et al.* (2010) Investigation of acidic pharmaceuticals in river water and sediment by microwave-assisted extraction and gas chromatography-mass spectrometry. *Microchemical Journal*, 95, 353–358.

Vigh, K., Singh, D.K., Agarwal, H.C. *et al.* (2001) Insecticide residues in cotton crop soil. *Journal of Environmental Science and Health B*, 36, 421–434.

de Vos, J., Dixon, R., Vermeulen, G. *et al.* (2011) Comprehensive two-dimensional gas chromatography time of flight mass spectrometry (GCxGC-TOFMS) for environmental forensic investigations in developing countries. *Chemosphere*, 82(9), 1230–1239.

Záray, G. & Kántor, T. (1995) Direct determination of As, Cd, Pb and Zn in soils and sediments by electrothermal vaporization and inductively coupled plasma excitation spectrometry. *Spectrochimica Acta B*, 50, 489–500.

# Chapter 7

# Bioaccessibility and bioavailability in risk assessment

*Cs. Hajdu & K. Gruiz*
*Department of Applied Biotechnology and Food Science*
*Budapest University of Technology and Economics, Budapest, Hungary*

## ABSTRACT

Contaminants in the environment occur in various physical and chemical forms and in close interaction with the environmental phases. The combination of environmental and contaminant characteristics is responsible for this interaction, which determines the contaminant's accessibility. This interaction may be extremely strong in soil, sediment and other solid phase compartments. Mobility, i.e. volatility, water solubility and sorbability of a contaminant molecule and the bonding/sorption capacity of solid matrices together determine bioaccessibility, i.e. the probability of encountering and interacting with living organisms. Potential adverse effects on an ecosystem member are determined by another interaction between the environmental contaminant and the organism. A contaminant that can be taken up by an organism is called 'bioavailable'. Bioavailability or biological availability determines the potential of a chemical substance to be absorbed by an organism. The potential of the organism to absorb, distribute and metabolize the chemical substance plays an important role in environmental risk. As far as humans are concerned, the gastrointestinal tract belongs to the 'environment', and digestion modifies the contaminant's bioaccessibility by liberating the bond contaminant from the matrix.

Low bioavailability results in low probability of being absorbed by living organisms and lower risk compared to a bioavailable substance with the same rate of toxicity. Bioavailability is limited by bioaccessibility: an organism which has no access to a substance neither encounters nor interacts with it. Environmental risk management of chemicals should place great emphasis on the chemicals' mobility and bioavailability in the environment because they can overwrite the risk determined by known adverse effects.

This chapter aims to summarize the importance of mobility and bioavailability of chemicals in risk management and introduces some studies and methods to use bioavailability as a dynamic tool for obtaining experimental results. These results directly relate to the environmental risk posed by chemicals simulating worst-case situations and mimicking interactions between contaminants, environmental matrices and living organisms.

## 1 INTRODUCTION

Environmental compartments function as complex and dynamic systems incorporating three physical phases and several physical, chemical and biological interactions among the physical phases, the living and non-living compartments of the environment at molecular, cellular and organism levels. When a substance, an agent or any living entity, or simply energy enters this dynamic system, it changes the rate and direction of the ongoing processes. No static balance exists in nature, and the outcome of the changes is determined by a hypothetical balance, which, in turn, is determined by the interactions between the living and non-living actors.

Soil is the compartment most exposed to the interactions between physical phases because gaseous, liquid and solid phases play equally important roles in terms of both its structure and function. The soil microbiota represents a fourth, biological 'phase' which is extremely important. Soil microbiota and their habitat are a microcosm within the soil with special micro-surfaces and biofilms that significantly influence bioaccessibility and bioavailability of soil components and contaminants. This is the reason why this chapter mainly deals with soil, but the same bioavailability problems are of course present in sediments, water and air. Bioaccessibility is the prerequisite of bioavailability, which arises only when the organisms have access to the chemical substance. Bioaccessibility and bioavailability that influence environmental interactions are discussed with respect to i) soil as the habitat of the terrestrial ecosystem and ii) humans exposed to soil and contaminants in soil via inhalation, ingestion and skin contact.

Figure 7.1 shows the direct interactions of an organic contaminant with the most relevant soil phases, and its indirect interactions with soil organisms. Contaminated soil gas is inhaled by microbes and animals, and pore water is used by facultative and obligate anaerobic microorganisms living in the saturated (two-phase) soil. Plant roots may also utilize pore water. Soil moisture in the micro- and mesopores of the vadose zone ensures the basic condition for soil life. Microorganisms and the food web relying on them live in the pores of the three-phase soil in biofilms, and in the root hair of the plants. The unsaturated soil structure includes a mixture of inorganic and organic matter. The organic matter – mainly colloidal humus components – plays a major role in the interactions with the organic contaminants while the clay minerals interact with the inorganic ones. The main process pairs that play a role in the actual mobility and bioaccessibility of chemical substances (both nutrients and contaminants) are the following:

- volatilization–condensation;
- solubilization–precipitation;
- sorption–desorption;
- oxidation–reduction;
- weathering–structure formation;
- loss of soil and soil components–gain of soil and components.

Each of these processes have further types and components, e.g. solubilization may include dissolution in water or acids, emulsification by biotensides, chelation, complexation, enzymatic transformation, partial biodegradation, suspension, etc. The process pairs are active both at the interface between the contaminant and physical soil phases and at the interface between living organisms and physical soil phases. The mutual influence between contaminants and living organisms is covered by the term bioavailability.

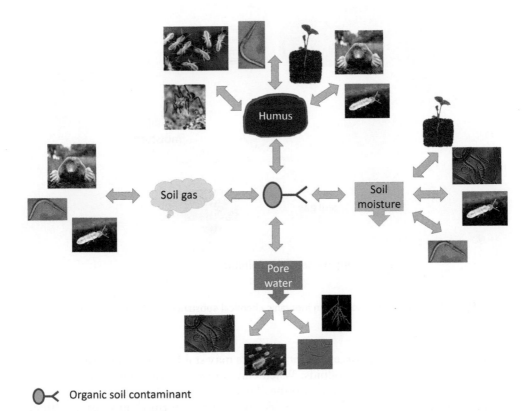

O< Organic soil contaminant

*Figure 7.1* Main interactions of an organic contaminant with soil and organisms living in soil.

Figure 7.2 demonstrates the interactions between inorganic substances and soil components. The variability is fairly large due to the variable affinity of inorganic substances to different inorganic and organic soil particles. The same processes play a role in nutrient and contaminant bioaccessibility: this makes soil health and fertility difficult to reconcile with soil which is contaminated with inorganics.

Soil contaminants are generally ***mixtures of chemicals***, and the interactions with soil phases, including the biota, may result in an infinite number of combinations. Changing environmental conditions or contamination events may trigger continuous changes in the physical, geochemical, nutrient and habitat status of the soil as well as in the density and diversity of the soil microbiota. The continuously changing non-equilibrium system is able to adapt to these conditions up to a certain limit.

***Age of soil contamination*** is a crucial characteristic that influences accessibility and availability and, as a consequence, the risk of the contaminant. Sorption of the contaminants to solid surfaces and the type of chemical bonds continuously change. The direction of these changes depends on the type and evolutionary phase of the soil as well as on the environmental conditions. When humus formation is the dominant

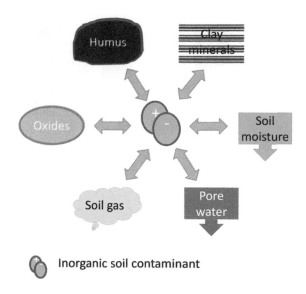

Inorganic soil contaminant

*Figure 7.2*  Main interactions of an inorganic chemical substance with soil components.

process in a soil, persistent organic contaminants may stably be built into the structure of the newly formed humus colloids. When clay minerals formation is the dominant process, metal ions are integrated into the clay structures. When sulfides are formed in anaerobic sediments, metal ions or oxides will be transformed into metal sulfides, even in crystalline forms. These processes result in biologically inaccessible forms of the contaminants. These kinds of stabilization processes occur in healthy soils and sediments, but these relatively stable forms are also part of the dynamic equilibrium system, meaning not completely irreversible binding. Regressive soil evolution processes, i.e. most soil degradation processes such as humus and clay disintegration or acidification, may mobilize the contaminants. The result in these cases is an increased risk not only in the soil, but also in the subsurface and surface waters in contact with the contaminated soil.

The effect of the environmental conditions on the soil contaminants' accessibility may also be significant. Organic contaminants go through oxidization, condensation, aggregation, and may form molecules of large size and low accessibility. In the case of metals the combinations of pH and redox potential (Eh) may result in completely different ratios of metal species such as ionic forms, oxides or hydroxides of different metal valences and sulfides. The stable mineral forms of metals in aqueous media are described by the so-called Eh–pH diagrams (also called Pourbaix diagrams after the name of its originator). These diagrams are constructed from calculations based on the Nernst equation and equilibrium solubility data of the metal in question and its species. The stability regions (read from the diagrams) are good for indication, but the actual, non-equilibrium values may differ from them given that metals always occur together with other molecules in the environment. Research is in progress on the Eh–pH diagrams of metals in natural aquatic systems and soils. Brookins (1988) developed

Eh–pH relationships based on thermodynamic data describing interactions between copper and potential inorganic ligands (oxides, hydroxides, carbonates, and sulfides) in natural aquatic systems (wetlands). Lin *et al.* (2011) characterized the stability of iron and manganese in saturated and unsaturated soils and groundwater by plotting Eh–pH diagrams. More knowledge on these relations would support hazard assessment of chemicals and upgrade the screening phase of environmental risk assessment of contaminated sites.

## 2 MANAGING BIOACCESSIBILITY AND BIOAVAILABILITY OF CONTAMINANTS IN THE ENVIRONMENT

The chemical approach in environmental risk management means that the risk is calculated based on generic environmental parameters and environmental fate characteristics of the chemical substances (regulatory risk management) or on the results of analytical measurements (contaminated sites). Analytical methodologies seek the exhaustive extraction of contaminants from environmental samples or try to differentiate between chemical species and extractability of the contaminating chemicals by partial and sequential extractions. Those chemical models which intend to model the "extraction" by biological systems (bioavailability) are called biomimetic extraction methods.

It has increasingly become evident that adverse effects of contaminants in the environment are not proportional to their concentrations measured by chemical analysis. When water and groundwater are the focus of environmental management, the water-extractable fraction of soil, sediment or solid waste represents the toxic fraction that poses risk to the aquatic ecosystem and water-consuming receptors. In contrast, water-extractable and biologically available fractions of the soil are not identical and both represent smaller portions than the total contaminant content in the soil as illustrated in Figure 7.3. The water-extractable fraction is the result of interactions between soil, contaminant and water. Water competes with soil solid to acquire the contaminant. The bioavailable fraction results from the interaction of the biological organization/system with the contaminant partitioned between solid and liquid phases of the soil. The biological system actively participates in this interaction by producing chelating agents, biotensides, degrading enzymes, etc. Both water extractability, bioaccessibility and biological availability are influenced by environmental conditions and the relative ratio of the participants.

The risk posed by hazardous contaminants in soil, sediment and solid waste is redefined by modeling or measuring the bioavailable fraction of contaminants. Test methods characterizing fate and behavior as well as adverse effects of chemicals in the environment are continuously developed and disseminated. The main aim of these tests is the correct estimation of the actual risk originating from a direct contact and material transport between contaminated environmental matrices and receptor organisms. A part of a contaminant is mobile and has the ability to be dispersed in the environment and exert an adverse effect, while another part is immobile and has no impact on the biotic and abiotic environment due to its stable chemical form and strong bond to matrices. This condition may be easy to modify or may be very stable. The dynamic behavior of contaminants in the real environment must be studied in order

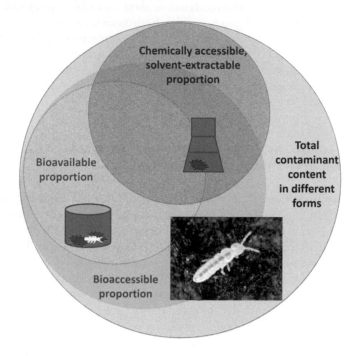

*Figure 7.3* Relationship between total, extractable, bioaccessible and bioavailable proportions of a soil contaminant.

to be able to differentiate between mobile, mobilizable and irreversibly immobile contaminants. Based on this information, bioaccessibility can be determined for hazard assessment purposes by "only" considering the chemical substance, the matrix and the environmental conditions but not the entire biological system.

Furthermore, soil remediation can be efficient in reducing risk when contaminant mobility and the resulting bioaccessibility in soil are hindered or eliminated. In such cases the immobilized contaminant is still present in the environment but does not pose any risk because it is not accessible, i.e. is not in contact with living organisms due to stable physical and chemical forms and not available, i.e. it cannot be taken up by living organisms. At the same time the physically or chemically stable/immobilized contaminants are not available to water, which lowers the risk of contaminant transport by water from soil. The risk-based approach considers contaminant stabilization/immobilization as an efficient risk reduction method. Unfortunately, many countries and authorities prefer the chemical model based approach, and do not accept immobilization as a full-value remedial measure. The crucial feature of contaminant immobilization in soil is the scale of reversibility: highly irreversible stabilization is a full-value solution for risk reduction.

The management of a complex and dynamic system of contaminated soil requires complex and dynamic engineering tools such as environmental monitoring, risk assessment methods and remediation technologies. When assessing the actual risk posed by contaminated soil, the mobility and bioavailability of the contaminants should be taken into consideration because the actual adverse effects of the contaminants greatly depend on their bioavailability. As risk is the probability of damage, it is clear that managing risk on the basis of a contaminant property such as the dynamically changing bioaccessibility in the environment can be a problem and it is not manageable mechanically. How should bioaccessibility and bioavailability be determined and taken into consideration to increase the certainty of risk assessment? There is no general answer, various modes can be and have been applied to assess the 'true' bioavailability as will be discussed below.

Risk may be overestimated if mobility and availability of environmental contaminants are not taken into consideration, thus leading to unnecessary expenditure on risk reduction. Underestimation due to momentary and short-term non-availability of contaminants is even more detrimental as it can result in a 'chemical time bomb', ready to 'explode' at any time. Risk management should maneuver between these two extremes to find the optimal conservative solution.

Risk and consequently the cost of risk reduction should reflect the real situation, i.e. the actual risk. For example, the risk posed by toxic metals in soil should not be evaluated based on analysis results from an *aqua regia* extract, but as a better option, based on an eluate or leachate resulting from the strongest possible acid rain, which is still a pessimistic scenario. This concept reflects the fact that environmental conditions influencing accessibility cannot vary within infinite limits, but within a certain range.

Bioaccessibility can be modeled by mathematical or chemical models and the results used for hazard assessment or contaminated site risk assessment in the screening phase. Bioavailability should be determined based on the results of tests that enable the interaction between contaminated soil and exposed organisms. Instead of ecotoxicity or toxicity tests one can apply mathematical models based on existing test results or biomimetic chemical tests, which give an answer close to the biological aspect. These results can be used for generic or site-specific risk assessment.

## 2.1  Mobility, bioaccessibility, bioavailability and risk assessment

When assessing a contaminated site, the results of direct chemical analyses carried out on the relevant samples may show poor correlation with actual effects. Chemical models based on direct sample analysis have little relevance to the real environment when the contaminant's bioavailability is limited. An integrated approach is needed to obtain a more realistic picture about the risk posed by pollutants in soil. This approach comprises complementary ecological or biological (ecotoxicity, mutagenicity, reprotoxicity, food-chain effects, diversity, etc.) tests and physico-chemical analysis of the contaminant and the soil.

Another possibility to move closer to environmental reality is dynamizing the chemical model system and combining it with the testing of adverse effects. Dynamization means in this context that the model should simulate the dynamic processes expected in the soil, e.g. possible changes in humidity, temperature, decrease or increase in pH and redox potential, changes in the nutrient supply, etc. Dynamization may involve gradual changes or impulse-like intervention (a push) to establish a realistic worst-case scenario in the test. The aim is to simulate the changing conditions and the maximum risk which may occur in the environment, but not more: a realistic maximum should be achieved and simulated by changing the conditions. Dynamic tests can be implemented in the laboratory or on the field. Monitoring of changes provides information for the evaluation and characterization of the probable worst case. In a dynamic test system, one can investigate the speed, intensity and duration of the response of the soil and the soil-living organisms. Dynamic testing and using its results for risk assessment is a tool which makes risk management optimally efficient both environmentally and economically. The results of dynamic tests may contribute to a correct risk assessment and to the planning of risk reduction.

## 2.2   Risk reduction in view of mobility and bioavailability

Reduction of the risk posed by soil contaminants may be based on physical, chemical, thermal or biological mobilization and removal of the contaminant or on the opposite: immobilization, stabilization and hindrance of its availability to water and living organisms. In the latter case, the contaminant remains in the soil or sediment, but in a stable and immobile form.

Reduced mobility and bioavailability of contaminants in the soil are equivalent to reduced short-term risk, but only dynamic tests can determine the contaminant's stability (irreversibility of stabilization) under varying environmental conditions. Temperature, pH, redox potential, humidity, ion concentrations, microbial activity and the equilibrium state of the soil processes influence the participating molecules' mobility. Over the long term, increased bioavailability may lead to biotransformation and biodegradation, which are generally risk-reducing processes.

The selection of the proper risk management measures depends to a large extent on mobility/bioavailability of the contaminants in soil:

–   If mobility and bioaccessibility of the contaminants are irreversibly reduced, both short- and long-term risks are low, and safety monitoring is the only necessary risk management measure.
–   If mobility and bioaccessibility are only reduced over the short term, the risk is temporarily reduced; therefore continuous monitoring and a final risk reduction measure are needed.
–   If the contaminants' mobility has not been and cannot be reduced, the risk is high; immediate risk reduction is necessary using remediation technologies other than stabilization.

In those cases when the goal is the removal of toxic contaminants, efficiency of remediation based on contaminant biodegradation is often limited by the contaminants' low bioavailability. Bioremediation of soils contaminated with high molecular-weight hydrocarbons such as polycyclic aromatic hydrocarbons (PAHs), chlorinated aliphatic and aromatic hydrocarbons, persistent pesticides or polychlorinated biphenyls (PCBs) may be slow, despite the presence of the microorganisms able to degrade these contaminants. This is one of the reasons why the more expensive and drastic chemical and thermal methods are preferred in their elimination. Limited bioaccessibility and bioavailability can be mitigated by pollutant-solubilizing (mobilizing) additives applied in the bioremediation technology. Several innovative technologies are based on the use of surfactants, cosolvents, nanomaterials and other mobilizing agents using molecular encapsulation and micelle formation. In general, these additives have proved to be effective bioaccessibility and bioavailability enhancing agents, but they may also be recalcitrant to biodegradation in the amended soils and toxic to higher soil-living organisms in charge of pollutant degradation. Fava *et al.* (2010) summed up the results of a number of studies documenting the advantages related to the use of commercially available biogenic products such as technical mixtures of hydroxypropyl beta-cyclodextrins, randomly methylated beta-cyclodextrins, soya lecithins or humic substances as pollutant bioavailability enhancing natural or naturally friendly agents in the bioremediation of soils historically contaminated by hydrophobic organic pollutants. Fava pointed out that cyclodextrins and saponins can be used as biological agents without causing additional environmental risk.

## 3   BIOAVAILABILITY AND BIOACCESSIBILITY – DEFINITIONS

Before giving a precise definition, the topic discussed above will be recapitulated. The fate and effect of toxic contaminants in soil and sediment are greatly influenced by their partitioning between physical phases, their physical and chemical forms, as well as by the environmental conditions. These together determine the pollutants' bioaccessibility. Bioavailability assumes the interaction with organisms. The attention of environmental scientists has been focused on these two terms in the last 15 years; but a clear definition and interpretation has still not been given and no uniform evaluation method has been established despite the fact that bioaccessibility and bioavailability of contaminants in soil and sediment constitute key parameters of their environmental risk.

For humans, pharmacology and toxicology have had a long-standing definition for the bioavailable part of a drug that reaches the circulatory system, while the bioaccessible fraction is the part that has the potential to reach the circulatory system once taken up by the organism/body. This definition can be modified/adapted to a case where a person swallows contaminated soil instead of a pure drug in a controlled matrix of the vehicle[1]. The problem is compounded if a mixture of contaminants or different chemical forms of the same contaminant are present in the soil.

In the context of environmental pollution a molecule is said to be bioavailable when it is ready to cross an organism's cellular membrane from the environment. It can only happen if the organism has access to the chemical (Kirk *et al.*, 2004), and is able to absorb it. The organism has access to those soil contaminants which are not very strongly bound to the solid matrix and are not located in a confined inclusion, or do not form a non-aqueous liquid phase. It depends on the physical and chemical form of the chemical substances as well as on the characteristics of the soil matrix and on the environmental conditions.

The contaminant in the soil is bioavailable when terrestrial organisms are able to take it up: the biophysical and biochemical interactions between the substance and the organism are able to ensure it. Bioavailability is organism-specific: what is

---

[1] A substance of no therapeutic value which conveys an administered drug.

available for a microorganism, is not necessarily available for a worm or a plant root and the opposite. When microbial processes are in the focus, e.g. biodegradation and soil microbiota are considered as a whole biodegrading entity, bioaccessibility and bioavailability may be overlapping to a large extent. Not, because knowledge is not enough to differentiate, but also because of the already mentioned flexibility and adaptability of the soil microbiota: when the contaminant is in an accessible form, it is highly probable that there are community members which can interact with it.

The topic can further be complicated by distinguishing between the 'ability', the potential access and the actual access to, as well as the potential availability and the actual uptake of a contaminant. For example large-size or non-motile microorganisms or soil-dwelling animals have no access to contaminants in very narrow soil capillaries, even if the contaminant is accessible (in terms of its chemical form and sorption) and available to bacteria (in terms of the existence of the interaction between the contaminant and microorganism). Plant roots which have access to and can take up available toxic metals, can also avoid metal-contaminated soil particles by chemotaxis, resulting in no actual uptake. Soil-living animals can alter soil structure and, as a result, the accessibility of contaminants. The same is valid for sediment-dwelling organisms, which actively contribute to changing the chemical forms of sediment contaminants by resuspending, mixing (bioturbation) and increasing the redox potential in surficial, sometimes also in the buried sediment layers.

Bioaccessibility and bioavailability in soil is a limiting factor in the removal of toxic substances by biodegradation, or in reducing the risk by advantageous biotransformation. Plant nutrient accessibility and bioavailability play an important role in crop production: hindered or blocked nutrient uptake, e.g. phosphorus deficiency can be traced back to soil acidification and the accompanying precipitation of phosphates in non-soluble chemical forms.

## 3.1   Definitions and mechanisms

*Bioaccessibility* can be defined as the potential of a chemical substance to encounter and interact with an organism. It is the characteristic of the substance, and depends on its physical and chemical properties in interaction with the environmental matrix and environmental conditions. Matrix effects can significantly decrease the bioaccessibility of a soil- or sediment-bound contaminant. It is not actively influenced by biological organisms. The actual access of specific organisms is mainly determined by physical conditions such as depths or pore size. Bioaccessibility is a precondition to bioavailability. Bioaccessibility can be determined and studied by chemical models or chemistry-based mathematical models. The bioaccessible part of contaminants is never less (maximum equal) than the bioavailable part.

*Bioavailability* specifies the amount of a chemical substance which reaches the site of physiological activity after adsorption/uptake. In this relationship the living organism and its physiology and anatomy as well as the exposure route and the pharmacokinetics play an important role. The actual adsorption is influenced by bioaccessibility and the organism's conditions and behavior. Bioavailability can be determined and studied by biological or biomimetic (chemical) models or mathematical statistical models based on them.

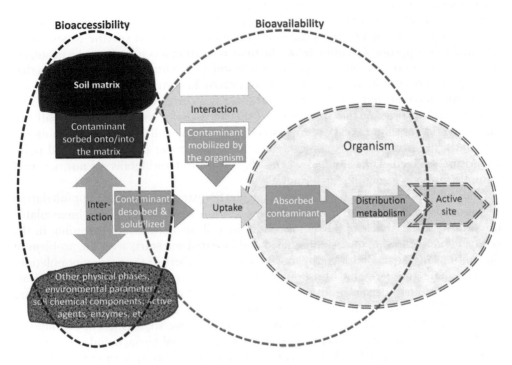

*Figure 7.4* Bioaccessibility and bioavailability of soil contaminants.

The bioaccessible fraction can be considered as the potentially bioavailable pool of the pollutant. In soils and sediments the difference between the bioaccessible and bioavailable amount can be significant, so that bioaccessibility is a preferred term in soil science, while aquatic ecotoxicology mainly uses the term of bioavailability assuming that a large part of the contaminant is in dissolved form (which may not be true when suspended solid matter is also present).

Several definitions have been established, mainly in pharmacological, but also in environmental context, e.g. Wiktionary (2014) or the IUPAC Glossary (2014). The selection introduced here may clarify the proper meaning and the differences between these terms. In soil, the bioavailable fraction of the contaminating compound is free to penetrate the organism's cell membrane at a given time (Semple *et al.*, 2004), and can be transferred into the circulatory system or other tissues where it has the ability to develop a biological response (Ruby *et al.*, 1993). Bioaccessibility is the integration of (i) desorption of the contaminant from the soil and (ii) its transportation to a biological membrane (Semenzin *et al.*, 2007; Puglisi *et al.*, 2009), called external availability by Caussy (2003). The final step, (iii), i.e. the uptake through the biological membrane is jointly determined by the bioavailability and by the previous step that limits bioaccessibility (see also Figure 7.4).

In the case of soil-living microorganisms, the three steps of the uptake process are not differentiated, and the situation is simplified considering the bioaccessible and the bioavailable fractions to be the same. In turn, soil-living organisms have a significant impact on the mobility of the contaminant in both directions: they are able to further mobilize the contaminants by acidic exudates, biotenzides and specific exoenzymes, or can do the opposite and immobilize the toxic contaminants in order to be protected by hindering uptake. Another simplification in soil and sediment ecotoxicology is that the exposure routes of the organisms are not exactly known. That is why they are not taken into account, but it is assumed that there is a direct contact between external and internal surfaces of the organism and the soil (similar to Daphnia or small fish and water). From the toxicity testing point of view it is still important to know the exposure routes of the test organism, otherwise the proper measurement end point cannot be selected. For example, worms digest soil but springtails do not, i.e. oral administration is negligible for springtails.

For higher organisms, including humans, the first step is the ingestion or inhalation (bioaccessibility), which, as a result of the digestion steps, provides the bioavailable fraction with the ability of passing through the cell membrane and spreading in the cell or tissue. Assessing the risk of inhaled and ingested soil starts with the problem of the amount ingested. Children, the most exposed members of the human population, ingest, inhale and dermally adsorb a significant amount of soil and dust. Chemicals which are inhaled, dermally adsorbed or ingested and mobilized in the gut fluid are bioaccessible (Caussy et al., 2003; Peijnenburg & Jager, 2003; Ren et al., 2006).

This definition does not hold for soil microorganisms, so, it is worth distinguishing between organisms contacting soil with their external surface and those which ingest the contaminated soil. One can expect, in both cases, a significant impact of the organism on contaminated soil and contaminants. For example, root exudates are able to dissolve non-ionic metals or exchange metal ions with hydrogen ions. Bacterial tenzides can form micelles from water-insoluble large-$K_{ow}$ substances and mobilize them. Higher animals ingest soil and digest it with a series of acidic, alkaline and enzyme-containing digestives. One could hardly create a chemical model that includes all these details, but well-chosen biotests or test batteries can incorporate all these availability-dependent data into the biological response of soil-dwelling organisms.

Further related terms are *availability* and *mobility* which mean similar behavior but in a broader sense and context: availability in the environment means that the chemical is available not only for living organisms but also for air or water and the term mobility considers the same fate characteristic from the chemical substance's point of view. *Availability* of a chemical substance means the extent to which this substance becomes soluble, disaggregates and becomes mobile via air, water or biological organisms. For metal availability, the extent to which the metal ion portion can disaggregate from the rest of the compound is the main influencing factor. Availability is not a prerequisite for bioavailability (GHS, 2011).

## 3.2  Contaminants' location and form in soil and the related accessibility and availability

Contaminants in soil represent a wide range of physical and chemical forms. The following overview in Table 7.1 summarizes the different occurrences, their bioaccessibility and bioavailability.

Table 7.1 Bioaccessibility and bioavailability of different type of soil contaminants.

| Contaminant | Location in soil | Physical, chemical form | Bio-accessibility | Bio-availability | Organism | Remarks |
|---|---|---|---|---|---|---|
| Volatile organic chemicals and metabolic products | Soil gas | Gas and vapor | ++ | ++ | Soil-living and terrestrial organisms | Transport to atmospheric air is a risk ($CO_2$, methane, chlorinated and fluorinated organics, etc.). |
| Organic chemicals | Groundwater pore water | Water-soluble | ++ | ++ | Facultative and obligate anaerobic microbes, plant roots, food web | Transport to surface waters and drinking water is a risk. Food chains/webs are endangered. |
| Organic chemicals | Soil moisture capillary waters | Water-soluble | ++ | ++ | Aerobic and facultative anaerobic microbes, plants, soil-dwelling animals, food web | Precipitation infiltrate or capillary upload from groundwater may differ. Significant adverse effect on food chains/webs. |
| Organic chemicals | Biofilm | Soluble | ++ | +++ | Biofilm organisms: bacteria and microbial food chain | Strong interaction between contaminant and biofilm organisms in the biofilm and efficient co-operation of the organisms may enhance biodegradation. |
| Organic chemicals | The surface of organic matter particles | Adsorbed | ++ | ++ | Microorganisms, animals | Become bioavailable after desorption. Microorganisms have versatile tools to increase substances' bioavailability (biotensides, complexing agents, enzymes, mucous fluids, etc.). Both in two- and three-phase soils. |
| Organic chemicals | The matrix of soil organic matter particles | Absorbed, covalently built in | (+) | + | Soil ingesting animals | Become accessible by weathering. Become bioavailable as a result of animal digestion. Mainly in three-phase soil. |
| LD NALP plume | Surface | Free layer | ++ | ++ | Aerobic, facultative anaerobic and obligate anaerobic microorganisms, plant roots, (animals) | The plume moves at the interface of the two- and the three-phase soil. Microbial degradation may occur on both sides of the plume. Soil-dwelling animals and some plant roots are able to avoid it. |
| LD NALP plume | Inside | Free layer | – | (+) | Physically not accessible, although the chemical itself may be bioavailable | The inside of contaminant plumes is highly persistent due to no access by the microbiota or to water, oxygen/electron acceptors, nutrients, etc. |
| HD NALP lens | Surface | Free layer/lens | + | + | Mainly anaerobic microorganisms | The lens is stable on the surface of the first or second impermeable layer (confining bed). |

(continued)

Table 7.1 Continued

| Contaminant | Location in soil | Physical, chemical form | Bio-accessibility | Bio-availability | Organism | Remarks |
|---|---|---|---|---|---|---|
| HD NALP lens | Inside | Free layer/lens | – | (+) | Physically not accessible, although the chemical itself may be bioavailable | The inside of the contaminant lens is highly persistent due to no access by the microbiota, or to water, electron acceptors, nutrients, etc. Risk of penetration of the water-impermeable layer. |
| Organic particulate matter | Soil organic matrix | Mixed in particle, part of soil texture | (+) | + | Only the surface is physically accessible, the inside not, although the chemical itself maybe bioavailable | It becomes accessible as a result of weathering. Weathering of the organic matter desintegrates/disaggregates the formerly hardly accessible particle. It becomes bioavailable as a result of animal digestion. Both in the two- and three-phase soil. |
| Volatile metal species | Soil gas | Gas or vapor | ++ | ++ | Soil-living and terrestrial organisms | Transport to atmospheric air is a risk (mercury, methyl mercury) |
| Metal ions | Groundwater, pore water | Dissolved | ++ | ++ | Microorganisms, plant roots, animals, food web | Transport to surface waters and drinking water is a risk. Food chains are endangered. |
| Metal ions | Soil moisture, capillary waters | Dissolved | ++ | ++ | Microorganisms, plant roots, animals, food web | Precipitation infiltrate or capillary upload from groundwater may differ. Significant adverse effects on food chains/webs may occur. |
| Metal ions | Biofilm | Dissolved | ++ | ++ | Biofilm organisms | Cooperation between biofilm organisms ensures efficient protection (resistance). |
| Metal ions | Surface of clay minerals and oxides | Adsorbed | + | + | Plant roots, microorganisms, animals | Desorption by ion exchange assisted and enhanced by interacting organisms: e.g. acidic root exudates, microbial complexing agents, etc. Both in two- and three-phase soil. |
| Metals (ions, oxides, silicates, sulfides) | Matrix of clay minerals and oxides | Absorbed or built in | (–) | + | Specialized microorganisms | Physically not accessible before weathering, soil deterioration or significant changes in environmental conditions (e.g. metal sulfides come to the surface by dredging of bottom sediment or mine-excavation of sulfide ores). May act as a chemical time bomb. |
| Metals/inorganic particulate matter | Soil texture | Mixed in particle, part of soil texture | (+) | + | The surface is physically accessible, the inside not. The chemical itself maybe bioavailable. | It becomes accessible due to weathering. Desintegration and disaggregation increase accessibility and availability. Mainly in two-phase soil. |

* LD NALP: low density non-aqueous liquid phase
** HD NALP: high density non-aqueous liquid phase

# 4  ASSESSING BIOAVAILABILITY OF CONTAMINANTS

Risk assessment should include accessibility and availability of the contaminants in soil because they have significant influence on the adverse effects and risks. Test methods which are expected to be environmentally realistic should ensure direct contact between the contaminant and the biological system or the chosen test organism. If the test concept cannot assure direct contact, availability should be accounted for by an empirical factor due to the accessibility of the substance.

Bioavailability suggests that the chemical substance spreads in the cytoplasm and reacts with the living cell. Organisms take up the bioavailable fraction of the contaminant and, depending on the metabolic potential of the organism, the contaminant can be

- accumulated in the organism with or without modification;
- secreted from the organism without or after modification;
- transformed or degraded by the organism's metabolic pathways.

The residence time of the contaminant in the organism's cell or body and the interactions with the morphological or functional part, i.e. receptor sites or molecules of the organism may result in changes in metabolism (enzyme activities) or adverse effects (toxicity, neurotoxicity, genotoxicity, reprotoxicity, etc.) in the body. The most pronounced changes as a result of the contaminant's effects can be used as measured end points in a toxicity test.

While bioavailability depends on both the chemical substance and the organisms e.g. a particular species exposed to the contaminant (Stokes *et al.*, 2006), the size of the **bioaccessible fraction** depends on the chemical substance and the environmental conditions (Hamelink *et al.*, 1994). If the requirement is to simulate the real environment, the test should ensure that the test organism and the environment do interact.

Bioavailability cannot be considered as physico-chemical partitioning like water extractability from soil, which is in direct relation to the soil–water partition coefficient and $K_{ow}$ of the chemical substance. The bioavailable fraction depends on the structure and activity of the living participant that interacts with the chemical substance. These living organisms with a dynamic adaptive behavior and active membrane systems which influence metabolic and physiological characteristics have a significant impact on the fate of chemicals in the environment. Standardized bioavailability tests on 'standardized' test organisms are suitable for the comparison of chemicals, but real bioavailability is always case-specific. Mathematical models or standardized simulation tests are used in general to determine 'bioavailability of a contaminant'. In simulation tests the pure chemical is 'mixed' into a standard or reference soil, sediment or waste before being tested. These simulated environments are considerably different from the real ones. However, real environmental bioavailability of a contaminant can only be tested via measurable adverse effects or other activities. However, it should be borne in mind that the results of a toxicity test include bioavailability in a non-distinguishable form from the adverse effect.

Bioavailability, biodegradability or the bioaccumulation potential of pure chemicals are fictions, their tests create a paradoxical situation: one tries to characterize an interaction-specific chemical property without the interacting partner or by a standardized substitute in these tests. It is clear that the results of such tests have a limited value, but they still provide useful information in a generic or screening step of a tiered risk assessment procedure. Ahlf *et al.* (2009) studied the incorporation of metal bioavailability into the European regulatory framework. They proposed the use of the biotic ligand model (BLM) to predict the bioavailability-dependent ecotoxicological effect of metals in the environment.

## 4.1   Bioaccessibility and bioavailability assessment methods

The methods for assessing bioaccessibility- and bioavailability-dependent adverse effects – similar to any other environment assessment methods – are models of the real environment. Some of them are very distant models, while others are closer to the environment. The basic methods for testing mobility and bioavailability of toxicants as well as the terms used are explained here.

*Mathematical models* must be based on a large number of data. Depending on the quality and quantity of data, they may truly represent reality and are very useful tools in regulatory and predictive risk assessment. When modeling bioaccessibility, the chemical speciation of the contaminant, its binding to the components of the soil matrix and the environmental conditions must be considered. In the context of bioavailability, geochemical conditions, contaminant and matrix interaction as well as the organisms' characteristics and activity must be taken into consideration.

*Physico-chemical models* are mainly based on the empirical correlation between the physico-chemical characteristics of a molecule and its fate and behavior in the environment. The most plausible example is establishing a link between the concentration of a chemical substance and the extent of its effect. A correlation measured *in vitro* between concentration and response of pure chemicals is not valid for the soil. In the context of bioaccessibility, one can mention the AVS concept developed for the bioavailability of metals in sediments: sulfide-bound metals are described by the ratio of simultaneously extracted metals and acid volatile sulfide. In their BLM model, Ahlf *et al.* (2009) considered the dissolved phase, metal complexes, dietary, and particle-bound metals. This model determines bioaccessible metal fraction, and based on that, predicts the probable toxicity in the environment.

*Simple biological models* are based on the contaminant's bonds to receptors or membranes, transport through the skin or the cell membrane, on the toxicity, bioaccumulation or biodegradation by single species. Biotests do not measure bioavailability directly, but the uptake or the effects, assuming that the degraded or bioaccumulated contaminant that shows an adverse effect, has been available to the test organism. If the response is positive, it is obvious that the effect has occurred, proving that the contaminant has been bioavailable. But if the response of the test organism is negative, it is hard to decide whether the contaminant is not available or the effect has not materialized due to the organism's characteristics. The test organism may have a protective tool and may not be sensitive to the toxicant. In the case of biodegradation, the test organism may have no enzymes for degrading the otherwise 'available' contaminant. Organisms secreting contaminants instead of accumulating them are not suitable for measuring bioavailability through testing their bioaccumulation. Summing up, confirming bioavailability by testing the fate and effect of the contaminants after interaction with the test organism(s) is a simple task. But the opposite, demonstration of the lack of bioavailability needs an integrated approach, e.g. i) testing several different toxicity end points on a number of test organisms or ii) the application of dynamic tests for the simulation of an increased or maximized (highest realistic scale) bioavailability. iii) Another way to obtain a clear picture is the stepwise application of a physico-chemical, e.g. BLM model for metals and the biological testing of the sample's toxic effect. An integrated evaluation can provide a larger number of combinations and refine the information on the assessed environment:

– bioavailable contaminants are toxic to the ecosystem;
– bioavailable contaminants are non-toxic due to the ecosystem's (or its members') low sensitivity or resistance;
– non-bioavailable contaminants are non-toxic to the ecosystem (members);
– non-bioavailable contaminants are not irreversibly immobile, and toxicity emerges after a certain time

   o   as a result of interactions with the ecosystem, e.g. the mobilizing effect of plants by root exudates or
   o   as a result of changes in environmental conditions and following unidentified transformations, e.g. acidification and consequent solubilization of precipitates, or resuspension and oxidization of anaerobic sediments, etc.

The conclusion is that biological models are closer to the real environment than other models but still cannot perfectly model the interactions in a complex system and to distinguish bioavailability from uptake and the effects of contaminants. To avoid misinterpreting the negative biological test results, and, as a consequence, underestimating risk, the actual state of the contaminated soil/sediment must be evaluated in an integrated way using joint physicochemical analyses and ecotoxicity tests.

*The triad approach* is an environmental characterization and monitoring tool that applies the combination of physico-chemical, biological/ecological and toxicological assessment tools and evaluates their results in an integrated way, relative to each other. Application of the 'triad' approach has several advantages in comparison to only chemical or only biological assessments. The integrated evaluation using the soil testing triad will be introduced for the interactive tests and the dynamic simulation tests in Section 7.6.

Testing bioavailability or bioavailability-dependent toxicity requires the application of *interactive tests* – also called contact or direct contact tests – which allow for the test organism to interact with the soil matrix and the contaminants. The test organism exerts an impact on the soil and on the contaminant by the produced acids, and other exudates, exoenzymes, biotensides, etc.

*Simulation tests* try to simulate real conditions, integrating all important components and actors of the environmental process in which environmental contaminants encounter and affect living organisms. A good example is the multi-step digestion of contaminated soil before human toxicity testing.

*Dynamic simulation* of contaminant bioavailability and bioaccessibility means creating test conditions which increase availability and accessibility compared to the original situation. The aim is to achieve maximum realistic mobilization/solubilization of contaminants under conditions totally different from normal natural conditions. This can be achieved by changing temperature, pH and redox conditions, quantity and quality of precipitation, flood on floodplains, exposure to light and the presence of organic matter in the soil or groundwater, etc.

Ensuring real-scale or greater bioavailability and bioaccessibility during the testing of adverse effects is an essential requirement, otherwise the effect measured in the test may become smaller than in reality leading to risk underestimation. Due to uncertainties, some overestimation is generally desirable meaning that accessibility and availability of contaminants should represent a *realistic maximum*, which is the realistic worst case from the point of view of risk assessment. Exhaustive extraction of soil contaminants represents a worst case which would never occur in the real environment and it would result an extreme overestimate in environmental toxicity and risk.

The same concept can be used to forecast the mobilization of a contaminant during soil remediation, e.g. soil washing, remediation based on chemical oxidation or biodegradation with the aim to enhance the removal of the contaminant from the soil using physical, chemical or biological technologies.

To model the realistic environmental scenario, the concept should include both the environment and the body of an organism as reactive media since both can change the contaminant and its availability. The environmental parameters such as temperature, pH, redox potential, the presence of other contaminants and the components of the solid matrix can change the physical and chemical form of the contaminant and, as a result, its mobility, accessibility and availability. The body of the organisms exposed to contaminated soil is even more reactive, able to switch to generic and contaminant-specific metabolic and transport mechanisms (interactions by the gastrointestinal tract, active transport through skin or inner membranes, secretion in a chemically mobilized form or accumulation in a chemically bound, immobile form) or create specific responses (toxicity or immune response). When measuring the potential adverse effects of a contaminant in soil, the test method should let the risk-increasing changes take place. A moderately pessimistic test concept simulates the realistic worst case. The results of such tests can fit into a tiered and iterative *risk assessment* procedure and used for *decision making*.

For environmental risk assessment, the test organism is placed into the contaminated soil and the response of the organism is measured. The response can be lethality, change in energy production or any other metabolic and enzyme activity, the absorption, accumulation or biodegradation of the contaminant (earthworm or plant bioaccumulation). Biodegradation is an advantageous interaction between soil microbiota and the contaminant which leads to risk reduction. Those soil contaminants are available which can reach the inside of the biodegrading cells, and those are biodegradable which are accepted by any of the cell's degrading enzymes and forwarded on a metabolic pathway. Bioavailability and biodegradation of contaminants in soil are strongly linked to each other; bioavailability may limit biodegradation. Microbial biodegradability assumes bioavailability, so that these two tests are often used and considered as alternatives.

## 5   MATHEMATICAL MODELS FOR CONTAMINANT BIOAVAILABILITY IN SOIL

Mathematical equations properly estimate toxicity or bioaccumulation when suitable quality and amount of data are available to create the QSAR (Quantitative

Structure-Activity Relationship) between chemical characteristics, e.g. $K_{ow}$ and the toxicity to a certain test organism (EU-TGD, 2003). Physico-chemical properties of chemical substances determine their water solubility, partitioning between phases, including biological phases and their affinity to air, water or solid environmental phases. The mathematical models use chemical properties and predict mobility and fate of the chemicals in a standard test scenario to calculate bioavailability or bioaccessibility of the compound. These calculated results are valid for pure chemicals. Some of the mathematical models are freely available in the form of software packages. EPI Suite™ (2014) for example needs only some easily available input data e.g. the contaminant's molecular weight, chemical structure, melting point and boiling point, from which the partition coefficients between physical phases such as octanol–water ($K_{ow}$), organic carbon–water ($K_{oc}$) and sediment–water ($K_p$) partition coefficient, the biodegradation half-time or even $EC_{50}$ or $EC_{20}$ values for fish species can be calculated.

## 6  CHEMICAL MODELS FOR CONTAMINANT MOBILITY AND AVAILABILITY IN SOIL

The target organisms in the environment are degrading microbes, plants or animals. The chemical substance in the soil is transformed physically and chemically before it reaches the target organism's active site, i.e. a receptor molecule, and becomes ready to act. The type and rate of contaminants' transformation depends on the environmental conditions. Risk management should focus on partitioning of chemicals between environmental phases and the balance of the process pairs determining the direction of contaminant transport in evaporation–condensation, dissolution–precipitation, suspension–sedimentation, sorption–desorption, oxidation–reduction, etc. These process pairs and their balance in a steady state specify the availability of chemical substances in the environment. Not only availability to the living organisms, but in a wider sense availability to air and water, interacting with chemicals and 'taking up' and transporting chemicals in space and time should be considered.

Special chemical extraction methods have been created to mimic the uptake by a participant of the environment or the leaching of the chemical substance from matrices. Some of these chemical extraction methods are extremely conservative tools for the calculation of environmental risk (hexane–acetone extraction of organic contaminants or *aqua regia* extraction of toxic metals), some others are closer to the realistic worst-case estimates (extraction of metals from soil by organic acids, or of organic contaminants by tenside-containing water).

### 6.1  Partition between n-octanol and water to predict accessibility of organic contaminants

The octanol-water partition coefficient ($K_{ow}$) of organic chemicals, an intrinsic molecular characteristic, correlates with their mobility and bioavailability. Each phase (step) of the interaction between the organism and the contaminant can be characterized by the $K_{ow}$: desorption from the soil material; transportation to the cell through the water phase of the soil and uptake by the cell.

As bioavailability suggests, the uptake of contaminants, measurement of toxic effects, biodegradation and bioaccumulation are suitable tools to predict bioavailability of contaminants. By now it has been proved that log $K_{ow}$ is in linear relationship with toxicity, biodegradation and bioaccumulation characterized by the following end points:

– EC$_x$, EC$_{50}$ or LC$_{50}$, (concentration having x% inhibitory effect, especially 50% effect, or causing lethality in 50% of the test organisms);
– half-life (DT$_{50}$) or biodegraded fraction (S$_{bi}$) of the contaminant;
– BCF, bioconcentration factor, the rate of contaminant concentration in the organism or in the environment.

Based on the same relationship, chemical methods can be used to predict the bioavailable fraction of contaminants in soil such as liquid–solvent extractions, solid phase and membrane-based devices. Some of the chemical agents which produce similar 'uptake' as a living organism are called **biomimetic** extractants or sorbents.

The distribution of **organic contaminants** between soil, water and living organisms can be modeled by moderately polar organic solvents e.g. butanol, methanol (1%, 50% methanol in water or 100% methanol), n-propanol, Tween 80, ethyl acetate and the aqueous solution of cyclodextrin (CD) (Kelsey et al., 1997; Tang & Alexander, 1999). These extraction methods are generally followed by the determination of the concentration using chemical analysis of the extract or the mass by gravimetry after evaporating the solvent. The basis of the model is the correlation found between the simultaneously measured amount of extracted contaminant and the bioaccumulated or biodegraded amount of contaminant. For example, the amount of an organic contaminant extracted from the soil by cyclodextrin solution was found to correlate with the biodegraded amount in the same soil measured by a biodegradability test (Reid et al., 2000; Stokes et al., 2005; Semple et al., 2007; Fenyvesi et al., 2008). Contaminant amounts extracted by cyclodextrin and butanol were compared to amounts accumulated by *Eisenia fetida* and *Lolium florum* from deuterated PAH (Gomez-Eyles et al., 2011) and correlation was found. For more details see Chapter 9, Volume 3.

Chemical models can be applied in predictive risk assessment but their results cannot guarantee that the contaminant will really be biodegraded, for example it may kill the degrading organisms if it is very toxic before biodegradation could commence. Nevertheless, these chemical models fit well to chemistry-based risk management and chemists alone can assess the risk, thus avoiding tests on living organisms.

## 6.2   Solid phase and membrane-based extractions – chemical bioavailability models

Solid phase and membrane-based extractions are chemical models for bioavailability estimation of organic contaminants. The solid phase and membrane-based systems move towards cell modeling: semi-permeable membranes or/and polymers inside of the devices are dedicated to mimicking the cell membrane (Huckins et al., 1990), able to simulate saturation and time-dependency of material uptake. These biomimetic sorbents were tested for soil contaminants such as PAHs, PCP and PCBs and compared to earthworm and plant root accumulation (Table 7.2) and the contaminant's log $K_{ow}$.

*Table 7.2* Solid phase and membrane-based extraction methods compared to solvent extraction.

| Extraction method and device | Bioassay or chemical model for comparison | Contaminant matrix | Result | Reference |
|---|---|---|---|---|
| Butanol Cyclodextrin Tenax TA beads extractions | Earthworm and rye grass root accumulation | PAHs from soil | Butanol was the best for earthworm. All of them correlated well for rye grass roots | Gomez-Eyles et al., 2011 |
| POM-55 55 μm thick POM membrane extraction | log $K_{ow}$ | PCB in sediment and water | POM-55–water partition coefficient correlates with log $K_{ow}$ | Cornellissen et al., 2008, 2009 |
| TECAM TECAM extraction | $K_{ow}$ Earthworm accumulation | PAHs in two phase soil | Good correlation between TECAM–water partition and both $K_{ow}$ and earthworm accumulation | Tao et al., 2008, 2009 |
| SPMD extraction HPBCD extraction using hydroxypropyl-beta-cyclodextrin | Earthworm accumulation | PAH in soil | SPMD absorbs higher proportion of small PAH molecules than earthworms. HPBCD correlates well with earthworm accumulation. | Bergknut et al., 2007 |
| TECAM triolein-embedded cellulose acetate membrane | Carrot root accumulation | DDT contaminated soil | TECAM after 6 hours shows good correlation with carrot accumulation | Yang et al., 2010 |
| XAD-2 polystyrene non-ionic beads | $CaCl_2$, butanol, dichloromethane sequential extraction earthworm accumulation | PCP in soil | XAD-2 correlates with both, using soil contaminated for a long time. | Wen et al., 2009 |
| POM-55 55 μm thick polyoxymethylene membrane | Earthworm accumulation | PAH contaminated sediment | POM overestimates the bioaccumulation of PAHs compared with the bioassay | Barthe et al., 2008 |

PAH: polycyclic aromatic hydrocarbons; PCB: polychlorinated biphenyls;
DDT: dichlorodiphenyltrichloroethane; PCP: pentachlorophenol;
Tenax TA beads: porous polymer resin based on 2,6-diphenylene oxide.

Another artificial cell model is the semi-permeable membrane device (SPMD) consisting of an external LDPE (low-density polyethylene) layer which can be filled with triolein, isooctane, trimethylpentane or with an ionic liquid e.g. organic salts (Esteve-Turillas *et al.*, 2008). SPMDs are widely used for sampling air and water, but a more permanent external layer is required for soil. Table 7.2 shows a few SPMD applications together with other solid phase and membrane-based extraction methods using TECAM, triolein-embedded cellulose acetate (Figure 7.5); XAD, polystyrene, POM, or polyoxymethylene membranes.

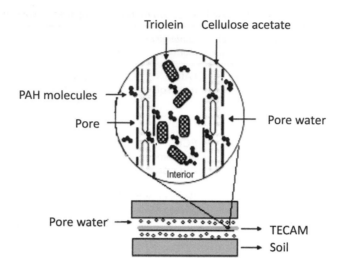

*Figure 7.5* Schematic diagram of TECAM accumulation of PAHs from soil (Tao *et al.*, 2008).

## 6.3 Liquid-phase extractions to predict accessibility of toxic metals

It has been accepted that neither the water-soluble, nor the total concentration of toxic metals in soil is in correlation with their accessibility and bioavailability and they are not suitable for the characterization of environmental risk. In addition, the chemical forms of toxic metals largely influence their toxicity, fate and transport, as well as partitioning between the soil phases. Various partial extraction methods such as extraction with salt solutions, acidic and alkaline solutes, chelating agents, etc. have been studied and routinely used to model toxic metal bioavailability. Extractions with solutes of various pH and ion concentrations were applied to mimic living organisms in interaction with the contaminated soil. The extractions are followed by chemical analysis, and the result is handled as the bioavailable fraction of the total metal content of the soil. Numerous studies proved that EDTA (ethylenediaminetetraacetic acid) is truly mimicking the metal accumulation by plants (Gupta & Sinha, 2007; Zhang *et al.*, 2010) and the metal uptake by benthic invertebrates (Tack & Verloo, 1995). The solution of ammonium lactate and acetic acid (LE) is also considered by some authors as a good imitator of plants (Lakanen & Erviö, 1971), but some others found that LE greatly overestimates plant uptake (Feigl, 2011). There is no generally suitable extractant, some solvents may correlate with certain plants in a certain soil for a certain element such as, in the case of phosphorous, 0.01 M $CaCl_2$ solution proved to be a true extractant for modeling plant-available phosphorous in soil (Hylander *et al.*, 1995). According to a more conservative approach, when simulation of a realistic worst case is the goal instead of true mimicking, other type of solvents such as diluted nitric acid can be used for environmental risk estimation.

An important conceptual question may arise when extracting contaminants from soil by different extractants if an environmental process is mimicked or a standardized

*Table 7.3* List of sequential multiple metal extraction methods and the separated fractions.

| Extraction method | Fractions | Reference |
|---|---|---|
| Tessier method | 1 Exchangeable<br>2 Associated with carbonates<br>3 Associated with Fe–Mn oxides<br>4 Associated with organic matter<br>5 Residual | Tessier et al., 1979 |
| Three-level extraction | 1 Mobile<br>2 Immobile<br>3 Pseudo total metal fraction | Gupta et al., 1996<br>Aten & Gupta, 1996 |
| BCR method | 1 Acid-extractable fraction<br>2 Reducible fraction<br>3 Oxidizable fraction<br>4 Residual | Geebelen, 2003 |
| Selective sequential extraction (SSE) procedure | 1 Water-soluble<br>2 Exchangeable<br>3 Acid-soluble<br>4 Bound to Mn oxides<br>5 Bound to amorphous Fe oxides<br>6 Bound to crystalline Fe oxides<br>7 Oxidizable fraction<br>8 Residual | Becquer et al., 2005 |
| Chemometric Identification and Substrate and Element Distribution (CISED) | Increasing strengths of simple mineral acids as the extractant, followed by chemometric data processing of the resulting multi-element data obtained from the extract analysis. | Cave et al., 2004<br>Wragg & Cave, 2012<br>Appleton et al., 2012 |
| Extracts approximating water availability, plant availability and total metal content | Parallel extraction with water<br>Artificial acid-rain acetate-buffer solution (pH = 4.0 or 4.5) ammonium acetate 0.5 M and EDTE (0.02 M) (pH = 4.65) *aqua regia* | Gruiz et al., 1998,<br>Gruiz, 2000<br>Feigl et al., 2009b |
| AVS and SEM procedure for anoxic sediments | Parallel determination of acid volatile sulfide (AVS) simultaneously extracted metals (SEM) SEM ($\mu$mol/g)/AVS ($\mu$mol/g) | US EPA, 1991 |
| Bioaccessibility of lead | Parallel determination of acidic glycine extractable and total extractable Pb and calculating their ratio | US EPA, 2012 |

chemical extraction proceeds. In the first case a geometrically valid leaching test is the best option with realistic infiltrating water quantity, no adjustment of pH and other conditions. In the second case the extractant should be buffered and standardized conditions are necessary during the extraction. The environmental trueness of such an extraction is very low, but it is useful for comparative studies.

Another widely used type of extraction is the multi-step sequential extraction (SME) method. The sequential extraction separates the metals into five fractions: exchangeable, bound to carbonates, bound to Fe-Mn oxides, bound to organic matter, and residual (Tessier *et al.*, 1979). Simplified methods of three or four sequential steps were also developed as introduced in Table 7.3. The extracts are generally analyzed by multi-element analysis methods such as ICP-AES (Inductively Coupled Plasma with Atomic Emission Spectroscopy), ICP-MS (Inductively Coupled Plasma with Mass Spectrometry) and XRF (X-ray fluorescence analysis). The results of SME enable the

*Table 7.4* In situ measurements and devices for the dynamic assessment of labile metal species.

| Method | Measured end point | Reference |
|---|---|---|
| *In situ* DGT device | Measuring the diffusive gradients in thin films | Davison et al., 1994; Zhang et al., 1995a and 1995b |
| Gel integrated microelectrode array (GIME) with voltammetric detection | Dynamic electroanalytical metal speciation | Buffle & Tercier, 2005 |
| Gel integrated stripping chronopotentiometry (GISCP) | Dynamic electroanalytical metal speciation | Town, 1998; van Leeuwen & Town, 2002; Town & van Leeuwen, 2004 |
| Hollow fiber permeation liquid membrane | | Buffle et al., 2000; Salaun & Buffle, 2004 |
| Donnan membrane technique | Can be used as equilibrium or as dynamic technique | Temminghoff et al., 2000; Weng et al., 2005; Kalis et al., 2006 |
| Competitive ligand exchange/ adsorption stripping voltammetry | Can be used as equilibrium or as dynamic technique | Xue & Sigg, 2002 |

estimation of mobility, bioaccessibility and leaching from soil's solid to soil's water phase and, to some extent, toxicity.

Anoxic sediments may contain a large amount of toxic metals in a non-mobile chemical form. As sulfide controls the mobility of metals in anoxic sediments, AVS, the acid volatile sulfide content of sediments ($H_2S$ is released when treated with strong acid) is important information on the amount of the sulfide-form metals. The relative amount of SEM (SEM = simultaneously extracted metals, which are solubilized in the acid-treatment step: Cd, Cu, Hg, Ni, Pb, and Zn) and AVS can be used for the prediction of metal bioavailability. If the molar ratio of SEM for bivalent metal ions to AVS exceeds one ($>1$), the metals in the sample are potentially bioavailable.

The US EPA (2012) method is typically applicable for the characterization of lead bioaccessibility in soil. It is intended to be used as reference for developing site-specific Quality Assurance Project Plans (QAPPs) and Sampling and Analysis Plans (SAPs). The amounts of lead in 0.4 molar glycine (pH = 1.5) extract and in the so-called total extract (both received from the same amount of soil) are compared to each other and it is called the '*in vitro* bioaccessibility' of Pb in the soil or solid waste:

$$\text{Lead } in \ vitro \text{ bioaccessibility} = Pb_{ext} \times V_{ext} \times 100/Pb_{total} \times M_{soil},$$

where

- $Pb_{ext}$ – lead concentration in the acidic glycine extract;
- $V_{ext}$ – volume of the extract;
- $Pb_{total}$ – Pb concentration in the solid sample;
- $M_{soil}$ – mass of the soil sample analyzed.

The *in situ* DGT device (Table 7.4), based on measuring the diffusive gradients in thin films was developed by Davison and Zhang between 1990 and 1995 (Davison *et al.*, 1994; Zhang *et al.*, 1995a; 1995b). It has become a popular tool in the last ten years for measuring 'labile' metals in water, sediments and soils (Han *et al.*, 2013).

In addition to ready-for-use measurement cells, separate components allowing self-assembly are supplied. The plastic device with a 2.0 cm diameter window is loaded with a three-layer system of i) a restricted pore size resin-impregnated hydrogel; ii) an open pore diffusive gel layer (polyacrylamide) and iii) a membrane filter contacting the environmental matrix. Specific gels are available for metals (Chelex gel), for phosphorus (Fe oxide gel), for cesium and for sulfides. It can be applied for *in situ* measurements (single measurement or time averaged concentrations), for measuring fluxes in sediments and soils, for the determination of kinetic and thermodynamic constants. Its pH tolerance is limited: it can work in the range of pH = 5–9.

DGT directly measures the mean flux of mobile/labile species (including free and kinetically labile metal species) to the device during the deployment. Where supply from sediment particles to solution is rapid, the interfacial concentration is the same as the concentration of metal in water or bulk pore water. For a given device and deployment time, the interfacial concentration can be related directly to the concentration of labile metals (Zhang *et al.*, 2001 and Zhang & Davison, 2001).

In the interpretation of DGT Research (2014) this concentration represents the supply of metal to any sink, be it DGT or an organism that comes from both diffusion in solution and release from the solid phase. It is proven by comparative studies that this concentration is suitable to approximate bioaccessibility, so it can be a simple surrogate measurement of the biologically potentially effective concentration (Søndergaard *et al.*, 2014). It supports well the presently accepted Free-Ion Activity Model (FIAM) which stipulates that the biological response of organisms to metals in water-based systems is proportional to the free-ion activity of the metals and not their total or dissolved concentrations (Zhang *et al.*, 2002). DGT–bioassay correlation experiments conducted demonstrate clearly that DGT provides a more representative measure of metal bioavailability, and hence toxicity, than conventional water quality parameters (i.e., total and/or dissolved metal concentrations). Specifically, the results indicate that metal uptake by DGT and aquatic biota (e.g., *Daphnia magna* and rainbow trout) are both reduced in the presence of metal-complexing ligands (INAP, 2002; Senila *et al.*, 2012). It means that metal-ligand complexes that are unavailable to aquatic biota are also undetected by DGT.

The theoretical background of the TGD method to using DGT in sediments and saturated soils can be found in the publications of Davison *et al.* (2000) and Zhang *et al.* (2001). A detailed description on the application and the mode of calculations of the DGT measured concentrations, the bioavailable concentration in the liquid medium and kinetic parameters is also written in the DGT handbook in details (DGT Research, 2014). The DGT measured concentration can be calculated as shown by the equation below:

$$C_{DGT} = M \times \Delta g / (D \times t \times A),$$

where
- $C_{DGT}$ – DGT measured concentration;
- $M$ – metal mass accumulated in the resin gel;
- $\Delta g$ – thickness of the diffusive gel;
- $D$ – diffusion coefficient of the metal in the gel;
- $t$ – time of depletion;
- $A$ – exposure area.

The real concentration in sediment or soil pore water depends on the rate of resupply (R) of solutes to the pore water. The resupply can be determined as the ratio of $C_{DTG}$ to C, the real concentration of labile metal forms in the pore water. R is the measure of the dynamic ability of the solid phase to resupply the pore water in response to the depletion induced by the DGT sink.

The *in situ* electroanalytical techniques combined with gel- and membrane-separation shown in Table 7.4 are mostly based on the application of microelectrodes immersed into gels or combined with membranes. The gels and membranes separate free metals from a complex matrix, which detect the electrochemically active forms, i.e. the charged metal ions from the electrodes. The electrodes should be calibrated before their *in situ* environmental application. Sigg *et al.* (2006) published a comparative study on six electrochemical methods applied to natural surface waters. They concluded that the free ion activity model, which makes the free ions responsible for biological activities and adverse effects, is a highly simplified model of a dynamic and complex situation, and the relative time scales and the rates of complex dissociations should also be taken into consideration, even in the case of water. The fast development in electroanalytical methods hopefully makes this tool suitable to contribute to the necessary dynamic tool battery, which will be able to produce reliable predictors for bioavailability, biological uptake and ecotoxicological risk in natural waters, sediments and soils.

## 7   COMPLEX MODELS

In environmental management, bioavailability and bioaccessibility must be determined when the actual risk posed by hazardous contaminants present in the environment (soil/sediment) should be evaluated or model experiments and field implementation of a remediation technology should be designed and monitored. One has to consider in both cases that the dynamic interaction of contaminated soil and the microbiota may change contaminant bioavailability. The detection of the interactions between contaminated soil and soil-living organisms require interactive tests, dynamic microcosms and their integrated monitoring using physico-chemical, biological and toxicological methods. Microcosms established for testing bioavailability and changes in bioavailability should model realistic average or worst-case scenarios. A laboratory lysimeter, for example, can be irrigated i) with average precipitation, ii) with an amount of precipitation representing the rainy season, or iii) with an artificially prepared acidic precipitation, simulating a worst case from the point of view of e.g. Zn and Cd mobilization.

### 7.1   Interactive laboratory tests

Interactive soil and sediment tests ensure direct contact between contaminated soil/sediment and test organisms, thus simulating a real environmental situation. Generic tests required by legislation to assess the hazards posed by chemicals create a standard test environment, while problem-specific tests intend to model site-specific transport and land-use-specific exposure scenarios. Testing the biological effects of contaminated soil can, if necessary, be achieved by performing chemical analysis of the concentration and composition of soil contaminants.

Most of the interactive tests measuring either toxicity, biodegradation or bioaccumulation as an end point are in strong relationship with bioavailability. The effect can only be measured when the contaminant is really available to the test organism: an interaction is established between the test organism and the contaminant and the contaminant is adsorbed by the test organism.

The following direct-contact laboratory tests can be used to characterize contaminant bioavailability in general and in real contaminated soil. The results of direct-contact test methods integrate the degree of mobility, bioavailability and the consequent effects of the contaminated soil:

- Biodegradation of a chemical substance mixed into a real soil with indigenous microflora in closed-bottle or aerated microcosms;
- Ongoing biodegradation of contaminants in a contaminated soil sample using indigenous soil microflora in closed-bottle or aerated microcosms;
- Bioleaching of the contaminant from contaminated soil using soil indigenous microflora in laboratory lysimeters;
- Bioaccumulation of the contaminants from contaminated environmental samples (soil, sediment, waste) using test plants or earthworms: pot experiments;
- Measuring acute and chronic toxicity of contaminated soil, sediment, waste or slurries by direct-contact bioassays using bacterial, fungal, animal or plant test organisms;
- Acute and chronic toxicity of contaminated soil extracts and leachates: liquid-phase bioassays using bacterial, fungal, animal or plant test organisms;
- Mutagenicity of contaminated soil, sediment, solid waste direct-contact mutagenicity assay using bacterial test organisms;
- Any other adverse effects (reprotoxicity, endocrine disruption, etc.) of the contaminated environmental samples.

In addition to laboratory tests, field ecosystem assessment, bioaccumulation of field-grown organisms, gene expression profile of the soil microbiota, and many other indicators of biodegradation, bioaccumulation or toxicity can be applied as an end point for the demonstration of the bioavailability of the contaminant occurring in the real environment.

## 7.2 Dynamic testing

A single measurement of a contaminant's mobility and bioavailability in a soil sample yields static information about one time point. However, no data are available about the future rate and direction of changes or any response to external e.g. environmental or climatic impacts. To obtain information on all these dynamic parameters, the microcosm must be monitored: sampling the microcosm regularly, analyzing the samples by parallel chemical analyses and biotesting, evaluating and plotting the results as a function of time as illustrated by the microcosm biodegradation example (Figure 7.6). The three cases shown in the graph are:

1. the contaminant's mobilization rate is equal to the biodegradation rate;
2. the mobilization rate is greater than the biodegradation rate;

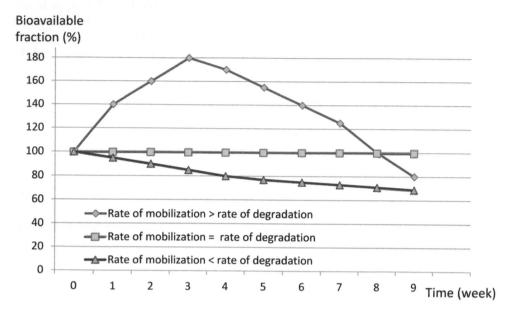

*Figure 7.6* Change in bioavailability in the initial phase of biodegradation. 100% = initially bioavailable amount.

3.  the mobilization rate is lower than the biodegradation rate. In the latter case (red line), bioavailability is the limiting factor of biodegradation, and the limitation can be eliminated by enhancing bioavailability.

    The impact of changes in external parameters on the contaminant's mobility and bioavailability in soil can dynamically be tested in several ways. i) When the starting point is an equilibrium-state environmental soil sample under realistic/average environmental conditions, the study may apply significantly different test conditions and monitor the behavior of the soil in the microcosm until it reaches the new equilibrium. ii) If the soil is not in equilibrium, first a controlled steady state has to be reached in the batch or flow-through soil microcosm. After the steady state has been achieved, the environmental parameters (temperature, pH, redox potential and humidity/water content, etc.) should be changed at a realistic or a slightly accelerated rate. The changes are monitored until the soil adapts to the new environmental conditions and reaches a new steady state which is different from the original one (Figure 7.7). The reaction of the soil to the changed environmental conditions can be:

1.  increased (blue line) or decreased (not shown in the graph) mobility and bioavailability, and a new steady state;
2.  temporary increase in mobility and bioavailability, then return to the original or a slightly different new steady state;

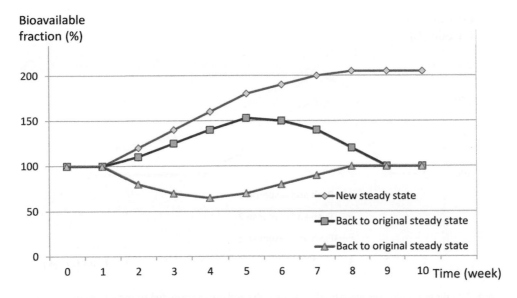

*Figure 7.7*  Bioavailability testing: typical changes in the soil microcosm between two different steady states.

3.  temporary decrease in bioavailability, then return to the original or a slightly different new steady state.

A microcosm or a field test can be made dynamic by stepwise adjusting new environmental conditions or applying a one-off, impulse-like impact, a push. The push can be changing pH, irrigation with 'acid rain', adding nutrients, injecting solubilizing agents, introducing more and special microorganisms, suddenly increasing temperature, or changing the redox potential by aeration, oxygen removal or injection of electron-donor molecules, etc. The microcosms should be in a steady state before the application of the push, and the response of the soil to the impulse should be monitored, e.g. by measuring the changes in the chemical forms, concentrations, partitions, mobility and bioavailability of the contaminants, respiration rate, metabolites, activity of the test organisms (Figure 7.8). The types of responses from the point of view of the contaminant's mobility and bioavailability are:

1.  Sudden increase or decrease (not shown in the graph) in mobility and bioavailability of the contaminant in soil and after a fast response, then reverting to the original steady state;
2.  Slow increase and very slow attenuation, returning slowly or not returning at all to the original steady state;
3.  Sudden increase, then slow decrease and return to a new steady state.

The effect of impulse-like technological interventions on the soil, on the contaminant and on the soil microbiota can be observed in the field by measuring air, water and

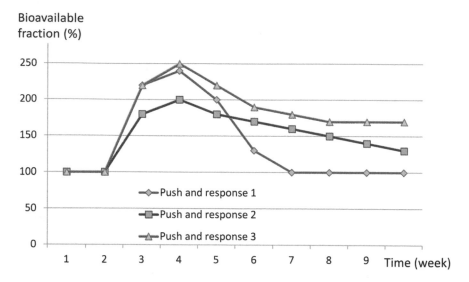

*Figure 7.8* Bioavailability testing: typical changes in the soil microcosm due to the effects of an impulse-like impact.

contaminant flows, contaminant concentrations in soil gas or groundwater, ion concentrations in soil water and soil moisture, and conductivity of soil water and soil moisture, or measuring the products of respiration and the biodegradation of microorganisms. The direct-push soil analysis technologies (see Volume 3) allow targeted studies also in deep soil layers (in 30 m or even 150 m depths).

## 7.3   Integrated evaluation

Both the interactive and dynamic tests should be evaluated by measuring the effect on single or multiple test organisms or by observing the process in a microcosm using integrated physico-chemical and biological monitoring: measuring the effects (lethality, biological and biochemical activity, changes in diversity, etc.) and chemical concentrations as a function of time. The physico-chemical, biological and ecotoxicity results are evaluated together in relation to each other. The consistency or the rate and type of inconsistency between physico-chemical and biological–ecotoxicological data give extensive information about the character of the risk, e.g. the biological availability of the contaminant; the presence of a chemical time bomb; the interaction between single contaminants (synergism or antagonism) and draws our attention to chemically not measured or non-measurable, but still dangerous components via their adverse effects (Gruiz *et al.*, 1999; Gruiz, 2005).

The integrated test methodology for contaminated soil has been called by the authors Soil Testing Triad (STT). It contains physico-chemical analytical, biological/ecological assessment methods and toxicity tests (Figure 7.9).

Physico-chemical analysis, biological (ecological) and toxicity tests are of similar significance. The methods are complementary. They provide information about the

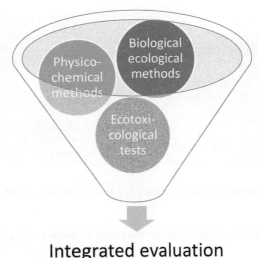

*Figure 7.9* Composition of the Soil Testing Triad (STT).

quality and quantity of the contaminant; the characteristics and the biological status of the soil; the activity, vitality and adaptive behavior of the soil-ecosystem; the effects, mobility, bioavailability and biodegradability of the contaminant; and the response of the soil to external effects.

The target of physico-chemical analyses may be the soil, the contaminant or the members of the ecosystem. The biological and ecological characteristics are used to determine the diversity and quality of the ecosystem and the dynamic nature of soil, e.g. its response to certain natural or provoked external effects. The answer gives information about the buffering capacity and the adaptive behavior of the soil and its biological system. The third party in the triad is toxicity testing: it provides information about the adverse effects of the contaminant or the contaminated soil, and thus the hazards posed to non-adapted ecosystems or humans. The STT plays a central role in site assessment, environmental risk assessment and environmental monitoring. Environmental toxicity testing indicates the actual adverse effect that is directly associated with the environmental risk posed by the soil. This value often differs from the results of chemical analyses.

## 8   EXAMPLES OF INTERACTIVE TESTING OF BIOAVAILABILITY IN SOIL

Simultaneously to chemical analysis, soil samples polluted with inorganic and organic contaminants have been directly tested using different test organisms to find the bioavailable portion. This was compared to the metal content which is available to

*Figure 7.10*   Grey flotation tailings covered by a thin soil layer: extensive water erosion.

*Table 7.5*  Toxicity of flotation tailings and cover soil tested by bacteria and plants.

| Test method | Azotobacter agile dehydrogenase | Sinapis alba root and shoot growth | Vibrio fischeri luminescence |
|---|---|---|---|
| Black soil layer | Very toxic | Toxic | Very toxic |
| Grey flotation tailings | Non-toxic | Slightly toxic | Non-toxic |

chemical extractants. A comparative evaluation of the results of physico-chemical analyses and toxicity tests enables conclusions on bioavailability, which can be used to calculate the risk posed by the polluted soil.

## 8.1   Toxic metal bioavailability in mine tailings – the chemical time bomb

A flotation tailings dump at an abandoned mining site in Hungary has been covered by a thin soil layer without isolating the waste from the soil to accelerate plant growth and improve the aesthetic value of the 4-million-tonne deposit located near a village. The pictures in Figure 7.10 were taken five years after placing the cover. The cover and the tailings were eroded by wind and water and, as Table 7.5 indicates, the soil became highly toxic. The soil cover could not fulfill its role: erosion, leaching and acidification as well as metal uptake by plants were in progress (Gruiz, 2000; Dobler *et al.*, 2001).

The metal content of the soil and of the tailings layers was analyzed using ICP–AES after *aqua regia* extraction. The results of the 'total metal content' were in sharp contrast to the results of toxicity. When artificial rainwater was used for extraction, it was found that the metal content of the tailings was not extractable by water (Table 7.6).

In spite of the relatively low metal content of the soil cover (two- to three-fold of the soil quality criteria) the metals were extractable by water and available to the test organisms. The metal content of the soil cover was water soluble, therefore the metals were taken up and easily transported into the food chain. Low water permeability of the tailings and the acidic character of the top layer resulted in continuous acidic

Table 7.6  Comparison of the total and the water-soluble metal contents in soil and tailings.

| Layer | pH | Total metal content (mg/kg) | | | Rain-water-soluble metal content (mg/kg) | | |
|---|---|---|---|---|---|---|---|
| | | Zn | Pb | Cu | Zn | Pb | Cu |
| Black soil layer | 4.7–5.2 | 603 | 186 | 72 | 42.2 | 1.9 | 0.5 |
| Grey flotation tailings | 7.0–8.0 | 31,858 | 4,971 | 2,450 | 3.4 | 1.2 | 0.6 |

Table 7.7  Metal content of some plants on the tailings dump.

| Species | Zn (mg/kg dry plant) | Cu (mg/kg dry plant) | Cd (mg/kg dry plant) |
|---|---|---|---|
| Achillea millefolium | 255.4 | 17.0 | 2.4 |
| Agrostis sp. | 409.8 | 31.9 | 6.3 |
| Carex sp. | 354.6 | 55.0 | 3.0 |
| Echium vulgare | 607.5 | 45.3 | 5.0 |
| Phalaris canadiensis | 144.9 | 4.1 | 0.5 |
| Phragmites australis | 767.5 | 40.5 | 0.7 |
| Populus sp. | 1158.5 | 12.9 | 19.5 |
| Silene alba | 693.5 | 50.4 | 2.6 |
| Silene vulgaris | 505.5 | 20.7 | 4.6 |
| Tussilago falifara | 568.5 | 38.9 | 8.8 |
| Maximum limit for forage plants*/vegetable** | 150–200/20 | 15–50/10 | 1.0/0.5 |

*17/1999 EüM; **44/2003 FVM (Hungarian regulations).

extraction of the metals from the tailings and their capillary transport into the soil cover, i.e. the habitat of the local ecosystem.

The plants that grow on the surface of the soil overlaying the tailings are highly contaminated with accumulated metals (Table 7.7) (Dobler et al., 2001).

Flotation tailings represent a typical chemical time bomb: their metal content is non-available until precipitation, acid rain, acidic infiltrates and plant root exudates are not in contact with the waste. When metals are extracted, leached or otherwise mobilized, they are transported by capillary forces into the top soil layer. Tailings in direct contact with the soil layer represent an infinite contaminant source. The long-term risk and the chemical time-bomb fate of these materials can only be confirmed by the integrated evaluation of the results of chemical analyses and toxicity tests. If this seemingly neutral waste material (no short-term toxicity was measured and the water extract showed low contamination due to the high pH) is diffusely dispersed by erosion, the sudden increase in its specific surface area accelerates weathering, acidification, leaching and mobilization of its toxic metal content.

## 8.2  Decreased bioavailability, lower toxicity – a soil remediation tool

Availability of toxic metals to water and plants is responsible both for the short-term risks to aquatic and terrestrial ecosystem, as well as for the long-term human

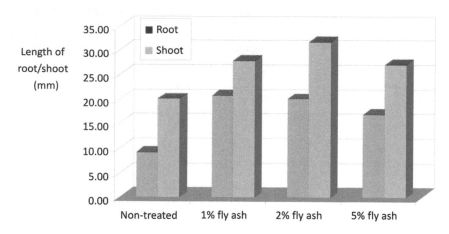

*Figure 7.11*   Root and shoot elongation of *Sinapis alba* in metal-contaminated soil with and without chemical stabilization by fly ash.

and ecological risks posed by contaminated soil through direct exposure and food chains. Weathering and leaching of the tailings material further increases the risk due to increased bioavailability of toxic metals. Mobility and bioavailability of toxic metals should be decreased to reduce the risk. Availability of metals in soil can be reduced by increasing their sorption rate, strengthening their bonds or conversion into stable chemical forms: e.g. from ionic to metal-hydroxide, -oxide or -silicate compounds. These transformations are controlled by chemical and mineralogical processes resulting new mineral-formation over the long term, considered the opposite of weathering.

Figures 7.11 and 7.12 illustrate the measured plant toxicity and plant-uptake results of contaminated soil before and after chemical stabilization by fly ash to decrease mobility and bioavailability of toxic metals (Feigl *et al.*, 2007, 2009b). Plant toxicity decreased significantly: root growth increased by 50% and shoot growth by 150%. Metal uptake of the plants decreased from $Zn = 740$ mg/kg and $Cd = 3$ mg/kg plant by up to 70%.

The total metal content of the soil with and without stabilization is the same: the fly-ash treatment has not reduced the total concentration but only the bioavailable fraction.

Water and ammonium acetate ($pH = 4.5$) extracts were analyzed using ICP–AES to obtain chemical data about the changes. The toxic metal contents of water and acetate extracts are shown in Figures 7.13 and 7.14.

Water can extract 180 mg/kg Zn from the soil, while acetate solution extracted 300 mg/kg Zn from the same soil, which contains a 'total' of 1800 mg/kg Zn concentration (extracted by *aqua regia*). Plants (*Sinapis alba*) 'extracted' 120 mg Zn per kg wet plant, which is equivalent to 740 mg/kg dry plant biomass.

Fly ash (2% and 5%) reduced Zn concentration in the water-extract from 180 to 10 and ~0 mg/kg whereas Cd concentration decreased from 1.1 to 0.2 and 0 mg/kg, respectively. The plant-accumulated amount in leaves decreased from 740 mg/kg to

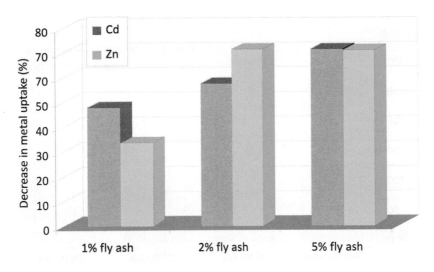

Figure 7.12 Decrease in metal uptake by plants (*Sinapis alba*) after soil stabilization with fly ash.

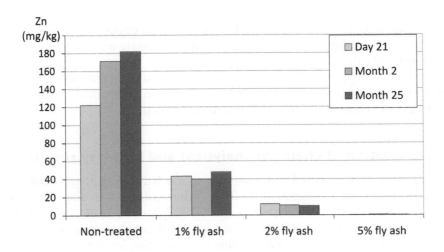

Figure 7.13 Zinc concentration in the water extract of the soil before and after stabilization.

220 mg/kg plant dry biomass. Plant toxicity (*Sinapis alba*) decreased to only 50% of the initial value. The decrease in toxicity is closer to the metal concentration reduction in the acetate extract (30–50%) than to the water extract (approx. 90%). The availability of the metals to plants is higher than to water due to acidic root exudates mobilizing metals locally. A comparison of the decrease in plant-accumulated and acetate-extracted amounts shows a maximum 50% decrease in the latter in contrast to 70% decrease in plant bioaccumulation.

The results of this bioavailability mitigation experiment showed that acetate extraction slightly overestimates the risk compared to plant bioassays. The rapid plant

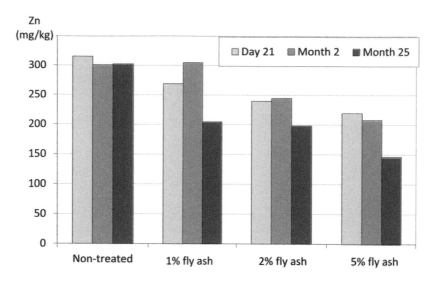

*Figure 7.14*  Zinc concentration of ammonium acetate extract before and after soil stabilization.

accumulation test, developed by the authors, is in itself a pessimistic biological model (using a type of seedling that tends to concentrate toxins), which overestimates the true metal uptake by plants.

It is worth noting that a moderate overestimate fits well a conservative risk assessment approach, and that the best possible set of tests for assessment and monitoring requires a wide selection of biological and chemical methods.

## 8.3  Correlation of chemical analytical and bioassay results

River sediments contaminated with mine waste have led to high toxic metal contents in the soil of flooded allotments in the Toka Valley, a former mining area in northern Hungary (Feigl *et al.*, 2009a; Gruiz *et al.*, 2009; Vaszita *et al.*, 2009). Chemical analysis of soils and wastes indicated that high metal content often failed to lead to toxicity (cf. Tables 7.4 and 7.5) and sometimes the opposite, i.e. low concentration was accompanied by high toxicity. The authors carried out several parallel assessments and comparisons between the results of toxicity bioassay and chemical analysis of different extracts from metal-contaminated soils. The results demonstrated that the flooded gardens in the relatively small Toka Valley show large variations and different contamination patterns. The results of chemical analyses and toxicity tests show a range between very good correlation and no correlation at all, depending on the type, origin, age and the proportion of the contaminating material and the characteristics of the recipient plot and its soil.

Figures 7.15 and 7.16 illustrate two extreme cases. Figure 7.15 shows the test results of soil samples from an allotment (Plot 1) homogeneously polluted by a regular (annual) high water level. The flooded soil is morphologically homogeneous, and plant toxicity and chemically measured metal concentration from an acetate buffer

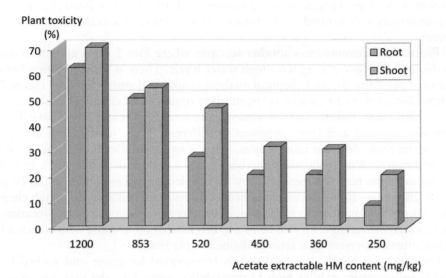

*Figure 7.15*  Plant toxicity is proportional to the concentration of the acetate-extracted toxic metals. Plot I is exposed to regular annual floods (HM: sum of all extractable metals As, Cu, Cd, Hg, Pb and Zn).

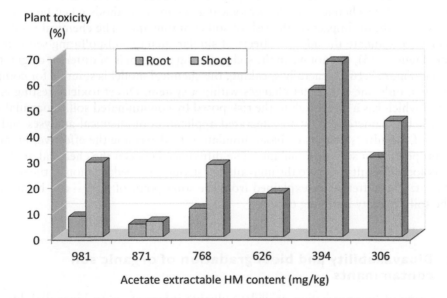

*Figure 7.16*  Plant toxicity is not proportional to the chemically measured mobile metal concentration. Plot 2 was exposed to sudden runoffs (HM: sum of all extractable metals As, Cu, Cd, Hg, Pb and Zn).

correlate well. The gradient visibly decreases with the distance from the creek. The proportionality is illustrated by the toxicity as a function of acetate-extractable metal content.

Figure 7.16 demonstrates another scenario where Plot 2 was exposed to sudden runoffs and subsequent long-term high water levels. There is no correlation between measured plant toxicity and chemical analysis results as demonstrated by Figure 7.16. A clear discrepancy can be seen between plant toxicity and chemical analytical data. The sudden heavy runoffs and subsequent long-term high water levels have disposed of eroded material and creek sediment of different origin (eroded soil and parent rock, waste rock, different mine wastes, other waste from illegal disposals upstream, etc.) and morphology (yellowish, grayish, stone-like, flocculent-like, etc.). In spite of the contradiction between metal content and toxicity, good explanation can be given for the differences found between the sediment-polluted soil samples. High chemical content and low plant toxicity indicate non-weathered waste rock or ore. Flotation tailings – as seen in Section 8.1 – also belong to this kind of mine waste with non-available metal content, representing a serious chemical time bomb.

On the other hand, dissolved metals transported by water and sorbed by the flooded soil particles exhibit high bioavailability, similar to the very fine rock and mine waste which show advanced weathering. Former anoxic bed sediments, with stable metal forms are easily transformed into labile, e.g. water-soluble ionic forms by chemical and microbiological oxidation under oxidative conditions once they reach the surface. This is why soil samples with low metal concentration may exhibit high toxicity and the opposite: low toxicity with high metal content.

The examples of the two flooded plots demonstrate the necessity of integrated application of the chemical and toxicological assessment methods when bioavailability has a significant impact on the risk of soil contaminants. The chemical results are sufficient to estimate the adverse effects of similar quality soil-polluting sediment or waste (Figure 7.15), but not when the soil-polluting material is of different origin, type and age (Figure 7.16). Generally speaking, the chemical model is suitable for comparison when only one parameter changes within a system. Direct toxicity testing is the method which has a direct link to the risk posed by contaminated soil. Bioavailability can only be characterized by an integrated application of chemical analyses and the testing of the adverse effects or bioaccumulation to determine the effective, bioavailable proportion of soil contaminants. The difference between the chemical analytical and biological results refers to the unavailable, strongly sorbed fraction of the soil contaminant. This difference – explained from the soil's point of view – can be assigned to the soil's toxicity buffering capacity.

## 8.4   Bioavailability and biodegradation of organic soil contaminants

Monitoring of bioremediation provides valuable information on bioavailability and biodegradation when chemical analyses and biological tests are applied in an integrated way. Only bioavailable contaminants can be biodegraded, but bioavailability and biodegradation are hard to distinguish. It is evident that the biodegraded fraction of the contaminant has become biologically available before the degrading attack by the microorganisms. But the opposite is not necessarily true: not all bioavailable contaminants are biodegraded. In addition, if the contaminant is biodegradable and only moderately bioavailable, limited bioavailability will in general be the bottleneck in the biodegradation-based soil treatment process. In this case, mobilization of the contaminants using solubilizing, mobilizing or bioavailability-enhancing agents provides the solution. Enhanced biodegradation confirms the effect of bioavailability-enhancing agents (Volumes 4 and 5).

Multiple interactions can make contaminants bioavailable in soil: the initial bioavailability – especially if it is not high enough to fulfill the substrate need of the soil microbiota (waste, wastewater, etc.) – can be increased microbiologically. Soil microorganisms can utilize the less available substrates such as litter and large organic molecules. Making these large molecules bioavailable is part of their metabolic activity and the necessary enzymes, biotensides, complexing agents are part of their tool battery. A complex community of microorganisms, specialized in 'common dining', so-called commensalism, has the task to mobilize organic molecules and make them bioavailable in soil. Under normal circumstances the process of producing bioavailable substrates and biodegrading them is harmonized to a great extent. However, in the case of a disturbed environment e.g. when unnatural substrates, i.e. xenobiotics or mixtures of various persistent organic pollutants are present, the natural harmonization within the soil community is not always successful.

In the following examples the temporary accumulation of the mobilized, but not yet degraded portion of the soil pollutant mixture will be detected by toxicity tests. Mobilization of substrates or biodegradable contaminants is a natural process, but it often does not take place in contaminated soils; therefore many soil remediation technologies apply mobilizing agents.

Figure 7.17 shows the typical trend in soil toxicity for transformer oil, a moderately biodegradable hydrocarbon mixture. Soil samples from the field were artificially contaminated with 30,000 mg/kg transformer oil and aged for two months, then studied in biodegradability studies in laboratory microcosms. Two different types, a sandy and a clayey soil were biovented (R0 = biovented only) and treated by the chemical mobilizing agent RAMEB, randomly methylated cyclodextrin (R1 = biovented and treated by RAMEB). Toxicity was measured by bacterial, protozoan, insect and plant test organisms. Figure 7.7 shows the bioluminescence inhibition of *Vibrio fischeri* during the study. The toxic equivalency (TEQ) is given in 4CP (see Section 7.9): the risk is acceptable under TEQ = 10 mg/kg, while a TEQ less than 5 mg/kg represents 'no risk'. The sandy soil's initial toxicity is low and the clayey soil is not toxic at all in spite of their contaminant content. After the beginning of remediation, the contaminant's mobility and bioavailability increased, as indicated by the extremely high toxicity after a one-week treatment. The contaminant must become available to the microorganisms for it is the prerequisite of biodegradation. This can be achieved spontaneously (activating mobilizing microorganisms) or artificially (applying mobilizing agents).

In the example shown, the concentration and toxicity decreased proportionally in both soils during biodegradation. Starting from the second week, the decrease in toxicity was in agreement with the chemical results but the sudden increase in the first week was hardly (sandy soil) or not detectable (clayey soil) using gas-chromatography. If the contaminant consists of a wide range of compounds such as in the case of mineral oil products, mobilization–degradation may occur in several consecutive steps.

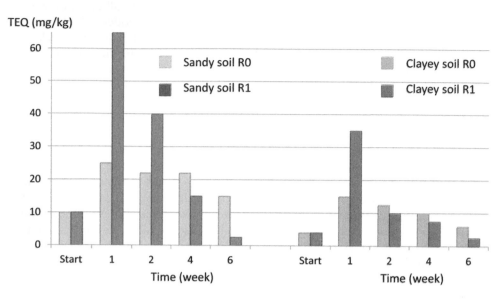

TEQ (mg/kg)

*Figure 7.17* Increased toxicity caused by the soil microbiota in the initial phase of biodegradation.

Figure 7.17 shows the first key step of the bioremediation model experiment on the soils contaminated with transformer oil.

Comparing the sandy and clayey soils, Figure 7.17 demonstrates well that the pollutant is less strongly sorbed and easier to mobilize in sandy soil (higher toxicity) while clayey soil is capable to buffer toxicity. As this clayey soil was a more suitable habitat for the degrading microorganisms than the sandy one, cell numbers were higher and the contaminant content lower after six weeks: remaining hydrocarbon content is 15,000 mg/kg in the sandy and 9,000 mg/kg in the clayey soil. A field demonstration of the same technology in a $10 \times 20 \times 5$ m soil volume resulted in 90% contaminant depletion, and no toxicity after 10 months of treatment.

Figure 7.18 shows another example, the changes in toxicity during a slurry phase bioremediation of an aged coal-tar polluted soil-like waste. Bioavailability and biodegradability of coal tar is known to be poor, but the results shown here do not support this belief. In the experiments presented, the hexane-acetone extractable and gravimetrically determined initial hydrocarbon content (total extractable: EPH) was 20,000 mg/kg and the gaschromatographically measurable content (GPH) only 8,000 mg/kg. The treatment covered aeration (M0 = only aeration) and the application of a self-made microbial inoculant to enhance mobilization and biodegradation (M1 = aeration and inoculation). According to the gaschromatographic analyses, about 50% of GPH was eliminated in 6 weeks and 80% in 10 weeks treatment. Toxicity shows a different picture: from the initial medium scale toxicity, it increases until the 6th week, reaching extremely high toxicity (measured by the highly sensitive marine luminobacterium), then it starts to decrease, until reaching a close-to-acceptable level of TEQ = 15 without a microbial inoculant and TEQ = 10 with the inoculant. Increased

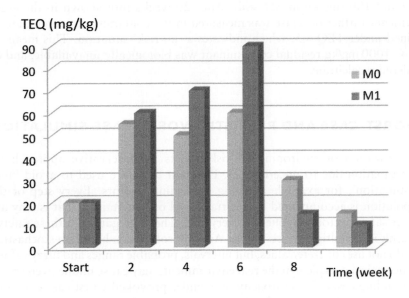

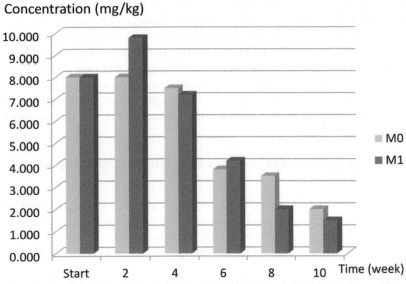

*Figure 7.18* Toxicity (TEQ) and concentration (EPH), in bioremediation of coal-tar contaminated soil.

biological activity in itself goes parallel to the increase of biological availability of relatively stable contaminants in an active soil. Certain members of the community produce biotensides and complexing agents which help all the other community members acquire the strongly sorbed contaminants (and other nutrients). The highest increase in toxicity was in the beginning, and in the presence of the inoculant. The residue, in spite of its low toxicity, was still present in a high concentration (GPH: 1800 mg/kg in M0 and 1000 mg/kg in M1 soil). After 20 weeks (not shown in the graphs), an insignificant further decrease was measured in the contaminant content, and toxicity remained under TEQ = 5, which indicates a 'no risk' situation. This means that the approx. 1000 mg/kg residual contaminant was biologically unavailable, and could be considered recalcitrant.

## 9 WORST-CASE AND REALISTIC WORST-CASE SIMULATION

Characterization of environmental risk adopts a conservative approach to avoid underestimation due to uncertainties. Different tools are used to avoid 'intentional overestimation', for example applying uncertainty factors. Every step of the serial extrapolation is accompanied by a certain level of uncertainty, that is why an uncertainty or safety factor is applied at every step. The aggregation of these safety factors may result in a multiple overestimate. Another solution is based on stochastic models which do not use concrete values, but intervals, probable ranges and fuzzy values. Complex uncertainties, typical in the real environment, may cause undue overestimates and corresponding costs. Direct toxicity tests under provoked worst-case conditions may produce results which fulfill the data requirements of a conservative risk assessment and can be used directly for decision making without risk posed by underestimation or failing to detect the problem (e.g. failing to include the toxic contaminants into the analytical program-based model). The application of pessimistic biological models will show – even in the most overestimated scenario – less risk than the chemical model (e.g. based on hexane-acetone extracted contaminant concentration). Chemical models cannot measure the biologically unavailable portion and cannot characterize the dynamic changes based on the interaction between the contaminants and the organisms.

### 9.1 Realistic worst-case models for dynamic testing of bioavailability

The difference between the actual and potential adverse effects, the short-term and long-term risks and the potential under- or overestimation of environmental risk posed by contaminants in soil, sediment or solid waste is in close relation with the bioavailability of contaminants and the possible changes in their mobility and availability.

Organic contaminants' bioavailability is in close correlation with their $K_{ow}$ value and with the corresponding partition among soil physical phases or their solubility in pore water. The interaction with the soil microbiota is an equally important influencing factor. Toxic metals' bioavailability depends on the chemical form of the metal, the pH, composition, sorption capacity of the soil and on the effect of plant root exudates and other biological agents. A pessimistic, also called realistic worst-case test scenario, changes the influencing factors to reach the worst case in the still living and actively working soil. One can induce mobilization, increased bioavailability of the contaminants before or during the test by lowering the $K_{ow}$ with tensides, cosolvents or complexing agents, and increasing or decreasing the pH. One may apply a series of tests with an increasing level of provoked changes, measuring the trend of the probable

*Table 7.8* Toxicity results of pessimistic models for toxic metal contaminated soil.

| Test scenario | Soil | No additive | Lime additive | Acid rain | EDTA addition | EDTA + acidic rain |
|---|---|---|---|---|---|---|
| Growth inhibition of | Non-contaminated soil | 0 | 9 | 9 | 0 | 0 |
| *Tetrahymena pyriformis* (%) | Soil contaminated with a mixture of toxic metals | 45 | 97 | 88 | 0 | 76 |
| Luminescence inhibition of | Non-contaminated soil | 0 | 0 | 0 | 0 | 30 |
| *Vibrio fischeri* (%) | Soil contaminated with a mixture of toxic metals | 64 | 42 | 64 | 34 | 90 |
| Reverse mutation of | Non-contaminated soil | 3 | 6 | 8 | 3 | 4 |
| *Salmonella typhimurium* TA 1535 (No. of revertant colonies/Petri dish) | Soil contaminated with a mixture of toxic metals | 39 | 20 | 28 | 2 | 73 |

risk increase, and thereby making the test more dynamic and predictive. The influencing factor can be applied as a one-off or a repeated impact and the effect can be static or impulse-like according to the test set-up.

The basic test set-up is a small-scale microcosm with direct contact between the soil and the test organism (Gruiz *et. al.*, 2001). Toxic metal mobility and bioavailability in the contaminated soils were increased by changing the pH (by applying lime and 'acid rain') and by adding EDTA, a chelate-complex-forming agent to the soil. The indicators of bioavailability in these demonstration cases were toxicity and mutagenicity. The results are shown in Table 7.8.

Acid rain alone has increased metal mobility, as shown by the effect on the unicellular animal, *Tetrahymena pyriformis*. Lime reduced both toxicity and mutagenicity measured by microorganisms, but surprisingly increased the toxicity to *Tetrahymena*. An EDTA and acid rain combination increased dramatically both toxic and mutagenic effects (point mutation on *Salmonella typhimurium* TA 1535). However, EDTA alone rather hides the toxic effects by chelate complexing. The results indicate high risk potential for the tested soil: both an increase and a decrease in the pH increased the toxicity in some of the indicator organisms. This is because the soil contained high concentrations of Cd, Zn, Cu, Pb and As although in a relatively stable form. Not only did these models increase the scale of response, they made a better differentiation possible.

$K_{ow}$ was changed by the addition of RAMEB (randomly methylated beta-cyclodextrin) – an inclusion-complex-forming solubilizing agent – to the contaminated soil. Changes in the bioavailability of pentachlorophenol in contaminated soil was investigated in a matrix experiment where increasing PCP concentrations and increasing RAMEB concentrations were applied and the changes in the availability were indicated by direct contact toxicity and mutagenicity tests (Table 7.9).

An interesting trend can be observed in tests on *Tetrahymena*: both RAMEB and the test organism interact with the toxic molecule (they compete with each other). 1% RAMEB increased the bioavailability of PCP, but higher RAMEB concentrations were able to decrease toxicity by encapsulating most of the PCP molecules and make them

*Table 7.9* Results from the pessimistic models of increased bioavailability of PCP in soil.

| Test | Soil PCP concentration | No additive | 1% RAMEB addition | 2.5% RAMEB addition | 5% RAMEB addition | 10% RAMEB addition |
|---|---|---|---|---|---|---|
| Growth inhibition | Non-contaminated soil | 0 | 0 | 0 | 0 | 0 |
| of *Tetrahymena* | 25 mg/kg PCP | 13 | 53 | 6 | 0 | 0 |
| *pyriformis* (%) | 50 mg/kg PCP | 17 | 83 | 49 | 30 | 41 |
| | 100 mg/kg PCP | 79 | 100 | 69 | 56 | 69 |
| | 200 mg/kg PCP | 99 | 100 | 100 | 100 | 100 |
| Luminescence | Non-contaminated soil | 0 | 0 | 0 | 0 | 0 |
| inhibition | 25 mg/kg PCP | 5 | 0 | 0 | 0 | 0 |
| of *Vibrio* | 50 mg/kg PCP | 46 | 29 | 29 | 11 | 0 |
| *fischeri* (%) | 100 mg/kg PCP | 81 | 69 | 42 | 32 | 22 |
| | 200 mg/kg PCP | 88 | 84 | 81 | 71 | 57 |
| Reverse mutation | Non-contaminated soil | 2 | ND | 3 | 3 | 2 |
| of *Salmonella* | 25 mg/kg PCP | 12 | ND | 18 | 16 | 11 |
| *typhimurium* TA 1535 | 50 mg/kg PCP | 0 | ND | 29 | 47 | 81 |
| (No. of revertant | 100 mg/kg PCP | 2 | ND | 42 | 219 | 115 |
| colonies/Petri dish) | 200 mg/kg PCP | 1 | ND | 517 | 411 | 396 |

less available to the test organisms. In terms of bacterial toxicity (*Vibrio fischeri*), RAMEB reduces toxicity by complexing and making PCP less available to the test bacterium proportionally to the applied concentration.

A mutagenicity study of PCP is particularly interesting because recently published results reveal a number of contradictions: some authors reported PCP as a positive, others as a negative mutagen (Seiler, 1991; Gopalaswamy & Nair, 1992; Sekine *et al.*, 1997). In some studies, Ames reverse-mutation assay proved PCP to be negative, either with or without metabolic activation. In this case the frameshift mutation occurred on *Salmonella typhimurium* TA 1538 after complex formation with RAMEB, which indicates that the contaminant's low availability may be the reason for its being negative. Other mechanisms may also play a role, but these have not been explored yet.

Artificially modified bioavailability indicates that the tested soil has the potential to increase its adverse effects when the environmental conditions enhance the contaminant's mobility and bioavailability. Contaminants are often in a non-available form in abandoned, undisturbed soils because the soil and soil microbiota itself provide protection against toxic contaminants. In other cases, mine waste, for example, has a mineral composition which binds contaminants strongly into the mineral structure. These stable forms are not always irreversible over the long term, but may be ready for degradation e.g. by natural weathering. This way the built-in hazardous metals can be mobilized. The 'chemical time bomb' phenomenon covers the presence and potential remobilization of latent, immobile and non-bioavailable contaminants. The study of toxicity or mutagenicity under the circumstances of enhanced bioavailability (e.g. due to additives) provides a realistic worst-case scenario, and the results can be used for decision making.

## 9.2 Effect of soil sorption capacity on bioavailability

Mobile forms of inorganic contaminants such as toxic metal ions have the affinity to bond to oxides and hydroxides or to the large specific surface of clay minerals. Later, during soil formation, they can be built into the molecular grid of the newly formed minerals. When the mineral structure of the soil is deteriorated, these metals can be mobilized again.

The mobility of organic contaminants in soil mainly depends on the organic material content (OM) of the soil, given that organic contaminants are sorbed onto the surface of the particles or built into the structure of stable (large size) humus molecules. Sorption and incorporation of the contaminants into the humus takes place during the long-term humus formation. The opposite process, humus degradation may also occur. Both humus formation and humus degradation largely influence the bioavailability of organic contaminants in the soil. This may be the key factor in the case of inherited polluted sites where these changes have already taken place.

Regardless of time and humus formation, humus influences the organic contaminants' mobility and bioavailability. Some regulatory screening values are higher for humic soil, which shows the rationality in risk management.

Dependence of OM on the bioavailability of soil contaminants was demonstrated on contaminated soils with different OM contents. Two different sandy soil types with 0.45% humus content and a brown forest soil with 1.3% humus content were contaminated with 4CP ($\log K_{ow} = 2.39$; Toxnet, 2014). Bioavailability was detected by toxicity tests: *Vibrio fischeri* bioluminescence inhibition, *Tetrahymena pyriformis* growth inhibition and *Folsomia candida* mortality tests.

A soil correction factor $f_{soil/water}$ is used to characterize the difference between toxicities in soil and water. This factor gives the ratio of a contaminant's toxicity in soil to the toxicity of the same concentration of the same toxicant in water (Eq. 7.1). It is calculated from the $EC_{50}$ values of 4CP in soils with different OM and in water. Figure 7.19 shows the toxicity results depending on soil type.

$$f_{soil/water} \left[\frac{L}{g}\right] = \frac{EC_{soil50}\left[\frac{mg}{g}\right]}{EC_{water50}\left[\frac{mg}{L}\right]} \tag{7.1}$$

The $f_{soil/water}$ values ($>1$ L/g) show how many times the toxicity of the same contaminant amount in soil is less than in water. A lower $f_{soil/water}$ for sandy soil compared to forest soil indicates that the toxicity in water is closer to sandy soil than to forest soil as forest soil with a higher OM content has greater sorption capacity which results in lower bioavailability.

## 10 BIOACCESSIBILITY AND BIOAVAILABILITY OF CONTAMINANTS FOR HUMANS

Bioaccessibility and bioavailability of environmental contaminants may be a dominant factor in exposures and human health risk. Microorganisms or other small animals interact with the contaminants through their entire inner and outer surfaces as they

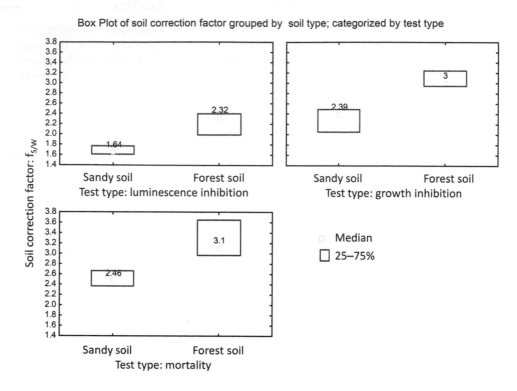

*Figure 7.19*   4CP toxicity in sandy and humic forest soils: soil toxicity compared to water.

live in and are in direct contact with the environmental matrices. It such cases bioaccessibility and bioavailability may be very close to each other. Human exposures are different: humans' interaction is restricted to a limited interface by inhalation, ingestion and skin contact during a limited time period and with a limited proportion of contaminants.

The most radical exposure from the viewpoint of bioavailability is the oral exposure, which endangers mostly children because of uncontrolled uptake and high sensitivity. Contaminants taken orally from soil are considered to be equal to the bioavailable fraction absorbed by digestion. Dermal uptake can be modeled *in vitro* by active animal skin, and the biologically available fraction of the contaminant which penetrates the skin can be measured (see Chapter 3). Bioavailability tests using respiration are mainly based on animal experiments: the large surface area of the lung and repeated exposures are assessed. Human oral bioavailability is the ratio of the internal (effective) dose to the orally administered dose. The orally administered dose is quantitatively characterized compared to the intravenously administered dose.

The applied mathematical, physico-chemical and biological models and their application in the assessment of human health risk via environment are discussed below.

## 10.1   Mathematical models for calculation of bioaccessibility- and bioavailability-dependent human risk

The pathway of chemical substances within the human body, into the blood and the organs is precisely described, but the pathway from the external environment into the human body is still not sufficiently known because of the large variety of scenarios. Human exposure models for pharmacological and toxicological purposes focus on the absorption and mobility of contaminants inside the human body: this specific scenario provides a practical model for the risk assessment of drugs. The mathematical models based on biochemical and physiological information are routinely used in human health and safety management and in the pharmaceutical industry. For the environment and for soil in particular, the only existing model is the gastrointestinal model. Absorption of a contaminant (toxicant) from the environment into the circulatory system is the critical kinetic component of bioavailability and environmental risk assessment of hazardous environmental contaminants. Statistical studies which model skin permeability showed that the efficacy of dermal absorption is related to the log $K_{ow}$ value (Cleek & Bunge, 1993; Roberts et al., 1995; Vecchia & Bunge, 2002), and availability via inhalation depends on the vapor pressure, molecular weight, and log $K_{ow}$ of organic contaminants. Calculation of dust inhalation is based on the bonds to and the size of soil dust particles. The size of the airborne particles is a crucial factor; fractions under 2.5 μm in diameter are completely absorbed by the lung tissue.

The most complex process occurs in the gastrointestinal tract after ingestion of soil. Toxicity of soil contaminants depends upon their absorption from the gut. Information on how well a contaminant is absorbed from the gut is important in determining how much of a contaminant humans can be exposed to before health effects occur. The difference between ingested and absorbed amounts is generally significant for soil contaminants due to their limited accessibility and availability. The subcellular mechanisms of absorption, and the influencing factors, as well as the transport from the gastrointestinal tract to the blood should be properly understood for individual contaminants Not only the amount, but also the rate of uptake is important, particularly under acute or subacute exposure conditions. The dose-dependent absorption of lead, for example, is influenced by several factors, which together result in the rate of absorption (Maddaloni et al., 1996; Ruby et al., 1996; US EPA, 1994 and US EPA, 2003):

–   interactions with other inorganic and organic substances, particularly such nutrients as calcium, iron, phosphate, vitamin D, fats, etc.
–   metal species and particle size may influence the solubility and because of that, the accessibility and bioavailability of lead;
–   physico-chemical complexity of the environmental matrices;
–   lead uptake is markedly lower when consumption is high than under fasting conditions;
–   children absorb more lead than adults do.

Several dynamic mechanisms can be identified behind these findings, which make the modeling of bioavailability difficult and the result uncertain. Chemical models and statistical evaluation of literature data can provide more relevant and practical estimates. The US EPA, for example, characterizes human risk of contaminated soil by the *relative bioavailability* (RBA) which gives the ratio of the amount of a contaminant absorbed from soil to the amount of that contaminant absorbed from food or water. In the case of lead absorption in humans, relative bioavailability in soil compared to water or food is about 60% on average (US EPA, 2009).

*In vivo* swine model was applied by Juhász *et al.* (2007, 2008) to establish the *in vitro* digestion model for arsenic by comparing adipose tissue, blood plasma concentrations and liver enzyme activities. As arsenic was studied by hundreds of studies, later on Juhász *et al.* (2011) worked out a method for predicting relative arsenic bioavailability in contaminated soils, using meta analysis and relative bioavailability–bioaccessibility regression models. They found that arsenic RBA from the source of chromated copper arsenate (CCA) posts, herbicide/pesticide, mining/smelting and weathered rock/iron cap[2] was 78.0, 78.4, 67.0 and 23.7%, respectively.

## 10.2 Chemical models for estimating accessibility of contaminants for humans

Contaminants in soil experience the most radical changes and mobilization effects when they enter the human gastrointestinal tract where extremely low pH prevails especially in the stomach, and the bile salts have solubilizing effects on compounds with high $K_{ow}$.

Two types of models are mainly used for bioaccessibility testing: chemical models (separation/speciation methods, such as extraction, membrane separation, complexation, etc.) and *in vitro* gastrointestinal digestion models.

Chemical analysis methods such as sequential extractions using extractants of different affinity to the metal species and growing extraction strength target the different solid matrices (humus, oxides, clay minerals) in the soil (Table 7.3). The interpretation of the results is not always clear, but the aim is always to find the bioaccessible fraction correspondent extract(s), e.g. Juhász *et al.* (2011).

There are more sophisticated analytical methods too, which can draw a true picture of chemical species and of the distribution of metals between different solid soil constituents: these are the synchrothron-based X-ray absorption fine structure (XAFS) and X-ray absorption near edge structure (XANEX) methods. Unfortunately they are extremely costly, time-consuming and of limited availability.

Numerous *in vitro* digestion methods can be found in the literature mainly originating from the pharmaceutical industry. In environmental science, the most relevant transformation of the contaminants is the focus when developing chemical/biochemical simulation models for digestion. These digestion models are well-studied and compared to real bioavailability data for the purpose of risk assessment. For example, playground soils contaminated with arsenic and PAH were extensively investigated using *in vitro* gastrointestinal models to assess the exposure of children. The results of the gastrointestinal model extraction from lead-contaminated soil were compared to the contaminant concentration in the blood serum of children playing on the playground by Ren *et al.* (2006). They have found a linear relationship between the bioaccessible

---

[2]Intensely oxidized, weathered or decomposed rock, usually the upper and exposed part of an ore deposit or mineral vein.

fraction mobilized from soil into the ingestion juice and the metal contamination of the ingested soil. Lead in the blood serum was also linearly related to the bioaccessible amount.

Determination of the bioaccessible fraction from contaminated soil is a major research topic all over the world. In the US the Solubility/Bioavailability Research Consortium (SBRC) worked out the SBET, also called SBRC method for testing lead and arsenic bioaccessibility and determining relative bioaccessibility.

The Bioaccessibility Research Group of Europe (BARGE, 2014a) is a network of European organizations for studying human bioaccessibility of priority contaminants in soil and disseminating oral data, mainly for regulatory purposes (BARGE, 2014b).

The Canadian working group for Bioaccessibility Research (BARC, 2014) has studied inorganic and organic contaminant bioaccessibility in soils on contaminated sites in Canada. A priority objective of BARC is to advance the scientific basis for incorporating bioaccessibility testing into site specific human health risk assessments.

Health Canada (2007) collected existing data from *in vivo* and *in vitro* bioaccessibility research and compared them to physico-chemical soil parameters that influence bioavailability. Some results on As, Pb and Cd are presented in this section.

A useful guide has been prepared for US defense facilities (Metals Bioavailability, 2003) to incorporate bioavailability adjustments into human health and ecological risk assessments. It has gathered current information on metals bioavailability, it explains concepts and identifies types of data necessary for assessing bioavailability and incorporating them into risk assessment. The first part of the guide gives a brief description on test methods suitable for human health and ecological risk assessment and deals with the acceptability of the results. In the second part, the technical background for metals bioavailability assessment, types of studies and connected protocols are discussed in details for every relevant metal.

### 10.2.1   Human bioaccessibility of toxic metals

Metal availability in soil is mainly determined by the dissolution–precipitation process pair. The same processes are present and influence human health effects by inorganic mineral components in soil. Dissolution kinetics controls the availability of metal ions, and basic rock minerals undergo continuous dissolution in natural soil. Other process pairs such as oxidation–reduction, sorption–desorption and ion exchange also play a crucial role in bioaccessibility of metals. Soil inhaled or ingested by humans contains metal species of different bioaccessibility. The basic bioaccessibility, valid in natural soil, will dramatically change as a result of new equilibrium of dissolution–precipitation, sorption–desorption, oxidation–reduction processes in the human body. The two together yield the actual metals bioaccessibility in the human body. This is compounded by the changing environment of the digestive system. The conditions and biochemistry of the organism (humans) determine the final bioavailable and taken-up proportion of the metal.

Bioaccessibility in the context of oral exposures means that contaminants are released from soil particles within the gastrointestinal tract during digestion. In addition to the chemical speciation of the contaminants, e.g. the chemical formula of a toxic metal, the co-occurring metals and metal species, their bonds to the soil's matrix are of crucial importance for bioaccessibility and in human and environmental toxicity (Figure 7.2).

For example arsenic may occur in the form of $Ca_3(AsO_4)_2$, $Mg_3(AsO_4)_2$, $As_2O_5$ in aerobic soil and as As or $As_2S_3$ in anoxic or anaerobic soil. Lead occurs as PbO, $PbCO_3$ or $Pb_3(CO_3)(OH)_2$ under aerobic and as Pb or PbS under anaerobic conditions. The chemical transformations during digestion increase bioaccessibility, depending on the original mineral speciation, chemical bonds and the actual parameters pH and redox potential of the digestive system.

The above described partial and limited description makes clear that modeling of metal bioaccessibility in the human body is accompanied by high uncertainty.

The chemical model of PBASE (Basta & Gradwohl, 2000) models the potentially bioavailable sequential extraction. It is based on the assumption that the acidic juice of the stomach (pH = 1–2 at fasting and pH = 2–4 when fed) is the main factor responsible for making metals accessible during digestion. PBASE extraction for metals applies a four-step extraction method using

1.   0.5 M $Ca(NO_3)_2$ for the exchangeable, readily soluble
2.   1 M NaOAc (pH = 5) for weak acid-soluble weak surface complexes
3.   0.1 M $Na_2$EDTA (pH = 7) for surface complexes and precipitates and
4.   4 M $HNO_3$ for very insoluble species.

The method of Mercier *et al.* (2002) is an even more simplified model based on the extraction of the soil samples at 37°C in HCl at pH = 2. This method was used to determine oral availability of antimony, arsenic, barium, cadmium, chromium, lead, mercury and selenium. In comparison to other bioaccessibility and bioavailability data, it resulted in a slight overestimate.

IVG (*In Vitro* Gastrointestinal) digestion method (Schroder *et al.*, 2004) is based on a leaching procedure using i) a gastric extraction fluid (NaCl and pepsin) at pH values of 1.5, 2.0, or 2.5 and ii) an intestinal fluid of pancreatin and bile extract (Rodrigues & Basta, 1999).

Total Migratable Screening, European Standard (EN 71-3, 2013) – also uses HCl to obtain the migratable portion of a tested object. The aim of the standard is a preliminary screening that looks at the total migration from a toy (Safety Toys Directive, 2009), but it is also useful for soil. The principle of the procedure is the extraction of soluble elements from solid materials under the conditions which simulate the material remaining in contact with stomach acid for a period of time after being swallowed. The following heavy metals are measured: aluminum (Al), antimony (Sb), arsenic (As), barium (Ba), boron (B), cadmium (Cd), chromium III ($Cr^{3+}$), chromium VI ($Cr^{6+}$), cobalt (Co), copper (Cu), lead (Pb), manganese (Mn), mercury (Hg), nickel (Ni), selenium (Se), strontium (Sr), tin (Sn), organic tin and zinc (Zn).

There are several digestion fluids in the gastrointestinal tract during digestion such as gastric juice with low pH and the intestinal fluids that can mobilize or immobilize toxic metals, which can also be bound to a protein molecule or accumulated in liver or fat tissues. Gastrointestinal models simulate one or several steps from the complex human digestion system applying a realistic residential time and using artificial biofluids.

–   Oral cavity: residential time: seconds–minutes, pH = 6.5;
–   Stomach when fasting residential time: 8–15 minutes, pH = 1–2; when fed residential time: 0.5–3 h, pH = 2–5;

- Small intestine has further sections:

  o duodenum: 0.5–0.75 h, pH = 4–5.5
  o jejunum: 1.5–2.0 h, pH = 5.5–7
  o ileum: 5–7 h, pH = 7–7.5

- Colon 15–60 h, pH = 6–7.5.

SBET – Simplified Bioaccessibility Extraction Test (Ruby *et al.*, 1996; Juhász *et al.*, 2008; Lamb *et al.*, 2009; Mingot *et al.*, 2011) is a routinely used rapid method which only comprises the gastric phase extraction.

PBET – Physiologically Based Extraction Test (Ooemen *et al.*, 2003) simulates the leaching of a solid matrix in the human gastrointestinal tract, and determines the bioaccessibility of a particular element that is available for adsorption during transit through the small intestine. The applied gastric component consists of pepsin, citrate, malate, lactic acid and acetic acid, and the intestinal component comprises pancreatin and bile salts.

Ruby *et al.* (1996) and Rodriguez *et al.* (1999) claimed in their studies that the PBET method showed a decrease in the bioaccessibility of lead due to the high pH in the intestinal phase where lead precipitated. Contradictory results were found on the bioaccessibility of arsenic, which became more accessible at high pH (Juhász *et al.*, 2008; Cave *et al.*, 2002). Juhász *et al.* (2008) compared the results of the five-step extraction method (Teisser method) (Wenzel *et al.*, 2001), those of the SBET digestion model and blood plasma from *in vivo* swine tests on animals kept on arsenic-contaminated soil. The comparison resulted in good correlation between the results of the SBET method and the *in vivo* model.

RBALP – The Relative Bioaccessibility Leaching Protocol was specifically developed for lead in soil. It uses physiologically relevant pH of the stomach but uses a glycine buffer as the extraction medium (Drexler & Brattin, 2007).

RIVM – Method of the Dutch National Institute for Public Health and the Environment. It mimics the chemical environment of the human gastrointestinal system using the mixture of pepsin, mucin and BSA as a gastric fluid and bile, duodenal juice, chime, pancreatin and lipase as intestinal secretion components (Versantvoort *et al.*, 2004).

SHIME – Simulator of the Human Intestinal Microbial Ecosystem. This *in vitro* model was developed according to De Boever *et al.* (2000) and Laird *et al.* (2007). The artificial gastric biofluid contains pectin, mucin, cellobiose, proteose peptone, starch and the intestinal one comprises pancreatin. This maintains a microbial community representative of that found in the human intestine. The SHIME is operated under anaerobic conditions with human gastrointestinal microorganisms in the colon stage since these microbes may influence metals bioaccessibility (Laird, 2010a and Laird, 2010b).

TIM – TNO Intestinal Model: it is a dynamic simulation for dissolution and bioaccessibility in the human gastrointestinal tract. It is used both for drugs, food and environmental samples, mainly soil. It is a multi-compartment method using lipase, pepsin as gastric fluid and bile, porcine, pancreatin as the intestinal component (Bosso & Enzweiler, 2008; Bosso *et al.*, 2008).

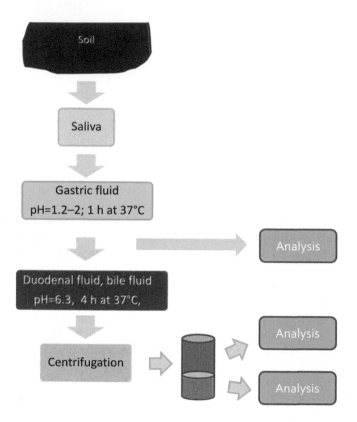

*Figure 7.20* Unified Bioaccessibility Method (UBM) gastrointestinal model (BARGE, 2014b).

GFE – Gastric Fluid Extraction methods is based on the PBET (Ruby *et al.*, 1996) and the IVG methods (Rodriguez *et al.*, 1999) using a liquid to solid ratio of 100:1, gastric pH of 1.8 and intestinal pH of 7.0. It applies only gastric secretion components: pepsin, citrate, malate and acetic acid.

UBM – Unified Bioaccessibility Method (BARGE, 2014b; Broadway *et al.* 2010) – is based on the Oomen *et al.* (2003) methodology and is recommended for inorganic substances. UBM contains four main digestion juices (saliva, gastric and duodenal juices as well as bile) and the samples are analyzed both after the gastric and intestinal treatment phase (Figure 7.20) to see the difference between the two.

According to the results found in the literature, the bioaccessibility of several toxic metals such as lead, cadmium (Tang *et al.*, 2006a), zinc or chromium (VI) is increased by gastric ingestion due to a low pH. The bioaccessibility peak of arsenic occurred after the duodenal juice treatment at a higher pH. The significantly more toxic and carcinogenic As(III) and As(V) are present in various ratios in the subsequent gastrointestinal tracts. That is why arsenic toxicity is hard to predict in the human body. Further time- and cost-intensive research is needed to explain and model the influence of the processes in the mouth on the bioaccessibility of toxic metals.

A comprehensive comparative study prepared by Health Canada (2007) yielded interesting results. Arsenic and lead content and their bioavailablity/bioaccessibility results were collected from several hundred soil samples, measured both by *in vivo* animal tests and *in vitro* applied artificial biofluids such as SBET, PBET, IVG, SHIME, TIM and GFE. Some important lessons learned are summarized below.

*In vivo* study results for arsenic-contaminated soil:

The average arsenic concentration of 55 soil samples was $1950 \pm 3760$ mg/kg, of which $23 \pm 20\%$ was bioavailable. Arsenic bioavailability was shown to be less than 5% in 13 of the 55 soils tested in *in vivo* studies. However, there were four soils where bioavailability was greater than 50%. Based on these results, Health Canada (2007) recommended the use of 25% as bioavailable for uptake into the bloodstream from the whole of soil arsenic in human health risk assessments. Statistical analysis indicates that the bioavailable percentage is independent of the total arsenic concentration in the soil samples. Rats contained significantly less bioavailable arsenic than other test species (monkey, rabbit, swine and rat were studied).

*In vitro* study results for arsenic-contaminated soil:

The average arsenic concentration of 102 soil samples was $1204 \pm 2775$ mg/kg. The bioaccessible percentage was $25.6 \pm 19.0\%$ from a total (100%) of arsenic. Therefore, based on *in vitro* studies, on average less than 26% of arsenic can be considered bioaccessible and available for uptake into the bloodstream. Given that soluble arsenic is almost 100% absorbed into the bloodstream, arsenic bioaccessibility and bioavailability (absorbed into the blood stream) are equal. An acidic medium (pH of 4.0) resulted in significantly lower arsenic bioaccessibility. An evaluation of accessible arsenic and liquid–solid ratio in the test indicated that there was no statistical difference across the varying ratios. The accessible fraction is independent of the total arsenic concentration in the soil sample.

*In vivo* study results for lead-contaminated soil:

soil samples were analyzed: the average lead concentration in the soils was $6067 \pm 3521$ mg/kg and the bioavailable percent was $46 \pm 27\%$. This means that on average less than 50% of total lead can be considered available for uptake into the bloodstream. The bioavailable portion is independent of the total lead concentration in the soil. Unlike arsenic, only 50% of soluble lead is absorbed across the intestinal wall.

*In vitro* study results for lead-contaminated soil:

Based on 86 discrete test results on soil lead bioaccessibility, the average lead concentration in soil was $3180 \pm 3285$ mg/kg ($N = 86$). The bioaccessible percentage in the stomach phase was $51 \pm 26\%$ and $4.2 \pm 5.0\%$ ($n = 45$) in the intestinal phase. There was some weak correlation between bioaccessible and total lead concentrations and with the pH of the soil.

The results introduced above based on the Health Canada survey can be utilized in generic predictions. Site-specific risk assessments need a different approach; site-specific bioaccessibility and bioavailability should be assessed case by case because the standard deviations are large due to soil types and environmental conditions.

### 10.2.2  Bioaccessibility of organic compounds in humans

While metals mobility is mainly influenced by the changing pH along the gastrointestinal tract, organic compounds' accessibility is determined by the mobilizing effects

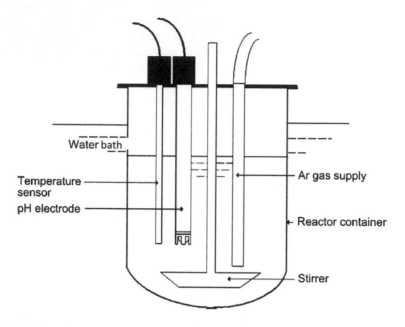

*Figure 7.21*   Equipment of an *in vitro* digestion test (Tang *et al.*, 2006b).

of bile salts. Among organic substances, PAH bioaccessibility is mostly studied by *in vitro* gastrointestinal models. Tang *et al.* (2006b) found increased bioaccessibility in the small intestinal phase, confirmed both by chemical analysis and effect testing. Figure 7.21 shows the set-up for the digestion test.

Otherwise negative phenanthrene toxicity occurred after intestinal digestion due to micelle formation with bile salts. *In vivo* bioavailability of dioxins and furans was determined in orally exposed minipigs and rats. Tissue sampling and analysis showed concentration in liver and adipose tissue (Wittsiepe *et al.*, 2007; Budinsky *et al.*, 2008). Chemical analysis failed to provide adequate information as to degradation products of organic contaminants are more toxic than the original substance.

### 10.2.3   Chemical models combined with biological models – measuring toxic effects after digestion

Pre-treatment before toxicity assessment applying a human digestion simulation model may result in a realistic scenario similar to the natural process. Theoretically, the toxicity (mutagenicity, reprotoxicity) tests with *in vitro* digested samples should provide highly realistic effect results similar to the actual effects in humans. On the other hand, toxicity tests may have several disadvantages, for example extensive dilution of *in vitro* digested samples. The digestion fluid (saliva:gastric:duodenal:bile = 1:1.5:3:1) has to be applied in an extremely high ratio to the soil – a minimum of 98:1 – to ensure the dissolution of metals (Ruby *et al.*, 1996). Before toxicity testing, further extraction and concentration of the digested sample is necessary.

In the case of organic contaminants, the detection of toxic metabolites in the digested fluid presents a problem unresolved as yet because tracing various unidentified metabolites is a difficult task for chemical analysis.

There are some developments to eliminate the disadvantages which occur in toxicity testing of contaminated soil and modeling maximum bioavailability:

1.  For organic contaminants, soil spiking with radioactive labeled contaminants can be a solution before *in vitro* digestion. It helps monitor the various transformations of the labeled compounds during digestion.
2.  The S9 enzyme mix represents a fairly simple gastrointestinal model. It includes the addition of the commercially available S9 enzyme mix that contains 5 mM glucose-6-phosphate, 4 mM NADPH and S9 from Spargue-Dawley rat liver oxidative enzymes (Maron & Ames, 1983). This method is suitable for further testing e.g. using the Ames test to examine mutagenic effect of soil (Hajdu *et al.*, 2008).
3.  Another promising but not yet practiced method for soils applies Caco-2 cell lines, which is a monolayered cell line originating from the small intestine and is routinely used in pharmacokinetics. It is suitable to estimate the absorption of pharmaceuticals and to test cytotoxicity by measuring cell layer integrity, enzyme activities or contaminant concentration in the harvested cells (Laparra *et al.*, 2005). As Caco-2 epithelial cells originate from the human large intestine, this *in vitro* test system can provide one of the most realistic absorption models of *in vitro* digested contaminants.

## II  CONCLUSIONS

Several methods are available for modeling and predicting bioavailability and bioaccessibility of hazardous environmental contaminants. The less environmentally relevant but simplest method is the determination of log $K_{ow}$ or various partial extractions and solvent distributions of the contaminants. These methods are often the cheapest but the results are only hazard estimates that fail to provide information on the actual environment and the target ecosystem or human receptors. The developments of bioavailability measurement methods as summarized in this chapter reflect the various concepts and introduce the mathematical, chemical and combined models that try to describe the phenomenon of bioavailability which itself has not yet been fully defined.

The first step to successfully manage this problem is to identify those points of the conceptual risk model (Figure 1.1 in Volume 1) where bioavailability may have an impact on the quantity or quality of risk. A thorough investigation of the environmental fate and behavior of contaminants should include the interactions with the living part of the ecosystems. Earthworm models, plant accumulation tests, *in vivo* swine experiments, and epidemiological data serve to address the contaminant–matrix–biota interactions and the representation of these processes in toxicity, mutagenicity, reprotoxicity tests. Another uncertainty is caused by the environmental conditions, which is handled by the 'pessimistic testing' concept introduced here or by realistic worst-case methods that can directly test the adverse effects exerted by contaminated matrices.

Since our knowledge lags behind what is required to understand contaminants' bioaccessibility and bioavailability in the environment, it is clear that this is one of the fields of environmental management that needs further development, innovative approaches, a uniform assessment and improved interpretation tools. These tools are necessary for efficient assessment methods and for treating uncertainties associated with bioavailability such that risk is neither unnecessarily nor unacceptably overestimated.

## REFERENCES

17/1999 (VI. 16.) EüM – Decree on the maximum permitted chemical contamination of food. Ministry of Health, Hungary.

44/2003 (IV.26.) FVM – Decree on the mandatory requirement of the Hungarian Feed Code. Ministry of Agriculture, Hungary.

Ahlf, W., Drost, W. & Heise, S (2009) Incorporation of metal bioavailability into regulatory frameworks—metal exposure in water and sediment. *Journal of Soils and Sediment*, 9, 411–419. [Online] Available from: dx.doi.org/10.1007/s11368-009-0109-6. [Accessed 15th August 2014].

Appleton, J.D., Cave, M.R. & Wragg, J. (2012) Modelling lead bioaccessibility in urban topsoils based on data from Glasgow, London, Northampton and Swansea, UK. *Environmental Pollution*, 171, 265–72. [Online] Available from: DOI: 10.1016/j.envpol.2012.06.018. [Accessed 15th August 2014].

Aten, C.F. & Gupta, S.K. (1996) On heavy metals in soil; rationalization of extractions by dilute salt solutions. *Science of the Total Environment*, 178, 45–53.

BARC (2014) *Bioaccessibility Research Canada*. [Online] Available from: http://www.bioavailabilityresearch.ca. [Accessed 15th January 2014].

BARGE (2014a) *Bioaccessibility Research Group of Europe*. [Online] Available from: http://www.bgs.ac.uk/barge/home.html. [Accessed 15th January 2014].

BARGE (2014b) *Unified bioaccessibility method*. Bioaccessibility Research Group of Europe. [Online] Available from: http://www.bgs.ac.uk/barge/ubm.html. [Accessed 15th January 2014].

Barthe, M., Pelletier, E., Breedveld, G.D. & Cornelissen, G. (2008) Passive samplers versus surfactant extraction for the evaluation of PAH availability in sediments with variable levels of contamination. *Chemosphere*, 71, 1486–1493.

Basta, N.T. & Gradwohl, R. (2000) Estimation of heavy metal bioavailability in smelter-contaminated soils by a sequential extraction procedure. *Journal of Soil Contamination*, 9, 149–164.

Becquer, T., Dai, J., Quantin, C. & Lavelle, P. (2005) Sources of bioavailable trace metals for earthworms from a Zn-, Pb- and Cd-contaminated soil. *Soil Biology & Biochemistry*, 37, 1564–1568.

Bergknut, M., Sehlin, E., Lundstedt, S., Andersson, P.L., Haglund, P. & Tysklind, M. (2007) Comparison of techniques for estimating PAH bioavailability: Uptake in Eisenia fetida, passive samplers and leaching using various solvents and additives. *Environmental Pollution*, 145, 154–160.

Bosso, S.T. & Enzweiler, J. (2008) Bioaccessible lead in soils, slag, and mine wastes from an abandoned mining district in Brazil. *Environmental Geochemistry & Health*, 30, 219–229.

Bosso, S.T., Enzweiler, J. & Angelica, R.S. (2008) Lead bioaccessibility in soil and mine wastes after immobilization with phosphate. *Water, Air & Soil Pollution*, 195, 257–273.

Broadway, A., Cave, M.R., Wragg, J., Fordyce, F.M., Bewley, R.J.F., Graham, M.C., Ngwenya, B.T. & Farmer, J.G. (2010) Determination of the bioaccessibility of chromium in Glasgow soil and the implications for human health risk assessment. *Science of the Total Environment*, 409, 267–277.

Brookins, D.J. (1988) Eh-pH Diagrams for Geochemistry. Berlin Heidelberg, New York, Springer-Verlag.

Budinsky, R.A., Rowlands, J.C. & Casteel, S. (2008) A pilot study of oral bioavailability of dioxins and furans from contaminated soils: Impact of differential hepatic enzyme activity and species differences. *Chemosphere*, 70(10), 1774–1786.

Buffle, J., Parthasarathy, N., Djane, N. K. & Matthiasson, L. (2000) Permeation Liquid Membrane for Field Analysis and Speciation of Trace Compounds in Waters. In: Buffle, J. & Horvai, G. (eds.) *In Situ Monitoring of Aquatic Systems; Chemical Analysis and Speciation.* Wiley.

Buffle, J. & Tercier-Waeber, M.L. (2005) Voltammetric environmental trace-metal analysis and speciation: from laboratory to in situ measurements. *Trends in Analytical Chemistry*, 24 (3), 172–191. [Online] Available from: dx.doi.org/10.1016/j.trac.2004.11.013. [Accessed 15th August 2014].

Caussy, D. (2003) Case studies of the impact of understanding bioavailability: arsenic. *Ecotoxicology and Environmental Safety*, 56, 164–173.

Caussy, D., Gochfeld, M., Gurzau, E., Neagu, C. & Ruedel, H. (2003) Lessons from case studies of metals: investigating exposure, bioavailability and risk. *Ecotoxicology and Environmental Safety*, 56, 45–51.

Cave, M.R., Wragg, J., Palumbo, B. & Klinck, B.A. (2002) *Measurement of the bioaccessibility of arsenic in UK soils.* Environment Agency R&D Technical Report, P5-062/TR002.

Cave, M.R., Milodowski, A.E. & Friel, E.N. (2004) Evaluation of a method for Identification of Host Physico-chemical Phases for Trace Metals and Measurement of their Solid-Phase Partitioning in Soil Samples by Nitric Acid Extraction and Chemometric Mixture Resolution. *Geochemistry: Exploration, Environment, Analysis*, 4, 71–86.

Cleek, R.L. & Bunge, A.L. (1993) A new method for estimating dermal absorption for chemical exposure 1. General approach. *Pharmaceutical Research*, 10, 497–506.

Cornelissen, G., Arp, H.P.H., Pettersen, A., Hauge, A. & Breedveld, G.D. (2008) Assessing PAH and PCB emissions from the relocation of harbour sediments using equilibrium passive samplers. *Chemosphere*, 72, 1581–1587.

Cornelissen, G., Okkenhaug, G., Breedveld, G.D. & Sørlie, J.E. (2009) Transport of polycyclic aromatic hydrocarbons and polychlorinated biphenyls in a landfill: A novel equilibrium passive sampler to determine free and total dissolved concentrations in leachate water. *Journal of Hydrology*, 369, 253–259.

Davison, W., Fones, G., Harper, M., Teasdale, P. & Zhang, H. (2000) Dialysis, DET and DGT: in Situ Diffusional Techniques for Studying Water, Sediments and Soils. In: (Eds.) Buffle, J. & Horvai, G. *In Situ Chemical Analysis in Aquatic Systems.* Wiley. pp. 495–569.

Davison, W., Zhang, H. & Miller, S. (1994) Developing and Applying New Techniques for Measuring Steep Chemical Gradients of Trace Metals and Inorganic Ions at the Sediment–Water Interface. *Mineralogical Magazine*, 58A, 497–498.

De Boever, P., Deplancke, B. & Verstraete, W. (2000) Fermentation by gut microbiota cultured in a simulator of the human intestinal microbial ecosystem is improved by supplementing a soygerm powder. *The Journal of Nutrition*, 130(10), 2599–2606.

DGT Research (2014) *DGT – for measurements in waters, soils and sediments.* [Online] Available from: www.dgtresearch.com/dgtresearch/dgtresearch.pdf. [Accessed 15th January 2014].

Dobler, R., Burri, P., Gruiz, K., Brandl, H. & Bachofen, R. (2001) Variability in microbial populations in soil highly polluted with heavy metals on the basis of substrate utilisation pattern analysis. *Journal of Soils & Sediments*, 1(3), 151–158.

Drexler, J.W. & Brattin, W.J. (2007) An *in vitro* procedure for estimation of lead relative bioavailability: With validation. *Human & Ecological Risk Assessment*, 13, 383–401.

EN 71-3 (2013) *European Standard on the Safety of Toys* – Part 3: Migration of certain elements. [Online] Available from: http://ec.europa.eu/enterprise/policies/european-standards/harmonised-standards/toys/index_en.htm. [Accessed 15th January 2014].

EPI Suite™ (2014) Estimation Program Interface (EPI) Suite Version 4.10. *Exposure Assessment Tools and Models.* [Online] Available from: http://www.epa.gov/oppt/exposure/pubs/episuite. htm. [Accessed 15th January 2014].

Esteve-Turrillas, F.A., Yusa, V., Pastor, A. & Guardia, M. (2008) New perspectives in the use of semipermeable membrane devices as passive samplers. *Talanta*, 74, 443–457.

EU-TGD (2003) Technical Guidance Document on Risk Assessment in support of Commission Directive 93/67/EEC on Risk Assessment for new notified substances. Commission Regulation (EC) No 1488/94 on Risk Assessment for existing substances Part 3. [Online] Available from: http://echa.europa.eu/documents/10162/16960216/tgdpart3_2ed_en.pdf. [Accessed 11th August 2014].

Fava, F., Berselli, S., Bertini, L., Di Gioia, D., Fenyvesi, É., Gruiz, K., Marchetti, L., Molnar, M., Szejtli, J. & Zannon, D. (2010) *Biogenic agents for improving the bioavailability and biodegradation of hydrophobic pollutants 2006. EUROGENE portal. February 2010.* [Online] Available from: http://eurogene.open.ac.uk/content/biogenic-agents-improving-bioavailability-and-biodegradation-hydrophobic-pollutants-2006. [Accessed 11th January 2014].

Feigl, V. (2011) Remediation of toxic metals contaminated soil and mine waste using combined chemical- and phytostabilization. Ph.D. Thesis (in Hungarian). Budapest University of Technology and Economics, Budapest, Hungary.

Feigl, V., Atkári, Á., Anton, A. & Gruiz, K. (2007) Chemical stabilisation combined with phytostabilisation applied to mine waste contaminated soil in Hungary. *Advanced Materials Research*, 20–21, 315–318.

Feigl, V., Anton, A., Fekete, F. & Gruiz, K. (2009a) Combined chemical and phytostabilization: field application. *Land Contamination & Reclamation*, 17 (3–4), 579–586.

Feigl, V., Uzinger, N. & Gruiz, K. (2009b) Chemical stabilisation of toxic metals in soil microcosms. *Land Contamination & Reclamation*, 17 (3–4), 483–494.

Fenyvesi, É., Molnár, M., Kánnai, P., Balogh, K., Illés, G. & Gruiz, K. (2010) Cyclodextrin-extraction of soils for modelling bioavailability of contaminants. In: Sarsby, R. W. & Meggyes, T. (eds.) *Abstract book of the International Conference of Construction for a Sustainable Environment, Vilnius, Lithuania, 1–4 July, 2008.* Leiden, CRC Press/Balkema. p. 34.

Geebelen, W., Adriano, D.C., van der Lelie, D., Mench, M., Carleer, R., Clijsters, H. & Vangronsveld, J. (2003) Selected bioavailability assays to test the efficacy of amendment-induced immobilization of lead in soils. *Plant and Soil*, 249, 217–228.

GHS (2011) *Globally Harmonized System* of classification and labelling of chemicals. 4th revised edition. New York & Geneva, UN. [Online] Available from: http://www.unece .org/fileadmin/DAM/trans/danger/publi/ghs/ghs_rev04/English/ST-SG-AC 10-30-Rev4e.pdf. [Accessed 10th December 2013].

Gomez-Eyles, J.L., Collins, C.D. & Hodson, M.E. (2011) Using deuterated PAH amendments to validate chemical extraction methods to predict PAH bioavailability in soil. *Environmental Pollution*, 159, 918–923.

Gopalaswamy, U.V. & Nair, C.K.K. (1992) DNA binding and mutagenicity of lindane and its metabolites. *Bulletin of Environmental Contamination and Toxicology*, 49(2), 300–305.

Gruiz, K. (2000) When the chemical time bomb explodes? – Chronic risk posed by toxic metals at a former mining site. *Proceedings of ConSoil 2000, Leipzig, Germany.* 662–670.

Gruiz, K. (2005) Biological tools for soil ecotoxicity evaluation: Soil testing triad and the interactive ecotoxicity tests for contaminated soil. Italy, INCA. *Soil Remediation Series*, 6, 45–70.

Gruiz, K., Murányi, A., Molnár, M. & Horváth, B. (1998) Risk Assessment of Heavy Metal Contamination in the Danube Sediments from Hungary. *Water Science & Technology*, 37(6–7), 273–281.

Gruiz, K., Molnár, M. & Bagó, T. (1999) Interactive bioassay for environmental risk assessment. *Proceedings of SECOTOX'99, Munich. March 15–17.*

Gruiz, K., Horváth, B. & Molnár, M. (2001) *Környezettoxikológia, Environmental toxicology.* Budapest, Műegyetem Publishing Company. (In Hungarian.)

Gruiz, K., Molnár, M. & Feigl, V. (2009) Measuring adverse effects of contaminated soil using interactive and dynamic test methods. *Land Contamination & Reclamation,* 17(3–4), 445–463.

Gupta, A.K. & Sinha, S. (2007) Assessment of single extraction methods for the prediction of bioavailability of metals to *Brassica juncea* L. Czern. (var. Vaibhav) grown on tannery waste contaminated soil. *Journal of Hazardous Materials,* 149(1), 144–150.

Gupta, S.K., Vollmer, M.K. & Krebs, R. (1996) The importance of mobile, mobilizable and pseudo total heavy metal fractions in soil for three-level risk assessment and risk management. *Science of the Total Environment,* 178, 11–20.

Hajdu, Cs., Gruiz, K. & Fenyvesi, É. (2010) Direct testing of soil mutagenicity. In: Sarsby, R. W. & Meggyes, T. (eds.) *Construction for a sustainable environment. Proceedings of the International Conference of Construction for a Sustainable Environment, Vilnius, Lithuania, 1–4 July, 2008.* Leiden. CRC Press/Balkema. 229–237.

Hajdu, Cs., Gruiz, K., Fenyvesi, É. & Tukacs, L. (2009) Bioassays for pessimistic risk assessment of soil. *ISTA 14 Conference, Metz, France, Augustus 30–September 4, 2009.*

Hamelink, J.L., Landrum, P.F., Bergman, H.L. & Benson, W.H. (1994) *Bioavailability: Physical, Chemical and Biological Interactions.* Boca Raton, Fl, CRC Press.

Han, S., Naito, W., Hanai, Y. & Masunaga, S. (2013) Evaluation of trace metals bioavailability in Japanese river waters using DGT and a chemical equilibrium model. *Water Research,* 47(14), 4880–92. [Online] Available from: DOI: 10.1016/j.watres.2013.05.025. [Accessed 11th January 2014].

Health Canada (2007) *Ingestion bioavailability of arsenic, lead and cadmium in human health risk assessments: critical review, and recommendations* – Report. Environmental Health Assessment Services, Safe Environments Program, Project No. No50604. [Online] Available from: http://www.bioavailabilityresearch.ca/Health%20 Canada%20Bioavailability.final.pdf. [Accessed 15th January 2014].

Huckins, J.N., Tubergena, M.W. & Manuweera, G.K. (1990) Semipermeable membrane devices containing model lipid: A new approach to monitoring the bioavailability of lipophilic contaminants and estimating their bioconcentration potential. *Chemosphere,* 10(5), 533–552.

Hylander, L.D., Svensson, H.I. & Simán, Gy. (1995) Extraction of soil phosphorus with calcium chloride solution for prediction of plant availability. *Communications in Soil Science and Plant Analysis,* 26, 7–8.

INAP (2002) *Diffusive Gradients in Thin-films (DGT)* – A Technique for Determining Bioavailable Metal Concentrations, INAP (International Network for Acide Prevention). [Online] Available from: http://www.inap.com.au/public_downloads/Research_Projects/Diffusive_ Gradients_in_Thin-films.pdf. [Accessed 15th January 2014].

IUPAC (2014) *Glossary of Terms Used in Toxicology.* [Online] Available from: http://sis.nlm.nih.gov/enviro/iupacglossary/glossaryb.html. [Accessed 15th January 2014].

Juhasz, A.L., Smith, E., Weber, J., Rees, M., Rofe, A., Kuchel, T., Sansom, L. & Naidu, R. (2007) Comparison of *in vivo* and *in vitro* methodologies for the assessment of arsenic bioavailability in contaminated soils. *Chemosphere,* 69, 961–966.

Juhasz, A.L., Smith, E., Weber, J., Rees, M., Rofe, A., Kuchel, T., Sansom, L. & Naidu, R. (2008) Effect of soil ageing on *in vivo* arsenic bioavailability in two dissimilar soils. *Chemosphere,* 71, 2180–2186.

Juhasz, A.L., Weber, J. & Smith, E. (2011) Predicting arsenic relative bioavailability in contaminated soils, using meta analysis and relative bioavailability–bioaccessibility regression models. *Environmental Science & Technology,* 45, 10676–10683. [Online] Available from: dx.doi.org/10.1021/es2018384. [Accessed 15th January 2014].

Kalis, E.J.J., Weng, L.P., Dousma, F., Temminghoff, E.J.M. & van Riemsdijk, W.H. (2006) Measuring free metal ion concentrations in-situ in natural waters using Donnan Membrane Technique. *Environmental Science & Technology*, 40, 955–961.

Kelsey, J.W., Kottler, B.D. & Alexander, M. (1997) Selective chemical extractants to predict bioavailability of soil-aged organic chemicals. *Environmental Science & Technology*, 31, 214–217.

Laird, B.D., Peak, D. & Siciliano, S.D. (2010a) The effect of residence time and fluid volume to soil mass (LS) ratio on *in vitro* arsenic bioaccessibility from poorly crystalline scorodite. *Journal of Environmental Science & Health, Part A. Toxic/Hazardous Substances and Environmental Engineering*, 45, 732–739.

Laird, B.D. (2010b) Evaluating Metal bioaccessibility of soils and foods using the SHIME – Ph.D. Thesis, University of Saskatchewan, Saskatoon, Canada. [Online] Available from: http://ecommons.usask.ca/bitstream/handle/10388/etd-11292010-165216/Laird_CGSR_Approved_Thesis.pdf?sequence=1. [Accessed 15th January 2014].

Laird, B.D., Van de Wiele, T.R., Corriveau, M.C., Jamieson, H.E., Parsons, M.B., Verstraete, W. & Siciliano, S.D. (2007) Gastrointestinal Microbes Increase Arsenic Bioaccessibility of Ingested Mine Tailings Using the Simulator of the Human Intestinal Microbial Ecosystem. *Environmental Science & Technology*, 41, 554–5547.

Lakanen, E. & Erviö, R. (1971) A comparison of eight extractants for the determination of plant available micronutrients in soil. *Acta Agricultural Fennica*, 123, 223–232.

Lamb, D.T., Minga, H., Megharaj, M. & Naidu, R. (2009) Heavy metal (Cu, Zn, Cd and Pb) partitioning and bioaccessibility in uncontaminated and long-term contaminated soils. *Journal of Hazardous Materials*, 171, 1150–1158.

Laparra, M.J., Velez, D., Barbera, R., Montoro, R. & Farre, R. (2005) An approach to As(III) and As(V) bioavailability studies with Caco-2 cells. *Toxicology in vitro*, 19, 1071–1078.

Lin, C.Y., Abdullah, M.H., Musta, B., Praveena, S.M. & Aris, A.Z. (2011) Stability Behavior and Thermodynamic States of Iron and Manganese in Sandy Soil Aquifer, Manukan Island, Malaysia. *Natural Resources Research*, 20 (1) 45–56. [Online] Available from: dx.doi.org/10.1007/s11053-011-9136-2. [Accessed 15th January 2014].

Maddaloni, M., Manton, W., Blum, C., LoIacono, N. & Graziano, J. (1996) Bioavailability of soil-borne lead in adults, by stable isotope dilution. *Environmental Health Perspectives*, 106, 1589–1594.

Maron, D.M. & Ames, B.N. (1983) Revised methods for the Salmonella mutagenicity test. *Mutation Research*, 113, 173–215.

Mercier, G., Duchesne, J. & Blackburn, D. (2002) Removal of metals from contaminated soils by mineral processing techniques followed by chemical leaching. *Water Air Soil Pollution*, 135, 105–130.

Metals bioavailability (2003) *Guide for Incorporating Bioavailability Adjustments into Human Health and Ecological Risk Assessments* at U. S. Department of Defense Facilities Part 1: Overview of Metals Bioavailability. Tri-Service Ecological Risk Assessment Workgroup. [Online] Available from: http://www.itrcweb.org/contseds-bioavailability/References/bioavailability01.pdf. [Accessed 15th January 2014].

Mingot, J., de Miguel, E. & Chacón, E. (2011) Assessment of oral bioaccessibility of arsenic in playground soil in Madrid (Spain): A three-method comparison and implications for risk assessment. *Chemosphere*, 84(10), 1386–91.

Oomen, A.G., Rompelberg, C.J.M., Bruil, M.A., Dobbe, C.J.G., Pereboom, D.P.K.H. & Sips, A.J.A.M. (2003) Development of an *in vitro* digestion model for estimating the bioaccessibility of soil contaminants. *Archives of Environmental Contamination & Toxicology*, 44(3), 0281–0287.

Peijnenburg, W.J.G.M. & Jager, T. (2003) Monitoring approaches to assess bioaccessibility and bioavailability of metals: Matrix issues. *Ecotoxicology & Environmental Safety*, 56, 63–77.

Puglisi, E., Vernile, P., Bari, G., Spagnuolo, M., Trevisan, M., de Lillo, E. & Ruggiero, P. (2009) Bioaccessibility, bioavailability and ecotoxity of pentachlorophenol in compost amended soils. *Chemosphere*, 77, 80–86.

Reid, B.J., Stokes, J.D., Jones, K.C. & Semple, K.T. (2000) Nonexhaustive cyclodextrin-based extraction technique for the evaluation of PAH bioavailability. *Environmental Science and Technology*, 34, 3174–3179.

Ren, H.M., Wang, J.D. & Zhang, X.L. (2006) Assessment of soil lead exposure in children in Shenyang, China. *Environmental Pollution*, 144, 327–335.

Roberts, M.S., Pugh, J.W., Hadgraft, J. & Watkinson, A.C. (1995) Epidermal permeability-penetrant structure relationships: 1. An analysis of methods of predicting penetration of monofunctional solutes from aqueous solutions. *International Journal of Pharmaceutics*, 126, 219–233.

Rodriguez, R.R. & Basta, N.T. (1999) An *in vitro* gastrointestinal method to estimate bioavailable arsenic in contaminated soils and solid media. *Environment Science & Technology*, 33, 642–649.

Rodriguez, R.R., Basta, N.T., Casteel, S.W., Pace, L.W. (1999) An in-vitro gastro-intestinal method to assess bioavailable arsenic in contaminated soils and solid media. *Environmental Science & Technology*, 33, 642–649.

Ruby, M.W., Davis, A., Link, T.E., Schoof, R., Chaney, R.L., Freeman, G.B. & Bergstrom, P. (1993) Development of an *in vitro* screening test to evaluate the *in vivo* bioaccessibility of ingested mine-waste lead. *Environmental Science & Technology*, 27(13), 2870–2877.

Ruby, M.W., Davis, A., Schoof, R., Eberle, S. & Sellstone, C.M. (1996) Estimation of lead and arsenic bioavailability using a physiologically based extraction test. *Environmental Science & Technology*, 30(2), 422–430.

Salaun, P. & Buffle, J. (2004) Integrated microanalytical system coupling permeable liquid membrane and voltammetric detection for trace metal speciation. Theory and applications. *Journal of Analytical Chemistry*, 76, 31–39.

Safety Toys Directive (2009) *Safety of toys. Directive* 2009/48/EC of the European Parliament and of the Council. [Online] Available from: http://eur-lex.europa.eu/LexUriServ/LexUriServ .do?uri=OJ:L:2009:170:0001:0037:en:PDF. [Accessed 15th January 2014].

Schroder, J.L., Basta, N.T., Casteel, S.W., Evans, T.J., Payton, M.E. & Si, J. (2004) Validation of the *in vitro* gastrointestinal (IVG) method to estimate relative bioavailable lead in contaminated soils. *Journal of Environmental Quality*, 33, 513–521.

Seiler, J.P. (1991) Pentachlorophenol. Mutation Research. *Reviews in Genetic Toxicology*, 257(1), 27–47.

Sekine, K., Watanabe, E., Nakamura, J., Takasuka, N., Kim, D.J., Asamoto, M., Krutovskikh, V., Baba-Toryhama, H., Ota, T., Moore, M.A., Masuda, M., Sugimoto, H., Nishino, H., Kakizoe, T. & Tsuda, H. (1997) Inhibition of azoxymethane-initiated color tumor by bovine lactoferrin administration in F344 rats. *Journal of Cancer Research*, 88(6), 523–526.

Semenzin, E., Critto, A., Carlon, C., Rutgers, M. & Marcomini, A. (2007) Development of a site-specific Ecological Risk Assessment for contaminated sites: Part II. A multi-criteria based system for the selection of bioavailability assessment tools. *Science of the Total Environment*, 379, 34–45.

Semple, K.T., Doick, K.J., Burauel, P., Craven, A., Harms, H. & Jones, K.C. (2004) Defining bioavailability and bioaccessibility of contaminated soil and sediment is complicated. *Environmental Science & Technology*, 38, 228A–231A. [Online] Available from: DOI: 10.1021/es040548w. [Accessed 15th January 2014].

Semple, K.T., Doick, K.J., Wick, L.Y. & Harms, H. (2007) Microbial interactions with organic contaminants in soil: Definitions, processes and measurement. *Environmental Pollution*, 150, 166–176.

Senila, M., Levei, E.A. & Senila, L.R. (2012) Assessment of metals bioavailability to vegetables under field conditions using DGT, single extractions and multivariate statistics. *Chemistry Central Journal*, 6, 119. [Online] Available from: DOI:10.1186/1752-153X-6-119. [Accessed 15th January 2014].

Sigg, L., Black, F., Buffle, J., Cao, J., Cleven, R., Davison, W., Galceran, J., Gunkel, P., Kalis, E., Kistler, D., Martin, M., Noël, S., Nur, Y., Odzak, N., Puy, J., van Riemsdijk, W., Temminghoff, E., Tercier-Waeber, M-L., Toepperwien, S., Town, R.M., Unsworth, E., Warnken, K.W., Weng, L., Xue, H. & Zhang, H. (2006) Comparison of analytical techniques for dynamic trace metal speciation in natural freshwaters. *Environmental Science & Technology*, 40(6), 1934–1941.

Søndergaard, J., Bach, L. & Gustavson, K. (2014) Measuring bioavailable metals using diffusive gradients in thin films (DGT) and transplanted seaweed (*Fucus vesiculosus*), blue mussels (*Mytilus edulis*) and sea snails (*Littorina saxatilis*) suspended from monitoring buoys near a former lead-zinc mine in West Greenland. *Marine Pollution Bulletin*, 78, 102–109. [Online] Available from: DOI: 10.1016/j.marpolbul.2013.10.054. [Accessed 15th January 2014].

Stokes, J.D., Wilkinson, A., Reid, B.J., Jones, K.C. & Semple, K.T. (2005) Prediction of polycyclic aromatic hydrocarbon biodegradation in contaminated soils using an aqueous hydroxypropyl-beta-cyclodextrin extraction technique. *Environmental Toxicology and Chemistry*, 24(6), 1325–1330.

Stokes, J.D., Paton, G.I. & Semple, K.T. (2006) Behaviour and assessment of bioavailability of organic contaminants in soil: relevance for risk assessment and remediation. *Soil Use Management*, 21, 475–486.

Tack, F.M. & Verloo, M.G. (1995) Chemical speciation and fractionation in soil and sediment heavy metal analysis: a review. *International Journal of Environmental Analytical Chemistry*, 59, 225–238.

Tang, X.Y., Zhu, Y.G., Cui, Y.S., Duan, J. & Tang, L. (2006a) The effect of ageing on the bioaccessibility and fractionation of cadmium in some typical soils of China. *Environment International*, 32, 682–689.

Tang, X.Y., Tang, L., Zhu, Y.G., Xing, B.S., Duan, J. & Zheng, M.H. (2006b) Assessment of the bioaccessibility of polycyclic aromatic hydrocarbons in soils from Beijing using an *in vitro* test. *Environmental Pollution*, 140, 279–285.

Tang, J. & Alexander, M. (1999) Mild extractability and bioavailability of aged and unaged polycylic aromatic hydrocarbons in soil. *Environmental Toxicology & Chemistry*, 18, 2711–2714.

Tao, Y., Zhang, S., Wang, Z., Ke, R., Shan, X. & Christie, P. (2008 ) Biomimetic accumulation of PAHs from soils by triolein-embedded cellulose acetate membranes (TECAMs) to estimate their bioavailability. *Water Research*, 42, 754–762.

Tao, Y., Zhang, S., Wang, Z. & Christie, P. (2009) Predicting bioavailability of PAHs in field-contaminated soils by passive sampling with triolein embedded cellulose acetate membranes. *Environmental Pollution*, 157, 545–551.

Temminghoff, E.J.M., Plette, A.C.C., Van Eck, R. & Van Riemsdijk, W.H. (2000) Determination of the chemical speciation of trace metals in aqueous systems by the Wageningen Donnan membrane technique. *Analytica Chimica Acta*, 417, 149–157.

Tessier, A., Campbell, P.G.C. & Bisson, M. (1979) Sequential extraction procedure for the speciation of particulate trace metals. *Analytical Chemistry*, 51(7), 844–851.

Town, R.M. (1998) Chronopotentiometric stripping analysis as a probe for copper(II) and lead(II) complexation by fulvic acid: Limitations and potentialities. *Analytica Chimica Acta*, 363, 31–43.

Town, R.M. & van Leeuwen, H.P. (2004) Depletive Stripping Chronopotentiometry: A Major Step Forward in Electrochemical Stripping Techniques for Metal Ion Speciation Analysis.

*Electroanalysis*, 16, 458–471. [Online] Available from: DOI: 10.1002/elan.200302844. [Accessed 10th December 2013].

TOXNET (2014) *Toxicology Data Network*. [Online] Available from: http://toxnet.nlm.nih.gov/. [Accessed 15th January 2014].

US EPA (1991) AVS and SEM Procedure. EPA-821-R-51-100. [Online] Available from: http://www.epa.gov/nscep/index.html. [Accessed 10th December 2013].

US EPA (1994) Guidance manual for the integrated exposure uptake biokinetic model for lead in children. Office of Emergency and Remedial Response, Washington, D.C. EPA/540/R-93/081, PB93-963510.

US EPA (2003) *Recommendations of the technical review workgroup for lead for an approach to assessing risks associated with adult exposures to lead in soil*. [Online] Available from: http://www.epa.gov/oswer/riskassessment/ragsa/pdf/rags_a.pdf. [Accessed 10th December 2013].

US EPA (2009) *Validation assessment of in vitro lead bioaccessibility assay for predicting relative bioavailability of lead in soils and soil-like materials at superfund sites*. OSWER 9200.3-51. [Online] Available from: http://www.epa.gov/superfund/bioavailability/lead_tsd_add09.pdf. [Accessed 10th December 2013].

US EPA (2012) *Standard operating procedure for an in vitro bioaccessibility assay for lead in soil*. Research Triangle Park (NC), US EPA. EPA 9200.2-86. [Online] Available from: www.epa.gov/superfund/bioavailability/pdfs/EPA_Pb_IVBA_SOP_040412_FINAL_SRC.pdf. [Accessed 10th December 2013].

van Leeuwen, H.P. & Town, R.M. (2002) Stripping chronopotentiometry at scanned deposition potential (SSCP). – Part 1. Fundamental features. *Journal of Electroanalytical Chemistry*, 536, 129–140.

Vaszita, E., Szabó, J. & Gruiz, K. (2009) Complex leaching of metal-sulfide-containing mine waste and soil in microcosms. *Land Contamination & Reclamation*, 17(3–4), 465–475.

Vecchia, B.E. & Bunge, A.L. (2002) Skin absorption databases and predictive equations. In: Guy, R.H. & Hadgraft, J. (eds.) *Transdermal Drug Delivery*. Chapter 3. Publisher Marcel Dekker.

Versantvoort, C.H.M., Van de Kamp, E. & Rompelberg, C.J.M. (2004) *Development and applicability of an in vitro digestion model in assessing the bioaccessibility of contaminants from food*. RIVM Report 320102002/2004, Bilthoven. [Online] Available from: http://www.rivm.nl/dsresource?objectid=rivmp:13066&type=org&disposition=inline&ns_nc=1. [Accessed 15th January 2014].

Wen, B., Li, R., Zhang, S., Shan, X., Fang, J., Xiao, S. & Khan, S.U. (2009) Immobilization of pentachlorophenol in soil using carbonaceous material amendments. *Environmental Pollution*, 157, 968–974.

Weng, L.P., Van Riemsdijk, W.H. & Temminghoff, E.J.M. (2005) Kinetic aspects of DMT for measuring free metal ion concentrations. *Analytical Chemistry*, 77, 2852–2861.

Wenzel, W.W., Kirchbaumer, N., Prohaska, T., Stingeder, G., Lombic, E. & Adriano, D.C. (2001) Arsenic fractionation in soils using an improved sequential extraction procedure. *Analytica Chimica Acta*, 436, 309–323.

Wiktionary (2014) *Bioaccessibility*. [Online] Available from: http://en.wiktionary.org/wiki/bioaccessibility. [Accessed 15th January 2014].

Wittsiepe, J., Erlenkamper, B., Welge, P., Hack, A. & Wilhelm, M. (2007) Bioavailability of PCDD/F from contaminated soil in young Goettingen minipigs. *Chemosphere*, 67, 355–364.

Wragg, J. & Cave, M. (2012) Assessment of a geochemical extraction procedure to determine the solid phase fractionation and bioaccessibility of potentially harmful elements in soils: A case study using the NIST 2710 reference soil. *Analytica Chimica Acta*, 722, 43–54.

Yang, X., Wang, F., Gu, C. & Jiang, X. (2010) Tenax TA extraction to assess the bioavailability of DDTs in cotton field soils. *Journal of Hazardous Materials*, 179, 676–683.

Xue, H.B. & Sigg, L. (2002) A review of competitive ligand exchange/voltammetric methods for speciation of trace metals in freshwater. In: Taillefert, M. &, Rozan, T.F. (eds.) *Environmental Electrochemistry: Analysis of Trace Element Bio-geochemistry.* Washington DC, USA, American Chemical Society Symposium Series No.811. pp. 336–370.

Zhang, H., Davison, W. & Grime, G.W. (1995a) New In-Situ Procedures for Measuring Trace Metals in Pore Waters, *ASTM STP Series*, 1293, 170–181.

Zhang, H., Davison, W., Miller, S. & Tych, W. (1995b) In situ high resolution measurements of fluxes of Ni, Cu, Fe and Mn and concentrations of Zn and Cd in porewaters by DGT. *Geochimica & Cosmochimica Acta*, 59, 4181–4192.

Zhang, H. & Davison, W. (2001) In situ speciation measuements. Using diffusive gradients in thin films (DGT) to determine inorganically and organically complexed metals. *Pure & Applied Chemistry*, 73, 9–15.

Zhang, H., Davison, W., Mortimer, R.J.G., Krom, M.D., Hayes, P.J. & Davies, I.M. (2002). Localised remobilisation of metals in a marine sediment. *Science of the Total Environment*, 296, 175–187.

Zhang, M.K., Liu, Z.Y. & Wang, H. (2010) Use of single extraction methods to predict bioavailability of heavy metals in polluted soils to rice communications in soil. *Science and Plant Analysis*, 4(7), 820–831.

Zhang, H., Zhao, F.J., Sun, B., Davison, B. & McGrath, S.P. (2001) A new method to measure effective soil solution concentration predicts Cu availability to plants. *Environmental Science & Technology*, 35, 2602–2607.

Chapter 8

# Microcosm models and technological experiments

*K. Gruiz, M. Molnár, V. Feigl, E. Vaszita & O. Klebercz*
Department of Applied Biotechnology and Food Science,
Budapest University of Technology and Economics, Budapest, Hungary

## ABSTRACT

Microcosms are easily controllable, small-size models of the environment. In spite of some shortcomings deriving from high surface/volume ratio and reduced diversity, microcosms provide highly realistic results, suitable for decision making. Microcosms are useful tools for learning and exploring nature, the water and soil ecosystems, for simulating natural or anthropogenic impacts on the ecosystems, for measuring toxicity by a multispecies, dynamic system and for the small-scale modeling of technologies.

This chapter introduces the practice and implementation of microcosm testing and modeling via examples from the authors' experience. The examples include microcosms used for modeling the fate and behavior of contaminants in the environment, their biodegradation and bioavailability, and microcosms for modeling bioleaching and stabilization. In addition to the experimental set-up of the small-scale models, this chapter discusses the integrated monitoring of microcosms in a general way by combining physico-chemical analysis and biological testing. It also discusses the evaluation and interpretation of the results' utilization in risk management and decision making. Applications and results of various microcosms are discussed in Chapters 4 and 5.

## I INTRODUCTION

In environmental toxicology, *microcosm* is defined as a small representative part of the real environment selected for the study of ecosystem response to environmental conditions or chemical contaminants (Chapters 4 and 5). Microcosm studies are usually performed in the laboratory, sometimes in the field.

Microcosms are also used as small-scale technological models for water treatment or soil remediation. These microcosms can simulate those natural environmental processes which are responsible for the hazard and risk posed by environmental contaminants or can provide a suitable basis for remediation and engineered interventions.

In both applications, microcosms simulate the real ecosystem, measure the effects of physical, chemical or environmental parameters and those of the pollutants on the environmental compartments, phases and, on the complex multispecies ecosystem at different trophic levels.

Microcosm and mesocosm studies can be considered as intermediates between bioassays and ecosystem studies. Fate processes and behavior of chemicals and biological changes on population, community or ecosystem level can be monitored by all kinds of physical, chemical, biological, ecological or toxicological tools and their combinations.

**Microcosm** in general means a 'little world', a 'world in miniature', for example a part of the environment, natural water, sediment or soil that is regarded as a miniaturized copy of the natural system. There is a wide range in the naturalness or artificiality of microcosms, from undisturbed natural aliquots (extracted from the real environment) under controlled conditions to a fully artificial controlled and regulated set-up of a system with desired complexity. It is a highly realistic model, but – since every single microcosm has its own history and evolution – the statistical evaluation of the changes in space and time needs special tools (Gruiz *et al.*, 2001).

The microcosm experiments do not only model the natural system but also allow for investigation of external interventions into the natural processes, thus challenging the system artificially. Microcosms enable the monitoring and control of natural processes and provide basic information for the design of biological and ecological technologies. The outcomes of technological processes and their complex effects on the physical, chemical and biological properties and activities can be measured and monitored in time. The series of technological parameters can be tested in order to find their optimum.

Microcosm applications can cover both water and soil ecosystems, and they can vary in size, duration and complexity (Gruiz *et al.*, 2001). They are key engineering tools in modern environmental management for both risk assessment and risk reduction and the results provided by them can be used for direct decision making.

The essence of microcosm testing – when modeling either natural or technological processes – is the appropriate selection of the determining factors and influential parameters. Once there is knowledge on or familiarity with the environment, the concept of the model should be created by transforming the complex environment into a simplified model. The main target of a microcosm concept is to maintain as much of the important and influencing environmental factors, and the diversity and activity of the community as possible. This means that the principle and the concept have priority when planning the microcosms, but the set-up and the implementation can be fairly basic. The concept of a microcosm needs an accurate conceptual risk model.

## 2 AQUATIC MICROCOSMS FOR SCREENING CHEMICAL SUBSTANCES AND TECHNOLOGIES

Open or closed batch reactors or flow-through set-ups in single or cascade arrangement can be used for water, wastewater, leachates and other liquid-form environmental matter. Living ecosystems can be compressed into a small-size ecosystem model or into a water-treatment technology model. Batch-type microcosms work with or without mixing, with inlets and outlets suitable for recycling or for injection of air, chemicals, nutrients, regulators and other chemical or biological additives as well as for the extracting of samples or aliquots. Physical factors such as heat or radiation can also be included in these microcosms. The aim of manipulations in the microcosm is the study of the contaminant and/or the community as well as the community's response to natural or simulated impacts.

Flow-through column and tube reactors are suitable for the simulation of natural transport behavior of chemicals, transport pathways in the water column, depths and the relevant redox potential as well as water treatment processes such as stripping, adsorption and absorption, filtration, flotation, various chemical reactions, or the intensification of community activity. They are popular set-ups for aquatic model systems and also for the simulation of water treatment including wastewater treatment technologies.

Realistic aquatic micro- and mesocosms contain sediment too, and can be used for testing the fate and the behavior of the chemical substance in the water–sediment system, the response of aquatic and benthic ecosystems to chemical substances. In addition, they can be used for general or targeted ecosystem studies and for the simulation of technological interventions in aquatic systems.

The American Society for Testing and Materials (ASTM International, 2014) standardized aquatic and terrestrial microcosms for regulatory purposes:

– Standard Practice for Standardized Aquatic Microcosms, fresh water (ASTM E1366, 2011)
– Standard Guide for Conducting a Terrestrial Soil-Core Microcosm Test (ASTM E1197, 2012)

United States Environmental Protection Agency (US EPA, 2014) uses standardized microcosms for generic and site-specific risk assessment:

– Generic Freshwater Microcosm Test (1996) is used for assessing the potential hazard of a chemical substance to freshwater ecosystems by acquiring data on the chemical fate and/or ecological effects of chemical substances and mixtures. Standardized aquatic microcosms, naturally derived mixed-flask culture microcosms, or naturally derived pond microcosms are used with and without sediment. The microcosms contain freshwater algae and zooplankton with an assortment of unidentified bacteria and fungi.
– Site-Specific Aquatic Microcosm Test (1996) is performed in a microcosm made of an indigenous water column and sediment core. This test system is capable of evaluating organic chemical substances, either soluble or insoluble, which may form either air–water surface films or aggregates which sink to bottom sediments. The acquired data can support hazard assessment of a chemical substance to a particular natural aquatic system.
– Sediment/Water Microcosm Biodegradation Test (1998) has been developed for testing pesticides, biocides and other toxic substances. Sediment and water with a representative sample of the natural microbial community from a test site of interest are required in this test.

Some uses of standardized microcosms and the results are:

– Use of microcosm data for regulatory decisions: fate and behavior of chemicals as well as the extent of their adverse effects at several trophic levels on competitive species, spontaneous detoxification and system recovery, etc.

–   Use of microcosm data for generic and site-specific risk management: quality and quantity of risk, effect of risk management options and risk reduction using technologies.
–   Determination of safe levels of chemical exposure.
–   Study of community changes and ecosystem functional complexity.
–   Study of community behavior after chemical stress, how ecosystems recover to resemble the reference state.
–   Measurement of microbial population and community dynamics, density, diversity, biochemical/enzyme activities, biodegradation and mineralization of toxicant molecules under various environmental conditions such as pH, redox potential, salinity, temperature and sediment/water interface conditions, etc.

The aim and size of aquatic micro-and mesocosms show a wide scope from a few milliliter jar or column through laboratory aquariums or water columns, to open air ponds, artificial lakes, koi ponds or huge-size aquariums or aquacultures.

Aquatic mesocosms integrate the experience of ecology, aquarium science, aquacultures and wastewater treatment. MESOCOSM (2014) is an information hub of mesocosm facilities in aquatic ecosystems world-wide, provided by the MESOAQUA project. The aim of most research on aquatic mesocosms is to better understand the ecological behavior of the aquatic component of the ecosystem and its utilization in ecological technologies such as

–   artificial ponds and wetlands for the treatment of contaminated runoff waters and leachates;
–   passive artificial systems for the neutralization and clean-up of acidic and/or contaminated mine drainage or leachates;
–   active subsurface soil zones for the filtration or detoxification of seepage, drainage or leachates;
–   rhizofiltration for the elimination of nutrients or other contaminants from runoff waters;
–   living machines for the remediation of contaminated lakes and reservoirs or just for the conservation of natural water bodies as well as for the treatment of wastewaters;
–   all possible ecotechnologies applied to conservation or sustainable use of waters.

From the point of view of evaluation and interpretation of the results it is important to understand the limitations of micro- and mesocosms due to the compression of the real environment into a small volume (Landis & Yu, 1999; Adey & Loveland, 1999).

The aim of testing and manipulating the ecosystem and the bio- or ecotechnologies determine the type and the set-up of the micro- and mesocosms. Aquatic micro- and mesocosms are supplied and available in a wide range for teaching and research purposes as well as for koi ponds or water gardens: these may be suitable for technological purposes, too. But if the ready-made ones do not fit the necessary set-up, they must be designed and built by ourselves.

The most common form of an aquatic mesocosm is a pond that contains natural fresh or marine water and sediment, and a community consisting of at least micro-organisms, plankton, macrophytes, macroinvertebrates and, if necessary, fish. The mesocosm may be the perfect tool for studying impacts on an ecosystem under field-relevant conditions. Technological mesocosms measure the impact of technological parameters such as flow rate, temperature, pH, redox potential, aeration, ion concentration, nutrient content, the effect of additives, special nutrients, chemicals of beneficial or adverse effects on the aquatic ecosystems, diversity and activity of the community, as well as element and energy cycling of the micro- or mesocosm system. Based on the results of technological studies, the real environmental system can be manipulated and controlled by engineering tools applying the technological parameters that proved to be optimal in the micro- and mesocosm experiment. The purpose of these engineering manipulations may only be to maintain the healthy state of our waters, i.e. conservation, or to remediate the unhealthy environment by reducing the risk of deterioration and reversing disadvantageous changes. Our natural environment can only be engineered in an eco-efficient and sustainable way if the actions are based on a deep insight into the structure and function of ecosystems, natural processes, trends and acceptable risk levels. Micro- and mesocosms greatly support the acquisition of such knowledge on natural diversity and processes such as element cycling, eutrophication, and the fate and behavior of contaminants in the environment, e.g. their biodegradation.

Figure 8.1 shows a simple laboratory aquatic microcosm vessel equipped with inlet and outlet fittings for aeration, water pumping (subtraction, flow-through or recycling), injection and sampling. The size can vary from a few milliliters to hundreds of liters.

In Figure 8.2 parallel aquatic microcosms are shown in a water bath for temperature control. The open-top columns are provided with plugs at the bottom and filled

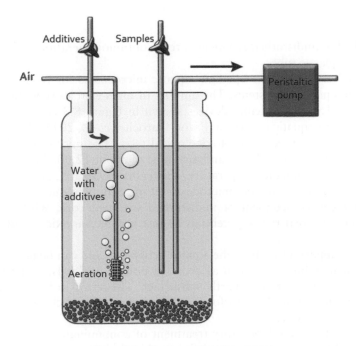

*Figure 8.1*  Open aquatic microcosm.

Aeration

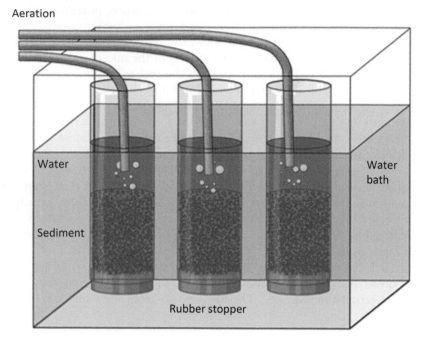

*Figure 8.2* Open laboratory microcosms with undisturbed sediment cores placed into a thermostat for testing biodegradation, eutrofication or toxicity.˜

with disturbed or undisturbed sediment cores. Continuous aeration and water cycling are applied by plastic tubes.

Figure 8.3 and 8.4 show open-air aquatic microcosms for research and testing of changes in aquatic ecosystems. The number of units enables repetitions and serial investigations. The experimental devices shown in Figure 8.3 can study the impact of warming on an aquatic ecosystem (Yvon-Durocher *et al.*, 2011). Figure 8.4 shows open-air microcosms used for ecotoxicological tests. The microcosms have a water depth of 1 m, a diameter of 3.9 m, and a volume of about 12,000 L. They are open to the elements, e.g. rain, light, temperature (University of Guelph, 2013).

Figure 8.5 shows a sediment microcosm set-up. Sediment microcosms can be run using a homogenized sediment or undisturbed sediment core, which can simulate original layers and real redox potentials to test aerobic, anoxic or fully anaerobic processes.

Plants are important parts of the aquatic ecosystem and can be utilized in natural water treatment technologies such as artificial ponds, wetlands, living machines or reactive zones on the surface or partly subsurface. The set-up in Figure 8.6 is suitable for modeling and testing the transforming, biodegrading and cleaning capacity of a plant-based aquatic system.

Figure 8.7 shows the laboratory treatment of contaminated water in a fluidization column. Different sorbents can be fluidized in the contaminated water to ensure

*Figure 8.3* Microcosms for investigating the impact of warming on a freshwater ecosystem (Yvon-Durocher, 2013).

intensive contact between the dissolved contaminant molecules and the sorbent. The same set-up was applied for the molecular encapsulation and removal of bisphenol-A using polymerized cyclodextrin beads (Gruiz *et al.*, 2010a).

## 3  SOIL MICRO- AND MESOCOSMS FOR MODELING ENVIRONMENTAL PROCESSES IN BIO- AND ECOTECHNOLOGIES

Soil micro- and mesocosms provide information under controlled conditions on the fate, natural changes and evolution within an actual representative part of the contaminated soil, and can model typical natural processes and remediation scenarios.

*Figure 8.4* Aquatic ecotoxicology microcosms (University of Guelph, 2013).

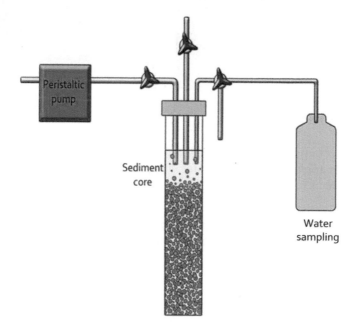

*Figure 8.5* Closed sediment microcosms.

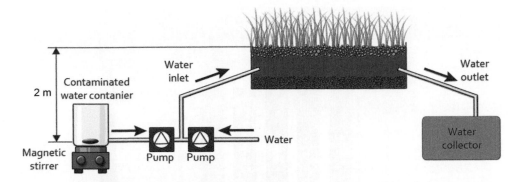

Figure 8.6 Aquatic microcosm with sediment, water and vegetation.

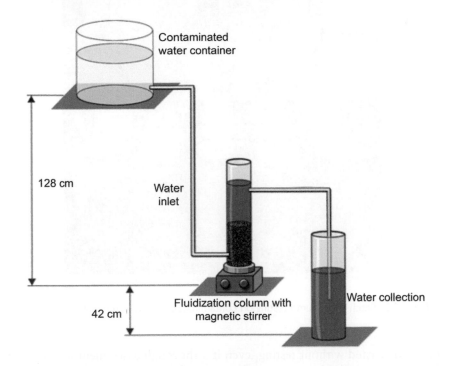

Figure 8.7 Microcosm for modeling wastewater fluidization.

General laboratory microcosms are being increasingly used for treatability testing of soil and groundwater. Microcosms are essential tools for pre-testing soil to help technology selection and planning because experience and routine cannot substitute site-specific information. Every site is unique and the number and combinations of biotic and abiotic parameters are so high that the impact of an additive or a technology

*Figure 8.8*  Batch and column testing on contaminated groundwater (Adventus, 2013).

cannot be forecasted without testing, even if a thorough assessment has been carried out. This is of particular importance when new additives and innovative technologies are being developed.

Figure 8.8 shows flow-through columns to mimic groundwater flow conditions. The columns contain soil from a contaminated site and a special inoculum for enhancing biodegradation. Groundwater is pumped through the column in an up-flow manner. The 2-liter glass jars filled with soil ensure longer residence time for natural attenuation of the contaminated groundwater by biodegrading the contaminants. Column and batch systems can be combined as shown in the photo: the outflow of the column is treated in the jars. The outflow from the jars is analyzed. The test system in the picture is used by the Laboratory Treatability Testing Services of the Adventus Company (Adventus, 2013).

The model ecosystems, or laboratory *microcosms* introduced here have become a key research tool in soil ecology and soil remediation. Due to the speed, statistical power and mechanistic insights attainable by laboratory-based microcosm experiments, they have made considerable contribution to today's knowledge about soil and the behavior of the soil microbiological community. Soil science and remediation have proved the value and importance of microcosm testing, but at the same time they drew attention to the fact that, given the small scale and artificiality, microcosm results must be validated by field model ecosystems (mesocosms), field tests or pilot applications to enable low-uncertainty extrapolation to field dimensions.

Various soil microcosm configurations/types have been developed to study soils contaminated with organic and inorganic pollutants.

- Open static (batch) reactors can be used to study the progress of biological and chemical processes, biodegradation, *in situ* natural attenuation, metal mobilization and immobilization (stabilization).
- Closed-reactor systems have been used for modeling anoxic conditions or for comparative studies of aerobic and anaerobic conditions and their effects on biodegradation or toxicity.
- Flow-through soil microcosms can be arranged horizontally or vertically and are used to study site-specific transport processes and fate, and the effects of risk reduction measures/interventions such as stabilization, filtering or the application of permeable reactive barriers.

Disadvantages of the microcosms are their small size and large surface–volume ratio as well as the lack of interaction with the natural environment. Their gradual application from the simplest set-up to the most complicated one supports the decision e.g. on the application of serial technological parameters (for example, aeration for one hour, 2, 6, 12 and 24 hours a day, or the application of 1%, 2%, 5% and 10% from an additive, etc.) and provides benefits or compensates for deficiencies. The possibility of repetitions and the 'unlimited' number of combinations or variations made it an essential tool for technology pre-experiments.

The following is an introduction to soil microcosms constructed and applied to monitoring natural or anthropogenic processes in the environment and to the measuring of the effect of contaminants and the results of remedial activities. Prior to the selection, planning and field application of an environmental technology, the technological parameters (pH, temperature, aeration, additives, etc.) must be examined and their applicable range determined. Technological microcosms proved to be the ideal tools for selecting and planning soil remedial technologies, as well as for the monitoring of the remediation process. Biodegradability of a mixture of soil contaminants, the effect and the proper amount of additives, the changes of mobility and the adverse effects of the contaminants, and the adaptive behavior of soil microbiota were successfully tested in microcosms by the authors of this chapter. These applications will be introduced below and their use demonstrated in Volumes 4 and 5 in the context of field applications (Molnár *et al.*, 2009a; Hajdu *et al.*, 2009; Molnár *et al.*, 2009b; Vaszita *et al.*, 2009a,b,c; Feigl *et al.*, 2009a).

*Figure 8.9* Test series for screening biodegradation in soil and the effect of enhancing additives.

## 3.1 Testing the effects of environmental and anthropogenic interventions in a small volume

50–500 g soil in closed, open or flow-through laboratory systems can be used for testing

- the effect of natural environmental and technological parameters (pH, temperature, moisture content, etc.) on contaminants' risk and microbial response;
- the adverse effects of pollutants and their different chemical forms;
- any adverse or beneficial effects on the natural soil biota or on controlled organisms placed into the soil.

The number, activity and diversity of the organisms or other end points characteristic of the soil ecosystem or test organisms are measured once (at the end) or at regular time intervals (Gruiz, 2005; Leitgib *et al.*, 2007). A large number of microcosms can be studied in parallel, and a series of variable technological parameters can be applied to ensure great statistical power and low uncertainty in deciding whether a parameter has an effect or not.

The soil sample can be placed into simple jars, pots, plastic containers (Figure 8.9) or more sophisticated tanks with built-in sensors, injectors or sampling devices.

The experimental pattern is based on the comparison of untreated and treated soils: the difference is measured and evaluated using statistical tools.

## 3.2 Testing biodegradation and bioavailability

Closed, open and flow-through reactors filled with soil (50–2000 g) are used to measure soil-specific biodegradation and bioavailability of pollutants. Contaminated soil is used or the substance to be tested is mixed with or infiltrated into the soil (Molnár, 2006; Molnár *et al.*, 2009a; Molnár *et al.*, 2009b). The simplest set-up is a pot or a jar filled with soil. The end points may be

– the decreasing amount/concentration or mobility of the substance,
– the increasing amount/concentration of the biodegradation products or
– any indicator of the metabolic activity of the soil microbiota.

Short- or long-term biodegradation, intensified and accelerated processes can be tested in static and dynamic modes. Dynamic mode in this case requires that the critical parameter is applied in an impulse mode and the response of the soil is monitored.

## 3.3  Testing long-term pollution processes in the environment

When studying soil pollution by simulating a polluting process, e.g. floods, precipitation infiltration or soil pollution caused by deposited contaminated sediment (Gruiz, 2005), it is important to understand those natural processes which influence the scale of risk and can serve as a basis for risk reduction. Transport and dissipation, spontaneous stabilization, physical, chemical or biological degradation may lead to rapid removal of or reduction in the active form of the contaminant. If the risk of a contaminant is reduced by transport in one place, but increased in another, the risk management measure should separate the first place from the other by controlling the transport. A typical environmental process is leaching, which can be very risky when it goes on in an uncontrolled way, but may serve as a biotechnology when the leachate is collected and treated. Complex leaching of mine waste stockpiled in the environment (Vaszita *et al.*, 2009a) was modeled in a long-term (3–4 years) experiment, observing and monitoring the processes in 2000–4000 g batch or flow-through microcosm vessels equipped with solid and leachate sampling possibilities. The results were integrated into the risk management plan of an abandoned mining site (see also Volume 5).

## 3.4  Testing microbial activity and plant growth in contaminated soil

Activating soil microorganisms and growing plants on contaminated soil may reduce risk due to dusting, runoff waters, leachates and erosion. Vegetation normalizes water balance, reduces or stops deflation and erosion, and increases organic matter content and microbial activity of the soil.

In the example shown in Figure 8.10 the soils in the pots were contaminated with incremental red mud doses/amounts. Plant growth (tolerance) was tested as a function of red mud content. The aim was to find those candidate plants which can grow on soils contaminated with various amounts of red mud and to find the upper red mud level candidate plants can tolerate. These simple tests give an answer within 1–2 weeks to questions which neither scientists nor practitioners can answer since nobody has experienced the special problem of an alkali red mud flood on agricultural soil that happened in the Ajka, Hungary, red mud disaster in 2010.

*Figure 8.10* Serial tests in static microcosms: soil contaminated with increasing concentrations of red mud.

## 3.5 Technological pre-experiments

Technology screening as part of remediation planning and scale-up prior to technology selection (Molnár *et al.*, 2009a; Feigl *et al.*, 2009a) provides useful information on the effect of engineering manipulations such as aeration, nutrient supply by injection or other ways, introduction of additives, chemically or biologically active agents, etc. A good example is the development of a CDT technology for biologically non-available soil pollutants (Molnár *et al.*, 2009a). Various technological versions were tested in 0.5–40 kg microcosms to determine the scale of aeration, nutrient addition and cyclodextrin application. Cyclodextrin is a bioavailability-enhancing agent: its mode of introduction and the necessary amount were also exactly specified in the small-scale microcosms. Another example for the technological application of soil microcosms is the development of *in situ* chemical stabilization of a soil contaminated with toxic metals. The optimum stabilizing agent and its necessary concentration range was identified in short- and long-term laboratory experiments in open-batch soil microcosms (Feigl *et al.*, 2009a). The findings of the microcosm experiment were validated in field experiments (see also Volume 5).

Some microcosm types are introduced below from the practical point of view and with regard to the aim of the microcosm application. The microcosm set-up as detailed in the next chapters accommodates the testing of three technologies based on the predominant natural processes:

– soil remediation based on natural attenuation and biodegradation;
– natural stabilization of metals in the soil and the application of combined chemical- and phytostabilization;

*Figure 8.11* Solubilization of groundwater contaminants using tensides and cosolvents in increasing concentration from left to right.

–  natural mobilization of metals in soil or waste rock and its utilization for bioleaching-based technologies;
–  simulation of a red-mud spill on agricultural soils;
–  simulation of secondary sodification due to inadequate irrigation and alkali intervention.

In addition to natural soil processes and technologies based on natural and biological activities, microcosms also enable the measurement of the result of physico-chemical treatments. Soil washing, injection of additives, chemical reactions such as ISCO (*in-situ* chemical oxidation) can be tested at small size to support the decision on treatability of contaminated soil and groundwater. Figures 8.11 and 8.12 show the very first experiment on treatability of contaminated groundwater using tensides and cosolvents to increase solubility of water-insoluble contaminants (Figure 8.11) and the application of oxidizing agents such as peroxide, persulfate or permanganate (Figure 8.12).

The tests on groundwater contaminated with trichloroethylene (TCE), shown in Figure 8.11, investigated the rate of solubilization, biodegradation and chemical oxidation at increasing cosolvent concentration. The test bottles contain a TCE lens which functions as an inexhaustible contaminant source and simulates the underground situation. The more dissolved the contaminants, the denser the solute is. Chemical analyses provide more details on quantities.

Laboratory batch experiments modeling ISCO with potassium permanganate are shown in Figure 8.12. The tests were set up in 250 mL glass vessels containing groundwater contaminated with 1000 mg/L TCE. The oxidizing agent was potassium permanganate (from left to right: control without $KMnO_4$, 1.2 g/L $KMnO_4$ and 6.0 g/L $KMnO_4$). Large amounts of precipitated $MnO_2$ can be seen in the reactor vessels. The permanganate-based ISCO was effective at laboratory scale, and the high

*Figure 8.12*   Chemical oxidation of dissolved contaminants in groundwater using potassium permanganate.

TCE (1000 mg/L) concentration of the groundwater dropped significantly. The applied
1.2 g/L KMnO$_4$ removed 92% of TCE within 24 hours. A disadvantage of the technology is the insoluble precipitate. A considerable amount of MnO$_2$ precipitated in
the reactors and the same can be expected in the real environment.

## 4   BIODEGRADATION AND BIODEGRADATION-BASED REMEDIATION STUDIES IN SOIL MICROCOSMS

The planning of soil bioremediation needs reliable methods for assessing and monitoring soil microbiota and its activity. Testing site-specific biodegradation of the pollutant
in soil microcosms is a prerequisite for a successful technology selection and planning.
Microcosm studies for testing natural biodegradation and performance assessment of
bioremediation are frequently used to investigate natural processes in contaminated
soil and to examine the adverse effects of organic contaminants and the beneficial
changes due to remedial activities (Alvarez & Illman, 2006; Fernández *et al.*, 2005;
Gruiz *et al.*, 2001; Leitgib *et al.*, 2007; Molnár *et al.*, 2005; Molnár *et al.*, 2009a;
Salminen *et al.*, 2002).

### 4.1   Testing natural and enhanced biodegradation

Static and dynamic microcosm systems were set up with three-phase soils artificially
contaminated with petroleum hydrocarbons to study the bioavailability and natural
biodegradation of the contaminants in a soil from a polluted industrial site (Molnár
*et al.*, 2009a). Experiments on contaminated soil spiked with diesel oil and transformer

*Figure 8.13* Soil biodegradation microcosms in jars to test enhanced biodegradation.

oil at 30,000 mg/kg concentration and reference soil were run for 10 weeks. Biodegradation and the activity of the indigenous soil microorganisms were characterized by chemical and biological methods at the very beginning (after spiking the soil), after the adaptation period (4 weeks), and after 6, 8 and 10 weeks.

Uncontaminated reference soil was used as a control to compare the effects of contaminants. The soils were amended with inorganic nutrients ($(NH_4)_2SO_4$, $KNO_3$, $KH_2PO_4$) to reach a final C:N:P ratio of 100:10:1, and were incubated at $25 \pm 2°C$.

### Static soil-microcosms for screening/checking natural biodegradation

Static open batch reactors containing 0.1–2.0 kg of contaminated and reference soils were incubated under atmospheric (aerobic) conditions between $10–30 \pm 2°C$. Optimal humidity (10–15% by mass) was maintained throughout the experiments. Duration varied typically from one week to one year. The number of tests can be large due to the small space requirement and low cost.

These simple test vessels filled with soil proved suitable for testing the biodegradation of aged petroleum hydrocarbons, transformer oil, mazout, pesticides, solvents or any organic soil contaminant of industrial origin. The influence of soil type, soil microbiology as well as of the physical phases present (two- and three-phase soils and soil slurries) can be tested. The aim of the tests is to decide whether biodegradation takes place, and whether the biodegradation process can be enhanced with additives or by changing environmental parameters.

To follow the processes, the initial and final states are evaluated and compared to reference or control. The evaluation needs an integrated approach applying physico-chemical, biological and toxicological methods and the comparison of their results.

The influence of cyclodextrin on enhancing bioavailability of contaminants was tested at different concentrations in three tests each: control, 0.1% cyclodextrin and 0.5% cyclodextrin from left to right in Figure 8.13.

### Static soil-microcosms for measuring biodegradation rate and finding the optimal scale of technological parameters

Technological studies can be performed in the laboratory in well-controlled static microcosms (batch reactors: solid phase and mixed slurry) or flow-through soil microcosms, where the mobile phases of air and/or water can 'flow through'. The size of these technological experiments may vary between 0.5–100 kg soil or beyond. Duration of the experiments may vary between two weeks or two years. The number of tests is defined by the statistical method used and its power. Generally, a combination of two or three previously selected parameters is tested and evaluated in comparison with each other and the reference.

Microcosms are easy tools not only to evaluate the "yes" or "no" responses of the soil microbiota but also to find a detailed explanation about the mechanism and rate of biodegradation. One can measure the effects of serially varying parameters of environmental technologies such as aeration, nutrient addition and the application of physical, chemical or biological additives (e.g. electron acceptors and agents that adjust the soil redox potential and influence contaminant mobility and bioavailability). Regular sampling, integrated analysis and testing and integrated data evaluation help monitor and assess the processes. The number of samples to be taken influences the microcosms' size to a large extent.

In the cyclodextrin-enhanced soil bioremediation (CDT) case study, four- to ten-week-long microcosm experiments were performed and evaluated to study the progress of biodegradation and to find the optimum parameters for aeration, nutrient and RAMEB (randomly methylated beta cyclodextrin) applications. Aeration of three different intensities, nutrient addition and cyclodextrin (type and amount) application were evaluated. The microcosms simulating real situations showed that a $2 \times 1$ hours-per-day aeration results in equally high biodegradation as one for 12 or 24 hours, and that bioavailability and biodegradation cannot be further increased by the addition of more than 0.5% CD. These findings have been validated later in the field.

*Dynamic soil microcosms for monitoring biodegradation and the effects of interventions*

Laboratory flow-through column reactors were designed for the problem-specific dynamic testing of the effects of technological parameters such as aeration and nutrient supply on soil remediation. Columns of a volume of 1–3 liters filled with 0.5–2 kg of contaminated soil were used for the experiments at $15–25 \pm 2°C$ (Figure 8.14). The reactors filled with the contaminated soils were aerated for 2 hours daily at a rate of 10 L/h air. Two flow-through traps filled with NaOH ensured a $CO_2$-free atmospheric air inflow. The $CO_2$-free air was sucked through the soil columns by vacuum produced by a water jet pump or ventilator. The humidity was maintained at between 10–15% by mass throughout the experiment. The microbial activities in the soil were characterized by measuring $CO_2$ production and $O_2$ consumption. The same microcosm set-up is suitable for the qualitative and quantitative characterization of biodegradability, bioavailability or toxic effects of organic contaminants. The effects of technological parameters, e.g. temperature and aeration rate, and those of additives such as nutrients and tensides or oxidizing, reducing, etc. agents were also measured.

## 4.2   Integrated monitoring and evaluation of the biodegradation experiments

An integrated methodology including physico-chemical analyses and biological tests is the most useful tool to evaluate soil microcosms because soil has a high potential to influence contaminant mobility and bioavailability. There are several physico-chemical and biological end points that can detect the biodegradation processes in air, water (moisture) and soil samples taken from the microcosm or by applying *in situ* sensors.

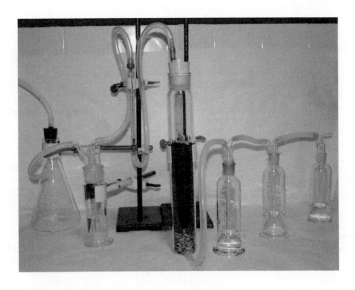

*Figure 8.14* Dynamic soil microcosms with flow-through column reactors.

Chemical analyses can determine the quality and quantity of the residual contaminants or the biodegradation product (metabolite) measured by *in situ* sensors, by using labeled substrates or complete or selective extraction. Water-based complexing agents or organic solvents can be used as extractants. The analytical method identifying the contaminant or contaminant mixture should be substance-specific. After extraction a separation or fractionation technique is applied such as ion chromatography, gas chromatography (GC), high performance liquid chromatography (HPLC) or capillary electrophoresis (CE, CZE), and gravimetry, flame-ionization detection (FID), mass spectrometry (MS), etc. for the detection of the contaminants depending on the material properties of the contaminating and biodegrading chemical substances.

Chemical methods alone cannot give a complete picture of the biodegradation process, in particular, for complex contaminant mixtures, aged and low-bioavailability contaminants. Chemical analysis should be supported by biological (the presence of biodegrading microorganisms or enzymes plays a role in the degradation process) and toxicological (decreasing toxicity or other adverse effects) methods. If chemical and biological/toxicological results are contradictory, the cause of this contradiction must be found. For example: the toxic substance disappears from the extract (it cannot be detected by chemical methods), but toxicity does not. The answer will be given by microbiological and toxicological test results. Is a more toxic metabolite – produced by the microbes – responsible for the still existing toxicity, or have the chemical properties of the original contaminants changed and is this why the applied analytical technique is not working any more, or something else? One can measure the concentration and the metabolic activities of the contaminant-degrading cells in the soil through enzyme activities such as the dehydrogenase enzyme activity and the changes in substrate specificity or diversity. The respiration rate can be

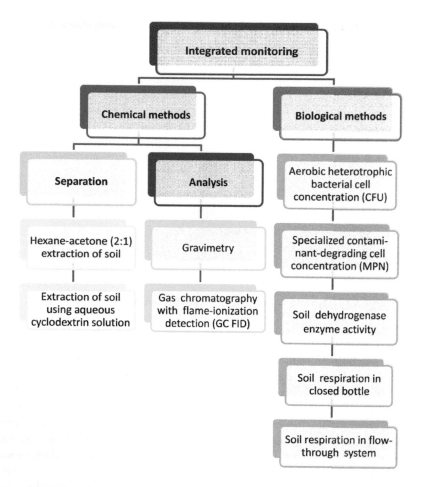

*Figure 8.15* Integrated methodology for the monitoring of biodegradation experiments.

continuously measured in the microcosm test vessel or on separate samples taken from the microcosm.

Dynamic testing of soil makes it possible to measure the intensity and emergence rate of the response to sudden changes in environmental conditions and to technological interventions.

Laboratory microcosms can be equipped with automatic precision pumps and dispensers which ensure that parallel tests are completely identical. The precise timing and scaling of the technological measures can also be ensured; thus the quality of the microcosm study results is similar to standardized biotests or chemical analyses. There is no guarantee that the communities will remain identical in the microcosms during the study.

The integrated monitoring of the micro- and mesocosms is illustrated by an example used for the remediation of a soil contaminated with transformer oil in an old inherited industrial site (Figure 8.15).

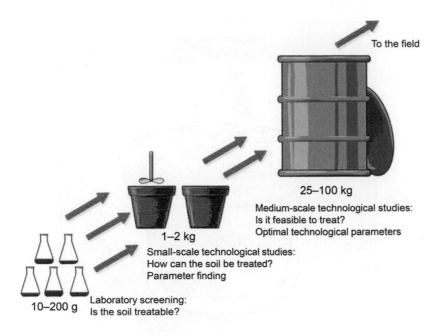

To the field

25–100 kg
Medium-scale technological studies:
Is it feasible to treat?
Optimal technological parameters

1–2 kg
Small-scale technological studies:
How can the soil be treated?
Parameter finding

10–200 g
Laboratory screening:
Is the soil treatable?

*Figure 8.16* Scale-up of laboratory soil microcosms.

## 4.3 Scaled-up technological micro- and mesocosms

Soil microcosms of different scales and configurations can be applied to testing natural and enhanced biodegradation with the aim to establish and develop innovative soil bioremediation technologies. Large-size (scaled-up) microcosms can be used to model the intensification of remediation technologies based on natural biodegradation by soil venting (bioventing), by the addition of nutrients, electron acceptors, bioavailability-enhancing agents such as cyclodextrins, cosolvents or tensides and microbes with special abilities (bioaugmentation).

Figure 8.16 shows that a large number of small-scale microcosms are performed and evaluated to find out which technological intervention has an effect at all. In the next step, the optimal value of technological parameters and their combinations should be determined by serially testing the effect of aeration, duration, nutrient or additive supplement such as the RAMEB in the innovative cyclodextrin technology (CDT), which was successfully used to enhance the biodegradation of a soil polluted with transformer oil (Fenyvesi *et al.*, 2003; Gruiz *et al.*, 1996; Leitgib *et al.*, 2003; Molnár *et al.*, 2003; Molnár *et al.*, 2005; Molnár *et al.*, 2009b).

The microcosm experiments with an incrementally increasing scale from laboratory- to field-scale (Figure 8.17) included four CDT development steps: small-scale laboratory screening tests, small- and large-scale laboratory microcosms for testing technological parameters and field tests (demonstration) on an inherited site contaminated with transformer oil at a transformer station. According to the experience of the technology developers, biodegradation and bioremediation tests in soil

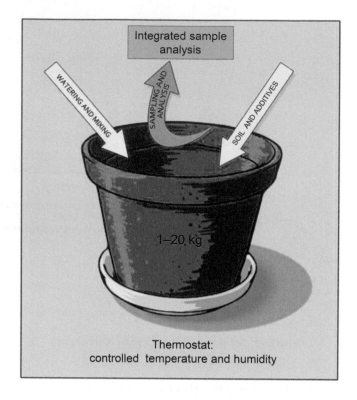

*Figure 8.17* Stabilization microcosm set-up.

microcosms under controlled conditions provided valuable information on the micro-biological processes in contaminated soil, and they supported decision making about the risk reduction options.

## 4.4 Summary of biodegradation testing for technological purposes

Biodegradation and biodegradation-based soil remediation tests in micro- and meso-cosm type experimental systems proved to be suitable for studying natural and enhanced biodegradation, and for finding the optimal technological parameters that support the development of scientifically established innovative technologies. In the case of already demonstrated technologies, their site-specific application and technological efficacy can be verified before new field applications.

## 5 TESTING TECHNOLOGIES BASED ON CONTAMINANT STABILIZATION

Microcosm models can be applied to technological pre-experiments, as part of the remediation scale-up procedure prior to technology selection and planning. In the case of *in situ* chemical stabilization or combined chemical and phytostabilization (see Volume 4), microcosm experiments provide useful information on the effect of the introduction of additives and plants to the soil (Gruiz & Vaszita, 2009).

Microcosm experiments are often used to assess the suitability and efficacy of chemical stabilizers applied for immobilizing soil contaminants. Other fields of application are effect and efficacy studies on stabilization-improving additives and process reversibility (Kiikkila *et al.*, 2002; Geebelen *et al.*, 2006; Feigl *et al.*, 2007). Microcosm assessments may be used to investigate the joint stabilizing effect of stabilizers and plants, which is a common solution in practice (Chlopecka & Adriano, 1997; Lombi *et al.*, 2002; Brown *et al.*, 2005).

Stabilization microcosm experiments comprise four steps:

- planning the experimental set-up and parameter combinations,
- assembling the microcosms and performing the experiment,
- monitoring and sampling individual microcosms,
- comparative evaluation of the results.

## 5.1   Experiment design

Stabilization microcosms were applied as the first tier of technological experiments to choose the best chemical stabilizer for soils contaminated with toxic metals and mine wastes in the Gyöngyösoroszi, Hungary, mining area and to identify its optimal concentration range (Volume 5). Table 8.1 shows the combinations of additives, soil and mine waste to be stabilized (Feigl *et al.*, 2009a, b; 2010a, b).

The microcosms that model toxic metal stabilization in soil and mine waste are arranged in as many parallel specimens as required or in a serial arrangement. 'Treated' microcosms are compared to a 'non-treated' control. The stabilization process is monitored via time-serial samples using an integrated methodology focused on the two main risk components, i.e. mobility and bioavailability of the toxic metals.

Another soil stabilization microcosm application assesses the stabilizing potential of lignite (Uzinger & Anton, 2008; Uzinger, 2010). A mathematical model providing information about a multidimensional response function has been applied to test samples that were influenced by a number of environmental and technological factors (Biczók *et al.*, 1994). The effects of four additives (lignite, Pb, Cr and Zn; the latter ones in the form of $Pb(NO_3)_2$, $Cr(NO_3)_3 \cdot 9H_2O$ and $ZnSO_4 \cdot 7H_2O$) were tested in three specimens. Five levels of each of the four factors considered were arranged in an orthogonal pattern based on the normalized matrix plan (Table 8.2). The plan enables the investigation of linear and quadratic effects and interactions of pairs of variables. Temperature and soil moisture were constant during the incubation period (4 weeks). The mathematical formula for the model is the following:

$$y = B_0 + B_1 * L + B_2 * Pb + B_3 * Zn + B_4 * Cr + B_5 * L * Pb + B_6 * L * Zn + B_7 * L * Cr + B_8 * Zn * Pb + B_9 * Cr * Pb + B_{10} * Cr * Zn + B_{11} * L^2 + B_{12} * Pb^2 + B_{13} * Cr^2,$$

where:
- $y$ = dependent variable;
- $B_0$ to $B_{13}$ = parameters of the model;

*Table 8.1* Experimental matrix for chemical stabilization: combination of metal-contaminated soils, additives and mine waste.

| Stabilizing additive | Start date | Pot size (kg) | Contaminated agricultural soil | Waste rock from the mine | Acidic mine waste |
|---|---|---|---|---|---|
| | | | Amount of additive (% by mass) | | |
| Fly ash A (Oroszlány) | 15.09.2003 | 2 | 1    2 5 | | |
| Fly ash B (Oroszlány) | 15.09.2003 | 2 | 1    2 5 | | |
| Hydrated lime | 10.05.2004 | 2 | 1 | | |
| Raw phosphate | 10.05.2004 | 2 | 1 | | |
| Alginite | 10.05.2004 | 2 | 1.5 | | |
| Lignite | 10.05.2004 | 2 | 10 | | |
| Lime + phosphate + alginite + lignite | 10.05.2004 | 2 | 1 + 1 + 1.5 + 10 | | |
| Fly ash from Tata | 17.02.2006 | 1.5 | 2      5 | 2      5 | |
| Red mud | 17.02.2006 | 1.5 | 2      5 | 2      5 | |
| Water treatment residue 1 | 17.02.2006 | 1.5 | 2      5 | 2      5 | |
| Water treatment residue 2 | 17.02.2006 | 1.5 | 2      5 | 2      5 | |
| Fly ash from Tata (T) | 23.02.2007 | 1.5 | 5 | 5 | |
| Fly ash from Visonta (V) | 23.02.2007 | 1.5 | 5 | 5 | |
| Fly ash from Tata + lime | 23.02.2007 | 1.5 | 2.5 + 2  5 + 2 | 2.5 + 2  5 + 2 | |
| Fly ash T + V + lime | 23.02.2007 | 1.5 | 2.5 + 2.5 + 2.0 | 2.5 + 2.5 + 2.0 | |
| Lime | 03.06.2008 | 1.5 | | 2 | 2 |
| Lime + steel shots | 03.06.2008 | 1.5 | | 2 + 1 | 2 + 1 |
| Fly ash T | 03.06.2008 | 1.5 | | 5 | 5 |
| Fly ash T + steel shots | 03.06.2008 | 1.5 | | 5 + 1 | 5 + 1 |
| Fly ash T + V + lime | 03.06.2008 | 1.5 | | 2.5 + 2.5 + 1 | 2.5 + 2.5 + 1 |
| Fly ash T + V + lime + steel shots | 03.06.2008 | 1.5 | | 2.5 + 2.5 + 2 + 1 | 2.5 + 2.5 + 2 + 1 |

- L = lignite concentration in %;
- Pb, Zn, Cr = toxic metal concentration in mg/kg.

This kind of experimental arrangement ensures high statistical power when looking for the relationship between simultaneous effects such as pH, three metals and lignite additives.

## 5.2 Microcosm set-up and implementation

The stabilization microcosms are simple, open batch reactors with a mass of 0–20 kg, placed into a thermostat to ensure controlled temperature and humidity. The authors applied such vessels for contaminated soil and mine waste stabilization. The vessels were filled with 1–2 kg soil or mine waste, mixed with different stabilizers at 1–10% by mass and were incubated at 25°C for 2–3 years. They were mixed and watered to 60% of their water holding capacity every month. The soil was sampled and analyzed at certain intervals to check the complex physico-chemical and microbiological processes in the microcosms (Figure 8.17) (Feigl *et al.*, 2009a).

Table 8.2  Experimental matrix for the assessment of
the stabilizing potential of lignite.

| Sample code | Lignite (%) | Pb (mg/kg) | Zn (mg/kg) | Cr (mg/kg) |
|---|---|---|---|---|
| Control | 0 | 0 | 0 | 0 |
| 1 | 7.5 | 1125 | 1125 | 1125 |
| 2 | 2.5 | 1125 | 1125 | 1125 |
| 3 | 7.5 | 375 | 1125 | 1125 |
| 4 | 2.5 | 375 | 1125 | 1125 |
| 5 | 7.5 | 1125 | 375 | 1125 |
| 6 | 2.5 | 1125 | 375 | 1125 |
| 7 | 7.5 | 375 | 375 | 1125 |
| 8 | 2.5 | 375 | 375 | 1125 |
| 9 | 7.5 | 1125 | 1125 | 375 |
| 10 | 2.5 | 1125 | 1125 | 375 |
| 11 | 7.5 | 375 | 1125 | 375 |
| 12 | 2.5 | 375 | 1125 | 375 |
| 13 | 7.5 | 1125 | 375 | 375 |
| 14 | 2.5 | 1125 | 375 | 375 |
| 15 | 7.5 | 375 | 375 | 375 |
| 16 | 2.5 | 375 | 375 | 375 |
| 17 | 10 | 750 | 750 | 750 |
| 18 | 0 | 750 | 750 | 750 |
| 19 | 5 | 1500 | 750 | 750 |
| 20 | 5 | 0 | 750 | 750 |
| 21 | 5 | 750 | 1500 | 750 |
| 22 | 5 | 750 | 0 | 750 |
| 23 | 5 | 750 | 750 | 1500 |
| 24 | 5 | 750 | 750 | 0 |
| 25 | 5 | 750 | 750 | 750 |
| 26 | 5 | 750 | 750 | 750 |
| 27 | 5 | 750 | 750 | 750 |
| 28 | 5 | 750 | 750 | 750 |
| 29 | 5 | 750 | 750 | 750 |
| 30 | 5 | 750 | 750 | 750 |
| 31 | 5 | 750 | 750 | 750 |
| 32 | 5 | 750 | 750 | 750 |
| 33 | 5 | 750 | 750 | 750 |
| 34 | 5 | 750 | 750 | 750 |
| 35 | 5 | 750 | 750 | 750 |
| 36 | 5 | 750 | 750 | 750 |

## 5.3  Monitoring of the microcosms

An integrated methodology was developed to monitor stabilization experiments which combined physico-chemical analysis with biological and ecotoxicity testing. Its purpose was to provide a realistic view of the risk posed by metal contaminants in the soil (Gruiz *et al.*, 2009). The outcome is that the risk of contaminating metals can be reduced by changing their mobility and accessibility, but not by removing them from the soil. Chemical stabilization can be based on

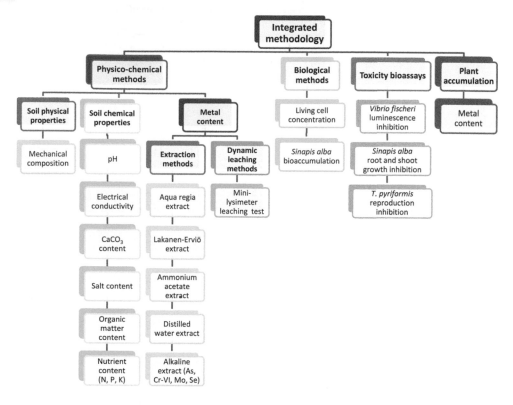

*Figure 8.18* Integrated methodology for the monitoring of chemical stabilization in soil microcosms.

– suppressing ionic metal forms,
– reducing dissolution by supporting precipitation,
– reducing desorption in favor of sorption,
– changing redox state in order to obtain a less water-accessible and bioavailable form,
– enhancing mineral formation by reducing weathering.

The risk posed by metal species of ionic–non-ionic, dissolved–precipitated, sorbed–desorbed, oxidized–reduced and incorporated minerals or free metal forms cannot be characterized using chemical analyses alone. The difference between adverse effects of the soil before and after stabilization is directly related to the metals' risk determined by mobility, biological availability and actual adverse effects, and thus biological and ecotoxicological methods would be preferred for technology monitoring.

The chemical analyses as part of the integrated methodology for the monitoring of chemical stabilization (shown in Figure 8.18) developed by Feigl *et al.* (2007) included the application of soil extractants with increasing acidity: i) distilled water, ii) aqueous solution of ammonium-acetate (pH = 4.5), iii) ammonium-acetate + EDTA (ethylene-diaminetetraacetic acid) and iv) *aqua regia*. The mobility of As, Cr-VI, Mo and Se was analyzed using an alkaline extract since these metal and metalloid species pose a threat in an alkaline environment. The metal content of the acidic and alkaline extracts was determined by ICP-AES (Inductively Coupled Plasma–Atomic Emission Spectrometry). The metal content of these extracts characterizes the extractability i.e. the 'chemical availability' of the metals. Chemical availability was not only measured in extracts but also in leachates using a dynamic leaching method in microlysimeter columns (developed for this purpose) at the end of the long-term stabilization experiments. The mini-lysimeters are also microcosms: they are filled with stabilized soil, average precipitation is applied and a realistic seepage or drainage is simulated in a soil column of optional height.

Besides these experiments other soil characteristics such as mechanical composition, texture, pH, EC, $CaCO_3$ content, salt content, organic matter content and nutrients (ammonium-acetate extractable P and K, total N) were also determined. According to the aim of the experiment, in addition to the above parameters, one can analyze for example the natural redox indicators such as oxygen, $NO_3^-$, $NO_2^-$, $SO_4^{2-}$, $SO_3^{2-}$, FeII, FeIII, MnII, MnIII, $CH_4$, or artificial redox indicators added to the soil such as thionine, toluidine-blue or cresyl violet in dissolved or immobilized forms. The results help better understand and interpret the redox-potential-dependent biological and chemical processes.

The actual risk of various metal species can be best characterized by direct effect assessment (Gruiz, 2005). Biological indicators and their changes can be monitored and used for the evaluation of the efficiency of remediation. Soil activity was characterized in the stabilization study by the concentration of the aerobic heterotrophic cells in the soil. The diversity and activities of the microbiota provides a huge variety of bioindicators. Soil toxicity measurements in the stabilization experiments included direct contact tests with test organisms from three trophic levels: bacteria (*Vibrio fisheri* luminescence inhibition test), plant (*Sinapis alba* root and shoot growth inhibition test) and animal (*Tetrahymena pyriformis* single cell animal reproduction inhibition test). A rapid (five-day) bioaccumulation test on *Sinapis alba* developed by the authors was used to obtain direct information on the suitability of the chemical stabilizer from the point of view of the phytostabilization process following chemical stabilization.

## 5.4   Evaluation, interpretation and use of the stabilization microcosm results

The results of the chemical analyses and toxicity measurements have to be evaluated together in order to obtain a detailed picture on the effect of stabilizing agents. The application of the integrated monitoring and evaluation methodology enables choosing the best stabilizing agent that should be applied in the large-scale technological experiments (Figure 8.19).

## 5.5   Summary and conclusions of stabilization microcosm application

The authors used the simple batch or lysimeter type microcosm to develop a technology based on chemical stabilization for the remediation of mine waste containing toxic metals. In addition to the effects of additives, long-term processes and the effects

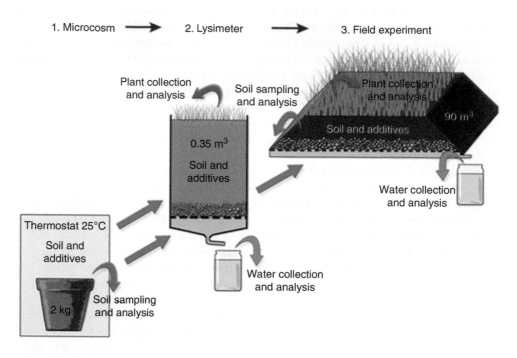

*Figure 8.19* Scaled-up technological experiments for combined chemical and phytostabilization.

of vegetation can successfully be modeled. The results of microcosm and mesocosm experiments were validated by field demonstration of the developed technology.

## 6  TESTING AND UTILIZING THE COMPLEX LEACHING PROCESS

Microcosm models and experiments can simulate site-specific conditions and provide information about parameters needed to improve risk assessment and enhance risk reduction (Gruiz *et al.*, 2001). For most applications the soil–groundwater pathway is considered, therefore the leaching behavior of contaminated soil and solid waste is of great importance (Kalbe *et al.*, 2008).

Complex leaching in the environment is the result of physico-chemical and biological processes playing a role in the weathering of rock and the formation of soil. In the case of contaminated rock and soil or uncontrolled waste deposits, the complex leaching process and its environmental risks can be studied in leaching microcosms. The technological aim is either to reduce the leachate amount and its contaminant concentration, or to remove the contaminant from the soil, rock or solid waste by leaching. Also, bioleaching as a basic process of biomining, which recovers valuable metals from ores or mine waste (biomining of copper, gold or uranium), can be studied in microcosms.

Laboratory leaching microcosms are lysimeters (Figure 8.20) which can be used for source characterization and pollution transport determination which serves as a

*Figure 8.20*  Principle of the lysimeter: water filtrating through the soil is collected and analyzed.

basis for risk assessment. Leaching experiments in laboratory microcosms provide information about the production and chemical composition of solutions derived for example from sulfide rocks present in a watershed (Munk *et al.*, 2006). The lysimeter's principle is that water filtrates through the three-phase soil sample, is collected at the bottom and then analyzed (Figure 8.20). The water balance of a site can be estimated with the help of a lysimeter. The amount of precipitation that the area receives and the amount lost through the soil are recorded (leachate measured in the lysimeter). The difference is the amount of water remaining in the soil and the proportion lost to evapotranspiration (Figure 8.21).

Lysimeters are used to collect leachates from contaminated soil and to measure the contaminant concentration in the leachate and thus the mobile or mobilizable contaminant fraction can be measured. The difference between the contaminant concentrations in soil and in leachate characterizes the interaction between contaminant and soil (solid phase) from which the retention of a contaminant or the sorption capacity of the soil can be estimated. Both are basic characteristics responsible for environmental risk of soil contaminants.

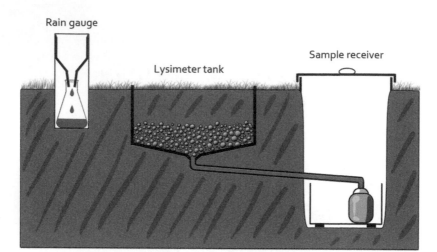

*Figure 8.21  In situ lysimeter set-up with natural precipitation and an underground collector system.*

Mine wastes containing metal sulfides are oxidized due to the weathering process which is accelerated by naturally occurring sulfur- and iron-oxidizing bacteria. This phenomenon is recognized as bioleaching and it results in acid mine (AMD) or acid rock drainage (ARD) (Bosecker, 2001). AMD/ARD is characterized by acidic, metal-rich waters and red-orange iron-bearing solids formed by the weathering of pyrite ($FeS_2$) and subsequent oxidation of soluble ferrous iron (Fe[II]) to insoluble ferric iron (Fe[III]). Iron-oxidizing bacteria present in AMD/ARD environments can accelerate the rate of ferrous oxidation by almost $10^6$ compared to inorganic mechanisms. Thus, iron-oxidizing bacteria can be responsible for nearly uninterrupted generation of AMD. Acid mine/rock drainage is the biggest environmental threat facing the mining industry. The environmental impact of AMD and ARD can be severe in both abandoned and active mining areas. The generation of low pH drainage enhances the dissolution of heavy metals in water. The dissolved metals contaminate soil, ground- and surface waters and can pose high risk to the ecosystem and human health (Gruiz *et al.*, 2007; Sipter *et al.*, 2008).

Interdisciplinary studies have resulted in substantial progress in understanding the processes that control the release of metals and acidic water from inactive mines and mineralized areas, the transport of metals and acidic water to streams (Gruiz *et al.*, 2009), and the fate and effect of metals and acidity on downstream ecosystems (Sipter *et al.*, 2008).

## 6.1   Flow-through soil microcosm for studying bioleaching

Vaszita *et al.* (2009a) conducted microcosm experiments in four flow-through soil microcosms which were operated for more than five years to model bioleaching of toxic metals from the Gyöngyösoroszi, Hungary, mine waste material in contact with rainwater. Two scenarios were modeled: 1. bioleaching within a large waste dump

*Figure 8.22* Simple leaching microcosms.

in microcosms containing only mine waste (M1, M2); and 2. bioleaching in a waste dump in contact with the surrounding soil in microcosms containing mine waste on top of a soil layer (T1, T2).

## 6.2 Microcosm set-up

Microcosm set-ups and monitoring depend on the aim of the test that can simulate

- natural leaching processes and the amount and quality of ARD;
- leachate production and reduction;
- the effect of permeable reactive barriers with different reactive fillings;
- an ore-processing technology applied to mine waste;
- soil remediation based on complex bioleaching of contaminated soil.

The microcosm set-up was arranged to simulate the natural leaching processes (dominated by bioleaching) of an acidic mine waste and the filtering ability and capacity of the soil layer beneath. The set-up included four 6-liter HDPE containers (Figure 8.22) filled with 4.5 kg homogenized, crushed metal-sulfide-containing mine waste placed on top of a 5-cm gravel layer to allow for the artificial precipitation to filtrate through. 6-mm diameter holes were drilled in the bottom of the HDPE containers to collect the resulting leachate in the underlying HDPE tray. In addition, the T1 and T2 microcosms contained 1 kg of single-layer control soil between the mine waste and the gravel layer (Figure 8.23). This uniform 8-cm soil layer was wrapped in a polyamide membrane to prevent the soil particles from being washed out by the water flowing through.

The microcosms were manually irrigated at pre-arranged time intervals simulating four low-intensity events and one heavy rain per month. The irrigation rate was based on an annual rainfall rate of 756 mm/m$^2$/year (OMSZ 2002) with an allowance for

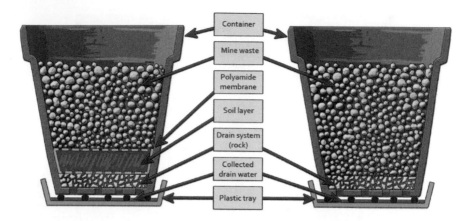

*Figure 8.23* Leaching microcosm containing mine waste on top of a thin soil layer and mine waste only.

evaporation. The low-intensity rain event was simulated by spraying 50–100 mL tap water on the surface of the soil microcosm. The heavy-rain event comprised about 400 mL of leachant flowing through the microcosm over twelve hours resulting in an output leachate.

A similar configuration can model and test the reactive filling of permeable reactive barriers and active alkali soil zones for the treatment of different leachate types.

## 6.3 Monitoring the leaching microcosms

Monitoring of the microcosms was based on physical, chemical and microbiological methodologies as shown in Figure 8.24.

As metal mobilization in the solid material is the focus of these experiments, metal content of the waste and soil was analyzed using ICP atomic emission spectroscopy (ICP-AES). The total metal content was determined from an *aqua regia* extract and the mobile fraction from an ammonium acetate + acetic acid + EDTA extract (Lakanen & Erviö, 1971). Cell concentration of the sulfide-oxidizing bacteria was measured according to Sand *et al.* 2007.

## 6.4 Evaluation and interpretation of the results

The evaluation of the results was focused on simulating various conditions of the process:

– Initial phase within which the steady-state equilibrium of the process can be reached and the annual water input is equal to the average annual rain.
– Dry condition phase within which water input is 1/3 of the average annual rain.
– Final phase within which the microcosms are depleted of metal sulfides, and the water input is ~2/3 of the average annual rain.
– Simulation of strong (5x average) precipitation and flood.

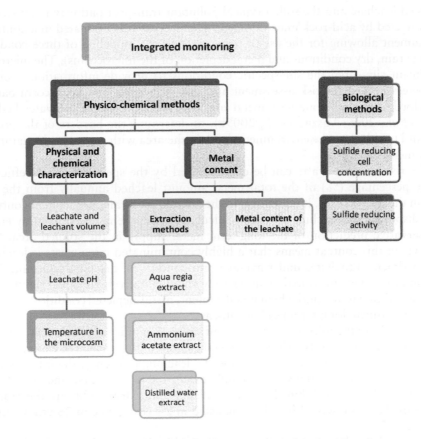

*Figure 8.24* Integrated methodology for the monitoring of the leaching microcosm.

The simultaneously measured physical, chemical and biological results were evaluated in an integrated way.

When modeling natural processes, the input parameters considered are the quantity and quality of the contamination source, the amount and distribution of precipitation, the quality and capacity of the soil layer. When the microcosm simulates a technology based on bioleaching, the input parameters considered can be the water flow, the geometry of the container and the additives. The results of the soil microcosm experiment were evaluated based on the following end points: input/output leachate amount, leachate pH, metal concentration of the leachate, amount of leached metal, leaching efficiency versus total and mobile metal content.

## 6.5  Summary and conclusions about leaching microcosm application

The laboratory soil microcosm experiment modeled, at laboratory level, complex leaching of mine waste that contained base-metal sulfides and a site-specific process non-measurable under field conditions. It also simulated the contact between the produced leachate and the soil, a typical pollutant transport pathway in an environment affected by acid-rock/mine drainage. The process was evaluated in a controlled environment allowing for the use of replicates and the modeling of three conditions (average rain, dry conditions and metal depletion of the microcosms). The microcosm experiment simulated the site-specific conditions to provide information about the parameters needed for risk assessment and risk reduction. The microcosm parameters helped to estimate long-term metal emissions from the sources at watershed level (Gruiz *et al.*, 2006; Vaszita *et al.*, 2009b), to determine the borders of the polluted areas and to calculate the environmental risk in the area with regard to the nature and fate of metal emissions.

The risk of mine waste can be characterized by the specific leaching efficiency, i.e. the percentage (%) of the total metal amount leached annually from the mine waste in the microcosms. It is identical to the emitted metal amount from uncontrolled waste disposal. The leaching efficiency calculated from the microcosm study enables a conservative prediction of the long-term metal supply in the studied area. 'Conservative' in this context means that a highly contaminated mine waste is leached in the experiment. Leachates under average rain conditions, dry conditions and heavy rain conditions were measured quantitatively and qualitatively to model weather variables. In addition to the initial high metal content, metal-depleted conditions were also studied to estimate long-term leaching efficacy.

The results of the microcosm experiment showed significant differences depending on the rain conditions and on the initial metal content of the waste. For example, under average rain conditions, based on the specific leaching efficiency, the total cadmium supply of the waste material would be fully exhausted in 8.5 years, zinc in 11.8 years and lead in 3,287 years. Taking leaching efficiency under dry conditions, the cadmium content of the waste would be fully exhausted in 24 years, zinc in 26 years and lead only in 24,657 years.

The results of laboratory lysimeter tests enable dynamic characterization of the environmental risk as a function of time, waste type and the weathering scale of the waste rock. Advanced weathering, for example, slowed down significantly the bioleaching process. The calculated time for full exhaustion of cadmium and zinc in the microcosms containing highly weathered waste material was 3–4-fold of the less weathered waste rock, but it was shorter for lead.

# 7  TRANSPORT PROCESSES STUDIED IN SOIL COLUMNS

Both 3- and 2-phase soils can be modeled in disturbed or undisturbed natural soil columns or in artificially assembled soil profiles. This model enables us to measure the infiltration rate and the effects of precipitation- and irrigation-transported ions, nutrients, and other agrochemical additives, liquid- and solid-phase-related contaminants and the physico-chemical and biological processes at various depths of the soil profile.

As a next level, technological interventions can also be studied to modify natural or contamination processes and reduce the risk. This kind of soil column is an ideal set-up to model technologies in combination with depths, infiltration and drainage. Soil columns with a height from 20 cm to 3 meters are described in the literature. Soil columns can be equipped with electrodes (measuring the changes *in situ*) and sampling ports and devices.

## 7.1  Test set-up

An assemblage designed for measuring the effect of red mud on agricultural soils is presented here to support decision making on remedial measures.

As a consequence of the Ajka, Hungary, aluminum factory disaster (October 2010), red mud flooded more than 800 hectares. In many areas the red mud layer remained on the surface for more than 3 months after the flood. Soils have not been sampled during the first risk management phase when the mitigation of the human health risk and of the damage to the residential area was the highest priority. Therefore field samples were not available that could have helped to unravel the infiltration process and adverse effects of the slurry containing NaOH. A column microcosm was designed to identify and possibly quantify the rate of infiltration and the depth-dependent changes of soil properties in a more abstract system compared to uncontrollable field situations. One of the main goals was to detect the mobilization and potential enrichment of toxic metals (especially As, Co, Cr, Ni, Pb), other potentially toxic elements (B, Ba, Cd, Cu, Hg, Mo, Se, Sn, V and Zn) and Na in the soil profile and the changes in the mobility of ions that play a role in plant nutrition. An integrated methodology consisting of physico-chemical, microbiological and ecotoxicological methods was applied for monitoring the transport and effects of red mud in the soil profile.

The columns – shown in Figure 8.25 – were 100-cm-long PVC tubes with a diameter of 15 cm, but similar 2.5 m long columns of 30 cm diameter were also used. The bottom of the columns was closed with a polyethylene textile. The soil filled into the column was neither completely mixed, nor undisturbed. Four characteristic layers were identified and separately removed from their original place in the soil and, after having been dried and homogenized, were filled into the column according to the original *in situ* sequence of the layers:

–   0–30 cm: sandy loam with gravel,
–   30–50 cm: sand with higher gravel content,
–   50–80 cm: sand with gravel and a coherent gravel layer at 60 cm,
–   80–100 cm: sand with gravel.

The bulk density of each soil layer in the columns was adjusted to the same value and the weight of the column was measured after every 5 cm soil layer. For this purpose, the soil had to be dried before filling it into the columns. Three periods were investigated: 30, 60 and 120 days. Each treatment had three repetitions, thus the test arrangement comprised nine columns. At the end of the given time period, the columns were sectioned into four pieces according to the soil layers to be analyzed.

Before the application of 10 cm red mud layer on the top, the soils were saturated up to the maximum water holding capacity from the bottom of the column to simulate the situation before the disaster. The water content of red mud (originating from the area of the disaster) was set at 38% according to the original value. During the time

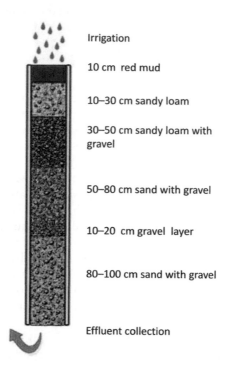

Irrigation

10 cm  red mud

10–30 cm sandy loam

30–50 cm sandy loam with gravel

50–80 cm sand with gravel

10–20  cm gravel  layer

80–100 cm sand with gravel

Effluent collection

*Figure 8.25*  Soil column experiment on red mud.

of the experiment, the columns were irrigated according to the precipitation observed in the disaster area.

## 7.2   Monitoring the soil column microcosm

Integrated monitoring was applied to observe the processes in the soil column (Figure 8.26). The effluent water was collected at the bottom of the column and analyzed at the end of each time period when the following parameters were measured from the soil samples:

– Microbiology: plate count test, microbial biomass, FDA analysis (fluorescein diacetate assay to estimate general microbial activity);
– Soil chemistry: pH, organic matter content and quality, $CaCO_3$ content, salt content, acid buffering capacity, cation exchange capacity (CEC), base saturation, partial extraction (*aqua regia* soluble element fraction; acetate buffer soluble fraction; water soluble fraction for modeling soil solution and leaching; ammonium acetate + EDTA soluble fraction for modeling plant availability);
– Soil physical properties: particle size distribution, upper limit of plasticity, water retention capacity, stability of soil aggregates, dispersity factor;

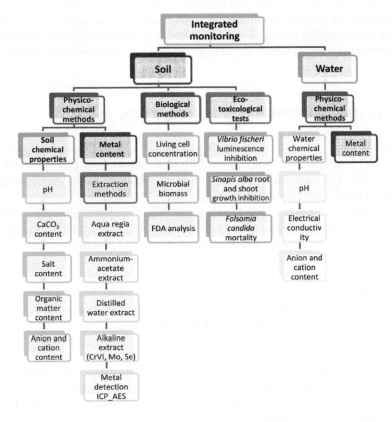

*Figure 8.26*  Integrated monitoring of the red mud column experiments.

– Ecotoxicology tests: *Vibrio fischeri* luminescence inhibition assay, *Sinapis alba* germination, root and shoot growth inhibition assay, *Folsomia candida* mortality test.
– The pH and element composition were measured also from the collected leachates.

## 7.3 Evaluation

Data were analyzed for treatment effects using one-way variance analysis (ANOVA). Variance was related to the main parameters (time, different layers). Significant differences among the parameter means were calculated by Duncan's post hoc test and by the LSD (least significant difference) test at $p < 0.05$.

## 7.4 Summary

The three- or two-phase laboratory soil column enables the investigation of depth-depending physico-chemical and biological processes. Undisturbed, completely mixed or semi-disturbed systems can be established: the red-mud column employed the semi-disturbed version. Natural and contaminating processes such as floods, water infiltration and desiccation, contaminant intake (deposition, infiltration, mixing into surface soil layers) and *in situ* applied technologies can be detected and modeled in these long-column microcosms, which are able to simulate depth-dependent characteristics, close to reality.

## 8 MODELING SECONDARY SODIFICATION

Salinization is the accumulation of water-soluble salts in the soil. These salts include potassium ($K^+$), magnesium ($Mg^{2+}$), calcium ($Ca^{2+}$), chloride ($Cl^-$), sulfate ($SO_4^{2-}$), carbonate ($CO_3^{2-}$), bicarbonate ($HCO_3^-$) and sodium ($Na^+$) ions. Sodium has a key role in salinization, also called sodification.

Salts dissolve and move around with water. When water evaporates, salts are left behind and an excess of sodium destructs the soil structure, which, due to the lack of oxygen, becomes incapable of sustaining either plant growth or animal life (SoCo, 2009).

Primary salinization involves salt accumulation through natural processes due to a high salt content of the parent material or in groundwater. Secondary salinization is caused by human interventions such as inappropriate irrigation practices, for example, using salt-rich irrigation water and/or insufficient drainage.

Sodification is a process of exchange between sodium present in the soil water and divalent cations (mainly $Ca^{2+}$) adsorbed on the exchange complex of the soil. It occurs when sodium is present in higher concentration than the divalent cations. The exchangeable sodium content of the soil increases and some minerals precipitate in the order of their own solubility from the less soluble ones (e.g. calcite, $CaCO_3$) to the most soluble ones (e.g. NaCl).

## 8.1 Modeling sodification in microcosms

Its dependence on geochemistry, hydrogeology, precipitation, groundwater flow, soil texture, soil water balance and several other environmental contextual parameters sets limits to the extent the complex sodification process can be modeled.

Since sodification is based on the transport by precipitation/rain water/runoff or groundwater flow and accumulation of metal ions (most frequently $Na^+$), the soil's water balance is the key aspect to be considered when setting up the sodification microcosm.

Secondary sodification is generally caused by relatively short-term anthropogenic effects such as irrigation with salty or alkali water or by inadequate agrotechnologies. The authors of this chapter looked into the topic of secondary sodification in the context of the red-mud flood problem in Hungary where Na intake of the soil was extremely high.

Three different microcosm set-ups containing 5 kg soil in 50–100 cm high columns were assembled to model worst-case scenarios (Figure 8.27). The microcosm tests were also aimed at screening the sodification potential of various soils and developing techniques to prevent this disadvantageous process by additives or water regulation. Because the natural sodification process is fairly slow, saline-sodic soils develop during

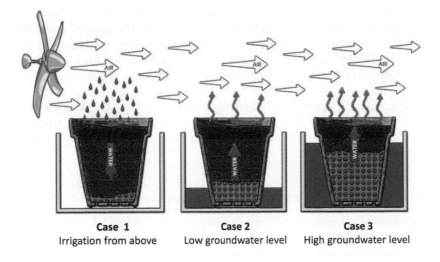

**Case 1**
Irrigation from above

**Case 2**
Low groundwater level

**Case 3**
High groundwater level

*Figure 8.27* Water balance scenarios in microcosms used for modeling sodification.

several years. The process tested in the soil microcosm was accelerated by increasing the evaporation rate (applying air flow and heating). Thus a 1–2-year-long sodification process was modeled within one–two months. Sodification microcosm study results may fill a gap because the direct effect of NaOH on soil has not yet been studied, therefore our knowledge on the effect of $Na^+$ ions originating from NaOH is limited.

## 8.2 Sodification microcosm set-up

The microcosms were aimed at testing the sodification of red-mud flooded soils. The scenario of tilling (mixing) residual red-mud into the soil was modeled. The reason behind this technology option was that the removal of the thin (1–2 cm) residual red mud deposits and of the upper contaminated soil layer after the red mud spill was not feasible due to the large affected area, therefore it was tilled into the agricultural soil (see also Chapter 5).

The red-mud sodification microcosms modeled three cases in which the distance of the groundwater level to the soil layer was different:

– Case 1 models a deep groundwater level scenario where the soil's water balance is not affected by the groundwater. In this case the ion-loaded infiltrated water flows vertically toward deeper layers. This causes sodification only if the surface layers are rich in salts or the soil is irrigated with salt-containing water.
– Case 2 models a shallow 3-phase soil scenario where the water table is at medium depth and only rises occasionally when capillary forces transport water and ions from the deeper layers to the surface. This type of sodification occurs if the groundwater itself or the deep soil layers contain an abundant stock of $Na^+$ or other ions.

–   Case 3 models a high, near-surface groundwater level scenario where neither precipitation nor capillary forces have a considerable effect on $Na^+$ ion transport. Instead, ion transport is chiefly determined by the groundwater flow, and if ion-containing precipitation occurs, groundwater flow may dilute the ion concentration. However, if the groundwater is contaminated with ions, the capillary forces transport the ions continuously to the surface where they are concentrated.

It should be noted that any of these three cases rarely occur solely in nature. Since groundwater level and rainfall rates are always fluctuating, ion flows also change with time, and eventually the sodification process is determined by the superposition of all of these processes.

## 8.3  Technological microcosms for reducing risk of sodification

The same extreme microcosm set-ups can support technological experiments aimed at preventing or reversing sodification in different soils, and at assessing the effect of different soil additives to reduce sodification.

## 8.4  Evaluation and interpretation of results

The changes can be monitored by analyzing separately the vertical layers of the microcosms. The most important indicators of sodification are $Na^+$ and other exchangeable ions. Soil geochemistry and water balance indicators such as soil texture, porosity and rate of capillary rise can show the direct effect of ion accumulation on soil, while biological and toxicological results (diversity of soil microflora, plant growth) can complete the monitoring methodology.

## 8.5  Summary of sodification modeling

In spite of the simplification and problems of modeling, the design introduced above was able to distinguish between the effects of different NaOH concentrations simulating a worst-case scenario. It was possible to draw attention to the risk of sodification when NaOH containing red mud remains on the soil surface or is mixed into the soil at more than 5% by mass.

## REFERENCES

Adey, W.H. & Loveland, K. (1999) *Dynamic Aquaria: Building living ecosystems*. 2nd revised edition. Academic Press. ISBN10: 0120437929 ISBN13: 9780120437924.
Adventus (2013) Laboratory treatability testing services of the Company Adventus. [Online] Available from:   http://www.adventusgroup.com/solutions/lab_treat_test_services.shtml. [Accessed 23rd February 2014].
Alvarez, P.J.J. & Illman, W.A. (2006) *Bioremediation and natural attenuation*. New Jersey, Wiley-Interscience.
ASTM (2014) Website of ASTM International. [Online] Available from: http://www.astm.org. [Accessed 23rd February 2014].

ASTM E1366 (2011) Standard Practice for Standardized Aquatic Microcosms: Fresh Water. [Online] Available from: DOI: 10.1520/E1366-11. [Accessed 23rd February 2014].

ASTM E1197 (2012) Standard Guide for Conducting a Terrestrial Soil-Core Microcosm Test. [Online] Available from: DOI: 10.1520/E1197-12. [Accessed 23rd February 2014].

Biczók, Gy., Tolner, L. & Simán, Gy. (1994) Method for the determination of multivariate response functions. *Bulletin of the University of Agricultural Science*, 1993–1994, 5–16.

Bosecker, K. (2001) Microbial leaching in environmental clean-up programmes. *Hydrometallurgy*, 59, 245–248.

Brown, S., Christensen, B., Lombi, E., McLaughlin, M., McGrath, S., Colpaert, J. & Vangronsveld, J. (2005) An inter-laboratory study to test the ability of amendments to reduce the availability of Cd, Pb, and Zn in situ. *Environmental Pollution*, 138, 34–45.

Chlopecka, A. & Adriano, D.C. (1997) Influence of zeolite, apatite and Fe-oxide on Cd and Pb uptake by crops. *Science of Total Environment*, 207, 195–206.

Feigl, V., Atkári, Á., Anton, A. & Gruiz, K. (2007) Chemical stabilization combined with phytostabilisation applied to mine waste contaminated soils in Hungary. *Advanced Materials Research*, 20–21, 315–318.

Feigl, V., Uzinger, N. & Gruiz, K. (2009a) Chemical stabilization of toxic metals in soil microcosms. *Land Contamination & Reclamation*, 17(3–4), 483–494.

Feigl, V., Uzinger, N., Gruiz, K. & Anton, A. (2009b) Reduction of abiotic stress in a metal polluted agricultural area by combined chemical and phytostabilisation. *Cereal Research Communications*, 37(Supplement), 465–468.

Feigl, V., Gruiz, K. & Anton, A. (2010a) Remediation of metal ore mine waste using combined chemical- and phytostabilisation. *Periodica Polytechnica*, 54(2), 71–80.

Feigl, V., Anton, A. & Gruiz, K. (2010b) An innovative technology for metal polluted soil – combined chemical and phytostabilisation, In: Sarsby, R. W. & Meggyes, T. (eds.) *Construction for a sustainable environment. Proceedings of the International Conference of Construction for a Sustainable Environment, Vilnius, Lithuania, 1–4 July, 2008.* Leiden, CRC Press/Balkema, 187–195.

Fenyvesi, É., Csabai, K., Molnár, M., Gruiz, K., Murányi, A. & Szejtli, J. (2003) Quantitative and qualitative analysis of RAMEB in soil. *Journal of Inclusion Phenomena and Macrocyclic Chemistry*, 44, 413–416.

Fernández, M.D., Cagigal, E., Vega, M.M., Urzelai, A., Babin, M., Pro, J. & Tarazona, J.V. (2005) Ecological risk assessment of contaminated soils through direct toxicity assessment. *Ecotoxicology and Environmental Safety*, 62, 174–184.

Geebelen, W., Sappin-Didier, V., Ruttens, A., Carleer, R., Yperman, J., Bongué-Boma, K., Mench, M., van der Lelie, N. & Vangronsveld, J. (2006) Evaluation of cyclonic ash, commercial Na-silicates, lime and phosphoric acid for metal immobilisation purposes in contaminated soils in Flanders (Belgium). *Environmental Pollution*, 144, 32–39.

Generic Freshwater Microcosm Test (1996) Laboratory Microcosm – Ecological Effects Test Guidelines OPPTS 850.1900. Prevention, Pesticides and Toxic Substances (7101), EPA 712–C–96–134.

Gruiz, K. (2005) Biological tools for soil ecotoxicity evaluation: Soil testing triad and the interactive ecotoxicity tests for contaminated soil – In: Fava, F. & Canepa, P. (eds.) *Soil Remediation Series, no. 6.* Italy, INCA. pp. 45–70.

Gruiz, K. & Vaszita, E. (2009) Microcosm models and experiment: types and applications. *Land Contamination & Reclamation*, 17(3–4), 461–462.

Gruiz, K., Fenyvesi, É., Kriston, É., Molnár, M. & Horváth, B. (1996) Potential use of cyclodextrins in soil bioremediation. *Journal of Inclusion Phenomena and Macrocyclic Chemistry*, 25, 233–236.

Gruiz, K., Horváth, B. & Molnár, M. (2001) *Környezettoxikológia. Environmental Toxicology* (in Hungarian). Budapest, Müegyetem Publishing Company.

Gruiz, K., Vaszita, E. & Siki, Z. (2006) Quantitative risk assessment as part of the GIS based environmental risk management of diffuse pollution of mining origin. In: *Conference Proceedings CD. Difpolmine Conference, 12–14 December 2006, Montpellier, France.*

Gruiz, K., Vaszita, E. & Siki, Z. (2007) Environmental toxicity testing in the risk assessment of a metal contaminated mining site in Hungary. *Advanced Materials Research*, 20–21, 193–196.

Gruiz, K., Molnár, M. & Feigl, V. (2009) Measuring adverse effects of contaminated soil using interactive and dynamic test methods. *Land Contamination & Reclamation*, 17(3–4), 445–462.

Gruiz, K., Vaszita, E., Siki, Z., Feigl, V. & Fekete, F. (2009) Complex environmental risk management of a former mining site. *Land Contamination & Reclamation*, 17(3–4), 357–372.

Gruiz, K., Molnár, M., Fenyvesi, É., Hajdu, Cs., Atkari, Á. & Barkács, K. (2010a) Cyclodextrins in Innovative Engineering Tools for Risk-based Environmental Management. *Journal of Inclusion Phenomena and Macrocyclic Chemistry*, 70(3–4), 299–306.

Gruiz, K., Feigl, V., Hajdu, Cs. & Tolner, M. (2010b) Environmental toxicity testing of contaminated soil based on microcalorimetry. *Environmental Toxicology, Special Issue: 14th International Symposium on Toxicity Assessment*, 25(5), 479–486.

Hajdu, Cs., Fenyvesi, É. & Gruiz, K. (2009) Cyclodextrins in Environmental Bioassays. *Cyclodextrin News*, (published by Cyclolab ISSN 0951-256X), 23(10), 1–10.

Kalbe, U., Berger, W., Eckardt, J. & Simon, F.G. (2008) Evaluation of leaching and extraction procedures for soil and waste. *Waste Management*, 28(6), 1027–1038.

Kiikkilä, O., Pennanen, T., Perkiömäki, J., Derome, J. & Fritze, H. (2002) Organic material as a copper immobilising agent: a microcosm study on remediation. *Basic and Applied Ecology*, 3, 245–253.

Lakanen, E. & Erviö, R. (1971) A comparison of eight extractants for the determination of plant available micronutrients in soils. *Acta Agralia Fennica*, 123, 223–232.

Landis, W.G. & Yu, M.H. (1999) *Introduction to environmental toxicology: impact of chemicals upon ecological systems.* New York, Boca Raton, Florida, CRC Press LLC.

Leitgib, L., Gruiz, K., Molnár, M. & Fenyvesi, É. (2003) Bioremediation of Transformer Oil Contaminated Soil. In: Annokkée, G.J., Arendt, F. & Uhlmann, O. (eds.) *Wissenschaftliche Berichte*, FZKA 6943, pp. 2762–2771. Karlsruhe, Germany, Forschungszentrum Karlsruhe GmbH Publisher.

Leitgib, L., Kálmán, J. & Gruiz, K. (2007) Comparison of bioassays by testing whole soil and their water extract from contaminated sites. *Chemosphere*, 66, 428–434.

Lombi, E., Zhao, F., Zhang, G., Sun, B., Fitz, W., Zhang, H. & McGrath, S. (2002) In situ fixation of metals in soils using bauxite residue: chemical assessment. *Environmental Pollution*, 118, 435–443.

MESOCOSM (2014) A portal of information on mesocosm facilities worldwide. [Online] Available from: http://www.mesocosm.eu. [Accessed 23rd February 2014].

Molnár, M. (2006) Intensification of soil bioremediation by cyclodextrin – from the laboratory to the field. *PhD thesis*, Budapest University of Technology and Economics, Budapest.

Molnár, M., Fenyvesi, É., Gruiz, K., Leitgib, L., Balogh, G., Murányi, A. & Szejtli, J. (2003) Effects of RAMEB on Bioremediation of Different Soils Contaminated with Hydrocarbons. *Journal of Inclusion Phenomena and Macrocyclic Chemistry*, 44, 447–452.

Molnár, M., Leitgib, L., Gruiz, K., Fenyvesi, É., Szejtli, J. & Fava, F. (2005) Enhanced biodegradation of transformer oil in soils with cyclodextrin – from the laboratory to the field. *Biodegradation*, 16, 159–168.

Molnár, M., Fenyvesi, É., Gruiz, K., Illés, G., Hajdú, Cs. & Kánnai, P. (2009a) Laboratory testing of biodegradation in soil: a comparative study on five methods. *Land Contamination & Reclamation*, 17(3–4), 495–506.

Molnár, M., Leitgib, L., Fenyvesi, É. & Gruiz, K. (2009b) Development of cyclodextrin enhanced soil bioremediation: from laboratory to field. *Land Contamination & Reclamation*, 17(3–4), 599–610.

Munk, L.A., Faure, G. & Koski, R. (2006) Geochemical evolution of solutions derived from experimental weathering of sulfide-bearing rocks. *Applied Geochemistry*, 21, 1123–1134.

OMSZ (2002) Hungarian Meteorological Service, Meteorological Data.

Salminen, J., Liiri, M. & Haimi, J. (2002) Responses of microbial activity and decomposer organisms to contamination in microcosms containing coniferous forest soil. *Ecotoxicology and Environmental Safety*, 53, 93–103.

Sand, W., Jozsa, P.G., Kovacs, Zs.M., Săsăran, N. & Schippers, A. (2007) Long-term evaluation of acid rock drainage mitigation measures in large lysimeters. *Journal of Geochemical Exploration*, 92, 205–211.

Sediment/Water Microcosm Biodegradation Test (1998) Fate, Transport and Transformation Test Guidelines OPPTS 835.3180. Prevention, Pesticides and Toxic Substances (TS–788), EPA 712–C–98–083.

Sipter, E., Rózsa, E., Gruiz, K., Tátrai, E. & Morvai, V. (2008) Site-specific risk assessment in contaminated vegetable gardens. *Chemosphere*, 71(7), 1301–1307.

Site-Specific Aquatic Microcosm Test (1996) Laboratory Microcosm – Ecological Effects Test Guidelines OPPTS 850.1925. Prevention, Pesticides and Toxic Substances (7101), EPA 712–C–96–173.

SoCo (2009) *Sustainable agriculture and soil conservation – Soil degradation processes, Fact sheet no.4*. [Online] Available from: http://soco.jrc.ec.europa.eu/documents/ENFactSheet-04.pdf. [Accessed 23rd February 2013].

University of Guelph (2013) *Aquatic Ecotoxicological Microcosms* (photo) [Online] Available from: http://www.uoguelph.ca/ses/content/aquatic-ecotoxicology-microcosms. [Accessed 23rd February 2013].

Uzinger, N. (2010) Model experiments for heavy metal immobilization by lignite. *PhD thesis*. Keszthely, Hungary, Pannon University.

Uzinger, N. & Anton, A. (2008) Chemical stabilization of heavy metals on contaminated soils by lignite. *Cereal Research Communications*, 36(Suppl), 1911–1914.

Vaszita, E., Gruiz, K. & Szabó, J. (2009a) Complex leaching of metal sulphide containing mine waste and soil in microcosms. *Land Contamination & Reclamation*, 17(3–4), 465–474.

Vaszita, E., Siki, Z. & Gruiz, K. (2009b) GIS-based quantitative hazard and risk assessment of an abandoned mining site. *Land Contamination & Reclamation*, 17(3–4), 513–529.

Vaszita, E., Szabó, J. & Gruiz, K. (2009c) Complex leaching of metal sulphide containing mine waste and soil. *Land Contamination & Reclamation*, 17(3–4), 463–472.

Yvon-Durocher, G., Montoya, J.M., Trimmer, M. & Woodward, G. (2011) Warming alters the size spectrum and shifts the distribution of biomass in freshwater ecosystems. *Global Change Biology*, 17(4), 1681–1694.

Yvon-Durocher, G. (2013) Impact of warming on freshwater ecosystem. (photo) [Online] Available from: http://news.bbcimg.co.uk/media/images/49564000/jpg/_49564394_mesocosms 304gabriel.jpg. [Accessed 23rd February 2013].

Chapter 9

# Data evaluation and interpretation in environmental toxicology

*K. Gruiz, Cs. Hajdu & T. Meggyes*
Department of Applied Biotechnology and Food Science,
Budapest University of Technology and Economics, Budapest, Hungary

## ABSTRACT

Adverse effects of pure chemicals are usually characterized by the concentration–response or dose–response curve. The effective concentration or effective dose can be read from this curve, given that the concentration, mass or volume of the chemical substance used in the study is exactly known. The test results indicate the hazard posed by the chemical substance. When the environmental concentration is known, the risk can be calculated from the scale of the hazardous effect.

The assessment of adverse effects of environmental samples is different because the concentration of the contaminants in water or soil samples is not known, and in most cases an unidentified mixture of contaminants is present. Studies, tests, bioassays or microcosms can measure the toxicity quantitatively based on the actual adverse effects of environmental samples. The actual adverse effect is in direct relation to risk.

Direct toxicity assessment (DTA) ensures high environmental relevance representing all possible interactions between contaminants, ecosystem members and soil phases aggregating the effect of all contaminants present in the sample. In addition to this, DTA can simulate different water and soil uses and real, multiple exposures.

On the other hand, directly measured toxicity of environmental samples cannot be expressed in concentration thus it does not fit the chemical risk assessment model and concentration-based screening values applied to pure chemicals.

DTA provides the effective or non-effective soil doses or dilutions as test end point, read from the dose–response curve. This is a shortcoming not only from the regulatory point of view but also for most environmental professionals who think and act mechanically, according to the chemical model. The authors have introduced an option to bridge direct toxicity assessing methods with the chemical model of environmental risk assessment. The equivalency methodology applies the copper equivalency toxicity to waters and soils contaminated with inorganic and the 4-chlorophenol equivalent to organic chemicals.

A separate section is devoted to statistics in toxicology. An overview of statistical methods, the most frequently applied methods and the relevant IT tools are discussed in detail. A short summary is given on the use of toxicity results in risk assessment.

## I   INTRODUCTION

The use of chemical models in environmental toxicology, monitoring, risk assessment and risk management resulted in the chemical concentration-based assessment of environmental risk, threshold concentration values and other environmental quality criteria, expressed in concentration. However, if a mechanical way of thinking focuses attention only on concentration values in the environment, this handicaps the development of a more complex and realistic evaluation method including the site-specific fate of a contaminant and the scale of measured adverse effects which can truly characterize the actual risk posed by chemicals to the environment.

A "hazard" refers both to harmful properties of the contaminants, generalized by the term "toxicity", and the effective concentration of the chemical measured by standard toxicity tests (e.g. using fish, rats, etc.). The hazard scale is proportional to the hazardous properties and the amount/concentration of contaminants from which the hazardous levels can be determined. However, the risk scale is influenced by many other environmental factors and the individual sensitivities of the targeted ecosystem and human receptors. Environmental factors are responsible for the interactions and processes which determine the fate and effects of a contaminant in the real environment, e.g. temperature, moisture content, redox potential, microbial activity, bioavailability and accessibility. All the interactions are part of a dynamic system with spatial and temporal variations. Mathematical models and simulation tests can transform hazard to risk: from the potential of a chemical substance to damage into the actual probability to cause damage. A chemical substance is already hazardous in theory, but it is risky only in the environment. Generic risk lies between the two, given a regular environment and an average receptor organism (e.g. humans) in the models or in the tests.

Generic risk posed by a contaminant can be characterized by using standard matrices and standardized test conditions. These types of tests simulate the interactions between the pure contaminant, the artificial test medium and the properly controlled test organism. These test conditions are highly abstracted as they represent the intrinsic hazard of pure chemical substances and the environment is limited to a test medium. Due to the constancy of the experimental 'environment', the reproducibility of the measurement is good and the generic fate and behavior of the chemical substance and its environmental hazardousness based on the effects caused can be predicted from the measured results. In addition, a mathematical model can be established for each standardized test based on measured data. For instance the QSAR models are based on the quantitative relationship between the chemical structure and the activities, taking the form of environmental behavior and adverse effects.

The risk posed by the *contaminated environment* cannot be characterized by the contaminants alone. The adverse effects of environmental samples integrate the characteristics of the contaminants, the physicochemical characteristics as well as sensitivity and vulnerability of the environment and the interactions between contaminants, matrix compounds and ecosystem members. Measuring the toxicity of environmental samples is called *direct toxicity assessment*. It can be done *in situ* on site or in the laboratory. Based on the results, the risk can be quantitatively characterized without chemically identifying the contaminants. Unlike pure chemical substances, environmental samples may contain a mixture of contaminants interacting with each other, with the components of the environmental matrices, the microbiota and higher members of the ecosystem. During direct toxicity assessments neither the contaminant components, nor the matrix components, nor the microbiota members are known to or can be controlled by the assessor. It is only the response of some selected representatives of the ecosystem that is known. As a consequence, the contaminant content and the impact of the interactions cannot be expressed in terms of contaminant concentration. Instead, they can be specified as inhibition, lethality or abnormal behavior of the affected organism or community.

*Direct toxicity testing* (DTA) provides information for direct decision making based on the measured scale of adverse effects, because DTA

– measures the toxicity of the entire effluent;
– characterizes sediment and soil toxicity with a more realistic representation of the environment;
– accounts for the aggregated effect of chemical mixtures;
– provides results that include a combination of different exposure routes and effects;
– measures a response proportional to the bioavailable fraction;
– accounts for the effects of unknown chemicals and those for which the toxicity is not well known;
– increases safety, given that complex industrial effluents, industrially contaminated soils and sediments may remain toxic despite complying with chemical-based limits.

As described in Chapter 1 the test end points quantify environmental fate properties and adverse effects of chemicals. The test end points (a.k.a. study end points) may vary from the simplest inhibition rate, through dose–response function and pharmacokinetic characteristics to complex field assessment results of species density and diversity. The most widespread test end points for DTA are (i) the inhibition rate – calculated from the response scale of the test organism in the sample compared to the negative control or reference – and (ii) the dilution-series-based toxicity end points giving the volume or mass of the sample causing a certain percentage of inhibition, or the highest measured level without affect (NOEL), or the lowest one found to affect the test organism (LOEL).

In DTA the sample dose–response curve describes the dose-dependent response of the test organism or the test system. The measured effect values are plotted against the sample doses (mass or volume) represented by the dilutions of the 'dilution series' prepared from environmental samples of contaminated water, wastewater, sediment or soil. The test end points may be the amount or the dilution rate of the sample causing an arbitrary level (e.g. 10%, 20%, 50% or 90%) of inhibition of the measured response or the highest no-effect amounts read from the dose–response curve. In DTA the 'dose' of the environmental sample is the amount (volume or mass) of the original liquid or solid sample in the tested dilution. The sample dose in DTA is not the same as in animal tests (the amount administered by the test organisms) but the amount of the environmental sample(s) in the test medium. 'sD' stands for 'sample dose' in this chapter, to distinguish it from consumed or administered dose (D). The dose of the environmental sample corresponding to $EsD_{20}$ or $EsD_{50}$ and the no-effect or lowest-effect data can be calculated from the amount of the sample (in mL or g) used in the test and the rate of dilution. The sample amount (in mL or g) depends on the test methodology; there are microbiological tests using only 200 µg samples while other ones use 50 or 100 grams. As the effective amount defined above of an environmental sample is test-dependent, it cannot be used in general for the characterization of toxicity in this form. Instead, a comparable and convertible interpretation is needed, such as the equivalence method introduced by the authors in Section 9.

The *dilution rate* is a useful value that refers to a generally applied outcome of environmental toxicity testing, as it is in direct relation both to the sample and to the risk. Based on the chemical definition, the risk characterization ratio, RCR = PEC/PNEC, reflects how many times the predictable environmental concentration is greater than the likely acceptable concentration of a contaminant. In this sense, the dilution rate shows how many times the adverse effect of an environmental sample is higher than the no-effect dilution, the dilution of the sample not causing any adverse effects. This no-effect dilution is closely related to the acceptable exposure in the environment.

The no-effect dilution or the dilution needed to achieve a predetermined percent inhibition (e.g. less than 10%) seems to be a clear and easy end point, but there is still the problem of inconsistency with regulation and environmental quality criteria expressed as contaminant concentrations. Most of the decisions in environmental management are based on concentration values and on the comparison of chemically measured concentrations to threshold values. Since the disadvantages of this kind of simplification have been realized by practitioners and by regulators, a refined chemical model of differentiated limiting values has been introduced at the regulatory level, for example, special soil quality criteria for clayey, loamy and sandy soils, or aerobic and anaerobic sediments. Environmental regulation and management rarely apply dilution rates to define thresholds and obligations such as: 'The risk of the soil is unacceptable and should be reduced if the dilution rate necessary to reach the no-effect point is higher than 5 for any of three terrestrial test organisms representing three trophic levels'. This would be a correct but completely unusual criterion and hardly understandable for environmental professionals who grew up using chemical models. Despite all these some practical cases are known where environmental criteria are based on the results of direct toxicity assessment, e.g. treated wastewater discharge into specific surface water or remedied soil reuse for purposes of different intensity and sensitivity. In these cases 'the absence of toxicity' is the screening factor. Unfortunately, the results of direct toxicity assessment are not used when managing risk and planning risk reduction measures of complex pollution cases, although the scale of reduction or attenuation of adverse effects to reach a no-effect level could be equivalent to the scale of the necessary risk reduction regardless of the actual contaminants present. In current risk management practice, the directly measured adverse effects are only used for indication, and, based on positive toxicity screening results, chemical tools are involved to determine a concentration value. In spite of the limitations of the chemical approach – for example in the case of inherited mixed pollution, when the contaminants and metabolites present are highly uncertain – most of the decisions are made based on the measured concentrations of arbitrarily selected or regulated priority contaminants. The authors of this chapter prefer direct toxicity assessment as a routine tool in risk management in all cases when a mixture of contaminants is present, when bioavailability has great influence and causes uncertainties in exposure and when risk is reduced by biodegrading communities such as in the biological waste-water treatment and soil bioremediation processes. It is also highly recommended as a screening tool in the first step of a tiered site assessment and as an approval or verifying tool for risk reduction activities, as well as a general environmental monitoring tool.

It is a common experience that environmental toxicity results only make sense for eco-toxicologists because toxicity data are hard to understand for practitioners brought up on chemical models. This is one of the reasons for the actual gap between availability and practical application of environmental toxicology in contaminated land management, although everyone who manages the risk posed by environmental contamination knows that the adverse effects and their scale are and should be the core information supporting environmental decision making. Decisions based on the results of direct toxicity assessment could be an effective tool in environmental management, but today only a few examples can be given on the direct use of environmental toxicity data in regulation or decision making.

Another possibility for toxicity data interpretation is the verbal characterization of the measured toxicity. It is based on the creation of inhibition ranges for each test, assigning verbal characterization such as 'very toxic', 'moderately toxic', 'slightly toxic' or 'non toxic'. The interpretation is test-specific, which means that, for example, different ranges are applied to a bacterial and a plant toxicity test given that the shape of the dose–response curve is different for these two test organisms. When creating these ranges, the quantitative and differential character of the test results is lost.

A similar approach is applied to the regulatory classification and labeling of chemicals: based on detailed hazard assessment, the chemical substance, product or waste will be labeled and handled mechanically according to the 'label' or the group into which it has been classified. This makes the situation understandable for non-professionals and simplifies the management of chemicals, wastes and contaminated land, but it also leads to a loss of information. For example, it may happen that two environmental toxicants are in the same category and obtain the same label with an $EC_{50}$ of 11 mg/L and 99 mg/L because they fall into the same hazard category.

To make the adverse-effect results of bioassays better understandable and more applicable, we introduce the "equivalent evaluation and interpretation" and "equivalent calibration" tools for organic and inorganic contaminants. The essence of this evaluation and interpretation methodology is that the inhibition of environmental samples is compared to the toxicity of a reference (calibration) compound, which is the same for each measuring set, and finally the inhibition of samples is expressed as an equivalent toxicity value for the selected reference substance. These methods use a "calibration curve" for the conversion of the measured end point to the concentration of the calibrating (reference) substance (Gruiz, 2009). The toxicity equivalent is a technical tool which may bridge the chemical and biological models in risk management of contaminated land. It makes possible the comparison of the character of the impact of different contaminants with each other and of the contaminated environmental compartment with pure chemicals and other environmental samples.

The different concepts of environmental risk management are illustrated in Figure 9.1 that shows the application options of chemical, biological and ecosystem models and their relation to the risk management measure. Chemical models interpret ecosystem response as a chemical characteristic, typically as an effective concentration, then compare this concentration to a chemically based threshold value. Finally, the ecosystem response/characteristic is estimated based on the difference between the actual environmental and the threshold values. Chemical models reduce ecosystem

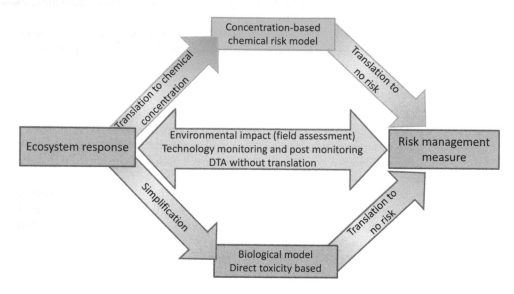

*Figure 9.1* The concept of environmental risk management applying chemical, biological and ecosystem models.

complexity to one or very few chemical characteristics. Consequently, they entail extreme uncertainties due to the discrepancy of the model from the real environment. These uncertainties are partly compensated for by a low measurement error. Biological models measuring toxicity directly from environmental samples are much closer to the real environmental situation, reflecting correctly the microenvironment and the environmental matrices, but not the whole environment. Measurement errors of toxicity tests are higher than those of the chemical analysis methods, but they can be reduced by well-established and precisely executed methods. Field assessment vindicates the role of closely modeling the environment, but the subset of the environment assessed (assessable) by the assessor corresponds only to a sample, which may or may not produce a sufficiently true value with low uncertainty.

The selection of the measurement end points and the evaluation methodology depend on test goals, the type of test, the sensitivity of the test organisms and on the quality and quantity of the toxic hazard. Expected capabilities of a proper measurement end point are summarized as follows:

–    A proper measurement end point serves the purpose of the assessment. The end point of the assessment should have a diagnostic value and should be in close relationship with the hazardous effect and risk;
–    The measured end point should be consistent with the main goal of the study in terms of preciseness, exhaustiveness and qualitativeness (e.g. sensitivity of the test organisms);
–    The end point allows for the measurement of direct and indirect effects such as genetic, metabolic (consumption and production), reproductive, growth and lethal effects;

– The measured end points should represent adequate sensitivity (average, higher or lower than average) and the response time is as short as possible;
– The measurement end point should provide a high level of the desired signal compared to the level of background, i.e. high signal-to-noise ratio;
– The implementation, evaluation and interpretation of the measurement should be easy and practical.

When testing contaminated soil, the selection of an adequate reference soil representing the uncontaminated soil is always a challenge. Artificial soils as reference such as the OECD-recommended mixture (5% peat, 20% clay with high kaolinite content and 75% sand with a dominant fraction of 50–200 μm particle size) need consistent material properties to be able to maintain constant quality. When using the same artificial soil for dilution, nonlinear changes may occur in the different soil characteristics influencing soil toxicity (OECD soil may inhibit growth due to low nutrient content, overcompensating for the toxicity based inhibition of a nutrient reaching the contaminated soil). If a natural soil is selected as reference, its storage, maintaining its equilibrium and microbial activity may be difficult. The best reference for contaminated soils taken from the field is an uncontaminated aliquot from the same or a very similar site.

The most frequently used evaluation methods applied in environmental toxicity assessments are introduced as follows. The evaluation of toxicity results of pure chemicals and environmental samples are discussed in parallel.


## 2  INHIBITION RATE

The rate of inhibition is the most popular measured value for laboratory bioassays, given that toxicity of a chemical substance or an environmental sample is generally compared to an uncontaminated reference, a negative control or blank, and the result is expressed as percentage.

The inhibition rate of *pure chemicals* dissolved in water results in a regular saturation curve (such as the dose–response curve) when plotted against concentration. Pure chemicals dissolved in organic solvents create a more complicated situation. The solvents may have (and generally do have) an effect on the test organism, so the dissolved chemical should be considered as a mixture with additive or non-additive effects. Pure chemicals mixed in soil in growing concentration for generating a concentration series create an even more complicated situation due to continuously changing contaminant speciation and non-equilibrium partitioning between three physical phases as well as to concentration-dependent bioavailability. The best practice for preparing a concentration series is mixing an exactly known mass (dose) of contaminant into every single aliquot of the standard soil. The inhibition percentages are plotted as a function of the contaminant concentration.

*Contaminated soils* and sediments are the most difficult target for toxicity assessment because the original sample should be diluted to get the mass dependent response and reach the 'no inhibition' point. The dilution of solid environmental samples raises a number of questions: what to dilute with, how to ensure equilibrium conditions, etc. Mixing two different soils is the beginning of the development of a new equilibrium

*Figure 9.2* Plant root and shoot growth inhibition is proportional to toxicity.

for all soil processes. A further common problem is that the reference or artificial soil used for the dilution of the contaminated soil may (and definitely does) generate further inhibitory or stimulatory effects, which would render the results ambiguous. The selection of the proper reference soil may be a problem in itself: it can be a site reference (similar soil from the same place but definitely not contaminated) or a general reference (artificial soil or laboratory reference). The result of a test carried out with contaminated soil is always the effective amount (mass) of soil causing a certain scale of inhibition (5%, 10%, 20%, 50%, 90%) or no inhibition (NOEC).

In soil studies, e.g. when measuring plant growth in contaminated soil (Figure 9.2), it is common that results from a complete dilution series or more than two effective concentrations are not available because toxicity disappears rapidly with the dilution of the contaminated soil. Another problem is that the uncontaminated reference soil or artificial soil used for the dilution of the contaminated soil generates additional inhibitory or stimulatory effects, which makes the evaluation ambiguous. Therefore, plant toxicity is often expressed just in the percentage of a relative inhibition compared to the negative control.

Inhibition of root and shoot growth is given as compared to the control, the growth on uncontaminated soil. Inhibition rate can be calculated in % by Equation 9.1:

$$X_L(\%) = 100 * (L_C - L_T)/L_C, \text{where} \tag{9.1}$$

– $X_L$ = relative growth of root/shoot compared to control, in %;
– L = seedlings' root or shoot length in millimeters;
– C = control, grown on reference soil and
– T = tested, grown on contaminated soil.

The rate of inhibition may consider not only the length but also the produced biomass (Equation 9.2).

$$X_M(\%) = 100 * (M_C - M_T)/M_C, \text{where} \tag{9.2}$$

– $X_M$ = relative growth in plant mass compared to control, in %;
– M = seedlings' mass in g or kg;

- C = control, grown on reference soil and
- T = tested, grown on contaminated soil.

Inhibition values can be assessed for the whole dilution series of the contaminated soil (or any other environmental sample) and the soil dose–response curve can be plotted from the inhibition rates: inhibition percentage as a function of the amount of the contaminated soil.

For statistical evaluation of the inhibition results, the most appropriate is to use the Student's $t$ statistic to test whether the means of the treated and the untreated control are different. The independent two-sample $t$-test can be applied if the sample sizes (the number of parallels) are equal and the two distributions have the same type of variance, e.g. if they are normal distributions.

## 3  CONCENTRATION/DOSE—RESPONSE RELATIONSHIP

The concentration–response curves of *pure chemicals* are sigmoid functions reflecting the fact that the response does not rise linearly as a function of the concentration, but a certain level is needed for triggering an effect and after a certain toxicant concentration the response cannot grow proportionally. The increasing concentration results in a growing response up to the inflection point, after which the response inhibition starts decreasing with the number of test organisms or with their activity in the test vessel.

Concentration–response curves are plotted from primary (measured) results or from derived values such as the inhibition rate (calculated from measured end points) or more complex derived values, such as the area under a growth curve, or any other significant characteristics of the response curve, e.g. the maximal response or the slope.

Figure 9.3 shows a typical concentration–response curve: the effects of the members of the concentration series made of PCP dissolved in water. The best sigmoid is fitted to the measurement points by using statistics-based methods supported by suitable software tools and the $EC_x$, NOEC and LOEC values are read from the curve. Statistical evaluation will be discussed in the next subchapter.

It is important to understand that there is a difference between the measured NOEC value, which means the highest measured concentration of the tested concentration series, which does not show a significant adverse effect, and the hypothetical NOEC, which is likely to be a larger value than the measured one. The difference between the measured and the hypothetical NOEC depends on the scale and spacing of the concentration series prepared for testing. This explains also the rule that the design of the study should be fitted to the goal, the tested end point and other requirements of the study.

In assessing the adverse effect (inhibition rate) of contaminated *environmental samples*, the above described method results in a sample dose–response sigmoid. The contaminant composition and concentration in the sample are not known. Instead of the contaminants' concentration, the sample mass or volume is applied as a variable.

As an example, Figure 9.4 shows the dose–response curve of the *Vibrio fischeri* culture measured with a serially increasing proportion of contaminated water (in reference water), given in mL. In other cases, e.g. when testing contaminated soil or

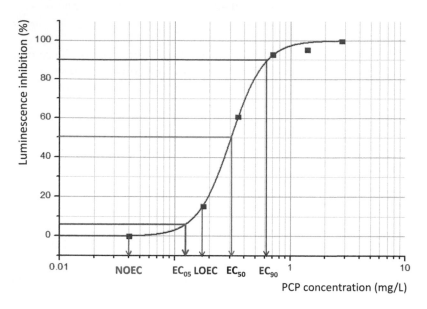

*Figure 9.3*  Test end points on the concentration–response curve of pentachlorophenol (PCP) dissolved
in water.
Legend to Figure 9.3:
Inhibition of the luminescent light emission (H%), a specific type of inhibition for the luminobacterium *Vibrio fischeri*
$EC_{90}$: concentration resulting in 90% inhibition
$EC_{50}$: concentration resulting in 50% inhibition
LOEC: lowest tested concentration exerting an effect
NOEC: highest tested concentration giving a no-effect response.

sediment the sample mass (in g) is used instead of the sample volume. Dilution rate
can also be applied as an alternative to mass/volume.

The inhibition rate of H% for luminescence is a special case of inhibition. H% is
a more complex value than I% in the plant example (Section 2), integrating a time-
correction factor based on the luminescent light measurement methodology (Equations
9.3–9.5):

$$H\% = 100 * (I_{calc} - I_{30})/I_{calc}, \text{where} \tag{9.3}$$

$$I_{calc} = f * I_0 \tag{9.4}$$

$$f = I_{c30}/I_{c0}, \text{with} \tag{9.5}$$

–  $I_0$ = initial luminescence of the bacterial suspension without sample;
–  $I_{30}$ = luminescence of the test suspension 30 minutes after sample addition;
–  $I_{c0}$ = initial luminescence of the negative control;
–  $I_{c30}$ = luminescence of the negative control after 30 minutes;
–  f = the ratio of $I_{c30}$ to $I_{c0}$, i.e. the negative control at the beginning and after
30 minutes;

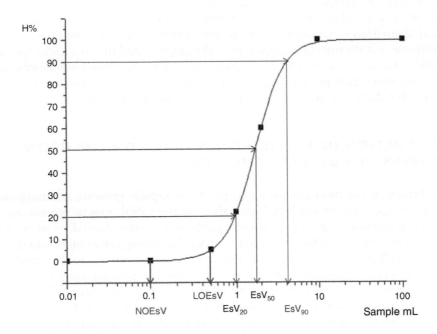

*Figure 9.4* Test end points on the sample volume–response curve of environmental water samples containing unidentified contaminants.
Legend to Figure 9.4:
H%: inhibition of the luminescent light emission (a specific type of inhibition)
$EsV_{90}$: sample volume resulting in 90% inhibition
$EsV_{50}$: sample volume resulting in 50% inhibition
$EsV_{20}$: sample volume resulting in 20% inhibition
LOEsV: lowest tested sample volume showing an effect
NOEsV: highest tested sample volume giving a no-effect response
s: sample whose toxicity is measured.

–  $I_{calc}$ = calculated light intensity, which would be emitted by the test suspension without toxic sample addition;
–  H% = luminescence inhibition of the sample.

The evaluation is based on the initial and the 30 minutes luminescence. As an alternative, one can continuously measure luminescent light emission for each sample and plot the time-dependent changes in the emitted light. After measuring the light emission–time curves, one has to characterize the typical shape of the curve and select the most characteristic features, which adequately characterize the response of the tested bacterial culture, and use this feature (slope, area under the curve, maximum of light emission, and so on) for the calculation of the inhibition rate.

A concentration/dose–response curve can be plotted from any measured end point which is proportional to the concentration/dose of the stressor and is able to quantify the response and thus differentiate it from a no-effect response or from effects of significantly lesser or greater values. Both directly measured end points (lethality, activities, reproduction) or derived values (inhibition rate, slope of the activity–time curve, etc.) can be utilized for characterizing the adverse effect of chemical compounds or contaminated environmental samples. The derived values may be the maximum value of the response and the time of its appearance, the slope of and the area under the curve, and other characteristic features of the time-curves for respiration, luminescence, any enzyme activities, heat production, growth and movement of the test-organism. Some examples for the use of derived end points are introduced in Section 5.

## 4  EVALUATION OF THE RESPONSE BASED ON THE GROWTH CURVES OF CULTURED ORGANISMS

Reproduction is the most adequate end point for rapidly growing microorganisms, bacteria, fungi, algae or single-cell animals. The microbial growth process includes consecutive reproduction cycles of microorganisms by repeated cell division according to the power of two, resulting in an exponentially increasing cell number and biomass. The initial cell divisions are slower (lag phase) due to adaptation to the growth substrate. The speed then increases steadily to a maximum (exponential phase) after which it slows down due to the lack of nutrients (assuming propagation in batch). Time dependence of the number of cells is shown by the growth curve in Figure 9.5).

*Tetrahymena* is a sensitive protozoan (single-cell animal) (light microscopic and electron microscopic images can be seen in Chapter 4). Its response to contaminants and contaminated environmental samples is proportional to the concentration of the contaminant or the amount of the contaminated environmental sample.

All phases and special points of the growth curve – the length of the lag-phase, the slope of the exponential phase or the maximum of the curve and the time of rise – are suitable indicators for the growth and growth inhibition. The integrative indicator of the area below the growth curve can be calculated by any mathematical tool, for example by approximate calculation, assuming that two evaluation points can be connected by a straight line (which is not entirely true), and the area under the growth curve can be mapped as right-angled trapezoids, e.g. using the formula below (Equation 9.6):

$$A = t_1 * (X_1 - X_0)/2 + (t_2 - t_1) * ((X_1 - X_0) + (X_2 - X_0))/2 + \cdots + (t_{n+1} - t_n)$$
$$* ((X_n - X_0) + (X_{n+1} - X_0))/2 \qquad (9.6)$$

- $A$ = area
- $X_0$ = nominal number of cells/mL at the beginning of test;
- $X_1$ = measured number of cells/mL at $t_1$;
- $X_n$ = measured number of cells/mL at time $t_n$;
- $t_1$ = time of first measurement after the beginning of test;
- $t_n$ = time of $n$th measurement after the beginning of test.

From **A**, as an indicator of cell reproduction (biomass), the inhibition values (I) can be calculated in percent, comparing the treated to the control. Then the individual I% values can be compared to each other or the serial I% values can be plotted vs. contaminant concentration or environmental sample mass/volume.

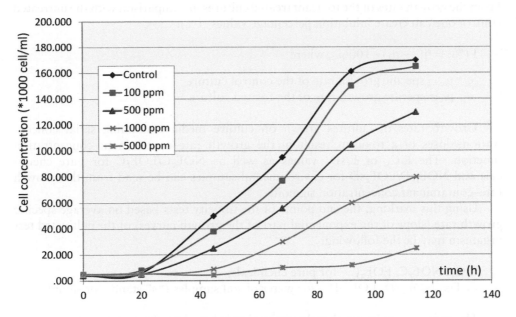

*Figure 9.5* Growth curves of *Tetrahymena pyriformis* at various contaminant concentrations.

The effect of pure chemicals can be measured this way and the $E_AC_{50}$, the $NOE_AC$ or $LOE_AC$ values (the "A" index indicates that the result is based on the area under the growth curve) can be read from the I%–concentration curve plotted from the inhibition values of the concentration series.

Environmental water and soil samples can also be characterized based on the growth curves by the effective dose of the environmental sample (the volume of the water or the mass of the soil). The sample amount causing a certain percentage of inhibition and the highest no-effect and lowest effect sample doses are symbolized by the following abbreviations for results obtained from the area under the curve:

–   $E_AsD_x$: effective dose for a specified percentage of inhibition rate, where x may range from 0% to 100%;
–   $NOE_AsD$: no-effect sample dose;
–   $LOE_AsD$: lowest effect sample dose.

Another possibility to evaluate the growth curves is the comparison of the growth rates. The average specific growth rate ($\mu$) for exponentially growing cultures can be read from the slope of the growth curve, plotted in the form of *lnX vs. time*, or calculated by the following formula:

–   $\mu = (\ln X_n - \ln X_1)/(t_n - t_1)$, where

    o   $\mu$ = average specific growth rate for exponentially growing cultures (logarithmic increase of biomass during the test period, expressed in $day^{-1}$);
    o   $t_n - t_1$ = time interval within the exponential growth phase.

From the growth rates of the toxicant treated cultures in comparison with the untreated control one can create inhibition percentage values (I%):

–   $I (\%) = (\mu_c - \mu_t) * 100/\mu_c$, where

   o   $\mu_c$ = specific growth rate of the control culture
   o   $\mu_t$ = specific growth rate of the treated culture

Growth rates of cultures grown on culture media containing serial concentrations/doses of a toxicant result in the growth rate–contaminant concentration function. The $E_rC_x$ or $E_rsD_x$ values as well as $NOE_rC/LOE_rC$ for pure chemicals and $NOE_rsD/LOE_rsD$ for environmental samples can be read from this growth rate–contaminant concentration sigmoid.

Using this marking, the end points of the toxicity tests based on average specific growth rate (slope of the exponential part of the growth curves) of the cultivated test organism may be the following:

–   $E_rC_x$, $NOE_rC$, $LOE_rC$ – for pure chemicals;
–   $E_rsDx$, $NOE_rSD$, $LOE_rSD$ – for water of soil samples ("s" stands for sample).

The index "r" indicates that the evaluation is based on the growth rate.

The yield (Y), the biomass increase during the test periods can also be used as a measurement end point. Its calculation is: biomass at the end of the exposure period minus biomass at the beginning.

Biomass and cell count may correlate with each other well but they can differ on a significant scale due to the deviations in cell size, which is why cell count is considered by many of the standardized methods to be a surrogate method compared to cell biomass analysis.

The same type of evaluation is used for any microbial and algal growth tests. The 'test organism' is a culture, the quasi-homogeneous culture of the microorganisms where the individual cells are not considered. The response of such a system depends mainly on the experimental parameters and is described typically by the log-normal or normal distributions. It is worth using all the data of the replicates (instead of the means). Data should be fitted to a functional equation by non-linear regression. Selection of the appropriate functional equation can be a problem: if it is not possible to fit to the whole of the curve, it is advisable to choose and fit different equations to the two distinct parts of the growth curve. From the fitted equation the characteristic point estimates of $E_rC_x$ values can be determined. More detailed discussion on statistics is found in the next subchapter.

## 5   EVALUATION OF THE EFFECT OF CONTAMINANTS ON HEAT PRODUCTION: A SPECIAL CASE

Toxicity evaluation based on the response–time function makes the tests and evaluation more complicated, but the information behind the numbers may be more important and necessary for making a good decision based on toxicity assessment.

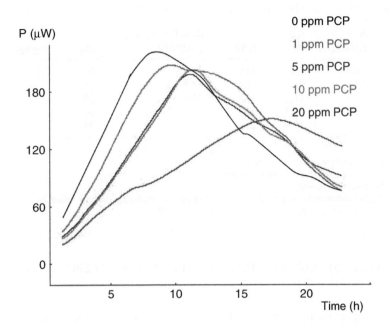

*Figure 9.6* Microcalorimetric power–time curve of *Azomonas agilis* in PCP-polluted soil.

An example for the evaluation of the heat production curve for quantitative toxicity assessment is explained here. Contaminated soil was tested in a microcalorimeter (see Chapter 5) adding an *Azomonas agilis* bacterial strain to the sterilized soil sample. The concentration series of PCP (pentachlorophenol, a pesticide and disinfectant) and DBNPA (2,2-dibromo-3-nitrilopropionamide, a quick-kill biocide) were assessed and the power–time curve was plotted.

The power-time curve of PCP-contaminated soil is shown in Figure 9.6. Theoretically, the area under the curve is proportional to the total heat produced during the test. This may be a good end point, but less capable of distinguishing among cases than some other characteristics of the curve. The results indicate that the higher the concentration of PCP in the soil, the smaller the initial rise ($R_i$) and the maximum ($P_{max}$) of the curve. A shift in time in the power maximum can be observed ($t_{max}$).

In the special case shown in Figure 9.6, instead of the total heat production (area under the curve), the most significant indicators of $R_i$, $P_{max}$ and $t_{max}$ were combined to establish the $R_i \times P_{max}/t_{max}$ factor, denoted $A_h$. $A_h$ integrates and magnifies the effects being in reciprocal relation with PCP concentration in the tested soil. The $R_i \times P_{max}/t_{max}$ value ($A_h$), the inhibition rate (I) and the $EC_{20}$ and $EC_{50}$ values are summarized in Table 9.1.

$EC_{20}$ and $EC_{50}$ are read from the concentration vs. $A_h$ (response) sigmoid although the $EC_{20}$ and $EC_{50}$ values can be determined from any of the end points (R, P, t, RxP/t) read from the curve, which is useful for making comparison. Other end points can also be read from the sigmoid such as NOEC or LOEC.

*Table 9.1*   Effect of PCP- and DBNPA-polluted soil on the heat production of *Azomonas agile*.

| PCP conc. (mg/kg) | $R_i$ ($\mu W/h$) | $t_{max}$ (h) | $P_{max}$ ($\mu W$) | $A_h = R_i \times P_m/t_{max}$ ($\mu W^2/h^2$) | $I_A$ (%) | $EC_{20}$ (mg/kg) | $EC_{50}$ (mg/kg) |
|---|---|---|---|---|---|---|---|
| 0 | 36.4 | 9.3 | 219 | 856 | 0 | 0.7 | 3.9 |
| 1 | 25.1 | 9.6 | 208 | 544 | 36 | | |
| 5 | 19.7 | 11.2 | 198 | 351 | 59 | | |
| 10 | 18.8 | 11.4 | 203 | 334 | 61 | | |
| 20 | 8.9 | 17.4 | 157 | 80 | 91 | | |

Legend to Table 9.1:
$R_i$ ($\mu W/h$): initial rise of the power–time curve;
$t_{max}$ (h): time of occurrence of the power maximum ($P_{max}$);
$P_{max}$ ($\mu W$): the maximum power;
$A_h = R_i \times P_m/t_m$: an indicator for the magnification of the positive responses, shortly called '$A_h$';
$I_A$(%): inhibition based on $I_A = (A_{hControl} - A_{hPCP})/A_{hControl}*100$.

## 6   EVALUATION OF BIODEGRADATION OF CHEMICALS IN WATER AND SOIL

Testing biodegradability and actual biodegradation in water, sediment, soil and wastewater or solid waste is based on similar methodologies: to utilize contaminants as energy substrates and to measure the consumption of the substrate or the production of inorganic end products or any metabolite or by-products of the degrading microbiota.

The biodegradation rate is the measure of the contaminants' tendency to be degraded or their effect on biodegrading organisms. Its measurement has the following aims:

– Characterizing the biodegradability of pure chemical substances or contaminants or mixtures of contaminants in the environment;
– Characterizing the biodegradation potential of the soil or wastewater community;
– Measuring the toxicity of a chemical substance or contaminated environmental samples on the microbiota or on other degrading members of the ecosystem of water, sediment, soil, or wastewater. The inhibition of biodegradation is measured in this case.

The evaluation of the biodegradation of chemicals is based on the mass balance, which is described by the reaction equation. A generic equation for organic compounds consisting of C (carbon), H (hydrogen) and N (nitrogen), i.e. the complete aerobic biodegradation (mineralization) in the presence of microorganisms and $O_2$, neglecting cell mass growth, can be described as (Equation 9.7):

$$C_XH_YN_Z + (X + Y/2 + 3Z/2)O_2 \rightarrow \text{microorganisms} \rightarrow$$
$$XCO_2 + Y/2H_2O + ZNO_3 + \text{biomass} \tag{9.7}$$

One can measure the decrease in the mass of the chemical substance being degraded, the oxygen consumed and the $CO_2$ or the $NO_3$ produced and calculate all the others based on the equation. If the microorganisms have the opportunity (e.g. time) to grow and propagate, the increase in cell mass should be added to the right side of the equation. If the product is different to $CO_2$, one has to use the molecular weight of the identified product in the calculation.

Depending on the naturally occurring degradation process other types of equations can be used for partial degradation and for anoxic degradations and artificial indicators (radioactive labels or color-markers) can also be applied to facilitate evaluation.

## 6.1   Monitoring the depletion of the chemical substance

Degradation of the substance is generally followed by chemical analyses. Sampling and analysis should be performed at least at the beginning and the end of the tests, but recording several time points may be beneficial. Isotope-labeled substances are easier to follow based on their activity. The result based on mass values is compared to the initial mass and given in percentage as a biodegraded proportion.

Toxic compounds may inhibit normal biodegradation in water and soil. In this case the biodegraded proportion should be evaluated in comparison with uncontaminated control water or soil to calculate inhibition rate in percent. The effect of a series of contaminant concentrations or of the contaminated wastewater volumes on the healthy biodegradation can also be measured. Based on the results, the dose/*concentration–response* curve can be drawn and *the no-effect* and various *effect* values can be read from the curve.

## 6.2   Evaluation of biodegradation based on $CO_2$ production

Biodegradability is most widely tested based on $CO_2$ production because the test equipment and set-up ensures that the only degradable substrate is the chemical substance/contaminant to be tested. The produced mass of $CO_2$ is equivalent to its degraded C-content. Soil without contaminant (without substrate) is used as a reference (negative control).

As an alternative methodology, impulse-like addition of the biodegradable substance to the steady-state aerated test system makes the test more dynamic. In this case the scale of the response and its time-dependent occurrence can be evaluated.

If the kinetics of biodegradation is known, the half-life time can be the end point of the biodegradability test.

The biodegradable fraction of the contaminants in wastewater or contaminated soil can be determined based on the produced $CO_2$, equivalent to BOD (biological oxygen demand) according to the carbon transformation formula (Equation 9.7).

Figures 9.7 and 9.8 show the respiration curves of artificially contaminated soil with growing concentrations of a mixture of toxic metals (Cd, Cu, Hg, Ni and Zn). The concentration series represents 0, 1x, 2x, 4x, 8x and 16x of the soil screening concentrations (SSC). Similar serial curves are created by contaminated soils added to healthy soil in growing percentage.

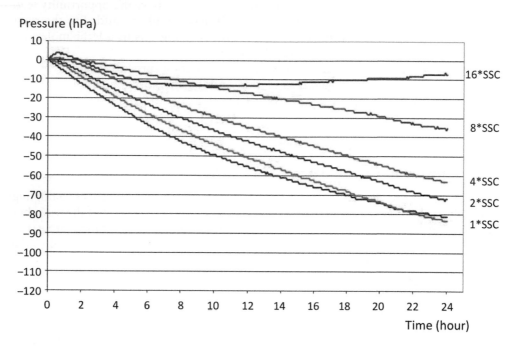

*Figure 9.7* Decreasing $CO_2$ production by the soil with increasing toxic metal concentration.

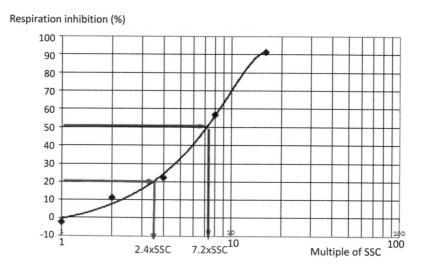

*Figure 9.8* The concentration–respiration inhibition curve from the 24h respiration rate.

## 6.3 Substrate induction

To characterize soil health, activity and its ability to biodegrade, induction of ready-to-biodegrade additives such as glucose is a usual test (Torstensson, 1994). The same

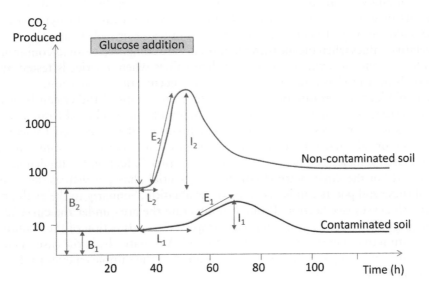

*Figure 9.9* Glucose-induced respiration rates for contaminated and uncontaminated soils.
Legend to Figure 9.9:
B: Soil respiration baseline value, before glucose addition
– $B_2$: uncontaminated soil: intensive respiration rate
– $B_1$: contaminated soil: low respiration rate
I: Soil respiration rate after the addition of easy-to-utilize glucose
– $I_2$: uncontaminated soil: higher increase
– $I_1$: contaminated soil: lower increase
L: Lag phase: adaptation to the new situation, before the exponential phase
– $L_2$: uncontaminated soil: short lag phase
– $L_1$: contaminated soil: long lag phase
E: Exponential phase with constant respiration rate
– $E_2$: uncontaminated soil: steep slope
– $E_1$: contaminated soil: less steep slope.

test is suitable to test inhibitory or adverse effects of contaminants based on the decreased biodegrading activity due to poisoning of the water or soil microbiota.

Substrate-induced respiration (SIR) as well as substrate-induced nitrification (SIN) bioassays can measure a response proportional to adverse effects using standardized and non-standardized test methods. Measuring the glucose-induced respiration rate (or that of any other adequate substrate) after the incubation of the soil and the suspected contaminating substance, the response to the easy-to-degrade substance (glucose) will be inversely proportional to the toxicity.

Using biodegradation measurements as an ecotoxicity test method, the effects of a series of concentrations (of the chemical substance or the contaminated water/soil) on carbon transformation can be determined. The data from these tests are used to prepare a dose–response curve and to calculate $EC_x$ or $EsD_x$ values.

Adding glucose to the soil, a sudden increase in respiration rate can be measured (Figure 9.9). The increase (I) is proportional to soil health, adaptive behavior, and the presence (if any) of contaminants with inhibitory effects. Test design and data evaluation and interpretation depend on the aim of testing: biodegradability of contaminants or any other substrates to degrade, respiration rate and dynamics of the wastewater/soil or ecotoxicity of contaminants in water or soil. Evaluation can provide relative values when the measurement results are compared to an uncontaminated, untreated or reference water/soil, or absolute values when a series is tested and the response is measured as a function of growing concentration or amount.

The biodegradation rate in substrate-induced respiration studies can be used for toxicity testing. The concentration–response sigmoid can be plotted from any of the characteristics of the respiration curve as a function of the applied toxicant concentrations or, alternatively, of the doses (mass or volume) or the dilutions of the contaminated sample. The height of the increase after induction (I), the length of the lag phase (L) or the slope factor (E) of the exponential phase, as well as the combinations of these end points can be used for the evaluation. Similarly to the evaluation of the growth curves (see Section 4) one can integrate the area under the curve and use the integrated $CO_2$ production values as "response" as a function of the contaminant concentration to calculate $EC_X$ or LOEC/NOEC. Alternatively, in the case of contaminated water/soil samples, the mass or the dilution of the sample enables to determine $EsD_x$ and LOEsD/NOEsD.

## 7   ATTENUATION RATE METHOD FOR ENVIRONMENTAL SAMPLES

From the *dose/concentration–response* curve, amongst others NOEC, NOEsV (no-effect volume of the water sample) or NOEsD (no-effect mass of the soil sample) values can be read. They are the highest concentration and the sample volume or mass which do not cause any adverse effect. This refers to the no-risk state of the water or soil in contaminated land management and can be considered as the target of risk reduction. The management measure should ensure a risk reduction rate needed to achieve the no-risk situation in the field. This is supposed to be equivalent to the no-effect dilution of the DTA sample in terms of toxicity. It is reasonable to introduce a safety factor based on the uncertainties accompanying the assessment.

The tested serial volume or mass of environmental samples is created during the test by diluting the original sample with uncontaminated water or soil. In these cases the dilution rate to reach the no-effect point on the dose–response curve is identical with the necessary risk reduction rate. This does not mean that the risk is reduced by a dilution process in the real environment; it means rather the attenuation due to natural or technological processes responsible for reaching an acceptable risk value.

Liquid-phase wastes (i.e. wastewaters) contain unidentifiable contaminants, as hundreds and thousands of toxicants enter the wastewater treatment plants and a chemical analysis is not practiced as a daily routine. Residual organic matter and toxicity can be measured easily and the decision on when wastewater can be discharged into effluents can be based on measured toxicity results. The toxicity of the treated wastewater poses a risk to the ecosystem of the recipient water body; the wastewater should not be discharged from the treatment plant, but further treated. The water of inadequate quality is stored in a buffer pool until recycling or further treatment. The ecosystem of the recipient water body in question can be represented by selected

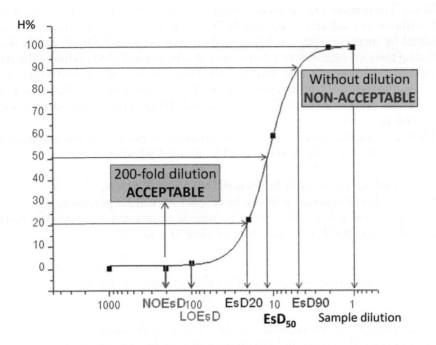

*Figure 9.10* Graphic determination of the dilution rate of a contaminated soil sample to reach the no-effect volume measured by the collembolan test.

aquatic organisms and can be used as site-specific test organisms. If the measured ecotoxicity exceeds the NOEsV values, it requires interventions by the operator, e.g. changes in the technological parameters, use of additives or any other treatment to reach the required lower toxicity. For properly representing an aquatic ecosystem, a minimum of three organisms from different trophic levels should be tested and the results used for decision making in a way that can take uncertainties into account.

When using measured ecotoxicity data for direct decision making, managing uncertainties should have priority. Uncertainty derives from several details: the selection of the test organisms, their representativeness, their rate of sensitivity, the composition of the wastewater, adaptation of the ecosystem, etc. Depending on the scale of the expected uncertainties, more test organisms and safety factors should be applied.

The dilution rate read from the dose–response curve should be considered as an attenuation rate from the technological point of view which should be reached by the waste water treatment technology.

If the contaminated groundwater or leachate shows 12 times higher toxicity than the no-effect volume of the groundwater or leachate, the applied risk reduction measure (treatment in ponds, artificial wetlands or permeable reactive barriers, etc.) should ensure a minimum of 12-fold reduction in toxicity according to the uncertainty (safety) factor used.

From Figure 9.10 the initial toxicity of the sample and the no-effect value can be read in dilution or in gram units because a sample of known mass is used in the test (100 g in the case shown). Dilution of the original sample is 1 and it caused 100%

inhibition. The original sample needed 200-fold dilution to reach the first measurement point without any adverse effect (NOEsD = 0.5 g soil). This approach can further be refined by increasing the resolution of the curve between NOEsD and LOEsD or calculating their average. A safety factor may also be applied. After refining the system, the necessary dilution rate decreased to 150. This way, the risk characterization ratio of a contaminated soil can be defined as RCR = the dilution rate that produces no adverse effect (150) = mass of the sample tested (100 g)/mass of the sample with no effect (0.66 g).

There are practical cases when decision making in environmental management is based on toxicity, for example if it should be decided whether or not

–    the treated wastewater can be released into natural waters,
–    the treated contaminated soils can be re-used for certain purposes,
–    the application of fertilizers, nutrients, organic wastes or any other waste materials on soils is acceptable from the point of view of toxic risk.

Using the attenuation-rate approach, one can make a decision not only by saying 'yes' or 'no' but by determining the allowable quantity of the discharged water or disposed waste. The same approach can be used to create and control remedial target values based on toxicity.

A stepwise approach is recommended to determine the target of remediation: after a rapid screening, the extent and distribution of toxicity at a contaminated site can be evaluated. If the screening tests indicate either mortality or a sub-lethal effect that is greater than a certain percentage (for example 20% or 25%) of that of the reference soil, the contaminated soils toxicity should be investigated in more detail by applying standardized or site-specific test batteries. The test battery for soil should include plants (monocot and dicot), soil-dwelling animals (arthropods and earthworms) and/or soil microorganisms and their activities such as respiration or nitrification rates. The last two integrate several microbial species' activity, representing the soil as a whole.

Results acquired from laboratory bioassays have shortcomings similar to those expected from any oversimplified model: their environmental reality is generally poor. A larger number of tests on numerous test organisms can provide a better estimation. Monitoring the species distribution in microcosms, mesocosms or field tests may serve as a scientifically established basis for decision making. Unfortunately, utilization of the results of species distribution to calculate a risk characterization ratio (RCR) and to integrate it into a rapid decision support tool is not yet practiced, mainly due to the time requirement. Instead of measuring species distribution, the application of more test organisms and end points is the solution for a statistically correct and still rapid decision making. The selection of the most restrictive responses or other statistically based criteria may increase efficiency and lower time requirement. The "omics" technologies may improve the situation in this field too.

In tests on non-site-specific, but generally applicable and high-sensitivity test organisms such as *Vibrio fischeri*, i.e. luminobacterium, the comparative strategy may help to characterize the tested soil in comparison to an uncontaminated local soil. In this case the target toxicity may be the same as the reference, more precisely 'statistically the same': it must not exceed the toxicity of the reference soil by more than e.g. 10 or 20%.

Another aspect of the decision is that, in addition to toxicity, other environmental parameters should also be considered in the decision making: pH and redox potential of the recipient environment or the intensity of light in the case of emission of photodegradable contaminants, etc. Bioavailability and its inclusion in risk assessment is a recurring problem to be dealt with. Validation of toxicity based on the results of long term monitoring of the recipient's quality is necessary to prove that there is no increased risk in the environment due to additional toxicity load.

## 8   TOXIC EQUIVALENCY OF CONTAMINATED ENVIRONMENTAL SAMPLES FOR EXPLORATION AND SCREENING

The toxic equivalency method is not unknown in environmental practice: it is used to quantitatively characterize the toxicity of mixtures of similar chemical substances. The first and best known application was developed for dioxins and dioxin-like compounds in the context of regulatory toxicology. The aggregated adverse effect of mixtures can be expressed in the form of **toxic equivalency** (TEQ) for a similar toxicant with additive effects. TEQ is based on the TEF value of individual dioxin-type chemical substances in the mixture. TEF quantifies the toxicity compared to that of the most toxic dioxin compound, the 2,3,7,8-TCDD (tetrachlorodibenzo-p-dioxin, van den Berg *et al.*, 2006). For example the toxicity of the 1,2,3,7,8,9-HxCDD (hexachlorodibenzo-p-dioxin) is 0.1 because it is ten times less than the toxicity of the 2,3,7,8-TCDD. Based on the chemical composition of a mixture, one can calculate its aggregated toxicity from the TEFs and the proportions of the components. A similar scheme has been developed for PCBs. Toxic equivalency exclusively uses chemical data which supports risk assessment and risk management. Within this context, the risk of complicated mixtures of different dioxin components (with a different size of effects) can be determined, as with single contaminants. This model can be applied in those cases where dioxin-like compounds are the only contaminants. With regard to dioxins, the equivalency method solely applies the chemical model based on chemical analytical data and fits well into regulatory risk management which is also based on the chemical model. The latter relies on environmental quality criteria (expressed in concentrations) and environmental exposures (also expressed in concentrations). The TEF values are constant factors.

The authors applied the equivalency approach in an inverse fashion and used the effect results of direct toxicity assessment of environmental samples. DTA characterizes the effects of unknown contaminants, which are complicated mixtures of chemically and toxicologically different contaminants, with unknown speciation, bioavailability and interactions, as typical in the environment.

Given that the measured toxicity values of environmental samples cannot be fitted directly to the chemical model-based ERA, the equivalency method was used to convert toxicity values to concentrations of environmental samples with unknown composition, speciation and bioavailability. This method required only the measurement of the effects and not that of the concentrations or the modeling of the environmental fate processes. This is not a very precise method because the assessor cannot decide whether or not the components are additive, whether their dose–response curves run parallel (as will be seen, they often do not), or if they have the same or different effect mechanism since the contaminants have not yet been identified. Nevertheless, toxic equivalency can be very useful in a tiered risk assessment to screen large areas with similar contaminants and select and classify a great number of environmental samples tested by different tests and test organisms. An additional benefit is that the equivalency method enables the results of (eco)toxicity studies to be translated into the language of chemistry and make them understandable for non-ecotoxicologists.

The authors have almost 20 years' experience in the use of toxic equivalency (TEQ) for quantitative characterization of environmental sample toxicity (Gruiz et al., 2001, Molnár et al., 2005; 2009; Leitgib et al., 2007; Feigl et al., 2007). This approach was spontaneously applied along the line of 'calibration' useful in chemical analytical practice. It was used for many years (first in 1992) for the evaluation of soils in a large mining area contaminated with various metal mixtures (Gruiz & Vodicska, 1992; 1993). At that time multi-component analytical methods were not as widespread as today and the owners and the authorities selected two out of approximately 20 metals that contaminated a former mining site. The risk management of the site was based on the concentration of the two metals. No logical correlation between the observed toxicity and the measured metal concentrations was found; therefore, soil, sediment and leachate toxicity was measured on the abandoned mining site. Copper was selected by the authors for this purpose to control the sensitivity of a special test bacterium and thus it was logical to express toxicity ($EC_{50}$, $EC_{20}$, NOEC) in copper equivalent – as though copper had only caused toxicity. Toxicity of the samples tested on various test organisms was expressed in copper equivalent concentration, which helped evaluation, interpretation and communication with different stakeholders, owners, municipalities, authorities, NGOs and the mine's management. Later on the copper equivalent was also used to monitor the efficacy of bioremediation technologies (Molnár et al., 2005; 2009; Leitgib et al., 2008; Feigl et al., 2007). It proved to be useful in explaining and interpreting toxicity test results and integrating them into the quantitative risk assessment procedure (Molnár et al., 2007; Leitgib et al., 2007) during the initial phases (screening phase and preliminary risk assessment). The basic concept and the creation and interpretation of the toxic equivalency method and toxic equivalency factor are introduced below in this chapter.

*Equivalent toxicity,* expressed for example in mg Cu/kg soil ($TEQ_{Cu}$), integrates the effects and the actual bioavailability of all metals present in the environmental sample, the interactions between the metals present and the partition of contaminants between soil phases, as well as interactions between metals and metal forms that influence effects and bioavailability. TEQ is able to handle the differences between bioassay methodologies and the sensitivity of the test organism to a contaminant, resulting in a generally applicable equivalent value, a common benchmark. Copper (Cu) was selected as the reference substance for all metals toxic to the environment because it is toxic to many of the ecosystem members (including the test organisms studied) and because its salts dissolve in water but are not particularly toxic to humans under laboratory conditions.

Toxicity of soils contaminated with organic compounds, *per analogiam*, is compared to 4-chlorophenol (4CP) toxicity and expressed in 4CP equivalent: $TEQ_{4CP}$ (mg 4CP/kg soil). Measured toxicity values of a 4CP dilution series serve as a calibration curve, and the toxicity of soil contaminated with an organic chemical substance can be expressed as if it was caused by 4CP. The risk management is based on this toxicity. However, this does not mean that 4CP can model other characteristics of the substance than toxicity. The reason why 4CP was chosen, similar to copper, is that it is toxic to all of the most widely used test organisms, it is easy to work with (it is not volatile, it is stable and has adequate solubility in water), and the risks to humans can be managed under laboratory circumstances (protective equipment, ventilation).

To calculate TEQ, 4CP or Cu calibration series and the environmental sample with unknown contaminants are prepared and measured at the same time. This calibration series also serves to control sensitivity and behavior of test organisms. The calibration curve has to be measured in water when water samples are tested. When dealing with soils, it is important to use uncontaminated soil which is the same as or similar to the soil tested for toxicity. The calibration curve should be measured at the same time as contaminated environmental samples are measured. The different dilutions belonging to equal effect can be read from the curves and the sample dose is converted into copper or 4CP concentration. Any special point of the concentration–response curves, e.g. $EC_{20}$, $EC_{50}$, NOEC or LOEC, can be used. The selection depends on the aim and concept of risk management. Depending on the calibration concept, the biological response to any environmental sample can be converted to a hypothetical concentration and risk management methods based on a chemical model can be applied, e.g. the result can be compared to environmental quality criteria expressed in concentration.

Equivalency tools have been employed by the authors in their everyday practice, and experience shows that standardized calibration tools are able to make toxicity testing more practicable and independent of the actual measurement conditions. Not only can test organisms be controlled, their sensitivity to different contaminants can be better characterized. Different test organisms can also be compared with one another. The equivalency tool is similar to those used by many chemical analytical methods that need a calibration between a measured end point (e.g. color) and the concentration of a chemical substance. It should be noted that 4CP or copper might not be present in the sample, but handling the results in concentration units makes sense in practical toxicity management.

## 8.1  Toxic equivalency for organic and inorganic contaminants

Bioassays measuring toxicity and the calculation of TEQs are introduced below as an example of a direct-contact soil bioassay using:

– *Vibrio fischeri* bacterial luminescence inhibition test;
– *Tetrahymena pyriformis* protozoan growth inhibition test;
– *Folsomia candida*, collembolan mortality test;
– *Sinapis alba* plant root and shoot growth test and
– *Daphnia magna*, water flea immobilization test.

The main benefits of a TEQ value lie in its potential to characterize the toxicity and risk of a contaminated environment without specifying the actual contaminants, i.e. in the case of unknown contaminant and of contaminant mixtures. Copper and 4-CP 'calibration series' have other benefits as well: due to their positive control in toxicity tests, they can confirm test repeatability and good biological status of test organisms. 4CP or Cu equivalents of toxicity of known (specified) contaminants enable a quantitative comparison between the two toxicants and the relative sensitivity of the test

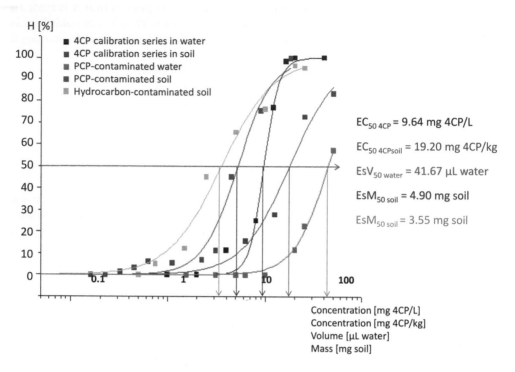

*Figure 9.11*  Calibration curve of 4CP and the dose–response curve of contaminated soils (mass) and water (volume).

organism. Comparing the adverse effects (to the same test organism) of different chemical substances or mixed contaminants in soil, the contaminant can be characterized by a toxic equivalency factor (TEF) with regard to 4CP (organic toxicants) or Cu (toxic metals). This shows how many more times it is toxic compared to 4CP or Cu. TEF can help in toxicant ranking and risk evaluation. TEF can be calculated from different test end points and it enables a dynamic comparative characterization and the aggregation of the results of different studies.

## 8.2   Graphical determination of equivalent toxic concentrations from measured data

The response of the test organism is plotted against the 4CP or Cu concentration series, and the sigmoid is fitted to the points providing the concentration–response curve. The same can be done to the dilution series of contaminated soil or water and the corresponding contaminated soil mass (M in mg) or contaminated water volume (V in mL) can be read from the curves obtained. The sigmoid is fitted to the measurement points (Figures 9.11 and 9.12) using statistical software (Origin 8.0 or Statistica 6.0). Any common end point (EC$_{50}$, EC$_{20}$, LOEC, NOEC, etc.) and the equivalencing chemical substances in terms of sample volume, sample mass, 4CP or Cu concentration, respectively, can be read from these curves of the samples. This is the basis for creating the

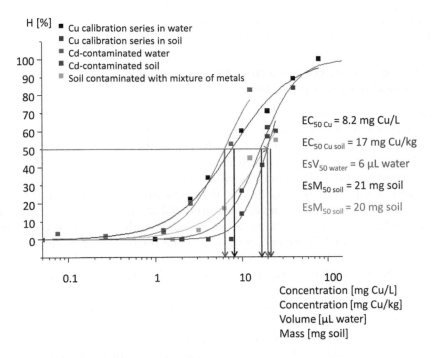

*Figure 9.12* Calibration curves of Cu equivalent and dose–response curves of Cd-contaminated soil (mass) and water (volume) and mixed metals-contaminated soil.

equivalent value, i.e. 'calibrating' the unknown contaminant. The same method can be applied for samples in dissolved form or in solid matrices.

The volume of the sample and the concentration–response curve of 4CP or Cu (Cu mg/L or 4CP mg/L) are plotted based on liquid sample tests. The value read from the sigmoid at a default effect (highest no-effect, lowest effective inhibition or the effects causing 5, 10, 20, 50, 90% inhibition) is the volume or mass of the sample equivalent to the amount of 4CP or Cu causing the same effect. The effective mass of the sample is labeled by the abbreviation $sM$ = sample mass/dose, or $sV$ = sample volume. $EsM_x$ (effective sample dose causing x% inhibition) values can be read from the curves and the $TEQ_x$ values calculated from the known 4CP or Cu values. The calibration curve examples of pure and mixed organic soil contaminants such as PCP and hydrocarbons are shown in Figure 9.11 and those of soils contaminated with toxic metals using the Cu equivalent in Figure 9.12. Both figures show the dose–response curve of *Vibrio fischeri* luminescence inhibition and the graphical determination of $EC_{50}$.

The graphically determined $EC_{50}$ values e.g. in Figure 9.12 can be directly interpreted as follows: $6 \mu L$ contaminated water causes 50% inhibition, which is the same level of toxicity as the one caused by a copper solution of 8.2 mg/L concentration. $6 \mu L$ from the unknown is equivalent (based on $EC_{50}$) to 8.2 mg/L copper. Further possible calculations: $6 \mu L$ (causing 50% inhibition) is the 41.66-fold dilution of the original sample ($250 \mu L$), consequently the original sample's toxicity (based on $EC_{50}$) is equivalent to 341.7 mg/L copper. In reality, the artificially contaminated Cd solution had 500 mg/L Cd concentration, the copper calibrating solution 400 mg/L Cu. Taking these into account, Cd is less toxic for *Vibrio fischeri* than Cu, the ratio of Cd/Cu toxicity is 0.68 based on the $EC_{50}$ value of *Vibrio fischeri* (See also Table 9.4).

## 8.3 Numerical determination of the toxicity equivalent concentration

The rate of mortality or inhibition is plotted against the volume/mass of the tested water/soil contaminated with unidentified contaminants: this gives the sigmoid of the sample dose–response curve. Simultaneously, the artificially contaminated water or soil with the reference contaminant (4CP or Cu) is also tested and its concentration–response curve is drawn. Both the reference chemicals' concentration ($EC_{50}$) and the tested water or soil dose ($EsV_{50}$ or $EsM_{50}$) that cause 50% inhibition can be read from the two curves, and these two values are considered as equivalent values causing the same level of toxicity (50%). The toxicity equivalent is the amount of the unknown water or soil expressed in the concentration of the reference contaminants: 4CP or Cu. The result means the equivalent $EC_{50}$ of the tested water or soil as though it were caused by the 4CP or Cu reference contaminants. By analogy of the dilution point causing 50% inhibition ($EC_{50}$), the same can be done using NOEC or LOEC.

The method is presented using the example of 4CP equivalent; the same can be used for calculating Cu equivalent.

If the liquid-phase sample is tested and the adverse effect is plotted against dilution in the graph, the effective volume can be calculated from the dilution rate read from the graph and from the test-specific sample volume. The equivalent toxicity of the liquid-phase sample can be calculated by equation (Equation 9.8). A similar equation is valid when no-effect volume (NOEsV) or lowest effect volume (LOEsV) are used as end points.

$$TEQ_{4CP(EC50)}\left[\frac{mg\ 4CP}{L}\right] = \frac{EC_{50(4CP)}[mg\ 4CP/L] \cdot V_{sample}[\mu L\ or\ mL]}{EsV_{50(sample)}[\mu L\ or\ mL\ sample]} \qquad (9.8)$$

where

- $TEQ_{4CP(EC50)}$: 4CP equivalent concentration of the sample with unknown contaminant(s) causing 50% inhibition (alternatively $TEQ_{4CP(LOEC)}$ and $TEQ_{4CP(NOEC)}$);
- $EC_{50(4CP)}$: effective concentration of 4CP solution causing 50% inhibition;
- $EsV_{50(sample)}$: effective volume of the liquid sample causing 50% inhibition;
- $V_{sample}$: tested volume according to the test protocol, e.g. the volume of the tested sample is 0.25 mL for *Vibrio fischeri* luminescence inhibition;
- $V_{sample}/EsV_{50(sample)}$: dilution rate of the unknown sample until reaching the 50% inhibition point in the graph.

If solid-phase sample such as soil or solid waste is tested, the equation should be modified according to (Equation 9.9). Similar equations are valid for LOEC and NOEC values when the basis of equivalencing is LOEC or NOEC.

$$TEQ_{4CP(EC50)}\left[\frac{mg\ 4CP}{kg}\right] = \frac{EC_{50(4CP)}[mg\,4CP/kg] \cdot M_{sample}[mg\ or\ g]}{EsM_{50(sample)}[mg\ or\ g\ sample]} \qquad (9.9)$$

where

- $TEQ_{4CP(EC50)}$: 4CP-equivalent concentration in soil of an unknown contaminant or mixture of contaminants causing 50% inhibition;
- $EC_{50(4CP)}$: effective concentration of 4CP solution causing 50% inhibition, lowest effect or no effect;
- $EsM_{50}$: effective mass of the solid sample causing 50% inhibition, lowest effect or no effect;
- $M_{sample}$: test mass for e. g. *Vibrio fischeri* luminescence inhibition; the volume of the tested sample is 0.25 mL;
- $M_{sample}/EsM_{50(sample)}$: dilution rate of the unknown sample up to reaching dilution resulting in 50% inhibition.

TEQ can be calculated for the no-effect and lowest effect points, too; in this case the volume or mass corresponding to no effect or lowest effect is used for the calculation by analogy. The toxicity of soil can be equivalenced compared to the calibrating water solution if the partition of the toxicity between soil and water is known or has been measured (See also Chapter 7).

## 8.4   Equivalent toxicity of contaminated water: examples and validation

The use of the equivalency tool in the case of water contaminated by single chemicals and mixtures of contaminants is introduced below by some examples. The unknown contaminant's toxicity is expressed in 4CP-equivalent toxicity based on the $EsV_{50}$ ($TEQ_{4CP(EC50)}$) or $EsV_{LOEL}$ or $EsV_{NOEL}$ and the relevant TEQ values. $EsV_{50}$ means the volume of contaminated water which causes 50% inhibition or 50% lethality; $EsV_{LOEL}$ is the lowest tested volume causing an observable effect, $EsV_{NOEL}$, the highest tested volume of contaminated water not causing measurable adverse effect.

The toxicity of the unknown samples expressed as if it were 4CP or Cu makes possible the comparison of the results to the screening values or other effect-based environmental quality criteria established for 4CP and Cu. Some of the existing screening, intervention, threshold limit and other guideline values are listed here:

- 4CP groundwater quality criterion: 1 µg/L (Hungarian soil regulation, 2000)
- Italian 4CP limit value for groundwater: 0.5 µg/L (Carlon *et al.*, 2007);
- 4CP soil quality criterion: 0.01 mg/kg (Hungarian soil regulation, 2000);
- 4CP intervention value for soil: 1–3 mg/kg (Hungarian soil regulation, 2000);
- 4CP soil intervention value: 10 mg/kg (Swartjes, 1999);
- Finnish 4CP soil guideline values for soil: lower: 10 mg/kg, upper: 40 mg/kg (Carlon *et al.*, 2007);
- Italian 4CP limit for public parks: 0.01 mg/kg, for industrial areas: 5 mg/kg (Carlon *et al.*, 2007);
- Cu groundwater quality criterion: 200 µg/L (Hungarian soil regulation, 2000);
- Italian Cu limit value for groundwater: 1000 µg/L (Carlon *et al.*, 2007);
- Cu groundwater intervention: 300–500 µg/L (Hungarian soil regulation, 2000);
- Cu soil quality criterion: 75 mg/kg (Hungarian soil regulation, 2000);

Table 9.2  4CP-equivalent toxicity of contaminated water based on 50% inhibition.

| Contaminants and tests | Measured $EsV_{50}$ mL sample | Calculated $TEQ_{4CP(EC50)}$ mg 4CP/L | Tested $V_{sample}$ mL | Analysis result mg/L | $EC_{50}$ mg/L | $TEF_{4CPEC50}$ – |
|---|---|---|---|---|---|---|
| PCP-contaminated water | | | | | | |
| Luminescence inhibition of Vibrio fischeri | 0.0417 | 57.79 | 0.25 | 1.8 | 0.30 | 32.1 |
| Growth inhibition of Tetrahymena pyriformis | 0.91 | 362.76 | 31.5 | 144 | 4.17 | 2.5 |
| Immobilization of Daphnia magna | 1.11 | 50.45 | 100 | 1.8 | 0.02 | 28.0 |
| BPA-contaminated water | | | | | | |
| Luminescence inhibition of Vibrio fischeri | 0.0205 | 117.55 | 0.25 | 156 | 12.76 | 0.76 |
| Immobilization of Daphnia magna | 0.69 | 81.16 | 100 | | 1.07 | 0.52 |
| DBP-contaminated water | | | | | | |
| Luminescence inhibition of Vibrio fischeri | 0.0233 | 103.43 | 0.25 | 30 000 | 2800 | 0.003 |
| Immobilization of Daphnia magna | >10* | N.A. | 100 | 30 000 | N.A. | N.A. |
| Mix of NSAIDs in water | | | | | | |
| Luminescence inhibition of Vibrio fischeri | >0.050* | N.A. | 0.25 | 7.5 | N.A. | N.A. |
| Immobilization of Daphnia magna | 9.10 | 6.16 | 100 | 7.5 | 0.68 | 0.8 |
| Nicotine in water | | | | | | |
| Immobilization of Daphnia magna | 0.40 | 140.35 | 100 | 20 000 | 79.8 | 0.007 |
| 4CP calibration series | | | | | | |
| Luminescence inhibition of Vibrio fischeri | 0.0121 | 200 | 0.25 | 200 | 9.64 | 1.0 |
| Growth inhibition of Tetrahymena pyriformis | 0.66 | 500 | 31.5 | 500 | 10.48 | 1.0 |
| Immobilization of Daphnia magna | 0.28 | 200 | 100 | 200 | 0.56 | 1.0 |

*Sample toxicity failed to reach the 50% inhibition level
N.A. not applicable

- Cu soil intervention: 200–300 mg/kg (Hungarian soil regulation, 2000);
- Cu soil intervention value: 100 mg/kg (Swartjes, 1999);
- Finnish Cu soil guideline values for soil: lower: 150 mg/kg, upper: 200 mg/kg, (Carlon et al., 2007);
- Italian Cu limit for public parks: 120 mg/kg, industrial areas 600 mg/kg (Carlon et al., 2007).

Table 9.2 shows the calculation details and the resulted $TEQ_{4CP}$ values of selected organic contaminants for Vibrio fischeri, Tetrahymena pyriformis and Daphnia magna. The basis of the TEQ calculation was the 50% inhibition. The following values are listed in the head of the table:

- The $EsV_{50}$, ($EsV_{LOEC}$ and $EsV_{NOEC}$ also can be used) is the volume of the contaminated water representing the theoretical point (extrapolated by fitting the sigmoid) of the serial dilution resulting in 50% inhibition. Its unit of measurement is *ml sample*.

- TEQ$_{4CP}$ is the calculated 4CP equivalent of the tested contaminated water: as if the water were only contaminated by 4CP. Its unit of measurement is *mg 4CP/L.*
- Sample volume is the volume prescribed/required by the test protocol (this data is necessary for the calculation of EC$_{50}$ and TEF);
- Result of the chemical analysis (only for identified contaminants) of the samples for the purpose of validating the toxic equivalency and calculating TEF) given in *mg contaminant in 1 liter sample.*
- EC$_{50}$ is the calculated toxicity, i.e. the concentration of the contaminated water causing 50% inhibition expressed in *mg/L sample* and determined from the analysis result and the measured dilution.
- TEF (Toxic Equivalency Factor) describes how many mg 4CP is equivalent to 1 mg contaminant. It is the equivalent mass of the contaminant causing the same toxicity, compared to 4CP. TEF = 1: equally toxic; TEF > 1: more toxic; TEF < 1: less toxic than 4CP. It can only be calculated when the contaminating substance has been identified and quantified by chemical analysis. The TEF is calculated from 50% inhibition, the NOEC or LOEC values are generally different due to the different shape of the concentration–response curve of the tested contaminant and 4CP.

### 8.4.1   4CP equivalent of selected organic contaminants in water: examples

In the demonstration cases, real or artificially contaminated water samples were used and their contaminant content was analyzed during the test for the validation of the equivalency method. The toxic equivalency factor (TEF) was calculated for concrete (identified) contaminants for comparison. Chemical contaminants tested and whose toxicities were expressed as 4CP equivalents, were pentachlorophenol (PCP), bisphenol-A (BPA), dibutyl phthalate (DBP) and nicotine. As a mixture, an artificial mix of NSAIDs (non-steroidal anti-inflammatory drugs) was prepared, containing diclofenac (645 mg/L), ketoprofen (456 mg/L), naproxen (438 mg/L) and ibuprofen (362 mg/L). The Cu-calibration has been demonstrated on cadmium-contaminated water and soil.

The dose–response curves in Figure 9.10 show that the 41.67 µL sample of PCP-contaminated water caused 50% inhibition in the *Vibrio fischeri* test, while only 12.1 µL of the 4CP solution with 200 mg/L concentration caused the same level of toxicity which corresponds to 9.64 mg/L 4CP in terms of concentration. Calculating back to the original solutions using Equation 9.8, the PCP solution is equivalent to a 57.79 mg/L 4CP solution. The TEF factor, which tells how many times PCP is more toxic to *Vibrio fischeri* than 4CP, can also be calculated. The concentration of the calibrating 4CP solution (200 mg/L) is *a priori* known and the unknown water sample contaminated with PCP has been analyzed: the result was a PCP concentration of 1.8 mg/L. The table shows that PCP is 32.1 times more toxic when EC$_{50}$ is the basis of the calculation (9.64/0.30). *Daphnia magna* shows a ratio similar to that shown by *Vibrio fischeri*, but *Tetrahymena* found PCP only 2.5-times as toxic as 4CP.

BPA based on its EC$_{50}$ value is less toxic than 4CP both to *Vibrio* and *Daphnia*, but a different TEF is obtained based on the NOEC values (Table 9.3) because *Vibrio* is sensitive to small BPA concentrations as well, while *Daphnia* is not.

Table 9.3  4CP-equivalent toxicity of contaminated water based on NOEC and LOEC values.

| Contaminants and test | Measured EsV_LOEC mL | Calculated TEQ_4CP(LOEQ) mg/L | LOEC mg/L | TEF_LOEC – | Measured EsV_NOEC mL | Calculated TEQ_4CPNOEC mg/L | NOEC mg/L | TEF_4CPNOEC – |
|---|---|---|---|---|---|---|---|---|
| PCP-contaminated water | | | | | | | | |
| Luminescence inhibition of Vibrio fischeri | 0.020 | 50.00 | 0.144 | 27.8 | 0.0100 | 50.0 | 0.072 | 27.8 |
| Growth inhibition of Tetrahymena | 0.134 | 559.48 | 0.61 | 3.9 | 0.067 | 559.5 | 0.30 | 3.9 |
| Immobilization of Daphnia magna | 0.25 | 100.00 | 0.0045 | 55.6 | 0.05 | 100.0 | 0.0009 | 55.6 |
| BPA-contaminated water | | | | | | | | |
| Luminescence inhibition of Vibrio fischeri | 0.00032 | 3,125 | 0.20 | 20.0 | 0.00008 | 6,250 | 0.05 | 40.1 |
| Immobilization of Daphnia magna | 0.75 | 33.33 | 1.17 | 0.2 | 0.5 | 10.0 | 0.78 | 0.06 |
| DBP-contaminated water | | | | | | | | |
| Luminescence inhibition of Vibrio fischeri | 0.0167 | 60,000 | 2.00 | 2.0 | 0.0000083 | 60,000 | 1.00 | 2.0 |
| Immobilization of Daphnia magna | 0.0167 | 15,000 | 0.50 | 0.5 | 0.000067 | 75,000 | 0.02 | 2.5 |
| Mix of NSAIDs | | | | | | | | |
| Luminescence inhibition of Vibrio fischeri | 0.0125 | 80.00 | 0.375 | 10.7 | 0.005 | 100.0 | 0.15 | 13.3 |
| Immobilization of Daphnia magna | 0.67 | 37.50 | 0.05 | 5.0 | 0.067 | 75.0 | 0.005 | 10.0 |
| Nicotine contaminated water | | | | | | | | |
| Immobilisation of Daphnia magna | 0.0125 | 2,000 | 2.5 | 1.0 | 0.00005 | 100,000 | 0.01 | 5.0 |
| 4CP calibration series | | | | | | | | |
| Luminescence inhibition of Vibrio fischeri | 0.0050 | 200 | 4.00 | 1.0 | 0.0025 | 200 | 2.00 | 1.0 |
| Growth inhibition of Tetrahymena | 0.15 | 500 | 2.38 | 1.0 | 0.075 | 500 | 1.19 | 1.0 |
| Immobilization of Daphnia magna | 0.125 | 200 | 0.25 | 1.0 | 0.025 | 200 | 0.05 | 1.0 |

The effect of nicotine is also interesting from this point of view: $TEF = 0.007$ based on $EC_{50}$, but $TEF = 1$ when calculated from LOEC and $TEF = 5.0$ from NOEC, characterizing nicotine as a slightly toxic substance (not causing lethality) in high concentrations, but also toxic in very low concentration ranges.

The toxicity based on $EC_{50}$ and LOEC or NOEC may be significantly different for a chemical substance, as we can see in Tables 9.2 and 9.3. It is best demonstrated by the $TEQ_{4CP}$ and $TEF_{4CP}$ values calculated from the different end points. The differences in TEQ and TEF calculated from $EC_{50}$, LOEC and NOEC are characteristic of the contaminants' interaction with the test organism and of the effect mechanisms. Growing TEF toward LOEC and NOEC indicates adverse effects of the chemical substance at very low concentrations or high dilution rates. This means that the dose–response curve declines slightly and does not reach zero inhibition.

A comparison of the NOEC, LOEC and $EC_{50}$ values with each other reflects the differences in the shape of the *concentration–response* curves. Equivalencing means that the contaminants' curve is compared to the curve of 4CP. 4CP gives a response proportional to the dilution (e.g. $EC_{50}$: 9.64; LOEC: 4.0; NOEC: 2.0) for all of the test organisms. The mixture of NSAIDs is a good example to demonstrate the differences in the shape of the dose–response curve: it hardly reaches the 50% inhibition, but a very high dilution rate is still needed to reach the no-effect point. It results in a surprisingly high TEQ value when calculating it from NOEC and LOEC. $TEF_{NOEC}$ of 10.7 (10 times more toxic than 4CP) draws attention to the importance of the less drastic adverse effects of low contaminant concentrations. Thus NSAIDs are typically non-killing agents, but they may be inhibitory or disrupting for the endocrine and immune system in small concentrations.

### 8.4.2  Copper equivalent of cadmium-contaminated water

To demonstrate the equivalent method for inorganic contaminants, the copper equivalent of water artificially contaminated with cadmium (Cd) is introduced here. Cu-calibration series were used as a comparison and the adverse effects of the water are characterized by the copper equivalent toxicity (as if the toxicity were only caused by Cu). Table 9.4 contains the equivalents based on $EC_{50}$. Table 9.5 shows the LOEC and NOEC values.

Cadmium appears to be less toxic to the two bacteria and the plant than Cu. All TEF from LOEC and NOEC are <1.

## 8.5  Toxicity equivalent of soil: examples and validation

Soil, sediment and solid waste toxicity is difficult to model therefore direct toxicity assessment of solid-phase samples provide useful information for decision making. The interaction of contaminants with each other, with the solid matrix and with thousands of living organisms in soil, sediments and waste greatly influence the actual adverse effect of a hazardous chemical or the aggregated effect of mixtures of chemicals. Direct toxicity tests on samples containing solid phase integrate the interactions within the solid matrix, the partition, mobility, bioavailability etc. Direct toxicity tests on soil, in spite of providing highly realistic results, are still rarely used in contaminated land management practice because latter is still based on the chemical model.

*Table 9.4* Copper-equivalent toxicity of cadmium-contaminated water based on $EC_{50}$.

| Contaminants and tests | Measured $EsV_{50}$ mL | $TEQ_{Cu(EC50)}$ mg/L | Tested $V_{sample}$ mL | Analysis results mg/L | $EC_{50}$ mg/L | $TEF_{Cu(EC50)}$ – |
|---|---|---|---|---|---|---|
| Cd-contaminated water | | | | | | |
| Luminescence inhibition of *Vibrio fischeri* | 0.006 | 341.7 | 0.25 | 500 | 12.0 | 0.68 |
| Dehydrogenase activity of *Azomonas agilis* | 0.09 | 80.0 | 2.5 | | 18.0 | 0.16 |
| Shoot growth of *Sinapis alba* | 2.19 | 285 | 5.0 | | 219.0 | 0.57 |
| Root growth of *Sinapis alba* | 1.44 | 433 | 5.0 | | 144 | 0.86 |
| Cu calibration series | | | | | | |
| Luminescence inhibition of *Vibrio fischeri* | 0.005 | 400 | 0.25 | 400 | 8.2 | 1.0 |
| Dehydrogenase activity of *Azomonas agilis* | 0.018 | 400 | 2.5 | | 2.9 | 1.0 |
| Shoot growth of *Sinapis alba* | 1.56 | 400 | 5.0 | | 125 | 1.0 |
| Root growth of *Sinapis alba* | 1.56 | 400 | 5.0 | | 125 | 1.0 |

### 8.5.1   4CP equivalent of selected organic contaminants in soil: examples

To demonstrate the use of 4CP equivalent, we compared the toxicity of soil contaminated with pentachlorophenol (PCP) of two transformer oils and unknown soils contaminated with benzenes, alkyl benzenes (BTEX), polycyclic aromatic hydrocarbons (PAHs) and extractable petroleum hydrocarbons (EPH). For demonstration purposes and for verifying the equivalent toxicity quantification method, the contaminant content was determined by chemical analysis and the 4CP equivalent (TEF) was calculated. In the case of PCP contamination, uncontaminated reference soil (from the same site) spiked with 4CP was used for comparison.

In this experiment series, soil was tested using the luminescent bacterium, the protozoan and *Folsomia candida,* the soil-dwelling springtail. Similar to water samples, PCP showed higher toxicity compared to 4CP in the case of microorganisms, but the collembolan did 'feel' PCP less toxic than 4CP; TEF = 0.77 (Table 9.6). This can be explained by the difference between 4CP and PCP in volatility and availability for the springtail, which is mainly impacted via respiration.

Soils polluted with hydrocarbon mixtures demonstrate that the chemical analysis (fulfilling regulatory obligations, but still only analyzing selected components) cannot accurately reflect the risk of contaminated soils. The analytically measured 780 and 7,600 mg/kg EPH (extractable petroleum hydrocarbons) content of soil samples contaminated with (PCB-free) transformer oil – differs significantly, almost tenfold. With respect to toxicity results, the difference between B and A soils is only sevenfold.

Soil contaminated with the mixture of non-volatile contaminants is characterized by 277, 143 and 100 (mg 4CP/L) of TEQ expressed as 4CP for the three test organisms. These TEQ values are 27-, 14- and 10-fold greater than the intervention value of 4CP (10 mg/kg) while the analyzed concentration of the three contaminant groups is significantly under the lowest Hungarian intervention value (EPH: 300; BTEX: 0.5 mg/kg; PAH: 5 mg/kg) (HU soil regulation, 2000). Trusting the response of the test organisms rather than the chemical model, this soil should be further assessed and remedied.

Table 9.5  Cu-equivalent toxicity of metal-contaminated water based on NOEC and LOEC values.

| Contaminants and tests | Measured EsV_LOEC mL | Calculated TEQ_CuLOEC mg/L | LOEC mg/L | TEF_Cu(LOEC) – | Measured EsV_NOEC mL | Calculated TEQ_Cu NOEC mg/L | NOEC mg/L | TE_Cu(NOEC) – |
|---|---|---|---|---|---|---|---|---|
| Cd-contaminated water | | | | | | | | |
| Luminescence inhibition/Vibrio fischeri | 0.000625 | 32.00 | 1.25 | 0.06 | 0.0000625 | 40 | 0.125 | 0.08 |
| Dehydrogenase activity/Azomonas agilis | 0.050 | 50.00 | 10.00 | 0.10 | 0.025 | 20 | 5.0 | 0.04 |
| Shoot growth of Sinapis alba | 1.0 | 250.00 | 100.00 | 0.50 | 0.50 | 100 | 50 | 0.2 |
| Root growth of Sinapis alba | 0.2 | 250.00 | 20.00 | 0.50 | 0.10 | 250 | 10.0 | 0.5 |
| Cu calibration series | | | | | | | | |
| Luminescence inhibition/Vibrio fischeri | 0.00005 | 400 | 0.08 | 1.0 | 0.0000062 | 400 | 0.01 | 1.0 |
| Dehydrogenase activity/Azomonas agilis | 0.0125 | 400 | 2.00 | 1.0 | 0.00125 | 400 | 0.20 | 1.0 |
| Shoot growth of Sinapis alba | 0.625 | 400 | 50.00 | 1.0 | 0.125 | 400 | 10.0 | 1.0 |
| Root growth of Sinapis alba | 0.125 | 400 | 10.00 | 1.0 | 0.0625 | 400 | 5.0 | 1.0 |

*Table 9.6* 4CP-equivalent toxicity of contaminated soil samples based on $EC_{50}$ values.

| Contaminants and tests | Measured $EsM_{50}$ g | Calculated $TEQ_{4CP50}$ mg/kg | Tested $M_{sample}$ g | Result of the analysis mg/kg | $EC_{50}$ mg/kg | $TEF_{4CPEC50}$ – |
|---|---|---|---|---|---|---|
| PCP-contaminated soil | | | | | | |
| Luminescence inhibition of Vibrio fischeri | 0.0049 | 981 | 0.25 | 191 | 3.74 | 5.1 |
| Growth inhibition of Tetrahymena pyriformis | 0.14 | 1,537 | 31.5 | | 0.85 | 3.2 |
| Mortality of Folsomia candida | 8.0 | 146 | 20 | | 76.4 | 0.77 |
| Transformer oil contaminated soil A | | | | | | |
| Luminescence inh. of Vibrio fischeri | 0.025 | 194 | 0.25 | EPH:780 | 77.5 | 0.24 |
| Growth inhibition of Tetrahymena pyriformis | 1.0 | 215 | 31.5 | | 24.6 | 0.27 |
| Mortality of Folsomia candida | 2.54 | 461 | 20 | | 99.0 | 0.59 |
| Transformer oil contaminated soil B | | | | | | |
| Luminescence inhibition of Vibrio fischeri | 0.0036 | 1,353 | 0.25 | EPH: 7,600 | 107.7 | 0.18 |
| Growth inhibition of Tetrahymena pyriformis | 0.35 | 615 | 31.5 | | 84.7 | 0.08 |
| Mortality of Folsomia candida | 1.1 | 1,060 | 20 | | 419 | 0.14 |
| Non-volatile hydrocarbon-contaminated soil | | | | | | |
| Luminescence inhibition of Vibrio fischeri | 0.0173 | 277 | 0.25 | EPH: 120 | N.A. | N.A. |
| Growth inhibition of Tetrahymena pyriformis | 1.5 | 143 | 31.5 | BTEX: 0.51 | N.A. | N.A. |
| Mortality of Folsomia candida | 11.6 | 100 | 20 | PAH: 1.68 | N.A. | N.A. |
| 4CP calibration series | | | | | | |
| Luminescence inhibition of Vibrio fischeri | 0.024 | 200 | 0.25 | 200 | 19.20 | 1.0 |
| Growth inhibition of Tetrahymena pyriformis | 0.43 | 500 | 31.5 | 500 | 6.83 | 1.0 |
| Mortality of Folsomia candida | 5.85 | 200 | 20 | 200 | 58.50 | 1.0 |

N.A.: not applicable.

Using LOEC and NOEC for equivalencing a completely different impression is obtained compared to the $EC_{50}$-based evaluation (Table 9.7).

Hydrocarbons of petroleum origin are non-proportionately losing their toxicity when diluted, compared to 4CP. The explanation may be that they are – strictly speaking – natural compounds.

### 8.5.2 Copper equivalent of soils contaminated with cadmium and a mixture of metals

Mixtures of metals are typical soil contaminants for example in industrially contaminated sites. The interactions in the soil contaminated with metal mixtures are enhanced

Table 9.7 4CP-equivalent toxicity of contaminated soil samples based on NOEC and LOEC values.

| Contaminants and tests | Measured EsM$_{LOEC}$ g | TEQ$_{4CP(LOEC)}$ mg/kg | LOEC mg/kg | TEF$_{4CP(LOEC)}$ — | Measured EsM$_{NOEC}$ g | TEQ$_{4CP(NOEC)}$ mg/kg | NOEC mg/kg | TEF$_{4CP(NOEC)}$ — |
|---|---|---|---|---|---|---|---|---|
| **PCP-contaminated forest soil** | | | | | | | | |
| Luminescence inh. of *Vibrio fischeri* | 0.000625 | 4.8 | 0.48 | 0.025 | 0.0003125 | 2 | 0.24 | 0.01 |
| Growth inh. of *Tetrahymena pyriformis* | 0.012 | 1625 | 0.072 | 8.6 | 0.006 | 1583 | 0.036 | 8.3 |
| Mortality of *Folsomia candida* | 2.5 | 100 | 23.88 | 0.52 | 1.25 | 100 | 11.94 | 0.52 |
| **Transformer oil-contaminated soil A** | | | | | | | | |
| Luminescence inh. of *Vibrio fischeri* | 0.0200 | 1.5 | 62.56 | 0.0019 | 0.0100 | 0.0625 | 31.28 | 0.00008 |
| Growth inh. of *Tetrahymena pyriformis* | 1.00 | 19.5 | 24.80 | 0.025 | 0.50 | 18.9 | 12.40 | 0.0242 |
| Mortality of *Folsomia candida* | 10.00 | 25 | 391 | 0.032 | 5.00 | 25 | 195.5 | 0.032 |
| **Transformer oil-contaminated soil B** | | | | | | | | |
| Luminescence inh. of *Vibrio fischeri* | 0.0005 | 6 | 15.2 | 0.0013 | 0.00025 | 2.5 | 7.59 | 0.000329 |
| Growth inh. of *Tetrahymena pyriformis* | 0.12 | 162.5 | 28.90 | 0.021 | 0.06 | 158 | 14.50 | 0.02 |
| Mortality of *Folsomia candida* | 2.0 | 125 | 760 | 0.0165 | 1.00 | 125 | 380 | 0.0165 |
| **Non-volatile hydrocarbons-contaminated soil** | | | | | | | | |
| Luminescence inh. of *Vibrio fischeri* | 0.003125 | 0.96 | N.A. | N.A. | 0.0016125 | 0.39 | N.A. | N.A. |
| Growth inh. of *Tetrahymena pyriformis* | 1.0 | 19.5 | N.A. | N.A. | 0.50 | 18.9 | N.A. | N.A. |
| Mortality of *Folsomia candida* | 2.5 | 100 | N.A. | N.A. | 1.00 | 125 | N.A. | N.A. |
| **4CP calibration series** | | | | | | | | |
| Luminescence inh. of *Vibrio fischeri* | 0.000015 | 200 | 0.012 | 1.0 | 0.0000031 | 200 | 0.0025 | 1.0 |
| Growth inh. of *Tetrahymena pyriformis* | 0.039 | 500 | 0.62 | 1.0 | 0.019 | 500 | 0.30 | 1.0 |
| Mortality of *Folsomia candida* | 1.25 | 200 | 12.5 | 1.0 | 0.625 | 200 | 6.25 | 1.0 |

N.A.: not applicable.

*Table 9.8* Copper-equivalent toxicity of metal-contaminated soil based on $EC_{50}$ values.

| Contaminants and tests | Measured $EsM_{50}$ g | $TEQ_{Cu\,50}$ mg/kg | $V_{sample}$ g or mL | Analysis result mg/kg | $EC_{50}$ mg/kg | $TEF_{CuEC50}$ – |
|---|---|---|---|---|---|---|
| Cd-contaminated soil | | | | 500 | | |
| Luminescence inhibition of *Vibrio fischeri* | 0.021 | 202 | 0.25 | | 42 | 0.41 |
| Dehydrogenase enzyme activity of *Azomonas agilis* | 0.325 | 86 | 2.5 | | 65 | 0.17 |
| Shoot growth of *Sinapis alba* | >5 | >139 | 5 | | >500 | <0.28 |
| Root growth of *Sinapis alba* | 2.13 | 366 | 5 | | 213 | 0.73 |
| Soil contaminated with a mixture of toxic metals | | | | As: 228 | | |
| Luminescence inhibition of *Vibrio fischeri* | 0.020 | 212.5 | 0.25 | Cd: 0.84 | 266 | 0.064 |
| Dehydrogenase enzyme activity of *Azomonas agilis* | 0.363 | 77.1 | 2.5 | Cu: 54.7 | 482 | 0.023 |
| Shoot growth of *Sinapis alba* | 3.78 | 184 | 5 | Pb: 2748 | 2511 | 0.055 |
| Root growth of *Sinapis alba* | 5.0 | 156 | 5 | Zn: 294 | 3325 | 0.047 |
| Cu calibration series | | | | 400 | | |
| Luminescence inhibition of *Vibrio fischeri* | 0.0106 | 400 | 0.25 | | 17 | 1.0 |
| Dehydrogenase enzyme activity of *Azomonas agilis* | 0.07 | 400 | 2.5 | | 11.2 | 1.0 |
| Shoot growth of *Sinapis alba* | 1.74 | 400 | 5 | | 139 | 1.0 |
| Root growth of *Sinapis alba* | 1.95 | 400 | 5 | | 156 | 1.0 |

by the variety of their chemical forms (chemical speciation) and changing environmental conditions. In such situations actual toxicity is essential information and direct toxicity assessment is a tool to fill the knowledge gap.

In this example soils artificially contaminated with cadmium and a mixture of metals stemming from contaminated land are tested and subjected to equivalencing. The validation after chemical analyses and TEF calculation helps to understand the results.

Similar to the previous cases, the results in Table 9.8 confirm that cadmium is less toxic to microorganisms and plants than copper; TEF is less than one in all tested cases. The mixture of arsenic, cadmium, copper, lead and zinc in a soil from an abandoned mining site with extremely high lead content showed somewhat less than 400 mg/kg copper in the soil. Considering the analyzed lead content itself (2,748 mg/kg), or total metal content together (3,325 mg/kg), this soil has a TEF of 0.02–0.06 compared to Cu. It can be explained by the immobile (stable) form of the metals in this soil and their limited biological availability.

As the above tables (Table 9.8 and 9.9) show, it often occurs that $EsM_{50}$ cannot be detected (Cd-contaminated soil root growth inhibition) because the inhibition does not reach 50%. In these cases long-term LOEC and NOEC values provide information about the adverse effects.

The TEQ values can interpret the differences due to matrix effects occurring in solid environmental samples and due to the differences in the test organisms' responses. They may help to understand the nature of organic contaminant toxicity in soil and integrate the toxicity of different and unknown contaminating substances and their interactions with the soil matrix and soil organisms.

Table 9.9 Equivalent toxicity of metal-contaminated soil based on NOEC and LOEC values.

| Contaminants and tests | Measured EsM$_{LOEC}$ g | TEQ$_{Cu(LOEC)}$ mg/kg | LOEC mg/kg | TEF$_{Cu(LOEC)}$ — | Measured EsM$_{NOEC}$ g | NOEC mg/kg | TEQ$_{Cu(NOEC)}$ mg/kg |
|---|---|---|---|---|---|---|---|
| Cadmium-contaminated soil | | | | | | | |
| Luminescence inhibition of *Vibrio fischeri* | 0.0025 | 250 | 5.0 | 0.5 | 0.001 | 2.0 | 250 |
| Dehydrogenase enzyme act. of *A. agilis* | 0.10 | 100 | 20 | 0.2 | 0.05 | 10 | 100 |
| Shoot growth of *Sinapis alba* | 5.0 | 100 | 500 | 0.2 | 2.50 | 250 | 100 |
| Root growth of *Sinapis alba* | 0.02 | 25,000 | 2.0 | 50 | 0.01 | 1.0 | 25,000 |
| Toxic metal mixture-contaminated soil | | | | | | | |
| Luminescence inhibition of *Vibrio fischeri* | 0.00625 | 100 | 83.13 | 0.03 | 0.003125 | 41.56 | 80 |
| Dehydrogenase enzyme act. of *A. agilis* | 0.0625 | 160 | 83.14 | 0.05 | 0.03125 | 41.57 | 160 |
| Shoot growth of *Sinapis alba* | 1.25 | 400 | 831 | 0.12 | 0.625 | 415 | 400 |
| Root growth of *Sinapis alba* | 2.5 | 200 | 1663 | 0.06 | 1.25 | 831 | 200 |
| Cu calibration series | | | | | | | |
| Luminescence inhibition of *Vibrio fischeri* | 0.00156 | 400 | 2.5 | 1.0 | 0.000625 | 1.0 | 400 |
| Dehydrogenase enzyme act. of *A. agilis* | 0.025 | 400 | 4.0 | 1.0 | 0.0125 | 2.0 | 400 |
| Shoot growth of *Sinapis alba* | 1.25 | 400 | 100 | 1.0 | 0.625 | 50 | 400 |
| Root growth of *Sinapis alba* | 1.25 | 400 | 100 | 1.0 | 0.625 | 50 | 400 |

## 9 STATISTICAL EVALUATION OF TOXICITY DATA

Uncertainties in toxicology should be managed by statistical evaluation and analysis of the dose–response function and in the risk assessment procedure, i.e. quantifying the probability and size of an adverse effect and determining the safe exposure limits. Biostatistics has been developed for the analysis of toxicological data, but it is equally important in the planning and execution of studies and in the analysis and interpretation of the results.

Statistical evaluation of toxicity data generally entails the problem of selecting the proper mathematical model and the most appropriate software. It is typical that several different statistical tools can be used for the evaluation of an experiment. The quality of the resulting test end point e.g. NOEC or $EC_x$ does not primarily depend on the statistical method but on the design and precision of the study. There are two pragmatic ways to select the proper evaluation method: (i) accepting the recommendation of a qualified statistician, or (ii) understanding the capacity of all potential statistical tools and selecting the most appropriate one for the purpose of evaluation and the dataset to be evaluated. When using a standard test organism and a standard test method, verified statistical methods can be applied as a routine, but the mathematical–statistical model should be chosen after thorough preparatory work and cautious circumspection in all other cases. To support the method selection, some general information on statistics and the models and methods applied in environmental toxicology are discussed below. Dose and concentration are used as interchangeable terms in this chapter.

### 9.1 Statistics in general

The general objectives and functions of statistical analysis are summarized below under the keywords of describing, identifying, comparing, predicting and explaining, based on the plausible overview of a professional blogger (Stats with cats, 2014):

- *Describing* the characteristics of biological entities, ecosystems, samples, areas etc. using default qualitative or quantitative descriptors, statistical intervals, correlation coefficients, graphics or maps;
- *Identifying* or classifying a known or hypothesized entity or group of entities using descriptive statistics; statistical intervals and tests, graphics and multivariate techniques such as cluster analysis;
- *Comparing* and detecting differences between statistical populations or reference values using simple hypothesis tests, and analysis of variance and covariance;
- *Predicting* measurements using regression and neural networks, forecasting using time-series modeling techniques, interpolating spatial data;
- *Explaining* latent aspects of phenomena using regression, cluster analysis, discriminant analysis, factor analysis, and data mining techniques.

The basic statistical parameters are as follows (Larson *et al.*, 2001):

- Statistics are numerical values used to summarize and compare sets of data.
- The following three statistics are measures of central tendency:
  - o The *mean*, or *average*, of $n$ numbers is the sum of the numbers divided by $n$. The mean is denoted by $\overline{x}$, which is read as "x-bar". For the data, $x_1$, $x_2, \ldots, x_n$, the mean is $\overline{x} = \frac{x_1 + x_2 + \ldots + x_n}{n}$.

- o The *median* of *n* numbers is the middle number when the numbers are written in order. If *n* is even, the median is the mean of the two middle numbers.
- o The *mode* of *n* numbers is the number or numbers that occur most frequently. There may be one mode, no mode, or more than one mode.

- Measures of dispersion tell how spread out the data are:

  - o The *range* is the difference between the greatest and least data values.
  - o The *standard deviation* describes the typical difference (or deviation) between the mean and a data value. The standard deviation σ of $x_1, x_2, ..., x_n$ is:

  $$\sigma = \sqrt{\frac{(x_1 - \overline{x})^2 + (x_2 - \overline{x})^2 + \cdots + (x_n - \overline{x})^2}{n}}.$$

  - o *Variance* also describes the difference between the mean and a data value. It can be considered the square of standard deviation:

  $$V = \sigma^2 = \frac{(x_1 - \overline{x})^2 + (x_2 - \overline{x})^2 + \cdots + (x - \overline{x})^2}{n}.$$

- Statistical graphs

  - o *Box-and-whisker* plot: The "box" encloses the middle half of the data set and the "whiskers" extend to the minimum and maximum data values. The plot itself comprises a rectangle (the "box") between the lower quartile and the upper quartile, and a straight line (the "whiskers") between the minimum and maximum data values, running through the box. The median divides the data set into two halves. The *lower quartile* is the median of the lower half, and the *upper quartile* is the median of the upper half.
  - o In a *histogram* data are grouped into intervals of equal width. The number of data values in each interval is the *frequency* of the interval.

A number of web pages (e.g. idre, 2014; Statistics, 2014; MathWorks, 2014; StatTrek, 2014; UDEL, 2014; Statpages, 2014) give advice for finding the best fitting statistical method for the above-mentioned cases in the form of lists, algorithms and tables, but all these cannot help in choosing the proper statistics if the evaluator/assessor does not understand the aim, the data type and its variations. The priority information needed for proper statistical method selection comprises:

- The objective of the evaluation;
- The number of variables;
- The scale of the variables;
- Whether the variable is dependent and/or independent;
- If the samples are autocorrelated and
- if the data are quantal or continuous.

The difference between quantal and continuous data is one of the main issues in toxicity evaluation. Quantal response data reflect incidences. Frequently used quantal end points are lethality (how many test organisms die or survive), immobilization (how many daphnids are mobile or immobile) or genetic anomalies. Continuous data are growth, enzyme activities, etc. data, which can take a continuously changing value during the study (body or organ weight, respiration or enzyme activities etc.). In toxicity studies the route and duration of the exposure are essential information and should be comparable to those occurring in nature. These two independent variables get less emphasis when standardized test evaluations are discussed. The statistics of time-dependence, exposure routes and exposure repetition deals with cases when more species are tested together or other interactions are expected. However, the biological changes under the impact of the assessed stressor (e.g. resistance, adaptation), as well as the extrapolation from laboratory to field are not discussed in this chapter, only attention is drawn to the importance of these aspects.

The mathematical basis of the statistics in toxicology and other biological studies goes back to the fifties and has not changed much (Brown, 1978), but new possibilities opened up with the expansion of information technologies. Nowadays universities and statistics courses offer descriptions and helpdesks for self-made statisticians and several software are available for free or for small subscription fees (StatPages, 2014; Unistat, 2014; Usablestat, 2014; Statistics, 2014; Statistical Solutions, 2014 and the large software providers such as BMDP, 2014; SAS, 2014 and SPSS, 2014; PROAST, 2014; Slob, 2003; etc.). Some of the complex software tools offer tool batteries for toxicity tests and bioassay evaluation (Unistat – Bioassay, 2014; Origin, 2014; Toxstat, 2014). Several high-quality tutorials/lectures are also available on YouTube as presentations of highly qualified professionals and from the software providers, e.g. the STATISTICA user guide (StatSoft, Getting started, 2014) or the excel introduction to XLSTAT (XLSTAT Introduction, 2014).

*Study design* in environmental toxicology plays a crucial role, for it is more important than the analysis itself. Design and analysis mutually determine each other: design of a study governs the necessary data analysis, e.g. ensuring satisfactory power or confidence interval may influence study design. When using standardized test methods, one should not deal with the design of the study only if input and output data fulfil the prerequisites of the standard method. The same is true for the test duration: duration, treatment and sampling schedule are key factors which strongly influence statistics in newly designed (not standardized) i.e. problem-specific studies. The majority of environmental toxicology studies are so-called prospective studies, applying randomized controlled trial and covering one or more treatments and one control. The importance of randomization is that the known and unknown prognostic factors are completely balanced in the different treatment groups. Treatment and control should occur in the same period of time.

Some well-known statistical tests and procedures are listed below. Most of them have importance in toxicology. For the definitions and explanations below, the following sources have been used: Berkeley (2014); Doncaster and Davey (2007); OECD (2006); Statistics (2014); Upton and Cook (2014) and Wikipedia (2014).

– *Akaike information criterion* characterizes the relative quality of a statistical model for a given set of data, so it is suitable for model selection.
– *Analysis of variance (ANOVA):* differences between group means and their variation among and between groups. The ANOVA model can be described as:

Response = Factor(s) + ε where the response refers to the data that require explaining, the factor or factors are the putative explanatory variables contributing to the observed pattern of variation in the response, and ε is the residual variation in the response left unexplained by the factor(s). For the confirmation of the model the residuals should be analyzed. Follow-up testing is necessary in order to assess which groups are different from which other groups or to test other hypotheses. If the test method applies a control, Dunnett's test (a modification of the *t*-test) can be applied for saying whether each of the other treatment groups has the same mean as the control.

- *Anderson–Darling test* is a powerful tool for detecting most departures from normality or deciding if the normal distribution adequately describes a set of data.
- *Bartlett test* is used to test if the samples are from populations with equal variances, called homogeneity of variances.
- *Bonferroni correction* is used in multiple comparisons to control the family-wise error rate.
- *Bootstrapping* measures accuracy to sample estimates. This technique allows estimation of the sampling distribution of almost any statistic. It uses simple methods such as resampling. The same technique can be used for the validation of the regression model by bootstrapping the residuals.
- *Box-Cox transformation* is often used for stabilizing variance and transforming data to normality, or normal distribution-like data. It is a rank-preserving transformation of data using power functions.
- *Chi-squared test*: a hypothesis test in which the sampling distribution of the test statistic is a chi-squared distribution when the null hypothesis (in toxicology, it is typically the assumption that the treated does not differ from the control) is true.
- *Cramér–von Mises criterion* is used for judging the goodness of fit of a cumulative distribution function compared to a given empirical distribution function, or for comparing two empirical distributions.
- *Correlation*: any kind of statistical relationships involving dependence.
- *Dunnett's test* is a multiple comparison procedure to compare each of a number of treatments with a single control. In contrast to the Bonferroni correction it exploits the correlation between the test statistics.
- *Factor analysis*: describes variability among observed and correlated variables in terms of a potentially lower number of unobserved variables called factors.
- *F-test*: is used for testing *F*-distribution under the null hypothesis to find the best fit. Exact "F-test" is used when the model has been fitted to the data using least squares.
- *Hosmer–Lemeshow test* is applied for goodness of fit for logistic regression models.
- *Kolmogorov–Smirnov test* is a non-parametric test of the equality of continuous, one-dimensional probability distributions.
- *Mann–Whitney U-test*: a non-parametric test of the null hypothesis that two populations are the same against an alternative hypothesis, e.g. that a particular population tends to have larger/smaller values than the other.
- *Multivariate analysis of variance* (MANOVA) is used when there is more than one response variable. In this case simultaneous observation and analysis of more than one outcome variable is necessary. The application of multivariate statistics is multivariate analysis. Multivariate statistics is concerned with multivariate probability

distributions of the observed/measured data and the interpretation of the results. Several types of multivariate analysis are known and applied such as: multivariate analysis of covariance (MANCOVA), multivariate regression analysis, principal components analysis (PCA), canonical correlation analysis and redundancy analysis (RDA), correspondence analysis (CA), multidimensional scaling, discriminant analysis and clustering. Artificial neural networks extend regression and clustering methods to non-linear multivariate models. Statistical graphics can be used to explore multivariate data.

- *Pearson's chi-squared test* or chi-squared goodness-of-fit test is used to assess two types of comparison, tests of goodness of fit and tests of independence. It is for studying/assessing any type of correlation with an arbitrary model. It has priority for discrete variables. If data are continuous, it is worthwhile to transform them into a discrete form and then apply the Pearson's chi-squared test. It applies for non-parametric models.

- *Pearson product–moment correlation coefficient* is the measure of the linear correlation between two variables, giving a value between +1 and −1. The value of 1 means total positive correlation, 0 means no correlation and −1 corresponds to total negative correlation. It is obtained by dividing the covariance of the two variables by the product of their standard deviations. It is for characterizing the degree of linear dependence between two variables. Permutation tests or bootstrapping can be used for constructing confidence intervals for Pearson correlation coefficients. To evaluate the distribution of the coefficient one can calculate the Student's distribution or the real distribution.

- *Permutation test* is for testing the hypothesis that two or more samples might belong to the same population. It combines the observations from all the samples first, than shuffles them and redistributes and resamples them 2–3 times (using the same resample sizes as the original samples). Finally it determines how often the resampled statistic gives the same values as the originally observed value.

- *Regression analysis* is the statistical evaluation of the regression and estimates the relationships among variables. The focus is on the relationship between a dependent variable and one or more independent variables. It demonstrates how the typical value of the dependent variable changes when any one of the independent variables is varied, while the other independent variables are held fixed. For validation of the model, confirmation of the goodness of fit one can use R-squared, the analysis of residuals or hypothesis testing. For checking statistical significance an F-test (for the fit), and t-tests (for individual parameters) are recommended.

- *Shapiro–Wilk's test* utilizes the null hypothesis principle to check whether a sample came from a normally distributed population.

- *Spearman's rank correlation coefficient* is a non-parametric test used to measure the strength of association between two variables where the value $r = 1$ means a perfect positive correlation and $r = -1$ a perfect negative correlation. It can be applied not only to data with normal distribution, but also to unevenly distributed data. It is an additional possibility when ANOVA and similar probes fail due to normality and homogeneity issues.

- *Student's t-test* is used to determine if two sets of data are significantly different from each other, and is most commonly used when the test statistic would follow a normal distribution.

Some statistics-related terms are defined as follow (Berkeley, 2014; Doncaster & Davey, 2007; OECD, 2006; Statistics, 2014; Upton & Cook, 2014; Wikipedia, 2014):

- *Accuracy* is a measure of how close the estimate is to the (unknown) true value of a parameter.
- *Precision* is a measure of the amount of variability in the estimate. This estimate is quantified by the standard error or the confidence interval. Precision can be increased by increasing the number of samples and by decreasing experimental variation.
- *Confidence interval* is an interval of parameter values that covers the true value of the estimated parameter with a certain % of confidence. In environmental toxicology 90% or 95% confidence level is typically used.
- Correlation or correlation coefficient, a numeric measure of the strength of linear relationship between two random variables, e.g. the Pearson product-moment correlation coefficient. Independent variables have a correlation of 0.
- *Data types* may be *quantal (binary), continuous* or *discrete*. Quantal data are recorded as yes/present or no/absent (also denoted by all or nothing), or dead or alive when lethality is the end point in toxicology. Quantal data can change between 0 and the total number of test organisms. Continuous data can take any value in an open interval, e.g. size of the organism, body weight and activity. Continuous data should be measured and usually have a dimension. The range in practice can be specified. Discrete data (to be counted such as number of certain developmental failures, number of heart rate, etc.) may be nominal, ordinal or interval. Categorical data – data sorted or divided into categories – are always nominal. If a variable can take on any value between two specified values, it is a continuous variable and the values follow a continuous distribution. However, if the value can only take on a finite number of values, the values follow a discrete (Bernoulli, binomial or Poisson) distribution.
- *Dependent variable* is substituted by response variable (as the pair of an explanatory variable), regressand, measured variable, explained variable, outcome variable, experimental variable or output variable.
- *Independent variable* in statistics is also named as a predictor variable, regressor, controlled variable, explanatory variable, exposure variable, risk factor, input variable. Explanatory variable is preferred when the quantities treated as "independent variables" may not be statistically independent.
- *Null hypothesis* as a general statement or default position refers to no relationship between two measured phenomena. In toxicology it means that the treated sample does not differ from the control. Rejecting null hypothesis is equal to concluding that there is a relationship between two phenomena or a potential treatment has a measurable effect. Both Type I and Type II error (see below) may occur.
- *Outliers* are inconsistent or questionable data points, out from the general trends. Statistical evaluation of the data is recommended to be executed both with and without the outliers.
- Parameter *i.e. statistical parameter* can be a population parameter, a distribution parameter, an unobserved parameter and is often a quantity to be estimated.
- *Non-parametric statistics and methods* are not based on parameterized (by mean or variance) probability distributions, so non-parametric methods make no

assumptions about the probability distributions of the variables being assessed. A *non-parametric model* is based on weak assumptions. This model is chosen when the assessor is not willing to assume any distribution of the parameters and it is based on the rank order of the observations and used mainly to test the null hypothesis. It is mainly used for quantal and discrete data in toxicology. With the application of non-parametric methods the problem of failing the prerequisites of the chosen probe can be excluded.

– *Parametric statistics* means that data have come from a type of probability distribution and make inferences about the parameters of the distribution. The model is fully specified, except the value of the parameter. Most of the statistical methods are parametric, and parametric methods are easier to handle than non-parametric ones. For parametric methods the population is generally known as approximately normal. Parametric methods are primarily applied to continuous data in toxicology.

– *Population* in statistics is a complete set of items that share at least one property in common that is the subject of a statistical analysis.

– *Randomization* ensures statistical independence among observations.

– *Statistical significance* is the low probability that an observed effect would have occurred due to chance.

– *Type I error* in statistics means false positives, occurring when the null hypothesis is the truth but the hypothesis test results in a rejection of the null hypothesis in favor of the alternative hypothesis. The probability of the occurrence of a Type I error is referred to as $\alpha$ and generally specified at 0.05, or 5%.

– *Type II errors* are false negatives occurring when the alternative hypothesis is true but the test fails to reject the null hypothesis (insufficient evidence to support the alternative hypothesis). The probability of making a Type II error is referred to as $\beta$ (1–power).

– *Power* is the probability of rejecting the null hypothesis ($H_0$) in favor of the alternative hypothesis ($H_A$), given that the alternative hypothesis is the true one. The power of a test varies with sample size, with the variance of the measured response and the size of the detectable effect. Power to detect differences can be increased by increasing the sample size and reducing variation in the measured responses (minimize experimental residual error) and using powerful statistics.

## 9.2 Statistical evaluation and analysis in environmental toxicology

Evaluation of the measured response on a minimum of two or a series of concentrations or doses may apply hypothesis testing, regression models or biology-based methods to the determination of the targeted test end points of traditional environmental toxicology such as I%, $EC_x$, (BMD), LOEC and NOEC.

Three main statistical methods will be discussed in detail in this chapter:

– hypothesis testing,
– concentration–response modeling with regression methods and
– biology-based methods.

The various statistics and statistical analyses include general tasks and rules. The toxicology-related steps are summarized in the following:

- *Experimental design* of the independent variables such as the concentration of the substance to test, the duration of exposure, number of replicates to control experimental variation, times of observation, etc. The goal of the test can be to determine I%, $EC_x$, LOEC and NOEC, and according to these test end points the study design will differ in demands. $EC_x$ needs sufficient number of concentrations for good curve fitting, NOEC needs warranted statistical power.
- *Randomization* ensures independence of errors from the statistical point of view. When carrying out a test, randomization may include experimental materials, test conditions, test organisms, treatments, choosing the place of containers and, finally, measuring the responses (e.g. the measurement order of the units).
- *Replication* works against noise and increases statistical power in hypothesis testing and confidence limits of parameter estimates.
- *Multiple controls* are included in the experimental design in those cases when the chemical substance to be tested is not water soluble and a solvent/vehicle should be used for its solubilization. In this case two controls belong to the study design: one without and another one with solvent. If the two do not differ, the results can be averaged and used as one control. If the difference is statistically significant, the solvent-free control should be eliminated from the evaluation. General rule is not to use solvent/vehicles if possible.
- *Process of data analysis* always follows the same pattern:

  o Data inspection and elimination of outliers.
  o Data inspection and assumptions about the *general pattern of the scatter of data* to decide if variance is homogeneous or heterogeneous. Heterogeneity of variances may be eliminated by the right transformation, which is able to approximate/convert data to normal distribution.
  o *Transformation* of data aims at stabilizing the variance so that the standard ANOVA tests or regression analysis allow for application. Non-normal distribution can be transformed e.g. by Box-Cox transformation.
  o *Parametric and non-parametric methods*: if the scatter is symmetric and homogeneous after transformation, parametric methods based on normality can be used for the analysis. If the normality is violated, generalized linearization model (GLM) or non-parametric method based on rank order of the observation should be used.
  o *Quantal and continuous data* should be distinguished. Quantal data are categorical data – in toxicology the end points of mortality/survival or immobilization are such – denoting the facts of dead or alive as well as mobile or immobile with a *yes* or *no* characterization. Continuous test end points such as weight, growth, activities, light emission, etc. are based on responses taking values on a continuous scale, e.g. time-scale. The values can be monotonic or non-monotonic. The statistical method should be selected according to these data characteristics as shown later in Table 9.10.
  o *Pre-treatment of data* other than transformation for model fitting: e.g. logit and probit transformation, which should be applied with caution, due to the

problem of the logarithm of a zero concentration. To apply a very small value instead of zero is not recommended.

o   *Model fitting* follows a general principle: the model contains specific parameters, and the goal of data analysis is to estimate these parameters. The parameters are estimated by fitting the model to the data.

o   *Model checking and validation* covers the evaluation of the appropriateness of the fitted model, controlling if the data indeed comply with the model assumptions. It is done by evaluation of residuals for checking if the variances are homogenous, if the deviations from the fitted model are systematic, and if the experimental factors deviate or not.

o   *Reporting the results* in a statistical analysis of toxicological results should cover not only the $EC_x$ and NOEC values and the summary report of the statistics, but all details which are critical and may influence the results: starting with the experimental design, the tests used for hypothesis validation and model fitting. A detailed list of the required information in the report will be given at the end of the chapter on statistical methods.

## 9.3   Hypothesis testing

*Hypothesis testing* is the statistical evaluation of the hypothesis on the distribution of variables. It is also called confirmatory data analysis: checking if the hypothesis was correct. It is a method for identifying statistically significant study results which are predicted as unlikely to have occurred by chance alone. Significance is judged according to a pre-determined threshold probability. It is based on distinguishing null hypothesis (i.e. no difference between control and treated) and the alternative hypothesis (the treated significantly differs from the control). Hypothesis testing includes several methods which should be selected from among the most suitable ones for the concept of the study to be evaluated. Parametric, sometimes non-parametric, one-sided and multiple comparison tests are generally recommended for the evaluation of toxicity study results. Multiple comparisons refer to simultaneous statistical inferences e.g. in the case of Dunnett's test (1955) or the Williams method (1971, 1972) for comparing a number of treatments to a single control group.

Multiple comparison procedures are commonly used in an analysis of variance after obtaining a significant test result. A significant ANOVA result suggests rejecting the global null hypothesis that the means are the same across the groups being compared. Multiple comparison procedures determine the means differing. A one-way ANOVA involves pair-wise comparisons: having 4 treatments including a control, the number of these pair-wise comparisons is 4*3/2; having *n* groups, the combination of treatments gives n*(n–1)/2 pairs.

Non-parametric methods can also be used for the evaluation of concentration–response data, e.g. the Kruskal–Wallis test (1952) or the Jonckheere–Terpstra test (Jonckheere, 1954; Terpstra, 1952), which are non-parametric alternatives to ANOVA. Multiple comparisons can be done using pair-wise comparisons and using a correction – for example a Bonferroni correction (Abdi, 2007) – to determine if the post-hoc comparison tests are significant.

*Post-hoc comparison tests* are applied in the second step of ANOVA or MANOVA, after the null-hypothesis has been rejected. The testing procedures may be the

Bonferroni adjustment or the Dunn test. When the hypothesis has been rejected, post hoc is still able to find (otherwise existing and visible) associations. This is possible because ANOVA is sometimes too stringent for biological data. A post-hoc test gives one more chance for clarification and objective evaluation.

ANOVA has no precise definition because it is the combination of ideas, serving very different purposes and adapted to the analyses of a wide variety of experimental designs, for example for prospective evaluation without knowledge of effect sizes (Doncaster *et al.*, 2013).

*Multiple comparisons ANOVA* F-test in one-way analysis of variance is the standard statistical tool to determine those concentrations which exert an effect significantly different from the untreated control. The null hypothesis is that the means of the responses on the effect of different concentrations are the same. The ANOVA F-test is for assessing whether any of the concentrations caused an effect on average superior, or inferior to the others versus the null hypothesis. F-test is able to perform multiple comparisons instead of the comparison of the responses on single concentrations to each other, using several $t$-tests. When only two groups are evaluated by a one-way ANOVA F-test, $F = t^2$ where $t$ is Student's $t$ statistic. The definition of F is: F = between-group variance/within-group variance.

Considering the concentration–response curve, LOEC, $EC_{20}$ and $EC_{50}$ are significantly different from the control, and NOEC is the same as the control within the confidence interval.

Hypothesis testing has many different uses in ecotoxicology, ranging from detecting whether there is a significant difference in the measured response between the control and a given concentration to establishing a LOEC or NOEC value. In the OECD guidelines for testing the effect of pure chemicals and determination of NOEC, hypothesis testing is the recommended priority method (OECD, 2006).

Despite being the main statistical tool, hypothesis testing may have significant limitations in toxicology, e.g. it does not consider mechanisms of the toxicant and the biology of the organism. If the goal is the determination of a NOEC, one will face further shortcomings:

- No confidence interval can be assessed for NOEC since it does not estimate a model parameter.
- NOEC is not a tested but a hypothetical highest concentration with no significant effect; any tested concentrations may be different to NOEC.
- In the case of low statistical power the biologically important differences between the control and treatment groups may not be identified as significantly different. If the power is high, it may occur that biologically unimportant differences are found to be statistically significantly different.

### 9.3.1  Hypothesis testing for the determination of NOEC

NOEC can be determined by using single-concentration (yes/no or pass/failed) toxicity test results or multi-concentration test results.

- The steps for the statistical evaluation of single-concentration test results are:
    o transformation of response data to get the proportion survived (log, square root, arc sine transformations);

- o testing normality assumption (e.g. Shapiro-Wilk's test);
- o testing the homogeneity of variance (F-test) and a $t$-test for significance. In the case of failure (unequal variance) a modified $t$-test should be used;
- o is the difference compared to control significant (e.g. in survival)? Yes or no?
- The steps for the statistical evaluation of multi-concentration test results are:
  - o transformation of data to proportion survived (log, square root, arc sine transformations);
  - o testing normality assumption (e.g. Shapiro-Wilk's test);
  - o testing the homogeneity of variance (e.g. Bartlett's test);
    - yes: $t$-test with Bonferroni adjustment (equal number of replicates) or Dunnett's test (not equal number of replicates),
    - no: Steels many-one (equal number of replicates) or Wilcoxon rank sum tests (non-equal number of replicates).
  - o NOEC estimation.

*The OECD (2006) guideline* gives detailed explanation and instructions for the experimental design and statistical evaluation of toxicity studies and for the selection of the proper statistical method. The scope of the guideline is "restricted" to the testing of pure chemicals by standard test methods, so the user of the guidance document cannot provide a recipe for every type of experiment, yet the guideline is very useful for understanding the statistics, statistical analyses, and the importance of consistency in statistics as well as the design of toxicity measuring experiments. Hypothesis testing in toxicology is summarized below based on the OECD guideline.

*The NOEC is defined* as the highest test concentration without significant effect below the lowest concentration that did result in a significant effect in a toxicity study. The hypothesis that is tested when determining the NOEC reflects the question: "Which is the concentration that is likely to have no effect on the test organism?" The true answer depends on several test-specific conditions; some typical ones are listed:

- Has solvent been used? Have both solvent and non-solvent controls been tested?
- Is the study a dose–response type test?
- Is the dose–response function monotonic?
- Are there more than two concentrations tested?
- Are data measured normally distributed and homogeneous?

Having all this information one can choose the proper statistical method.
*The one-sided hypothesis* – also called one-tailed test – is appropriate when NOEC is a concern in one direction only:

- $\mu_0 > \mu_i$, where $\mu_0$ = the mean of the control; $\mu_i$ = the mean of the treated samples
- Null hypothesis of $H_0$: $\mu_0 = \mu_1 = \mu_2 = \cdots = \mu_k$
- Alternative hypothesis of $H_1$: $\mu_0 > \mu_i$, for at least one i where

  - o $\mu_0$ denotes the mean of the control
  - o $\mu_i$ is the mean of the test populations, and $i = 0, 1, 2, 3, \ldots, k$.

*A two-sided form of the hypothesis* symbolized by $\mu_0 \neq \mu_i$ requires a two-sided trend or pair-wise test. In toxicity studies the model may assume monotonicity because the treatment group only differs in the exposure concentration/dose from the control, and growing exposure will tend to increase the effect, which shows up as an increase or a decrease in the measured end point:

- $\mu_0 \geq \mu_1 \geq \mu_2 \geq \mu_3 \geq \cdots \geq \mu_k$
- Null hypothesis: $H_{0*}$: $\mu_0 = \mu_1 = \mu_2 = \cdots = \mu_k$
- Alternative hypothesis: $H_{1*}$: $\mu_0 \geq \mu_1 \geq \mu_2 \geq \mu_3 \geq \cdots \geq \mu_k$, with $\mu_0 > \mu_k$.

The trend test is recommended for dose–response experiments where a monotonic response is expected and when data distribution is normal and the variance is homogeneous. If the one-sided approach also fulfils the requirement of the assessor, it should be preferred, given that the statistical tests of a one-sided hypothesis are more powerful than tests of the two-sided hypothesis.

Comparing single-step (pair-wise comparisons) and step-down trend tests to determine the NOEC, the step-down trend test should be given priority if monotonicity is biologically justifiable: e.g. in the case of plant tests hormesis[1] may appear i.e. positive effect of the toxicants at low concentration, so the response will not be monotonic. Monotonicity can be accepted as a rule or tested formally: in the latter case either monotonicity or non-monotonicity can be tested.

*The power of a toxicity test* can be calculated if the size of the effect to be detected, the variability of the end point measured, the number of treatment groups and the number of replicates in each treatment group are known. Power is important for the selection of a process, but it is not the only condition. The main goal should be finding the most appropriate method for the data and the end results: statistically insignificant concentrations, which are large enough to cause damage, must not be overlooked and, on the other hand, minor concentrations must not be characterized as statistically significant. Statistical significance should agree with biological significance or with regulatory thresholds.

*Experimental design and evaluation* is based on the initial variance estimates which are calculated based on historical control data. Post-hoc power should be compared to the design power and reported. Other important experimental factors are the number and spacing of exposure levels (concentrations, doses), the number of test subjects in each group, and the type of subgroups if adequate. These factors will determine the power which should be designed to be able to identify the biologically important effect levels. Substance concentration is the most important issue from this point of view. It should be close to the NOEC when the aim is NOEC determination. The number of the organisms tested may also be crucial. According to the allocation rule of Dunnett (1955), the number of organisms in the control group ($n_0$) should be calculated by the *square-root rule* to get a test design optimizing the power of Dunnett's test. By this rule, the value of $n_0$ is the solution of the equation

$$N = k\, n + n\sqrt{k}, \text{ and } n_0 = N - kn \rightarrow n_0 = n\sqrt{k}, \text{ where}$$

- $N =$ the total number of subjects tested;
- $k =$ the total number of treatments;

---

[1] Adaptive response of cells and organisms to a moderate (usually intermittent) stress.

*Table 9.10* Statistical methods for hypothesis testing in environmental toxicology (OECD, 2006).

| For quantal data | Parametric methods | Non-parametric methods |
|---|---|---|
| Single-step (pair-wise) | Dunnet | Mann-Whitney with Bonferroni-Holm adjustments |
| | Poisson comparisons | Chi-squared with Bonferroni-Holm adjustment |
| | | Steel's Many-to-One |
| | | Fisher's exact test with Bonferroni-Holm adjustment |
| Step-down (trend based) | Poisson Trend Williams | Cochran-Armitage |
| | Bartholomew | Jonckheere-Terpstra test |
| | Welch | Mantel-Haenszel |
| | Brown-Forsythe | |
| | Sequences of linear contrasts | |

| For continuous data | Parametric methods | Non-parametric methods |
|---|---|---|
| Single-step (pair-wise) | Dunnett | Dunn |
| | Tamhane-Dunnett | Mann-Whitney with Bonferroni correction |
| Step-down (trend based) | Williams | Jonckheere-Terpstra |
| | Bartholomew | Shirley |
| | Welch trend | |
| | Brown-Forsythe trend | |
| | Sequences of linear contrasts | |

– $n$ = equal number of subjects in all treatment groups;
– $n_0$ = number of subjects in the control group.

The argument for having more tested subjects in the control group is that the same control is applied for comparison in every single treatment group.

Covariates that influence the conclusion should be included in the evaluation, and the analysis should be adjusted to these covariates such as age and size or other characteristics of the test organism at the beginning of the test. For continuous, normally distributed responses with homogeneous variances, analysis of covariance (ANCOVA) is well developed. For continuous responses that do not meet the normality or homogeneity requirements, non-parametric ANCOVA is available.

In Table 9.10 some OECD (2006) recommended statistical methods are given for hypothesis testing of toxicity results as the combination of stepwise and trend-based, as well as parametric and non-parametric cases, with quantal and continuous data types distinguished.

*The problem of outliers* can be solved by a simple procedure: the first step is visual observation of dose–response data. Experience of professionals is an adequate basis for a personal judgment and decision making about outliers that should be excluded from the evaluation. For continuous data, the individual responses can be plotted in addition to the group means as a function of dose. For quantal data, the observed frequencies of response as a function of dose may be conclusive. Several outlier rules can be used for identifying undesirable data such as Tukey's rule (Tukey, 1977). It is worth

*Table 9.11*   Statistics-related information recommended by OECD (2006) to be included in the toxicity study report.

| NOEC: quantal study result | NOEC: continuous study result |
|---|---|
| | Description of the statistical methods used |
| Test end point assessed | Test end point assessed |
| Number of test groups | Number of test groups |
| Number of subgroups within each group/way of handling | Number of subgroups within each group/way of handling |
| Identification of the experimental unit | Identification of the experimental unit |
| Nominal and measured concentrations (if available) for each test group | Nominal and measured concentrations (if available) for each test group |
| Number of those exposed in each treatment group | The concentration/dose metric used (actual, log, equally spaced scores) |
| Number of those affected in each treatment group | Number of those exposed in each treatment group (or subgroup) |
| Proportion affected in each treatment group | Group means (median, if a non-parametric test was used) and standard deviations |
| Confidence interval for the percent effect at the NOEC, provided that the basis for the calculation is consistent with the distribution of observed responses | Confidence interval for the percentage effect at the NOEC, provided that the basis for the calculation is consistent with the distribution of observed responses |
| P value for test of homogeneity if performed | The NOEC value |
| Name of the statistical method used to determine the NOEC | P value at the LOEC (if applicable) |
| The concentration/dose metric used (actual, log, equally spaced scores) | Results of power analysis |
| The NOEC | |
| P value at the LOEC | Plot of response versus concentration |
| Design power of the test to detect an effect of biological importance based on historical control background and variability | |

distinguishing between outliers relating to entire treatment groups and individuals because the first one is typical when experimental factors are different in the group, which should be excluded by proper design and randomization.

### 9.3.2   Reporting hypothesis testing

The information required in the study report by the OECD (2006) methodology for testing pure chemicals is summarized as an example in Table 9.11, both for quantal and continuous data.

## 9.4   Regression and regression analysis

In *regression analysis* the estimation target is a function, a relationship between a dependent variable and one or more independent variables. When using standardized test methods, regression analysis helps to understand how the response – measured end point – changes when the concentration or dose of the contaminant is varied (in standardized tests, duration and uncontrolled interactions are excluded from the independent variables). The variation of the dependent variable in the regression function can be described by a probability distribution.

The most frequently applied regression models are as follows:

– *Linear interpolation of means* is the estimation of an intermediate value of the independent variable of a function. This may be achieved by curve fitting or regression analysis.
– *Linear regression* is modeling the relationship between a scalar dependent variable and one or more explanatory variables. In linear regression, data are modeled using linear predictor functions, and unknown model parameters are estimated from the data.
– *Polynomial regression* is the generalization of linear regression in which the relationship between the independent variable and the dependent variable is modeled by a polynomial.
– *Logit or logistic regression* is a type of probabilistic statistical classification model for binary response variables. The probabilities describing the possible outcomes of a single trial are modeled, as a function of the explanatory (predictor) variables using a logistic function.
– *Probit model* is a binary response model, employing a probit link function and uses the standard maximum likelihood procedure. It treats the same set of problems as logistic regression does.
– *Spearman-Karber* is a non-parametric method, which does not require the assumption of normality, compared to Probit. It is known as a method, resulting in close estimates to $EC_{50}$ even in the case of small datasets. It is only applicable for $EC_{50}$, and not for other end points.
– *Weibull* is a proportional hazard model which means that the effect of an independent variable on the hazard rate is assumed to be multiplicative. It is mainly used for modeling survival data in toxicology.

### 9.4.1   The use of regression and regression analysis in toxicology

Regression models are especially useful in concentration/dose–response curve fitting and in the determination of $EC_x$. $F$-test is suitable for studying the hypothesis that a proposed regression model fits the data well based on the comparison of two fits and the determination of the *significantly* better fit to the measured data. The larger the $F$-statistic, the more useful the model is.

The OECD (2006) guidance on statistical tools describes the theory and practice as well as the steps of curve fitting to determine $C_x$, the concentration causing x% effect in the measured end point, the latter typically being lethality or inhibition of any biological function. Several suitable functions are able to describe measured data properly, but there are favored mathematical models accepted widely and used frequently. Regression methods cover the analyses for quantal data and continuous data as well as parametric approaches (when a specific underlying distribution is assumed) and some non-parametric ones (not typical in toxicity). They are for estimating the arbitrary x% effect in the biological response variable and the associated confidence bounds (limits).

The concentration/dose–response curve can be symbolized by the function of $y = f(x)$, where

– $x$ is the independent variable: concentration or dose, and
– $y$ is the function of $x$, i.e. the dependent variable or response.

Other independent variables are not included in a standardized test with strict conditions and a one-off evaluation. Otherwise duration of the exposure is typically an important determinant: even a simple dose results in a growing cumulative value when measuring the response (e.g. the number of responding subjects) in several time points.

Dose–response data are handled by regression statistics as a whole, not point by point. The variables of the function are determined by the measured data. The number of parameters is proportional to the flexibility of the model: the larger the number of parameters, the more flexible the model is. On the other hand, rationality of the number of the parameters says that inclusion of an additional parameter into the model is recommended only if it leads to a significantly better fit.

The model can be fitted to the data in various ways. The *sum of squares* method (SS), for example, optimizes the fit by minimizing the sum of squared residuals. Residuals in this context are the distances between the data and the model. The fit may involve the maximization of the likelihood based on the assumed distribution of data: normal or log-normal for continuous data, binomial distribution for quantal data and Poisson distribution for count data.

The term power of the statistics is only used in hypothesis testing, in regression models the *confidence interval* is used for characterizing the goodness of the model. Sample size and variation in the response are important characteristics: small sample sizes and large variability in the response within groups will increase the width of the confidence interval of the parameters, and the fitted model may not reflect the true concentration–response relationship. The number of replicates, the location and number of concentrations can be increased for better statistics (smaller confidence interval). The design of the experiment depends on the x value of $C_x$ because $C_{90}$, $C_{50}$ and $C_{05}$ require different designs, mainly in the number and spacing of concentrations.

Basic assumptions in regression for toxicology are as follows:

– In the case of continuous data, the number of dose groups showing significantly different response levels is – by experience – a minimum of four including the control. It can be greater than four in practice;
– In the case of quantal data, the minimum number of partial responses should be two: almost complete survival and complete mortality. The test design should exclude the inappropriate extrapolation for determining the desired $EC_x$;
– A monotonic concentration–response relationship;
– The fitted curve is close to the true concentration–response relationship;
– There are weak assumptions about the mechanisms of the toxicant or the biology of the organism and interaction between more stressors and more species.

Some further limitations of concentration–response modeling should be mentioned:

– Extrapolated $EC_x$ values (outside the measured concentration range) are rather uncertain;
– Too large gaps between consecutive response levels lead to uncertain interpolation and make the fitting uncertain (allowing many different models to be fitted).

Quantal and continuous data are evaluated along the same steps:

– Fitting various models;
– Evaluating the fit in comparison;
– If more or all models fit well and result in the same $EC_x$, the determination of $EC_x$ has come to an end. If some models do not fit, more additional models should be applied and outliers and precision of the test-method should be checked.
– The goodness of the regression is investigated through the confidence intervals among $EC_x$ estimates and the lowest confidence bounds are chosen.

### 9.4.2   Evaluation of quantal data

As OECD (2006) describes in its guideline for standardized tests, the choice of the model is crucial, and is governed by the data. For quantal data the dose–response function ranges between 0 and 1 (0% and 100%), the response normally is monotone. Cumulative distribution functions (normal, logistic, Weibull) are the primary candidates for dose-response modeling. Cumulative distribution functions are popular because they can be interpreted as individual tolerance distributions within the population. Plotting the tolerance distribution cumulatively results in the quantal dose–response relationship. A predicted response of 20% lethality at a concentration of 100 mg/L can be interpreted as 20% of the individuals having a tolerance lower than 100 mg/L. Tolerance distribution of the individuals in the population is equal to the probability of a response from each individual, which means that the chance of each individual is 1:5 to give response to the 100 mg/L concentration.

The dose–response model for quantal data is a function of the concentration $y = f(x)$ where y is the true quantal response and x is the concentration. The function $f(x)$ cannot strictly equal zero at concentration zero, so the model should include a background incidence parameter (a):
   $y = f(x) = a + (1 − a)g(x)$, where

– $a$ denotes the true background probability of response;
– $g(x)$ is a function increasing from 0 to 1 when $x$ increases from zero to infinity.
– In this formulation the response at infinite concentration is $g(x) = 1$.

### 9.4.3   Choice of the models

– The *probit model* is the cumulative normal distribution function, usually applied to the log-concentrations, implying that a lognormal tolerance distribution is assumed.
– $y = a + (1 − a) * \text{pnorm}(b * \log 10(x/EC_{50}))$
– The *logit model* is the cumulative logistic distribution function. The logit model is also applied usually to the log-concentrations.
– $y = a + (1 − a)/1 + \exp(b * \log 10(EC_{50}/x))$
– Both the probit and the logit model have two parameters: the $EC_{50}$ and the slope (b), all other $EC_x$ values (the quantal response of y) are determined based on these two parameters. $EC_x$ should be adjusted to the background value. When using the additional risk model, the background value (let say 3%) will be added to

x%, when determining $EC_x$ (i.e. $EC_{20}$ should be read at $20 + 3 = 23\%$) from the dose–response curve.

- The *Weibull distribution* is not necessarily symmetrical, and is usually applied to the concentrations themselves and not their logs. $EC_{50}$ is related to two parameters in Weibull: the location and the slope.

$$y = a + (1 - a) * (1 - \exp(-(x/b)c))$$

- Simplified *multi-stage models* are often used for describing tumor dose–response data.

### 9.4.4  Evaluation of continuous data

A continuous response has a certain amount of scatter, depending on the homogeneity of the treatment group. This scatter follows a certain distribution such as normal, lognormal or Poisson distribution.

$EC_x$ in continuous responses relates to the change in the degree of the effect. This interpretation differs from the one for quantal responses where $EC_x$ relates to a change in response rate, such as growth rate, respiration rate, enzyme activity rate and any kind and any percentage of inhibition.

The function is formally the same as for quantal data
$y = f(x)$, where

- y covers here continuous data and
- $x\% = 100 \ ((y \times EC_x/y_0) - 1)\%$ the degree of effect.

### 9.4.5  Choice of the models

The observed/measured dose–response outcomes are smoothened by the regression model and analyzed for estimating the most correct $EC_x$ and assessing the confidence intervals. As the regression model has no meaning in biological sense (it has no biological interpretation), and the selection of the available models is wide, the choice of model is largely arbitrary; the most popular families of models are as follows:

- *Linear regression models* may be nonlinear in respect to the independent variable, they can be higher-order polynomial or quadratic. Nested models can easily be modified regarding the incorporated parameters or turning the linear model into quadratic and evaluate the fit of these modified models (applying *F*-test) until getting the best fit. The function is:

  o  $y = a + bx$   or   $y = a + bx + cx^2$   or   $y = a + bx + cx^2 + dx^3$,

- *Threshold models* contain an additional parameter reflecting a dose threshold, below which the change in the end point is zero. The function describing this situation is

  o  $y = a$           if $x < c$, where $c =$ the threshold concentration and
  o  $y = a + f(x - c)$   if $x > c$ – for any function
  o  $y = a + b(x - c)$   if $x > c$ – for linear function.

– *Additive vs. multiplicative models* are supposed to solve the problem of background noise, meaning that the response is not zero when the exposition is zero. The background level of a can be incorporated into the model as an additive or as a multiplicative value. Priority is given to the latter one because in addition to the logical solution that the additional value should be a percentage similar to the response, it is very useful in the case of testing more populations with different background levels. The two alternative formulae are:

  o $f(x) = a + g(x)$ – additive way of incorporating background level;
  o $f(x) = a \times g(x)$ – multiplicative way of incorporating background level.

– *Models based on "quantal" models* adjust continuous data applicable for quantal models.
– *Nested nonlinear models* were proposed by Slob (2002) including 5 models:

  o model 1   $y = a$   $a > 0$;
  o model 2   $y = a \times \exp(x/b)$   $a > 0$;
  o model 3   $y = a \times \exp(\pm(x/b) * d)$   $a > 0, b > 0, d \geq 1$;
  o model 4   $y = a \times [c - (c–1) \exp(–x/b)]$   $a > 0, b > 0, c > 0$;
  o model 5   $y = a \times [c - (c–1) \exp(–(x/b) * d)]$   $a > 0, b > 0, c > 0, d \geq 1$ where
    • $y$ = any continuous end point
    • $x$ = concentration or dose.

– *Hill model* dates back to 1910 and was elaborated for receptor binding and enzyme kinetics. The value of c=1 in the following function results in the Michealis-Menten equation.

  o $y = a \times x^c/b + x^c$, where
  o c is the Hill parameter

– *The Michaelis-Menten equation* has a widely known form:

  o $v = V_{max} \times [S]/K_m + [S]$, where
    • $v$ = reaction rate
    • $V_{max}$ = maximum rate
    • $[S]$ = substrate concentration
    • $K_m$ = is the substrate concentration, at which the reaction rate is half of the maximum.

The model is fitted to the dose–response data and the parameters estimated by applying the suitable software. It is necessary to be aware of the assumptions underlying the fit algorithm. An iterative algorithm tries to find better parameter values in an evaluation process if the fit can be improved by changing the parameter values. The solution is to find the maximum likelihood or minimum SS in the function. The iteration stops when the software has found a clear maximum in the log-likelihood function, or in a minimum in the SS.

Several assumptions are used in choosing and fitting the model such as experimental design ensuring randomized test conditions, no dependence between tested organisms, no systematic difference between dose groups, exactly known concentration/dose, normal distribution of the data (before or after transformation), homogeneity of variance and goodness of the fit. These assumptions should be tested by normality tests, goodness of fit tests, etc. and a confidence interval should be determined. Confidence may be assessed by:

– The delta method: plus or minus twice the standard error as estimated by the second derivative of the likelihood function;
– Based on the profile of the log-likelihood function, using the Chi-square approximation of the log-likelihood;
– Bootstrap methods;
– Bayesian methods, having preliminary knowledge on the range of the parameter value(s).

When can the fitted model be accepted? The fit is good when the fitted model incorporates all the measurement points and gives a true estimate for all not measured responses to all not applied concentrations/doses between two measurement points. Statistical evaluation should cover both: the measured and the estimated responses. The best way to check the goodness of fit is the visual check. Absolute and relative tests can simultaneously be executed to characterize the goodness of fit. Absolute tests are able to quantify the deviation of data from the model curve; relative tests compare the different test results with each other. The most comforting answer is when different models (e.g. logit, probit and Weibull) result in the same or very close parameters and the same $EC_x$ values. Good-quality measured data make it suitable to apply different models, but excessively high uncertainties and imprecise test implementation, lack of randomization or outlier exclusion etc. violate the initial assumptions and may cause disagreement of the model with the data and the incorrect rejection of the model. But the opposite may also happen: the goodness of fit looks perfect by standard testing because the test execution is very precise. However, the data do not follow the chosen model e.g. due to biological reasons.

### 9.4.6  Reporting regression statistics

Table 9.12 shows the summary of reporting regression statistics.

*Table 9.12* Regression statistics-related information recommended by OECD (2006) to be included in the toxicity study report.

| Quantal data | Continuous data |
| --- | --- |
| Test end point assessed | Test end point assessed |
| Number of test groups | Number of test groups |
| Number of subgroups within each group (if applicable) | Number of subgroups within each group (if applicable) |
| Identification of the experimental unit | Identification of the experimental unit |
| Nominal and measured concentrations (if available) for each test group | Nominal and measured concentrations (if available) for each test group |
| Number exposed in each treatment group (or subgroup if appropriate) | Number exposed in each treatment group (or subgroup if appropriate) |

(continued)

*Table 9.12* Continued

| Quantal data | Continuous data |
|---|---|
| Number affected in each treatment group (or subgroup if appropriate) | |
| Proportion affected in each treatment group (or subgroup if appropriate) | Arithmetic group means and standard deviations, but geometric group means and standard deviation if lognormality was assumed |
| The dose metric used | The dose metric used |
| The model function chosen for deriving the $EC_{50}$ ($EC_x$) | The model function chosen for deriving the EC |
| Plot of dose-response data with fitted model, including the point estimates of the model parameters and the log-likelihood (or residual SS) | Plot of dose-response data with fitted model, including the point estimates of the model parameters and the log-likelihood (or residual SS) |
| Fit criteria for other fitted models | Fit criteria for other fitted models |
| The $EC_{50}$ together with its 90%-confidence interval | The $EC_x$ (CED) together with its 90%-confidence interval |
| If required: the $EC_x$ together with its 90%-confidence interval | |
| Method used for deriving confidence intervals | Method used for deriving confidence intervals |

## 9.5  A comparative study on statistical evaluation of dose–response data

Statistical models are becoming more common with the spread of IT techniques and their importance has greatly increased in toxicology. OECD recommended the application of regression models for the evaluation of response vs. concentration/dose data and also decided to phase out NOEC as a chronic end point being uncertain due to the arbitrary selection of the lowest tested concentration and the difference between two adjacent concentration points. The research group of Isnard *et al.* (2001) prepared a comparative study on the statistical evaluation of 27 chronic ecotoxicity datasets of algae, daphnia and fish. The same study results were evaluated by ANOVA-type hypothesis testing (Dunnett, Williams and Jonkheere-Terpstra) and several different regression models (linear interpolation, polynomial regression, logit, probit and Weibull models). The confidence interval was estimated using the bootstrap resampling technique. In addition to a general comparison their aim was to find the best regression model and to substitute NOEC with an $EC_x$ close to NOEC, obviously $EC_{05}$.

The most important conclusions from the comparative study are:

- Different hypothesis testing methods lead to different results from the same dataset.
- Both $EC_{50}$ and $EC_{05}$ values were very close by linear interpolation, polynomial regression and logit, in the case of good quality data and good fitting.
- Probit and Weibull models were excluded in an early stage of the study, proven not to be useful in the evaluation of the selected datasets.
- Bootstrap-simulated confidence intervals were found better and more realistic than asymptotic calculations.

- The paired comparison of NOEC and $EC_x$ gave the result that there is no significant difference between NOEC and $EC_{05}$, allowing the pragmatic approach of substituting NOEC with $EC_{05}$.
- Logit provided suitable fitting in most studied cases, but probit did not converge in all cases.
- Experimental design and precise test implementation are crucial prerequisites.
- Fisher test is recommended for testing both ANOVA and regression models.

Kooijman (1993 and 1996) proposed a threshold model based on Dynamic Energy Budget (DEB), giving directly a no-effect result, but it is rather complicated and not validated yet. The DEB model looks useful also in biology based toxicology and toxicokinetics (Kooijman *et al.*, 2009).

## 9.6  Biology-based methods

Biology-based methods aim to explain not only the result itself, but the underlying processes. Human toxicology focusing on toxicokinetics and reproduction (Kooijman, 1996) specifies a "response surface" which is a function of concentration and exposure time and includes the chemistry and biology of the system. Several parameters determine the response surface. After estimating these parameters from data, the same parameters can be used for calculating the time dependence of $EC_x$, the effect of changing concentrations, the effect on growing populations or other, more complex changes in the field. Biology-based models enable to consider the time-dependence of the internal concentration (increasing in time), or the changing concentration in the test medium which directly determines the effect (Kooijman, 1983; Gerritsen, 1997; Péry *et al.*, 2002). The changes in the slope in time can be modeled by the response surface of biology-based methods. Biology-based methods are able to include more datasets, such as activity, lethality and the changes of internal concentration (accumulation in and elimination from the body) in time. The effective concentration in complex organizations depends on the internal concentration, and the target parameter is assumed to be linear to this. In single-cell microorganisms or very small organisms the internal and external concentration (in the medium) can be considered equal.

Parameter estimation of biology-based methods apply Maximum Likelihood (ML) tests e.g. the least squared deviation method in cases when the scatter is independently normally distributed and has a constant variance. This means that the deviation is independent for the different data sets such as various toxicity end points and the internal concentrations. For different datasets a composite likelihood function is used that contains all parameters for all models and can handle different types of distributions. The profile likelihood function is used to obtain confidence intervals for parameters such as the No-Effect Concentration (NEC). Biology-based methods use NEC as a free parameter.

*The eco-physiological model* differentiates three steps:

- Change in the internal concentration: from the local environment to the concentration in the test organism.
- Change in a physiological target parameter: from an internal concentration to a hazard rate such as assimilation rate, specific maintenance rate, etc.

- Change in an end point: the step from the hazard rate to a change in an end point such as reproduction rate, total number of offspring during an exposure period, etc.

The model applies three concentration ranges from where it is obvious that the occurrence of the no-effect case is very common:

- Effects due to shortage: in this range the chemical may function as a stimulant;
- No-effect range: activities are seemingly independent of the concentration;
- Toxic effects: inhibition arises in this range not only for toxicants but also for substances with a wide range of no effect, e.g. glucose.

The toxico-kinetic model is based on the accumulation flux, which is proportional to the concentration in the external environment and the elimination flux, which is proportional to the internal concentration (inside the organism). This result is a first-order kinetic model. The growth of the organism during the experiment causes deviation from first order kinetics. This deviation can be predicted and taken into consideration.

The **Dynamic Energy Budget model** (DEB) is not only based on the material fluxes which are described by the external and the internal dynamic concentrations (resulting from biaccumulation and elimination), but also includes energy fluxes. The DEB theory unifies the commonalities between organisms as prescribed by the implications of energetics which link different levels of biological organization i.e. cells, organisms and populations (DEB, 2014). The theory presents simple mechanistic rules that describe the uptake and use of energy and nutrients (substrates, food, light) and the consequences for the physiological organization throughout an organism's life cycle, including the relationships of energetics with aging and effects of toxicants. The model of Kooijman (1993, 1996, 2010) introduces the factor of reserve (food is converted to feces and reserve), and the reserve is used for maintenance and synthesis (energy) and allocated to structures known as growth; to maturity as development and to gametes as reproduction. The rate of the reserve use is the catabolic rate. The DEB model describes the full material balance, which can be very useful in biology-based modeling.

The DEB model specifies how changes in one or more target parameters translate into changes in a specified end point. Reproduction rates for example depend on age (the first few offsprings contribute much more to population growth than later offsprings). There is no need to study all ages of the test organism once its DEB parameters are known. The use of the DEB models are detailed in the OECD Guideline (OECD, 2006), a short summary is given here.

- The DEB survival model: the effects on the survival probability of individuals are specified via the hazard rate. The hazard rate (probability per time) is also known as the instantaneous death rate. The hazard rate $h(t)$ relates to the survival probability $q(t)$ as:

  $$h(t) = -q(t)^{-1} d/dt\ q(t).$$

- The DEB body growth model allows for three routes affecting body growth: (i) decrease in the assimilation rate; (ii) increase in the somatic maintenance costs; (iii) increase in the specific costs for growth. In the case of fish growth, as an example, the growth is described by the equation below (applied mainly in fisheries science):

  $$L_t = L_\infty - (L_\infty - L_0)\exp\{-r_b \times t\}, \text{where}$$

o L(t) is the length at time t,
o $L_0$ is the initial length,
o $L_\infty$ is the ultimate length (the asymptotic length at which growth is zero)
o $r_b$ is the von Bertalanffy growth rate.

Effects on growth are determined by an increase of the maintenance costs and by a decrease of assimilation in the DEB model. This equation can be applied if, for example, the fish follow a *von Bertalanffy* (1969) growth curve in the control and the chemical substance follows first order kinetics. Further preconditions are that the toxicokinetic parameters (assimilation, maintenance, cost for growth) should increase linearly with the internal concentration and the concentration of the test compound is constant.

– Reproduction allows indirect routes similar to those for growth (i) decreased assimilation rate; (ii) increased maintenance rate; and (iii) increase in cost for growth. In addition, direct routes are specified: (iv) an increase in the costs per offspring (an effect on the transformation from reserves of the mother to that of the embryo) and (v) death of early embryos which are not counted.

– Population growth is a complex process, but in the case of algal or duckweed growth it can be simplified (under certain assumptions) and considered as growth and division (one cell into two) in the context of the DEB model.

### 9.6.1 Parameters

– NEC, the no-effect concentration (internal concentration of the chemical substance), a 0% effect level at very long exposure times;
– Killing rate: effect on survival or tolerance concentration (external concentration of the chemical substance);
– Elimination rate (a dynamic parameter);
– The hazard rate = hazard rate of the control + killing rate × (internal concentration/BCF – NEC);
– Stress value = 1/tolerance value × (internal concentration/BCF – NEC);
– Eco-physiological parameters: are test-species-specific values and should be measured in advance for the DEB model. These are for growth and reproduction: $r_b$ = von Bertalanffy growth rate; $L_0$ = initial body length; scaled length of puberty and energy investment ratio. For population growth: inoculum size and specific population growth rate.

The fitted data should be checked by using goodness-of-fit methods. The DEBtox software includes these functions. Experimental design and the applied statistics should always be harmonized. DEBtox has been designed for the analysis of the results of specified OECD standard test methods: so, when using these ones, no harmonization is necessary from the users' point of view.

The IT tools necessary for the evaluation of the above-mentioned biology-based tests are the DEBtox and DEBtool packages. The latest versions can be downloaded free of charge from the electronic DEB laboratory (2014). The Hill parameters can be estimated using IBM SPSS nonlinear regression software and others (see next section).

## 9.7  IT tools for statistical evaluation

The development of information technology made mathematical calculation easy and eliminated the barrier of manual work and time requirement. Several pieces of software are available as mathematical tools for statistical evaluation using hypothesis testing or regression analysis. On the other hand, the wide selection of ready-made statistical methods and software make the correct decision difficult. Except for some routine cases advice from professionals is necessary when choosing the most appropriate statistics. On-line support is also available, e.g. the website of UCLA (2014). These support materials should not be considered as strict rules, but as general guidelines since data can be analyzed in multiple ways and most of them may yield a correct answer. These web-based decision support tools are categorized according to the nature of the independent variables and the number and nature (nominal or categorical, ordinal, interval) of the dependent variables. The type of data distribution also determines the choice of statistical methods. The same statistical methods are available from different sources in the form of free or paid software. Software packages are offered by the big providers such as BMDP, SAS, SPSS, R, STATA, OriginLab, etc.

For education and information on statistics see Statistics (2014), The Institute for Statistics Education. Statistics software information is also available from Statistics (2014). A broad overview can be found on the website of StatPages (2014), providing information on:

– General Packages: support a wide variety of statistical analyses;
– Subset Packages: deal with a specific area of analysis or a limited set of tests;
– Curve Fitting and Modeling: handling complex, nonlinear models and systems;
– Biostatistics and Epidemiology: especially useful in life sciences;
– Surveys, Testing and Measurement: especially useful in business and social sciences;
– Excel Spreadsheets and Add-ins: recent version of Excel is needed (see later in details);
– Programming Languages and Subroutine Libraries;
– Scripts and Macros: for scriptable packages such as SAS, SPSS, R, etc.;
– Other collections of links to free software.

GNU (2014) General Public License (GPL) applies to most of the Free Software Foundation's (2014) software and to any other program whose authors are committed to using it. GNU gives its users complete freedom. Free Software Foundation is a nonprofit organization with a world-wide mission to promote computer user freedom and to defend the rights of all free software users.

Some well-known statistics tools and web-pages also useful in toxicology are gathered and listed below in alphabetic order:

*Benchmark Dose* (BMD 2014) software, developed by US EPA is the IT support of the benchmark dose approach for deriving a 'Point of Departure' for risk assessment. It is a scientifically more advanced method compared to the No Observed Adverse Effect Level (NOAEL). The BMD method pre-defines a specific effect, referred to as the Benchmark Response (BMR) and estimates the dose (BMD) associated with the specified effect. The BMD is estimated from the complete dose response dataset by fitting dose response models. Statistical uncertainties in the data are taken into account by the confidence interval around the BMD, the lower limit of which (denoted by BMDL) is the Point of Departure for deriving exposure limits. The software package is suitable for dose-response analysis and deriving a BMDL from dose-response data. The software is available from the Integrated Risk Information System (IRIS, 2014) website.

*BMDP* (2014) is a statistical package developed in 1965 at UCLA. Based on the older BIMED program, developed in 1960 for biomedical applications, it used key-word parameters in the input instead of fixed-format cards, so the letter P was added to the letters BMD, although the name was later defined as being an abbreviation for Biomedical Package. BMDP was originally distributed free of charge. It is now marketed by Statistical Solutions. The software package of BMDP is distributed by Statistical Solutions (2014).

*IBM SPSS* Statistics (2014) is a software package used for statistical analysis. Long produced by SPSS Inc., it was acquired by IBM in 2009. The current versions (2014) are officially named IBM SPSS Statistics which stands for Statistical Package for the Social Sciences (SPSS) reflecting the original market, although the software is now popular in other fields as well, including health sciences and marketing:

- Descriptive statistics: cross tabulation, frequencies, descriptives, descriptive ratio statistics;
- Bivariate statistics: means, *t*-test, ANOVA, correlation (bivariate, partial, distances), non-parametric tests;
- Prediction for numerical outcomes: linear regression;
- Prediction for identifying groups: factor analysis, cluster analysis (two-step, K-means, hierarchical), discriminant;
- Companion products in the same family are used for survey authoring and deployment (IBM SPSS Data Collection), data mining (IBM SPSS Modeler).

JMP (2014) the Statistical Discovery software from SAS is an easy-to-use data analysis and graphics tool. The statistical analysis is linked with interactive graphics, in memory and on the desktop.

Microsoft Excel Add-ins (2014) provides several categories and a free trial version is also available (XLSTAT, 2014). Statistical tools package is part of the latest EXCEL version, but it is not included in the basic functions, but it should be downloaded from *Excel* by using *Options* within the *File* menu, and selecting *Add Ins* and *Analysis ToolPak*. Detailed information can be found on YouTube: How to Get Excel 2010 data analysis tool (Excel data analysis tool, 2014). A statistical functions list (Excel statistical functions, 2014) and detailed description about the statistical analysis tools is also available (Excel statistical analysis tools, 2014).

*MINITAB* (2014) is a general statistical and graphical analysis package. It can do various general analyses including time series. An interactive Assistant helps through every step of the analysis.

*Origin* (2014) is graphing and analysis software that includes regression and curve fitting tools for linear, polynomial, and nonlinear curve fitting along with validation and goodness-of-fit tests.

*PROAST* (2014) is a software package that has been developed by the Dutch National Institute for Public Health and the Environment for the statistical analysis of dose–response or concentration–response data and nonlinear regression. It can be used for (i) dose–response modeling, (ii) deriving a BMD in human risk assessment, and (iii) deriving an effect concentration in ecotoxicological risk assessment. It is suitable for an in depth analysis of a single dataset, but also for a quick (automated) analysis of a whole series of response end points, which may be useful for analyzing complete studies. It allows for comparing dose–responses among various subgroups, e.g. among sexes, study durations or among replicate studies. PROAST not only indicates if the various dose–response relationships differ among the subgroups, but gives information if the difference is in the background response, in sensitivity to the chemical, or in dose-response shape. The present version (38.9) has the possibility to run PROAST in a user-friendly Graphical User Interface (GUI) for standard applications. The GUI was developed in collaboration with the Health and Safety Laboratory (2014) in Buxton, United Kingdom.

R (2014) is a free software environment for statistical computing and graphics. It compiles and runs on a wide variety of UNIX platforms, Windows and MacOS. Free software can be obtained from the R Project for Statistical Computing website. It does basic statistics, resampling, regression, logistic regression, GLM (generalized linear models for logistic regression) and GEE (generalized estimating equations, taking into account correlation between measurements at multiple time points). User-written routines are also available.

SAS (2014) Statistical Analysis System is a software suite developed by SAS Institute for advanced analytics, business intelligence, data management, and predictive analytics. The compartments of SAS are:

– Base SAS – Basic procedures and data management;
– SAS/STAT – Statistical analysis;
– SAS/GRAPH – Graphics and presentation;
– SAS/OR – Operations research;
– SAS/ETS – Econometrics and Time Series Analysis;
– SAS/INSIGHT – Data mining;
– SAS/PH – Clinical trial analysis.

SAS University (2014) is a special edition for students and universities. It needs a virtual machine that is a file or folder that contains an entire computer with operating system and program in software. A virtual machine player is needed to access the SAS virtual machine. Two recommended virtual machine players are Oracle's VirtualBox (runs on Windows, Mac or Linux) and VMWare's VMWare Player (runs on Windows and Linux). Both are free of charge.

Stata 13 (2014) Statistical package designed for researchers of all discipline. Applicable for treatment of effects, multilevel mixed-effects GLM, power and sample size, multilevel SEM (structural equation modeling) with generalized outcomes, forecasting, long strings and BLOBs (binary large objects). A trial copy of Stata contains standard statistics, resampling, time series, regression, logistic regression, GLM and GEE.

StatCrunch (2014) is statistical software available online that allows users to perform complex analyses, share data sets, and generate compelling reports of their data. Interactive graphics help users understand statistical concepts and are available for export to enrich reports with visual representations of data.

*STATGRAPHICS* (2014) is an easy-to-learn, easy-to-use personal computer software package designed for experts and non-experts alike. There is no need to download any software since all calculations are done on a remote server and the statistical analysis is done through the web browser. Figures are handled and images and area calculations are based on pixels. It provides simple and multiple regressions.

*STATISTICA* (2014) is a statistics and analytics software package developed by StatSoft. STATISTICA provides data analysis, data management, statistics, data mining, and data visualization procedures. Its techniques include the widest selection of predictive modeling, clustering, classification, and exploratory techniques in one software platform. It offers a free STATISTICA trial (2014) version.

Several pieces of free statistical software are available online or can be downloaded. Free Software (2014) offers several free IT tools. On the page of StatPages free statistical software with short descriptions and links to the sources are introduced (StatPages free, 2014). In the following, a list of free trials or completely free statistical software is shown:

- *AM* (2014) serves analyzing data from complex samples, especially large-scale assessments, as well as non-assessment survey data. It has sophisticated stats, easy drag & drop interface, and an integrated help system that explains the statistics as well as how to use the system. It can estimate models via marginal maximum likelihood (MML) and automatically provides appropriate standard errors for complex samples via Taylor-series approximation, jackknife & other replication techniques.
- *Dataplot* (2014) is a tool for scientific visualization, statistical analysis and non-linear modeling. It has extensive mathematical and graphical capabilities. It is closely integrated with the Engineering Statistics Handbook (2014) from NIST/SEMATECH.
- *Develve* (2014) a stats package for fast and easy interpretation of experimental data: statistical testing, design of experiments and sample size calculation modes are available. Everything is directly accessible and the results are directly visible with no hidden menus.
- *Epi Info Version 7* (2014) Classic public health and epidemiology application, with free download, developed by Centers for Disease Control and Prevention (CDC, 2014) in Atlanta, Georgia (USA). The program allows for data entry and analysis including t-tests, ANOVA, non-parametric statistics, cross tabulations and stratification with estimates of odds ratios, risk ratios, and risk differences, logistic regression, survival analysis and analysis of complex survey data. See also OpenEpi.
- *ezANOVA* (2014) is a free program for analyzing data. It has been developed for statistics courses. It is not a particularly powerful tool, but it is useful for illustrating the basics of the Analysis of Variance (ANOVA).
- *GNU PSPP* (2014) is a program for statistical analysis of sampled data. It is a free replacement for the proprietary program SPSS, and appears very similar to it with a few exceptions. Its main characteristics are:

  o Choice of terminal or graphical user interface;
  o Choice of text, postscript or html output formats;

- o Inter-operates with Gnumeric, OpenOffice.org and other free software;
- o Easy data import from spreadsheets, text files and database sources;
- o Fast statistical procedures, even on very large data sets;
- o No license fees; no expiration period; no unethical "end user license agreements";
- o Fully indexed user manual;
- o Cross platform;
- o Runs on many different computers and many different operating systems.

– *ICRISTAT* (2014) serves data management and basic statistical analysis of experimental data. It is primarily used for analysis of agricultural field trials. It includes: data management with a spreadsheet, text editor, ANOVA, regression, genotype & environment interaction analysis, quantitative trait analysis, single site analysis, pattern analysis, graphics, utilities for randomization and layout, general factorial EMS (Expected Means Squares) and orthogonal polynomials.

– *MicrOsiris* (2014) statistical and data management package for Windows, derived from the OSIRIS IV package, developed at the University of Michigan. It provides extensive statistics: univariate, scatter plot, cross-tabs, ANOVA/MANOVA, log-linear, correlation/regression MCA (*multiple correspondence analysis*), MNA (*Mean Number of Class Attributes*), binary segmentation, cluster, factor, MINISSA (smallest space analysis), item analysis, survival analysis, internal consistency.

– *MIX* (2014) Meta-analysis with Interactive eXplanations is a statistical add-in for Excel. Recommended for learning meta-analysis.

– *OpenEpi* (2014) provides a number of epidemiologic and statistical tools.

– *OpenStat* (2014) a general stats package for all Windows versions and for Linux systems. Developed by Bill Miller with a very broad range of data manipulation and analysis capabilities and an SPSS-like user interface.

– *PAST* (2014) an easy-to-use data analysis package: common statistical, plotting and modelling functions, curve fitting, significance tests (F, t, permutation t, Chi-squared w. permutation test, Kolmogorov-Smirnov, Mann-Whitney, Shapiro-Wilk, Spearman's Rho and Kendall's Tau tests, correlation, covariance, contingency tables, one-way ANOVA, Kruskal-Wallis test), diversity and similarity indices & profiles, abundance model fitting, multivariate statistics, time series analysis, geometrical analysis, parsimony analysis (cladistics), and biostratigraphy.

– *SalStat-2* (2014) provides tools for data management, statistical calculations such as descriptive summaries, probability functions, chi-square, *t*-tests, 1-way ANOVA, regression, correlation, non-parametric tests, Six-Sigma and graphic system (inherited from matplotlib).

– *SISA* (2014) is a Simple Interactive Statistical Analysis tool for PC (DOS), a collection of individual DOS modules for several statistical calculations including some analyses not readily available elsewhere.

– *SOFA* (2014) Statistics Open For All is a user-friendly, open-source statistics, analysis, and reporting package.

– *Statext* (2014) has basic statistical tests such as rearrange, transpose, tabulate and count data; random sample; basic descriptives; text-plots for dot, box-and-whiskers, stem-and-leaf, histogram, scatterplot; find z-values, confidence interval for means, one- and two-group and paired t-test; one- and two-way

ANOVA; Pearson, Spearman and Kendall correlation; linear regression, Chi-square goodness-of-fit test and independence tests; sign test, Mann-Whitney U and Kruskal-Wallis H tests, probability tables (z, t, Chi-square, F, U); random number generator; Central Limit Theorem, Chi-square distribution.

- *Statist* (2014) is a compact, portable program that provides most basic statistical capabilities: data manipulation (recoding, transforming, selecting), descriptive stats (including histograms, box & whisker plots), correlation & regression, and the common significance tests (chi-square, t-test, etc.).

- *Statistical Software* (2014) by Paul W. Mielke Jr. is a large collection of executable DOS programs (and Fortran source). It includes: matrix occupancy, exact g-sample empirical coverage test, interactions of exact analyses, spectral decomposition analysis, exact mrpp (multi-response permutation procedure) and exact mrbp (analysis of multivariate data for the randomized block design, based on permutation procedures), Fisher's exact test for cross-classification and goodness-of-fit, Fisher's combined p-values (meta analysis), largest part's proportion, Pearson-Zelterman, Greenwood-Moran and Kendall-Sherman goodness-of-fit, runs tests, multivariate Hotelling's test, least-absolute-deviation regression, sequential permutation procedures, LAD (least sum of absolute deviations) regression, principal component analysis, matched pair permutation, contingency tables and Jonkheere-Terpstra.

- *STPLAN* (2014) performs power, sample size, calculations needed to study design. Includes binomial, Poisson, normal and log-normal distributions, survival times and correlation factors.

- *Tanagra* (2014) includes data mining, descriptive statistics (cross-tab, ANOVA, correlation), instance selection (sampling, stratified), regression (multiple linear), factorial analysis (PCA, MCA), clustering (kMeans, self-organizing map=SOM, hierarchical agglomerative clustering=HAC).

- *ViSta* (2014), the Visual Statistics System, features highly dynamic and interactive statistical visualizations.

- *WinIDAMS* (2014) is suitable for numerical information processing and statistical analysis from UNESCO. Provides classical and advanced statistical techniques including interactive construction of multidimensional tables, graphical exploration of data, time series analysis, and a large number of multivariate techniques.

## 10   ENVIRONMENTAL HAZARD AND RISK ASSESSMENT USING TOXICITY DATA

Statistically evaluated toxicity test results can be used for generating environmental hazard and risk values that form the basis of decision making in environmental risk management. Figure 9.13 shows the pathway of the measured toxicity data from acquisition to their use in risk assessment.

### 10.1   Extrapolation

Available or measured physico-chemical and toxicity information/data (or those of other adverse effects) should be extrapolated to estimate the extent and calculate

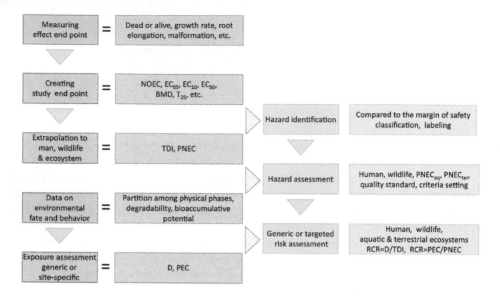

*Figure 9.13* The pathway of measured effect-data from acquisition to application in risk assessment.

the probability of future adverse environmental impacts. The *extrapolation* may include:

– from chemical properties to environmental fate and behavior of a contaminant, e.g.,

   o from $K_{ow}$ to partition among physical phases;
   o from $K_{ow}$ to degradability;
   o from $K_{ow}$ to the bioaccumulative potential,

– from chemical properties to biological effects:

   o from $K_{ow}$ to toxicity;
   o from the chemistry of mixtures to the effects of mixtures;

– from one response end point to another:

   o from $EC_{20}$ to $EC_{50}$;
   o from acute to chronic toxicity end points;
   o from effect doses to no-effect doses;

– from taxa to taxa:

   o from the tested species to another, e.g. an endangered species;
   o from the tested species to humans;
   o from more tested species to entire ecosystems;
   o from micro- and mesocosms to the field;
   o from a tested field to the real field, etc.;

– from species level to higher organizational levels:

  o from species abundance to food chains or food webs;
  o from short term damage of one or several species to the function of the entire the ecosystem (deterioration or recovery/sensitivity or resistance);
  o from direct effects to indirect effects and
  o from indirect effects to direct effects;

– spatial and temporal scales:

  o from one habitat to another;
  o from one region to another;
  o from an e.g. 10-year-old information to the current one, etc.

*Study end points* are calculated from measured end points taking into account uncertainties by applying a suitable statistics for evaluation (Section 9). Extrapolation steps are needed from the study end points of individual test organisms such as rats or mice in human toxicology or aquatic/terrestrial species in ecotoxicology. The extrapolation method is based on type, quality and quantity of the initial and target data. Most frequently factorial extrapolation is used with assessment factors, also called uncertainty or safety factors, but probabilistic methods based on known or estimated distributions are also widespread.

## 10.2 Hazard assessment

*Hazardousness* of a chemical substance is the result of its potential to pose an adverse effect on living organisms or non-living entities. Hazardousness is an inherent property of chemicals due to their molecular structure. Researchers know as early as in the planning phase, when the molecules do not yet exist, that the material will be hazardous. *Environmental hazard* is more extensive because the hazard is evaluated in an environmental context using information on the fate and behavior of the substance under environmental conditions and its effects occurring via the environment by air inhalation, water drinking, food consumption and dermal contact with environmental compartment. Toxicity data are used for hazard identification and hazard assessment.

### 10.2.1 Hazard identification

*Hazard identification* denotes the chemical substance's potential adverse effects (toxicity, mutagenicity, reprotoxicity) posed to organisms (humans, aquatic and terrestrial species, etc.), as well as the type and nature of the adverse effect. Toxicity is not the only threat which constitutes a hazard: several physical, chemical, biological hazards also play a role and act together. Integrated management of different hazards has not been fully achieved in practice. *Hazard characterization* is the qualitative and, wherever possible, quantitative description of the inherent property of the substance and/or the situation which has the potential to cause adverse effects. *Hazard assessment* includes the quantification of the hazard, a process which is required to determine the size and describe the characteristics of the possible adverse effects of a chemical substance and/or the situation to which an organism, population or the ecosystem may be exposed. These three steps are also parts of risk assessment, together with exposure assessment.

Several environmental management and regulatory tools are based on hazard identification and/or assessment, thus differing from risk assessment in its lack of exposure assessment. Regulation of chemical substances is largely based on their toxic hazard along with their environmental fate and behavior (degradation, transformation, accumulation, etc.). Most of the regulations integrate lists of chemicals of high concern, based only on the identified hazard. The *Globally Harmonized System of Classification and Labeling of Chemicals* (GHS, 2011) and its European counterpart, the CLH, under the regulation of *Classification, Labeling and Packaging of Substances and Mixtures* (CLP, 2008) are based on hazard identification only. Quality standards and environmental quality criteria (EQC) are created based on the quantified hazards of chemicals, e.g. the *European Quality Standards in the Field of Water Policy* (EQS Water, 2008).

Priority lists of chemicals or the lists of chemicals to be monitored are created mainly based on their hazard. Taking the substance's production volume into account is a step toward risk assessment. Tiered assessments apply hazard assessment as a preliminary tool.

### 10.2.2   Hazard quantification

**The quantified hazardous effect** may relate to individual organisms as a toxicity end point (NOEC, $EC_{50}$, etc.) or as an estimate from toxicity results for complete ecosystems. Individual *no-effect* concentrations can be used as a basis for the protection of the individual species. Several extrapolation methods are known to estimate the ecosystem's protective level from the measured toxicities posed to individual species. Two of the methods are explained here in detail: the factorial extrapolation and the species sensitivity distribution methods.

*Extrapolation from test results to humans*

**Extrapolating from animal test end points to humans**: interspecies and intraspecies as well as a duration factor can be applied as default, but information from literature or factual databases can also be used instead of default assessment factors. The frequency distributions of the assessment factors found can/should be evaluated by statistical methods.

– An *allometric scaling factor* (AS) is necessary because the greater size and life span of humans relative to experimental animals have a significant impact on the amount of chemical intake needed to provoke the same level of response: the human dose is 7-fold smaller than a mouse and 4-fold smaller than a rat dose. These AS factors are applied in toxicological studies often as a default.
– Another type of assessment factor is the default *duration extrapolation factor*: (i) factor 2 for sub-chronic to chronic; (ii) factor 6 for subacute to chronic; and (iii) factor 3 for subacute to sub-chronic.
– *Interspecies differences* in the human population show great variability so that a default factor of 10 is recommended for the protection of the most sensitive

Table 9.13 Assessment factors to derive a predicted concentration not affecting the ecosystem (PNEC).

| Aquatic data set | Terrestrial data set | Extrapolation factor |
|---|---|---|
| At least one short-term $L(E)C_{50}$ from each of three trophic levels of the base-set of fish, Daphnia and algae | $L(E)C_{50}$ from short-term toxicity test(s) on plants, earthworms, or microorganisms | 1000–100 |
| One long-term NOEC, either fish or Daphnia | NOEC for one long-term toxicity test (e.g. plant) | 100 |
| Two long-term NOECs from species representing two trophic levels such as fish and/or Daphnia and/or algae | NOEC for additional long-term toxicity tests of two trophic levels | 50 |
| Long-term NOECs from at least three species (normally fish, Daphnia and algae) representing three trophic levels | NOEC for additional long-term toxicity tests for three species of three trophic levels | 10 |
| Species sensitivity distribution (SSD method) | Species sensitivity distribution (SSD method) | 5–1 |
| Field data or model ecosystems | Field data/data of model ecosystems | case-by-case basis |

individuals. The assessor can depart from that when he has concrete information on deviations.

### Extrapolation from test results to the ecosystem

*In the extrapolation from single species tests to the ecosystem,* the factorial extrapolation method (FAME) has priority, and factors between 1000 and 1 are used depending on study type and duration. The study end point is generally the $EC_{50}$ or the *no-effect* concentration and, in a prognosis, the *predicted no-effect concentration.* A similar system is recommended for aquatic and terrestrial ecosystems (see Table 9.13).

Another example for factorial extrapolation is the secondary poisoning of birds or mammals. The recommended extrapolation factor by CSTEE (2000) decreases with the duration of the test as follows:

- $LC_{50}$ for birds, factor: 1000
- NOEC (28-day repeated dose test), factor: 100
- NOEC (90-day repeated dose test), factor: 30
- NOEC for chronic studies, factor: 10.

*Toxicity study end points* such as NOEC, $EC_{50}$, etc. reflect the responses of the test organism on the effect – generally a serial concentration or dose of a toxicant. Study end point is the result of statistical evaluation of hypothesis testing, regression analysis or other biology-based statistical methods (Section 9). Large enough uncertainties in the study results are superposed by the uncertainties of the extrapolation steps along the pathway of hazard or risk assessment, which ultimately produces the information that is used for environmental decision making. Risk managers should be aware of the scale of uncertainty; otherwise they may make the wrong decision.

Incorporation of the receptor-specific no-effect value into a hazard assessment procedure may yield environmental quality criteria (EQCs): $PNEC_{aquatic}$ is the basis for water quality criteria and $PNEC_{terrestrial}$ should be used for the generation of soil EQCs. These criteria form the basis of legally binding effect-based or risk-based environmental quality standards (EQS).

### PNEC derivation from acute and chronic toxicity data

Uniformly accepted methodologies are used for the derivation of predicted no-effect concentration (PNEC) on the respective ecosystem from effect data measured by single species tests. The most widespread tool is the factorial extrapolation method, the application of assessment or safety factors to compensate uncertainties (Garber *et al.*, 2010; EQS Water, 2008; ECHA, 2008). Assessment factors recommended by the EU-TGD (2003) for the derivation of freshwater $PNEC_{aquatic}$ and terrestrial ecosystems' $PNEC_{soil}$ values are shown in Table 9.10. If only water toxicity data are available, $PNEC_{soil}$ and $PNEC_{sediment}$ can be calculated from $PNEC_{water}$ by the equilibrium partitioning method, assuming that adverse effects are distributed between solid and water according to the equilibrium-partitioning coefficient $K_{soil-water}$ and $K_{susp-water}$.

### PNEC derivation from species sensitivity distribution

Instead of factorial extrapolation using fixed assessment factors, the *species sensitivity distribution* (SSD) can also be applied for extrapolation from single species to the entire ecosystems.

Sensitivity distribution can be defined as the distribution of sensitivity of different taxa of the aquatic or terrestrial ecosystem to the same contaminant. There are species with average or low sensitivity among the ecosystem members. Species highly resistant due to adaptation to contaminants should also be counted. The sensitivity of the selected test organisms should be fit to the aim of testing if only one or a few species are used. For example, an early indicator should be selected from the very sensitive ones. A low-sensitivity species can be used as a screening tool for hot spot identification. An average sensitivity may be used as a representative of the entire ecosystem. One organism can never represent the ecosystem (properly) so that species sensitivity distribution (SSD) has become the priority tool for this kind of representation. The level of representativeness of an ecosystem has been designated as 95%, i.e. 95% of the species should be represented. Not only species but higher taxonomic groups can also be targeted by sensitivity distribution if appropriate.

The SSD method is based on collecting ecotoxicity results (from literature and other data sources) of different test organisms from different taxa and fitting continuous distribution to the orderly structured toxicity data. One can read the concentration where the theoretical percentage of species exposed above their NOEC or $EC_{50}$ is less than e.g. five percent from the curve of *fraction-affected vs. effective concentration* (arbitrary end points such as NOEC or $EC_{50}$).

SSD may cover acute and chronic effects and the relevant test end points. The SSD curve will be plotted from $EC_x$ or $LC_x$ data for acute effects, the chronic one is obtained from NOEC, LOEC or MATC results. Both the measured and QSAR data can be applied for SSD assessment.
*Creation of a SSD curve:*

- As a first step, toxicity data are collected and selected.
- The minimum requirement for regulatory purposes is 10 toxicity data from 8 taxonomic groups;
- The toxicity data ($E_{Cx}$ or NOEC, LOEC, etc.) are log-transformed;
- $HC_x$ is estimated by ranking the species sensitivities and choosing the appropriate concentration. This non-parametric approach is less preferred than a parametric approach which applies a distribution fitted to the cumulative species sensitivities. The word "fit" means estimating the parameters of the distribution using statistical tools.
- The log-transformed toxicity data allow a cumulative distribution to be derived. Sensitivities are ranked from the most sensitive to the least sensitive.
- Several distribution functions have been recommended by different authors (OECD, 2005), e.g. log–triangular function, log–logistic and log–normal function. Aldenberg and Slob (1993) refined the way to estimate the uncertainty of the 95th percentile by introducing confidence levels, which was further refined by Aldenberg and Jaworska (2000). Placing confidence limits (cl) around the estimated $HC_5$ can be noted as follows: $HC_5^{cl}$, where "cl" denotes the confidence with which the estimated $HC_5$ is not higher than the true $HC_5$.

*Input data requirement of SSD for regulatory purposes:*

The evaluation should include as much $EC_{50}$ or NOECs as available from chronic/long-term studies, preferably on full life cycle or multi-generation studies. The minimum species requirements, for example according to the guidance for chemical safety assessment in the EU, are as follows (ECHA, 2008):

- Fish (salmonids, minnows, bluegill sunfish, channel catfish, etc.);
- A second family in the phylum Chordata (fish, amphibian, etc.);
- A crustacean (cladoceran, copepod, ostracod, isopod, amphipod, crayfish etc.);
- An insect (mayfly, dragonfly, damselfly, stonefly, caddisfly, mosquito, midge, etc.);
- A family in a phylum other than Arthropoda or Chordata (Rotifera, Annelida, Mollusca, etc.);
- A family in any order of insects or any phylum not already represented;
- Algae;
- Higher plants.

Minimal sample size (number of data), considering the statistical extrapolation methods necessary to apply when a PNEC is derived, is at least 10 different species' NOEC values (or more, preferably 15) covering at least 8 taxonomic groups (ECHA, 2008). Minimal requirements can also be specified otherwise; principally they should be based on statistical calculations. According to Newman *et al.* (2000) approximate optimal sample sizes based on statistical evaluation for $HC_5$ estimation ranged from 15 to 55 with a median of 30 species-sensitivity values. Similar sample sizes were needed for $HC_{10}$ and $HC_{20}$ estimation: estimates ranged from 10 to 75. No difference was apparent in ranges for $EC_{50}$–$LC_{50}$ or NOEC data. The study says that the statistically based sample sizes are much higher than those recommended as acceptable for regulatory purposes (i.e. four to eight species in different regulations).

*Derivation of PNEC from the SSD curve:*

–   The $HC_5$ or any other concentration belonging to arbitrary percentiles can be read from the SSD curve. The $HC_5$ concentration is the most frequently used hazard concentration for regulatory purposes.
–   $PNEC_{aquatic}$ can be created from $HC_5$ by dividing it by the assessment factor (AF): $PNEC_{aquatic} = HC_5/AF$
–   AF depends on the overall quality of the database (collected ecotoxicological data for SSD) and the diversity of the taxonomic groups covered by the database (organisms belonging to a minimum of 8 taxonomic groups). The recommended AF is generally 5, which can be lowered when results from much greater number of taxa and species are available than required as the minimum. An AF value less than 2 is only accepted in extraordinary cases. An AF of 2 was found most adequate, for example, for nickel (TGD EQS, 2011; Nickel EQS, 2011; Nickel RAR, 2008) because:

    o   The large size of the acute and long-term aquatic database covers all sensitive life stages.
    o   Acute aquatic toxicity results came from 65 freshwater and 21 marine species (together 86) of 12 taxonomic groups; exceeding the minimal requirements of 8 species from 8 taxonomic groups when employing the SSD approach.
    o   Chronic toxicity results were obtained from 31 freshwater and 15 marine species of 9 and 6 taxonomic groups, respectively.
    o   The representativeness of the taxonomic groups was based on 12 vertebrate, 7 invertebrate, one algal and one higher aquatic plant species.

Figure 9.14 shows the example of the graph taken from the nickel EQS dossier (Nickel EQS, 2011). Acute data were analyzed using ETX 2.0 (2014) from RIVM and Burrlioz (2014) software from CSIRO for deriving SSDs. The $HC_5$ using the ETX software was 0.065 mg/L (90% confidence interval = 0.0282–0.1289 mg/L). The result of Burrlioz of 0.069 mg/L is very close to ETX. An assessment factor of 2 yields 0.067 mg/L: $2 = 0.0335$ mg/L $\approx 0.034$ mg/L for the maximum allowable concentration (MAC) in the marine and freshwater ecosystem. The proposed MAC of 0.034 mg/L was below individual $EC_{50}$ values for the majority of tested species and below the most sensitive fish, invertebrate and plant species. Chronic marine toxicity input data and an AF of 2 provided a $PNEC_{marine}$ value of 8.6 µg Ni/L. It is used as an AA-EQS (annual average environmental quality criterion). Proposed AA-$EQS_{bioavailable}$ for freshwater ecosystem is 2 µg/L.

Creation of PNEC, MAC, or other EQC is based on statistical evaluation. Several statistical tools are recommended and compared by scientists and authorities. For example a CCME (2006) project evaluated several potential statistical models for freshwater SSD

(i)   for distribution: normal, logistic, lognormal, Burr-type, Weibull, and extreme value;

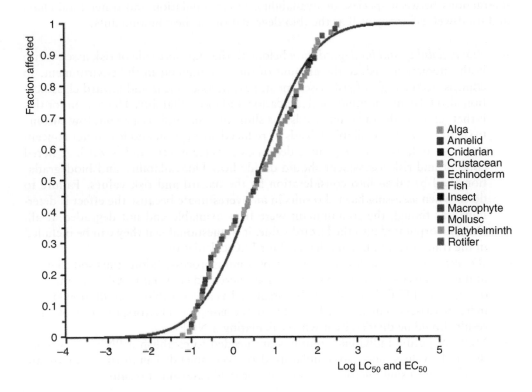

*Figure 9.14* SSD graph of nickel from 86 freshwater and marine toxicity data, taken over from the EQS dossier (Nickel EQS, 2011).

(ii)  for goodness of fit: Anderson-Darling, chi-square, Kolmogorov-Smirnov, Shapiro-Wilk's tests and several graphical techniques and
(iii)  for choosing the adequate statistical tool and software.

It is becoming more and more widespread to estimate $HC_5$ values for environmental risk assessment and regulatory applications of sensitivity distributions. Decision makers hope that this will refine information on the environment and that a less conservative risk value, and consequently less expensive risk reduction will be possible compared to $EC_{50}$ or the highly uncertain NOEC.

The end point of hazard assessment is DNEL or ADI in human toxicology and PNEC for the ecosystem. These predicted/derived no-effect values can be applied to:

– Hazard identification and assessment;
– Risk assessment by comparing DNEL/PNEC to the environmental concentration;
– Establishing environmental quality criteria (screening values, thresholds).

*Generation of measures for adverse effects from other than toxicity values* should also be included in an integrated risk assessment and decision-making procedure:

interactions between species, bioavailability, bioaccumulation and other food-chain and food-web effects as well as the (bio)degradation of the contaminants.

– *Bioavailability and biodegradation* belong to the exposure side of risk assessment: both interactions reduce the amount of the contaminant in the environment. In contrast to the logic of risk assessment, hazard assessment and hazard classification also take into account biodegradation and bioavailability. The reason for this is that ready biodegradability leads to a short life time and, as a result, low chronic toxicities. Low bioavailability leads to reduced adsorption and low inner concentrations, and, as a consequence, decreases adverse effects and hazard. A tiered concept and risk assessment should decide how bioavailability and biodegradation shall be taken into consideration by the hazard and risk values. Failure to do so when assessing hazard results in an overestimate because the effect is determined as though the contaminant were fully available and not degraded at all. Once incorporated into the hazard value, it is questionable if they can be included in the transport and fate model used for PEC calculation.
– *Bioaccumulation* and other secondary poisoning processes along the food chain or in the food web could be considered as an increased environmental concentration on the side of PEC, but it should be integrated as a factor (bioaccumulation factor) in hazard assessment and used similarly to the assessment factors: the toxicological result should be divided by it when calculating a NOEC.
– Method-specific individual solutions are needed to establish a PNEC or similar threshold values since no standardized or recommended uniform methods are available for using micro- or mesocosms or field assessment results.

## 10.3  Validation of toxicity tests

In addition to data evaluation using statistical methods, evaluation and validation of the test methods are equally important for decision making. Risk assessments should be based on the most reliable information available. Validation of the applied test methods is a science-based evaluation at the interface between test method development, application and acceptance for risk management or for regulatory purposes.

The statistics of toxicity data section demonstrated the number of uncertainties in effect and hazard assessment. Use of the most reliable and scientifically sound information on the effects of the chemical substances can reduce the extent of these uncertainties.

Standardized test guidelines primarily from US EPA, EU REACH and OECD, as well as quality assurance, quality control and good laboratory practice (GLP) are necessary to ensure safety of chemical substances and products. A standardized procedure for data evaluation will guarantee consistency and transparency of risk assessment which are key factors in public acceptance. The risk assessor needs a validation method to provide data of sufficient quality to underlie the hazard and risk assessments (Küster *et al.*, 2009; McCarty *et al.*, 2012).

Several organizations have developed validation tools such as ICCVAM and US EPA. The Klimish tool should also be mentioned as one of the most popular toxicological evaluation tools.

The ICCVAM criteria for test validation, including alternative test methods, are as follows (ICCVAM, 1997, 2003, 2014):

– Clear statement of the proposed use of the test results;
– The relationship of the test method's end point(s) to the biologic effect of interest;
– A detailed protocol of the test method;
– The extent of within-test variability, and reproducibility among laboratories;
– Proven performance of reference chemicals and representative test agents;
– Comparative data with competing or to be substituted tests;
– The limitations of the method;
– Reporting data to support the validity of a test method in accordance with Good Laboratory Practices (GLPs);
– Availability of data to support the assessment of the validity of the test method.

The main assessment criteria applied generally by US EPA are formulated as:

– Soundness: reasonable for the intended purpose;
– Applicability and utility;
– Clarity and completeness;
– Uncertainty and variability;
– Evaluation and review.

Klimish *et al.* (1977) elaborated an evaluation tool for assessing the reliability of toxicological studies, mainly for regulatory purposes. The Klimish score based on a systematic approach became a standard and uniform evaluation system in Europe and a software tool has been developed relying on it. It assesses reliability, relevance, and adequacy of data. The Klimish scoring method assigns studies to one of four reliability categories as follows:

– Reliable without restriction: "Studies or data generated according to generally valid and/or internationally accepted testing guidelines (preferably performed according to GLP) or in which the test parameters documented are based on a specific (national) testing guideline, or in which all parameters described are closely related/comparable to a guideline method."
– Reliable with restriction: "Studies or data (mostly not performed according to GLP), in which the test parameters documented do not totally comply with the specific testing guideline, but are sufficient to accept the data or in which investigations are described which cannot be subsumed under a testing guideline, but which are nevertheless well documented and scientifically acceptable."
– Not reliable: "Studies or data, in which there were interferences between the measuring system and the test substance, or in which organisms/test systems were used which are not relevant in relation to the exposure (e.g., unphysiologic pathways of application) or which were carried out or generated according to a method which is not acceptable, the documentation of which is not sufficient for assessment and which is not convincing for an expert judgment."
– Not assignable: "Studies or data which do not give sufficient experimental details and which are only listed in short abstracts or secondary literature (books, reviews, etc.).

All the main criteria are split into numerous very detailed items in the form of simple questions, which can be answered with yes or no when evaluating the study. For example the reliability of ecotoxicological studies is evaluated according to the following aspects:

- Clear description of the test procedure (complete documentation);
- Specification of the test substance (purity, by-products);
- Data on the test species and the number of individuals tested;
- Data on the measured parameters (including definitions);
- Data on exposure period;
- Use of solubilizers/emulgeators;
- Data on concentration control analysis;
- Data on neutralization of samples;
- Data on physical and chemical test conditions (pH value, conductivity, light intensity, temperature, hardness of water);
- Determined effect concentrations (EC/LC/NOEC/ LOEC);
- Data on the statistical evaluations (including the method);
- Data on dosing the test substance (static, semistatic, flow-through system);
- Additional items in the case of chronic studies: information about the investigated period of the life cycle of the test animals and data on feeding of test animals.

Relevance/adequacy of ecotoxicological studies is evaluated based on criteria such as:

- Testing strategy (organism, exposure scenario) should agree with the occurrence and the persistence of the test substance in the environment (target compartment);
- Data obtained from experiments on non-standard organisms (specialist, spread) should be converted into useful ecotoxicological information;
- Physico-chemical properties of the test substance (stability, volatility, solubility, sorbability) should be considered when planning the test.

Klimisch categories 1, 2 or 3 can be determined by the newly developed software tool, called ToxRTool (2014) (Schneider *et al.*, 2009). It is applicable to various types of experimental data, end points and studies to assess reliability of toxicological data. The tool consists of two parts, one to evaluate *in vivo* and another *in vitro* data. The final version, ToxRTool, is publicly available for tests and use (ToxRTool Instruction, 2014; ToxRTool Download, 2014).

## 10.4  Exposure assessment

Generic or site-specific exposure assessment is the next step in the approach toward risk assessment as shown by the scheme in Figure 9.13. The risk, as it is defined, depends on the probability of occurrence and the extent of the damage, which, in turn, depends on the adverse effects of the chemical(s) and the probability of any interaction with the receptors, e.g. their uptake. Probability of uptake is a function of the contaminant's biologically available proportion and the uptake mechanism of the receptor organism. The driving force of the potentially effective proportions is the concentration of the chemical substance in the environment, i.e. the exposure (PEC). Fully perspective risk assessment can calculate exposure relying on the planned volumes of production and use and the transport and fate modeling in a generic environment e.g. for authorization of a future product. The generic model parameters for transport and fate modeling are default values and are different from region to region. Site-specific transport parameters may differ from the generic parameters in both directions. The other terminal of exposure assessment may be a fully assessed contaminated site where both the measured chemical concentrations and the directly measured toxicity data are available. The prediction is restricted in this case to the contaminants' further spread and the time dependence of their transport and fate.

The result of the exposure assessment is a concentration both for a generic and a targeted case:

- It is a conservative estimate (due to model uncertainties and the variables in the environmental characteristics) in the generic situation, which may be applicable for the whole region (e.g. EU), but probably overestimated in every single local case. Generic PEC is useful for regulation but not for handling local problems. A local PEC may deviate from the generic PEC value in both positive and negative directions.
- A targeted, i.e. problem- or site-specific concentration may be in good correlation with the actual acute risk of the specific case. Long-term risks bear high uncertainty and additional physical, chemical and biological risks posed to the site may increase the uncertainties.

## 10.5  Risk assessment

Environmental risk is defined as the product of the probability (probable frequency) and the probable extent of damage due to adverse effects on humans and/or ecosystems. It is the occurrence of an uncertain event with a specific probability in general, which is determined by the extent of damage and the probability of the occurrence. Both can be quantified with large uncertainty, e.g. the scale and cost of ecosystem damage. This probability is expressed as the ratio of the predicted environmental concentration and the predicted highest no-effect concentration (on humans or the ecosystem), often called RCR or risk characterization ratio. This ratio can characterize the risk of chemicals in the environment by a quantitative value. RCR defines the probable concentration present in the environment and accessible by the users of the environment, and it determines if PEC is small or great, and how many times more than the acceptable (no-effect) concentration. Risk estimates may be greatly uncertain because not only can the value or the damage not be precisely estimated but the temporal and spatial extension is also uncertain.

Environmental risk assessment is split into two main branches, according to its integration into environmental management: (i) generic risk assessment and (ii) targeted, i.e. problem- or site-specific risk assessment. Generic risk assessment studies the hazardous chemical in a generic environment (e.g. Europe) for average receptors. Problem- or site-specific risk assessment looks into the chemicals in a certain environment (e.g. the Po River Delta or a garden close to a mine, etc.) and problems caused to local users of this environment.

Regulatory risk assessment and management mainly applies generic tools while targeted or problem-specific risk management relies on a tiered approach including both generic and specific information.

Both hazard and risk management include the steps of identification, analysis, assessment, control, and risk reduction by avoidance, minimization, or elimination of unacceptable hazard/risks. Environmental risk assessment relies on environmental monitoring, evaluation and interpretation of the acquired data and is governed by environmental law mainly concerning the fields of chemical substances, human health, water and soil, ecosystem/nature, waste and contaminated land. Terminology and definitions in risk assessment have been more or less harmonized in the last 10–15 years. IPCS (2004) collected and organized the relevant terms in a harmonization project in the context of chemicals regulatory management (i.e., notification/authorization, registration, and classification). Their guidelines serve as basis for the definitions in this chapter.

*Risk characterization* is the qualitative and, wherever possible, quantitative determination of the probability of occurrence of known and potential adverse effects of a chemical substance in a given organism, population or the ecosystem under defined exposure conditions. It includes uncertainties, and the TER (toxicity/exposure ratio: toxicity/PEC) and RCR (risk characterization ratio: PEC/PNEC) values are used (see below) for its characterization.

*Risk assessment* is the process intended to calculate/estimate the risk to a given target organism, population or system. It includes the exposure and the hazard, as well as uncertainties. The risk assessment process includes four steps: hazard identification, hazard characterization, exposure assessment and risk characterization.

*Risk evaluation* is a further step toward establishing the relationship between risks and benefits of exposure to a chemical substance or other agents. It is synonymous with risk–benefit evaluation.

*Validation* of the outcomes of the assessment is the most important final step in risk management. Reliability refers to the reproducibility of the outcome; and relevance can be defined as the meaningfulness and usefulness of the assessment for the defined purposes of environmental management.

Harmonized and standardized risk assessment methodologies are prescribed by several guidelines all over the world; in Europe EU-TGD (2003) is the basic Technical Guidance defining and laying down protocols and uniform tools for generic risk assessment of chemicals (industrial chemicals, pesticides and biocides) for regulatory purposes. Several other types of guidance, e.g. on REACH, CLP and biocides legislation are published on the website of ECHA (ECHA Guidance, 2014). US EPA (2014) developed guidance and tools (databases and models), for the assessment of human risks (US EPA – Human Risk, 2014), including developmental, carcinogenic, mutagenic, reproductive, neurotoxic risks; ecological risk (US EPA Eco Risk, 2014), as well as microbiological risks and the risk assessment of mixtures.

*Generic risk assessment* works with the assumption of a generic environment, e.g. a generic Europe, and is mainly used for regulatory purposes. It is a fully prospective activity forecasting a fully theoretical situation, e.g. the generic risk of a not yet notified and produced chemical substance in Europe. The generic risk assessment is based on the adverse effects of pure chemicals ($PNEC_{generic}$) and the results of transport modeling of the produced and used amount of the chemicals in the environment of a certain region. Study of the effects of pure chemicals and dissolved forms gives a well reproducible test situation. The standard study results are extrapolated to the hypothetical generic environment by relying on defaults and averages as well as on general experience. On the other hand, this 'sterile' system and the results of generic RA may greatly differ from reality. The application of uncertainty factors may reflect and handle the differences in space and time, but may also lead to undue overestimates. Simulation studies are advantageous in properly modeling multiple interactions and bioavailability, but have the same shortcomings as generic models: the spatial and temporal variations in the environment cause excessive uncertainty which should be compensated with large safety factors resulting in undue overestimation. An overestimate is accepted by the precautionary approaches (go beyond safety when scientific evidence on the safe conditions is lacking: "better safe than sorry"), but a feedback provided by environmental monitoring is always necessary. An additional uncertainty is that the theoretical environment may be generic, but the test organisms are not and thus the protection level of species or ecosystems remains a source of uncertainty.

*Generic risk values* are mainly used for decision making in risk management of chemicals, pesticides, biocides, waste, air, water and soil. The relevant national and European regulations such as the REACH regulation, pesticide and biocide directives, water framework and waste directives apply this tool for notification, authorizations, restrictions and for enacting generic risk reduction measures. Generic risk assessment can be placed between hazard assessment and targeted risk assessment. From many points of view its result is closer to the hazard than to the real risk values: actually the hazard values are evaluated in comparison with the same generic environment, so the generic risk values are hazards on the same denominator. The chapter on bioavailability (Chapter 7) introduces several examples where highly toxic substances have the same hazard, but taking the different environments into consideration, their risk will be small due to immobile and unavailable chemical forms or rapid biodegradation but may be significantly large due to complete availability and harmful transformations. Adaptation of the ecosystem, resistant species, and modified food webs may also contribute to a small risk situation even in the presence of significantly hazardous chemicals. In a workplace situation the risk of hazardous compounds can be reduced to a very small value by applying protective equipment.

*Targeted, i.e. problem- and site-specific risk* values are locally relevant and make decision making and risk management of contaminated sites, waste disposal, safe workplaces and other targeted tasks highly efficient within a certain time interval. Targeted risk assessment incorporates the past, the present and future. A retrospective assessment can be based on historical information covering the past up to the present, and a prospective assessment serves to predict the future situation. In this respect, risk assessment of a contaminated site will start with the collection of data on former and ongoing land uses, activities and all kinds of historical information on the chemicals used, measured concentrations and damage, and the adverse effects observed on human health and the ecosystem. Collection and evaluation of historical information will be followed by the acquisition of necessary new information and the evaluation of measured data on chemicals and adverse effects. The last step is the assessment of the present and the prediction of the future state of the environment, assuming that no risk reduction measures have been taken. Long-term data enable to identify the trends and increase the validity of the risk assessment. This topic is discussed in Volume 3 of the book series.

*Biology- and ecology-based* risk assessment/monitoring observes pharmacokinetic characteristics and epidemiological results for the human population and key species (individuals and populations), food chains, and species diversity for the ecosystem to find the point where changes are unacceptable. These assessment types are applied for cases when the adverse impact is already known. The time series of assessments (monitoring data) make the trends visible and enable a good prognosis. Biology-based assessment covers cumulative effects, so it is applicable to mixtures of chemicals and the combinations of chemical impact with other, e.g. meteorological or microbiological impacts. It may integrate the biomarkers approach, and early indication of potential damage. Using biomarkers in risk assessment means that the forecast is based on a nature-like biological and/or pharmacokinetic model and not on mathematical and chemical models only. The biology-based ecological risk assessment can also be tiered, applying methods from simple DTA to complete ecosystem assessments.

The comparison between measured contaminant concentrations and specific assessment criteria is the simplest way practiced in site-specific risk assessment that can decide whether a significant risk to human or ecosystem health (may) exist. The assessment criteria may be generic (laid down in national or regional regulations) or site-specific. A large number of studies and guidelines have been published in the last 20 years in connection with contaminated land management and remediation programs. These methodologies recommend a tiered assessment approach based on a conceptual risk model that integrates transport and exposure models, following the pathway of contaminants from the source to the receptors. The tiers of the assessment are as follows:

- First tier: a preliminary, largely desk-based assessment step to decide whether or not there is a potential exposure;
- Second tier: if the potential exposure cannot be excluded, a generic-type risk assessment is undertaken, which means a site-specific exposure assessment and comparison to generic no-effect criteria;
- Third tier: a refined risk assessment is executed using site-specific assessment criteria, i.e. site-specific target values;
- Fourth tier: site-specific risk management which includes a risk–benefit assessment and a comprehensive risk communication.

*Quantification of the risk*

There are several terms to be defined and quantified concerning the risk posed by chemicals to humans and the environment.

The risk posed to individual organisms can be characterized by the ratio of $TER = $ Toxicity/Exposure. The ratio of the effect on a living organism to the environmental exposure can be calculated as PEC using any of the toxicological end points ($LC_{50}$, $LD_{50}$ or NOEC/NOEL). TER quantifies the risk in terms of possibility of occurrence of an acute or toxic adverse effect ($EC_{50}$ or NOEC) posed to a certain organism.

The risk posed by a contaminant to the aquatic or terrestrial ecosystem (as a whole) can be expressed by a simplified risk assessment as a risk characterization ratio, abbreviated as **RCR**, which is the ratio of the predicted environmental exposure to the receptor-specific predicted effect or no-effect value (PEC/PNEC). Exposure is an estimate based on transport and fate modeling. The *no-effect* value is an estimate based on single species toxicity tests of the chemical substance or environmental samples, as well as results acquired from microcosms and field assessments. With the exception of the latter, all other estimates are expressed in concentration (or dose) and can be integrated into the RCR based risk assessment approach. RCR is widely used for screening of environmental chemicals and contaminated sites. The RCR approach is a strongly conservative one, which works with worst-case assumptions for both the exposure and the effects. It is applied by regulations (e.g. the European REACH) and regional risk assessment tools where generic environmental parameters are applied. Mathematical (QSAR, generic transport models), chemical (environmental fate and concentrations of chemicals, no-effect concentrations) and biological models (adverse effects) are applied in a problem-specific combination. The problem-specific method gradually refined generally reduces the conservatism of the assessment.

The human pendant of RCR is also called HQ, human hazard quotient designed by US EPA for the characterization of human risks posed by chemicals as the ratio between the ingested dose (D) calculated from measured or predicted environmental concentration and a human oral reference dose (RfD), both measured in mg/kg/day. The safe level, called target hazard quotient (THQ), is 1, a higher value indicates health risk. RfD is generated from NOEL, NOAEL or BMD (benchmark dose), e.g. $BMD_{10}$ or $BMDL_{10}$ of animal tests divided by the uncertainty factors. Instead of RfD, European legislation prefers DNEL. The derived no-effect level DNEL includes a NOAEL-specific uncertainty while TDI (tolerable daily intake) is calculated from the same animal test results by constant assessment factors. Instead of TDI, the term ADI (acceptable daily intake) is also used having the same meaning.

*Definitions*

- BMD, benchmark dose, is defined as the dose of a substance that is expected to result in a pre-specified level of effect, for example $BMD_{10}$ means the dose causing an inhibition rate of 10%.
- BMDL is the lower 95% confidence limit of BMD (the result of statistical evaluation).
- DNEL, the derived no-effect level defined by REACH (ECHA, 2008) as a human health-based limit value for threshold substances. The NOAEL dose descriptor may be used as a reference point. $DNEL = (NOAEL$ or $BMD)/(AF_1 \times AF_2 \times \cdots \times AF_n)$.
- RfD in human risk characterization is the daily oral exposure to the human population that is likely to be without an appreciable risk of deleterious effects during a lifetime. It is used by US EPA.
- TDI, ADI: tolerable/acceptable daily intake of a chemical substance, an amount that can be ingested (orally) on a daily basis over a lifetime without any appreciable health risk. Acceptable daily intake (ADI) was initially used for food additives, whereas the term TDI is preferred for contaminants. Their values are based on

animal experiments, and they are calculated as NOAEL divided by a multiple assessment factor of 100.

Human health risk professionals fight with the problem of non-threshold chemicals and their effects when assessing genotoxicity, carcinogenicity and endocrine disruption. The relevant board in the UK (IGHRC, 2003) and US EPA (2002) have recommended taking a common approach to risk assessment for both threshold and non-threshold types of substances (IOM, 2012), but other professionals still manage threshold and non-threshold chemicals differently. Many genotoxic agents and genotoxic carcinogens have theoretically no thresholds because there is a linear relationship between the dose and the number of DNA damage and even one molecule can trigger gene mutation. If the molecular mechanism is not a direct one, as in the case of gene mutations, a threshold should exist even theoretically because there is a minimum of damage to the mediating molecule (enzyme, hormone, etc.), which triggers the effect in an organism.

There are two semiquantitative methods available for the estimation of a no-effect level (US EPA, 1995; ECETOC, 2002; US EPA, 2005; EFSA, 2005; SCHER, 2009; ECHA, 2010; EFSA, 2012) for non-threshold genotoxic and carcinogenic toxicants:

-   The "linear dose–response" approach is based on a linear extrapolation from high to low doses, for example a straight line is drawn from the point of departure (typically the $BMDL_{10}$ or $T_{25}$) to the origin (zero dose with a zero response), and the slope of this line is used for estimation of cancer risk ($CR = exposure * slope\ factor$). The generated DMEL is compared to an acceptable cancer risk which is between $10^{-5}$ and $10^{-6}$ in different legislations. It is considered as the upper limit of risk;
-   The large extrapolation factor, also called margin of exposure (MOE) approach divides the estimated human exposure by the value of the reference point, usually the $BMDL_{10}$, but T25 can also be used. EFSA considers a MOE of 10,000 or higher, based on $BDML_{10}$, as a value of low concern.

### Definitions
-   DMEL is the derived minimal effect level used for non-threshold substances (ECHA, 2008). Both the $T_{25}$ and $BMDL_{10}$ dose descriptors may be used as reference points.

    o   DMEL at $10^{-5}$ (or $10^{-6}$) risk = $BMD_{10}/AF * 10,000$ for workers
    o   $BMD_{10}/AF * 100,000$ for general population
    o   $T_{25}/AF * 25,000$ for workers
    o   $T_{25}/AF * 250,000$ for general population.

-   MOE is the margin of exposure. MOE approach is useful for assessing exposures of genotoxic carcinogens and mutagens when carcinogenicity data are lacking. Both $T_{25}$ and $BMDL_{10}$ may be used as a reference point. $MOE = BMDL_{10}/human\ exposure$. The larger the MOE, the smaller the risk of exposure.
-   $T_{25}$ is a carcinogenicity potency estimate that is defined as the chronic dose rate which will cause tumors in 25% of the animals at a specific tissue site, after correction for spontaneous incidence, within the standard lifetime of that species.

– $TD_{50}$ is the standardized measure of carcinogenic potency, the daily dose rate in mg/kg body weight/day to induce tumors in half of the test animals that would have remained tumor-free at zero dose. Whenever there is more than one positive experiment in a species, the reported $TD_{50}$ value is a harmonic mean calculated using the $TD_{50}$ value from the most potent target site in each positive experiment.

The risk characterization ratio alone is not enough for decision making and efficient risk management. The problem of mixtures and the impact of the background concentrations should also be solved. Crommentuijn *et al.* (1997, 2001) suggested a Maximum Permissible Concentration (MPC) based on the addition of a background concentration (Cb) to a derived Maximum Permissible Addition (MPA). This way MPC = Cb + MPA. Long-term trends and wider scopes, as well as life cycles of connected chemicals, elements and energies should also be considered. A widespread database would be necessary to compare alternatives and substitute options of chemicals and technologies for the preparation of decisions. In addition to environmental and human health risks, the benefits of chemicals applications should also be considered and quantitatively assessed.

## 10.6 Summary comments on risk assessment and risk management based on toxicity data

The three risk assessment concepts – generic, targeted and biology-based – can be considered and applied as consecutive steps of a tiered methodology, and a lower-ranked step can be validated by using the higher-ranked one. PNEC can be refined or validated by a probabilistic no-effect value, and field assessment of actual impacts can validate risk modeling. Validation or other kind of comparison between the results of the different tiers needs careful analysis to understand discrepancies. For example one can include extremely sensitive species for SSD (instead of treating them as outliers). Another example is where one cannot distinguish between the environmental impacts of chemical contaminants and other simultaneous adverse effects, e.g., a virus infection, or micronutrient depletion.

The extension of risk in time and space is essential information in risk management: the assessment and the measure taken should cover the extension of the risk. Typically several different scopes should be managed at the same time so that scoping (determination of the scope or scopes) should always be the first step of the assessment and management activities.

Tiered risk assessment is fast and efficient; mathematical models can be widely applied, which reduces assessment costs by excluding negative cases at early tiers. A risk overestimate may lead to an unreasonable increase in risk reduction costs. Tiered assessment with a refined and increasingly precise assessment can minimize overestimation. Tiering can be applied to every risk-based management task, including planning the monitoring, generic and targeted risk characterization and assessment as well as the creation of environmental quality criteria and risk reduction target values.

The management steps in Figure 9.15 demonstrate the order of activities and the supportive relationship of environmental law (regulation) and monitoring to the management and the connected decision making. Hazard identification, quantification,

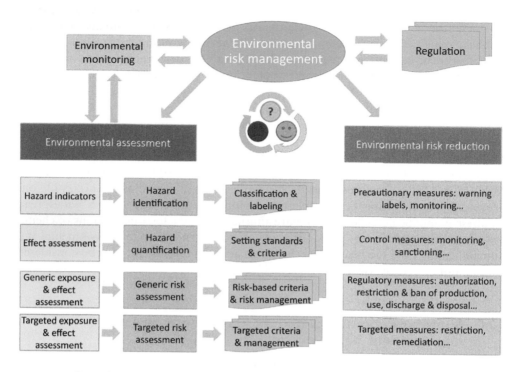

*Figure 9.15* Environmental risk management of environmental contaminants.

and generic and targeted risk assessment may be the consecutive steps of environmental risk assessment, but the results of the steps have their own usability in fulfilling legal requirements or special targets of environmental management. The reciprocal relationships between environmental risk assessment and monitoring indicate that monitoring data can be used for risk assessment, and monitoring design should be based on the chemicals' risk values (priority risk substances should be monitored). Environmental management should fulfill legal requirements and follow the relevant guidance, but experience and new knowledge from risk assessment and monitoring as well as from the application of risk reduction measures should be incorporated into the legal tool battery.

The most common applications of the assessment of environmental risk posed by adverse effects of chemicals to human and ecosystem health are as follows:

– Regulatory ERM of chemical substances (pesticides, biocides, industrial chemicals);
– Regulatory ERM of waters and soils;
– Regulatory ERM of wastes, waste disposal sites;
– Creation of generic environmental quality criteria;
– ERM of contaminated sites;
– Creation of site-specific EQC;
– Monitoring design based on risk values.

The environmental impact of chemical substances in air, water and soil has reached enormous global proportions. This impact cannot continuously be observed and assessed because of its large-scale dimensions and lack of knowledge about the healthy (non-deteriorated) environment, ideal references and the dynamic environmental processes. The risk of chemicals posed to the environment alone – even if the ecosystem/people of a subarea were perfectly known – could not generally be used for managing the regional or global environment. This is because several other risks and impacts (mixture of chemicals, hazardous effects other than toxicity, social and economic risks, etc.) must be taken into account and managed (see also Volume 1, Chapter 8).

The extrapolation steps are accompanied by large uncertainties depending on the type and quality of the information/data, and the variability within and the relationship between the entities from and to which it has been extrapolated. On the other hand, excessive uncertainty can result from the environment itself, meteorological and climatic conditions, changes of habitats, behavior of natural ecosystems, communities, ecosystem members, dynamic equilibrium processes, etc. The natural variability may create extreme uncertainties.

The consequences are that statistical evaluation and analysis should be applied not only to the determination of the true values of the toxicological end points (Section 9), but along the whole procedure of risk assessment and risk management. A wide selection of statistical tools can be applied in practice from uncertainty (extrapolation) factors through the analysis of distributions to the regression methods.

The quality of the extrapolated results depends on:

– The quality of the acquired data;
– The quantity of the available data;
– The proper selection of the test organism: does it represent the receptor ecosystem truly if not, what is the difference?
– The proper selection of the test type, the tested system and the end points;
– The realism of the model applied;
– The complexity of the situation, e.g. mixtures of contaminants, natural land loaded by anthropogenic impacts, climatic changes, unexpected meteorological conditions, extremely sensitive ecosystems, etc.
– The fit of the conceptual model to the scope and the aim of the management.

That is why impact prediction methods have been intensively developed in the last 25 years. Potential impacts, i.e. environmental risk posed by chemical substances to the ecosystem and man can be assessed using mathematical models, chemical analysis, biological and ecological testing and their combinations.

## 11  CONCLUSIONS

This chapter provides an overview of toxicity data evaluation and interpretation. Toxicity data should be used for risk assessment and decision making in risk management. The generic risk assessment of existing and new chemicals and the targeted risk assessment of contaminated environment are discussed, covering both human and ecological risks. Definitions of the most frequently applied terms are also given. The flow of toxicity data starts at data acquisition followed by data transformations and evaluations up to the use of the derived data in risk management.

A test end point can be determined by mathematical statistical methods from the measurement end points (e.g. lethality, inhibition rate, slope of the growth curve, etc.). The selection of the suitable test end point and the further evaluation and interpretation of the results depend on the aim and concept of the assessment. The test end points such as $EC_x$ or NOEC can be further extrapolated to other taxa or the entire ecosystem: from laboratory microcosms to field, from field to field, to other temporal or spatial conditions or to special cases of ecosystem adaptation or recovery.

Examples are used to introduce typical measurement and test end points and explain the evaluation and use of the toxicity results. Uncertainties as intrinsic part of hazard and risk and the essential statistical tools for their management are emphasized and discussed in detail. The extrapolation steps for generating hazard values for human populations and the ecosystem are explained and demonstrated by examples.

In addition to hazard and risk assessment based on the chemical model which measures the contaminant concentration and predicts the adverse effects from this value, direct toxicity assessment and the use of DTA results for decision making are introduced. Arguments are listed on decision making that is based on toxicity values measured directly on environmental samples. This helps avoid high interpretational uncertainty of chemical analysis.

The direct toxicity-based approach adopted nowadays is hard to integrate into the chemical-model-based environmental management theory and practice that uses concentrations, threshold values and other EQCs expressed in terms of concentration as well as monitoring by chemical analysis, etc. Direct toxicity assessment (laboratory or field tests) should have priority when the cumulative effect of several toxicants (including metabolites or degradation products) has to be measured and when the impact of the environmental matrices and their living part is significant. The complexity of the response of an ecosystem needs site-specific effect assessments; the puzzle of the environment cannot be compiled only from models and generic data. The relationship between a certain area's character and generic characteristics is also important information for environmental decision making. Response of key species must be known when identifying representative indicator organisms. Proper indicator organisms ensure efficient environmental monitoring from the point of view of information and expenses. Specific screening values or other quality criteria can be established based on the key species.

The calibration curve, i.e. any test end point plotted as a function of 4CP or Cu and the calculated *toxicity equivalent* (an environmental concentration expressed in terms of the equivalencing compounds 4CP and Cu) help to understand and interpret the toxicity-based characterization of environmental contaminants. On the other hand, the same test can provide the dilution rate necessary to reach the no-effect dose or concentration of an environmental sample. This supports decision making on the necessity and rate of environmental risk reduction.

According to the results discussed in this chapter, the calibration tool can characterize and integrate the effects of different contaminants and mixtures on test organisms of different sensitivity. Unidentified or unidentifiable contaminants can be handled with the help of this tool. Direct toxicity assessment can compare the response curves of the equivalencing chemical with that of the unknown contaminant and provide information about its nature and effect mechanism. The toxicity equivalent method enables the different bioassay results to be compared with one another and simultaneously evaluated. TEQ results can be used for quantitative risk assessment and decision making by calculating the quantitative risk and the target concentration.

Even if equivalent toxicity, risk value or target toxicity values do not contain more information than the measured toxicity data, information in the form of TEQ may be valuable. Comparability and aggregability enable the uniform use of directly measured toxicity results, similar to the effect data of known components. The understanding and handling of toxicity of environmental samples as an effective concentration (EC) value allow the integration of toxicity into the quantitative risk assessment procedure which is based on the PEC/PNEC of the contaminant. The equivalencing approach bridges the gap between the chemistry-based environmental management and the direct toxicity-testing-based approach and may support professionals in integrating scientific and practical/engineering knowledge.

This chapter provides the fundamentals of evaluation and use of toxicity data for environmental risk management. Best management practice can be achieved when decision makers harmonize the effect-based assessment with the management concept and select the best fitting test methods and evaluation tools.

## REFERENCES

Abdi, H. (2007) Bonferroni and Šidák corrections for multiple comparisons. In: Salkind, N. J. (ed.) *Encyclopedia of Measurement and Statistics*. Thousand Oaks, CA, Sage.

Aldenberg, T. & Jaworska, J. (2000) Uncertainty of the hazardous concentration and fraction affected for normal species sensitivity distributions. *Ecotoxicology and Environmental Safety*, 46, 1–18.

Aldenberg, T. & Slob, W. (1993) Confidence limits for hazardous concentrations based on logistically distributed NOEC toxicity data. *Ecotoxicology and Environmental Safety*, 25, 48–63.

AM (2014) *AM Statistical Software*. [Online] Available from: http://am.air.org. [Accessed 28th July 2014].

Berkeley (2014) *Glossary of statistical terms*. [Online] Available from: http://www.stat.berkeley.edu/~stark/SticiGui/Text/gloss.htm. [Accessed 28th May 2014].

Bertalanffy, L. von, (1969) *General System Theory*. New York, George Braziller.

BMD (2014) *Benchmark Dose*. [Online] Available from: http://www.epa.gov/raf/publications/benchmarkdose.htm and http://www.epa.gov/ncea/bmds. [Accessed 28th July 2014].

BMDP (2014) *BMDP Statistical Software* [Online] Available from: http://www.statsols.com/bmdp-statistical-software-in-cutting-edge-research. [Accessed 28th May 2014].

BMDS (2014) *Benchmark Dose Software*. [Online] Available from: http://www.epa.gov/ncea/bmds. [Accessed 28th May 2014].

Brown, C.C. (1978) *The Statistical Analysis of Dose-Effect Relationships*. In: Butler, G.C. (ed.) Chichester, New York, Brisbane, Toronto, John Wiley & Sons.

Burrlioz (2014) *Statistical software package to generate trigger values for local conditions within Australia*. CSIRO (http://www.csiro.au). [Online] Available from: www.csiro.au/Outcomes/Environment/Australian-Landscapes/BurrliOZ.aspx. [Accessed 8th August 2014].

Carlon, C., D'Alessandro, M. & Swartjes, F. (2007) *Derivation methods of soil screening values in Europe: a review of national procedures towards harmonisation*. Annex 3 Screening value. EUR 22805 EN – Joint Research Centre – Institute for Environment and Sustainability.

CCME (2006) *Potential Statistical Models for Describing Species Sensitivity Distributions.* CCME Project 382-2006. Prepared by Zajdlik & Associates Inc. for the Canadian Council of Ministers of the Environment. [Online] Available from: http://www.ccme.ca/assets/pdf/pn_1415_e.pdf. [Accessed 8th August 2014].

CDC (2014) *Centers for Disease Control and Prevention.* [Online] Available from: http://www.cdc.gov. [Accessed 28th July 2014].

CLP (2008) *Classification, Labelling and Packaging of Substances and Mixtures.* Regulation (EC) 1272/2008 of the European Parliament and of the Council and amending and repealing directives 67/548/EEC and 1999/45/EC, and amending Regulation (EC) 1907/2006. [Online] Available from: http://eur-lex.europa.eu/LexUriServ/LexUriServ.do?uri=CELEX:32008 R1272:EN:NOT [Accessed 29th July 2013].

Crommentuijn, T., Polder, M.D. & van de Plassche, E.J. (1997) *Maximum permissible concentrations and negligible concentrations for metals, taking background concentrations into account.* RIVM Report No. 601501001. Bilthoven, National Institute of Public Health and the Environment (RIVM).

Crommentuijn, T., Sijm, D., van de Guchte, C. & van de Plassche, E. (2001) Deriving ecotoxicological risk limits for water and sediment in the Netherlands. *Autralasian Journal of Ecotoxicology,* 7, 31–42. [Online] Available from: http://www.ecotox.org.au/aje/archives/vol7p31.pdf. [Accessed 29th July 2013].

CSTEE (2000) *The available scientific approaches to assess the potential effects and risk of chemicals on terrestrial ecosystems.* The scientific committee on toxicity, ecotoxicity and the environment (CSTEE). C2/JCD/csteeop/Ter91100/D(0). [Online] Available from: http://ec.europa.eu/food/fs/sc/sct/out83_en.pdf. [Accessed 8th August 2014].

Dataplot (2014) *Dataplot software system.* NIEST ITL. [Online] Available from: http://www.itl.nist.gov/div898/software/dataplot. [Accessed 28th July 2014].

DEB (2014) *Theoretical Biology.* Amsterdam, Vrije Universiteit. [Online] Available from: http://www.bio.vu.nl/thb/deb. [Accessed 28th July 2014].

DEB laboratory (2014) *DEB information page.* Theoretical Biology. [Online] Available from: http://www.bio.vu.nl/thb/deb. [Accessed 28th July 2014].

Develve (2014) *Statistical software.* [Online] Available from: http://develve.net. [Accessed 28th July 2014].

DIN 38412-34 (1991) Deutsche Einheitsverfahren zur Wasser-, Abwasser- und Schlammuntersuchung – Testverfahren mit Wasserorganismen (Gruppe L) – Teil 34: Bestimmung der Hemmwirkung von Abwasser auf die Lichtemission von *Photobacterium phosphoreum*; Leuchtbakterien-Abwassertest mit konservierten Bakterien (L 34) (German standard methods for the examination of water, waste water and sludge – Bio-assays (group L) – Part 34: Determination of the inhibitory effect of waste water on the light emission of *Photobacterium phosphoreum*; luminescent bacteria waste water test using conserved bacteria)

Doncaster, C.P. & Davey, A.J.H. (2007) *Analysis of Variance and Covariance: How to Choose and Construct Models for the Life Sciences.* Cambridge, Cambridge University Press. [Online] Available from: http://www.southampton.ac.uk/~cpd/anovas/datasets. [Accessed 28th May 2014].

Doncaster, C.P., Davey, A.J.H. & Dixon, P.M. (2013) Prospective evaluation of designs for analysis of variance without knowledge of effect sizes. *Environmental and Ecological Statistics,* 21, 239–261. [Online] Available from: dx.doi.org/10.1007/s10651-013-0253-4. [Accessed 28th May 2014].

Doncaster, C.P. & Davey, A.J.H. (2014) *Examples of Analysis of Variance and Covariance – Key to types of statistical models.* [Online] Available from: http://www.southampton.ac.uk/~cpd/anovas/datasets/index.htm. [Accessed 28th May 2014].

Dunnett, C.W. (1955) A multiple comparison procedure for comparing several treatments with a control. *Journal of the American Statistical Association,* 50, 1096–1121.

ECETOC (2002) The use of $T_{25}$ estimates and alternative methods in the regulatory risk assessment of non-threshold carcinogens in the European Union. Technical Report No. 83. [Online] Available from: http://library.wur.nl/WebQuery/clc/1645564. [Accessed 8th August 2014].

ECHA (2008) *Guidance on information requirements and chemical safety assessment.* Chapter R.10: Characterisation of dose [concentration]–response for environment. [Online] Available from: http://echa.europa.eu/documents/10162/13632/information_requirements_r10_en.pdf. [Accessed 8th August 2014].

ECHA (2010) *Guidance on information requirements and chemical safety assessment.* Chapter R.8: Characterisation of dose [concentration]-response for human health. Version 2. [Online] Available from: http://echa.europa.eu/documents/10162/13632/information_requirements_r8_en.pdf. [Accessed 8th August 2014].

ECHA Guidance (2014) *Guidance documents.* European Chemicals Agency. [Online] Available from: http://echa.europa.eu/support/guidance. [Accessed 8th August 2014].

EFSA (2005) *Harmonized approach for substances that are both genotoxic and carcinogenic.* [Online] Available from: http://www.efsa.europa.eu/etc/medialib/efsa/science/sc_commitee/sc_opinions/1201.Par.0002.File.dat/sc_op_ej282_gentox_en3.pdf. [Accessed 8th August 2014].

EFSA (2012) Guidance on selected default values to be used by the EFSA Scientific Committee, Scientific Panels and Units in the absence of actual measured data. *EFSA Journal*, 10(3), 2579. [Online] Available from: http://www.efsa.europa.eu/en/efsajournal/doc/2579.pdf. [Accessed 8th August 2014].

Engineering Statistics Handbook (2014) *e-Handbook of Statistical Methods.* NIST/SEMATECH. [Online] Available from: http://www.nist.gov/itl/sed/gsg/handbook_project.cfm. [Accessed 8th August 2014].

Epi Info (2014) *Software for epidemiologic studies.* [Online] Available from: http://www.cdc.gov/epiinfo. [Accessed 28th July 2014].

EQS Water (2008) *Environmental quality standards in the field of water policy.* Directive 2008/105/EC of the European Parliament and of the Council, amending and subsequently repealing Council Directives 82/176/EEC, 83/513/EEC, 84/156 /EEC, 84/491/EEC, 86/280/EEC and amending Directive 2000/60/EC of the European Parliament and of the Council. [Online] Available from: http://eur-lex.europa.eu/ LexUriServ/LexUriServ.do?uri=OJ:L:2008:348:0084:0097:EN:PDF. [Accessed 8th August 2014].

ETX 2.0 (2014) *Software for risk assessment and standard setting.* RIVM. [Online] Available from: http://www.rivm.nl/rvs/Risicobeoordeling/Modellen_voor_risicobeoordeling/ETX_2_0. [Accessed 8th August 2014].

EU-TGD (2003) *Technical Guidance Document on Risk Assessment* in support of Commission Directive 93/67/EEC on Risk Assessment for new notified substances; Commission Regulation (EC) No 1488/94 on Risk Assessment for existing substances; Directive 98/8/EC of the European Parliament and of the Council concerning the placing of biocidal products on the market. [Online] Available from: http://ihcp.jrc.ec.europa.eu/our_activities/public-health/risk_assessment_of_Biocides/doc/tgd. [Accessed 8th August 2014].

Everitt, B.S. & Skrondal, A.R. (2011) *The Cambridge Dictionary of Statistics* (4th Edition) – First published online: 5 Feb 2011. ISBN: 9780511784552. [Online] Available preview from: http://libro.eb20.net/Reader/rdr.aspx?b=554674. [Accessed 14th August 2014].

Excel data analysis tool (2014) [Online] Available from: https://www.youtube.com/watch?v=6tCDAKpGm00. [Accessed 4th August 2014].

Excel statistical analysis tools (2014) [Online] Available from: http://office.microsoft.com/en-us/excel-help/about-statistical-analysis-tools-HP005203873.aspx. [Accessed 4th August 2014].

Excel statitical functions (2014) [Online] Available from: http://office.microsoft.com/en-us/excel-help/statistical-functions-HP005203066.aspx. [Accessed 4th August 2014].

ezANOVA (2014) *Basics of Analysis of Variance*. [Online] Available from: http://www.cabiatl. com/mricro/ezanova. [Accessed 28th July 2014].

Feigl, V., Atkári, Á., Anton, A. & Gruiz, K. (2007) Chemical stabilisation combined with phytostabilisation applied to mine waste contaminated soil in Hungary. In: *Advanced Materials Research* 20–21, Trans Tech Publications, Switzerland, pp. 315–318. [Online] Available from: http://www.scientific.net/AMR.20-21.315 [Accessed 28th July 2014].

Free Sofware (2014) Free Software Foundation and GNU. [Online] Available from: http://www.fsf.org/ and http://www.gnu.org/software. [Accessed 10th September 2014].

Garber, K., Raimondo, S. & TenBrook, P. (2010) *Exploration of methods for characterizing effects of chemical stressors to aquatic animals*. US EPA. [Online] Available from: http://water. epa.gov/scitech/swguidance/standards/criteria/aqlife/upload/whitepaper_tools.pdf. [Accessed 8th August 2014].

Gerritsen, A. (1997) The influence of body size, life stage, and sex on the toxicity of alkylphenols to *Daphnia magna*. PhD thesis, University of Utrecht.

GHS (2011) *Globally Harmonized System of classification and labeling of chemicals*. 4th revised edition, Copyright © United Nations. [Online] Available from: http://www.unece.org/trans/danger/publi/ghs/ghs_welcome_e.html and http://live.unece.org/ trans/danger/publi/ghs/ghs_rev04/04files_e.html. [Accessed 29th July 2013].

GNU PSPP (2014) *Free statistical program*. Free Software Foundation. [Online] Available from: http://www.gnu.org/software/pspp. [Accessed 28th July 2014].

Gruiz, K. (2003) Interactive ecotoxicity tests for contaminated soil. In: ConSoil 2003, CD. *8th International FZK/TNO Conference on Contaminated Soil*, 12–16 May, 2003, Gent. Theme B, FZK, OVAM, TNO, pp. 267–275.

Gruiz, K. (2005) Biological tools for soil ecotoxicity evaluation: soil testing triad and the interactive ecotoxicity tests for contaminated soil – In: Fava, F. & Canepa, P. (eds.) *Soil Remediation Series* No.6. ISBN: 88-88214-33-X. Italy, INCA. pp. 45–70.

Gruiz, K. (2009) Early warning and monitoring in efficient environmental management. *Land Contamination & Reclamation*, 17(3–4), 387–406.

Gruiz, K. & Vodicska, M. (1992) Assessing Heavy Metal Contamination in Soil Using a Bacterial Biotest. In: *Preprints of Soil Decontamination Using Biological Processes*. International Symposium. Karlsruhe, 1992. pp. 848–855, Frankfurt am Main, Dechema.

Gruiz, K. & Vodicska, M. (1993) Assessing Heavy-metal Contamination in Soil Applying a Bacterial Biotest and X-ray Fluorescent Spectroscopy – In: *Contaminated Soil '93*. Arendt, F., Annokkée, G.J., Bosman, R. & van den Brink, W.J. (eds.) The Netherlands, Kluwer Academic Publ. pp. 931–932.

Gruiz, K., Horváth, B. & Molnár, M. (2001) Környezettoxikológia.*Environmental Toxicology* (in Hungarian). Budapest, Műegyetem Publishing Company.

Gruiz, K., Molnár, M. & Feigl, V. (2009) Measuring adverse effects of contaminated soil using interactive and dynamic test methods. *Land Contamination & Reclamation*, 17(3–4), 445–462.

Harju, M., Ravnum, S., Rundén Pran, E., Grossberndt, S., Fjellsbø, L.M., Dusinska, M. & Heimstad, E.S. (2011) *Alternative approaches to standard toxicity testing*. Scientific report. TQP ID 9 – OPTIO 257430181, NILU. [Online] Available from: http://www.gassnova.no/no/ Documents/Alternativeapproachestostandardtoxicitytestingoption_NILU.pdf. [Accessed 8th August 2014].

Health and Safety Laboratory (2014) *HSL website*. [Online] Available from: http://www.hsl.gov. uk. [Accessed 28th July 2014].

Hungarian soil regulation (2000) Quality standards for the protection of groundwater and the geological compartment. KöM-EüM-FVM-KHVM Decree 10/2000 (VI.2).

IBM SPSS (2014) Predictive analytics software and solutions. [Online] Available from: http://www-01.ibm.com/software/analytics/spss. [Accessed 28th July 2014].

ICCVAM (1997) Validation and Regulatory Acceptance of Toxicological Test Methods: A Report of the ad hoc Interagency Coordinating Committee on the Validation of Alternative Methods. NIH Publication No.: 97-3981. Research Triangle Park: National Institute of Environmental Health Sciences (NIH).

ICCVAM (2003) *ICCVAM Guidelines for the Nomination and Submission of New, Revised, and Alternative Test Methods*. NIH Publication No: 03-4508. Research Triangle Park: National Institute of Environmental Health Science (NIH)

ICCVAM (2014) Guidelines for the Nomination and Submission of New, Revised, and Alternative Test Methods D-1. Appendix D: ICCVAM validation and regulatory acceptance criteria. [Online] Available from: http://ntp.niehs.nih.gov/iccvam/suppdocs/subguidelines/sg034508/sgappd.pdf. [Accessed 14th August 2014].

ICRISTAT (2014) *International Crop Research Institute for the Semi-Arid Tropics*. [Online] Available from: http://www.icrisat.org/biometrics.htm. [Accessed 28th July 2014].

idre (2014) *Institute for Digital Research and Education*. UCLA. [Online] Available from: http://www.ats.ucla.edu/stat/mult_pkg/whatstat/choosestat.html. [Accessed 28th May 2014].

IGHRC (2003) *Uncertainty factors: Their use in human health risk assessment*. UK Government, the Interdepartmental Group on Health Risks from Chemicals (IGHRC), MRC Institute for Environmental Health. [Online] Available from: http://ieh.cranfield.ac.uk/ighrc/cr9.pdf. [Accessed 8th August 2014].

IOM (2012) *Review of methods to assess risk to human health from contaminated land*. The Institute of Occupational Medicine. Report prepared by: Searl, A. [Online] Available from: http://www.scotland.gov.uk/Resource/0041/00413136.pdf. [Accessed 8th August 2014].

IPCS (2004) IPCS/OECD *Key generic terms used in chemical hazard/risk assessment. IPCS Risk Assessment Terminology*. IOMC, Inter-organization programme for the sound management of chemicals. A cooperative agreement among UNEP, ILO, FAO, WHO, UNIDO, UNITAR and OECD. [Online] Available from: http://www.inchem.org/documents/harmproj/harmproj/harmproj1.pdf. [Accessed 8th August 2014].

IRIS (2014) *Integrated Risk Information System*. US EPA. [Online] Available from: http://www.epa.gov/iris/backgrd.html. [Accessed 28th July 2014].

Isnard, P., Flammarion, P., Roman, G., Babut, M., Bastien, P., Bintein, S., Esserméant, L., Férard, J.F., Gallotti-Schmitt, S., Saouter, E., Saroli, M., Thiébaud, H., Tomassone, R. & Vindimian, E. (2001) Statistical analysis of regulatory ecotoxicity tests. *Chemosphere*, 45(4–5), 659–69.

JMP (2014) *Statistical Discovery Software* from SAS. [Online] Available from: http://www.jmp.com. [Accessed 28th July 2014].

Jonckheere, A.R. (1954) A distribution-free $k$-sample test against ordered alternatives. *Biometrika*, 41, 133–145. [Online] Available from: http://dx.doi.org/10.2307/2333011. [Accessed 28th May 2014].

Klimisch, H.J., Andreae, M., & Tillmann, U. (1997) A systematic approach for evaluating the quality of experimental and toxicological and ecotoxicological data. *Regulatory Toxicology & Pharmacology*, 25, 1–5.

Kooijman, S.A.L.M. (1983) Statistical aspects of the determination of mortality rates in bioassays. *Water Research*, 17, 749–759.

Kooijman, S. (1993) *Dynamic Energy Budgets in Biological Systems—Theory and Applications in Ecotoxicology*. Cambridge, University Press.

Kooijman, S. (1996) An alternative for NOEC exists, but the standard model has to be abandoned first. *Oikos*, 75, 310–316.

Kooijman, S.A.L.M. (2010) *Dynamic Energy Budget theory for metabolic organisation*. [Online] Available from: http://www.bio.vu.nl/thb/deb/index.html. [Accessed 28th July 2014].

Kooijman, S.A.L.M., Baas, J., Bontje, D., Broerse, M., van Gestel, C.A.M. & Jager, T. (2009) Ecotoxicological Applications of Dynamic Energy Budget Theory. In: Devillers, J. (ed.) *Ecotoxicology Modeling, Emerging Topics in Ecotoxicology: Principles, Approaches and Perspectives*. 2. Springer Science+Business Media, LLC, US pp. 237–259. [Online] Available from: dx.doi.org/10.1007./978-1-4419-0197-2_9. [Accessed 28th July 2014].

Kruskal, W.H. & Wallis, W.A. (1952) Use of ranks in one-criterion variance analysis. *Journal of the American Statistical Association*, 47(260), 583–621. [Online] Available from: dx.doi.org/10.1080/01621459.1952.10483441. [Accessed 28th July 2014].

Küster, A., Bachmann, J., Brandt, U., Ebert, I., Hickmann, S., Klein-Goedicke, M.G., Schmitz, S., Thumm, E. & Rechenberg, B. (2009) Regulatory demands on data quality for the environmental risk assessment of pharmaceuticals. *Regulatory Toxicology & Pharmacology*, 55, 276–280.

Larson, R., Boswell, L., Kanold, T.D. & Stiff, L. (2001) *Algebra 2*. Evanston, IL, Boston, Dallas, McDougal Littell.

Leitgib, L., Kálmán, J. & Gruiz, K. (2007) Comparison of bioassays by testing whole soil and their water extract from contaminated sites. *Chemosphere*, 66, 428–434.

Leitgib, L., Gruiz, K., Fenyvesi, É., Balogh, G. & Murányi, A. (2008) Development of an innovative soil remediation: "Cyclodextrin-enhanced combined technology". *Science of the Total Environment*, 392, 12–21.

MathWorks (2014) *Statistics toolbox*. [Online] Available from: http://www.mathworks.com/products/statistics/?s_cid=sol_des_sub2_relprod3_statistics_toolbox. [Accessed 28th May 2014].

McCarty, L.S., Borgert, C.J. & Mihaich, E.M. (2012) Information quality in regulatory decision making: peer review versus good laboratory practice. *Environmental Health Perspectives*, 120, 927–934.

Microsoft Excel Add-ins (2014) *Add-ins website*. [Online] Available from: http://www.add-ins.com. [Accessed 28th July 2014].

MicrOsiris (2014) *Statistical Analysis and Data Management Software and Decision Tree*. [Online] Available from: http://www.microsiris.com. [Accessed 28th July 2014].

MINITAB (2014) *Statistical software*. [Online] Available from: http://www.minitab.com. [Accessed 28th July 2014].

MIX (2014) *Meta-analysis with Interactive eXplanations*. [Online] Available from: http://www.meta-analysis-made-easy.com. [Accessed 28th July 2014].

Molnár, M., Leitgib, L., Gruiz, K., Fenyvesi, É., Szaniszló, N., Szejtli, J. & Fava, F. (2005) Enhanced biodegradation of transformer oil in soils with cyclodextrin – from the laboratory to the field. *Biodegradation*, 16, 159–168.

Molnár, M., Gruiz, K. & Halász, M. (2007) Integrated methodology to evaluate bioremediation potential of creosote-contaminated soils. *Periodica Polytechnica, Chemical Engineering*, 51(1), 23–32.

Molnár, M., Leitgib, L., Fenyvesi, É. & Gruiz, K. (2009) Development of cyclodextrin-enhanced soil remediation: from the laboratory to the field. *Land Contamination & Reclamation*, 17(3–4), 599–610.

Newman, M.C., Ownby, D.R., Mézin, L.C.A., Powell, D.C., Christensen, T.R.L., Lerberg, S.B. & Anderson, B-A. (2000) Applying species-sensitivity distributions in ecological risk assessment: assumptions of distribution type and sufficient numbers of species. *Environmental Toxicology and Chemistry*, 19(2), 508–515. [Online] Available from: dx.doi.org/10.1002/etc.5620190233. [Accessed 28th July 2014].

Nickel EQS (2011). *Nickel and its compounds*. EQS dossier prepared by the Sub-Group on Review of the Priority Substances List (under Working Group E of the Common Implementation Strategy for the Water Framework Directive). [Online] Available from: https://circabc.europa.eu/sd/d/1e2ae66f-25dd-4fd7-828d-9fd5cf91f466/Nickel%20EQS%20dossier%202011.pdf. [Accessed 8th August 2014].

Nickel EQS Datasheet (2005) *Nickel and its Compounds.* Environmental Quality Standards (EQS) Substance Data Sheet. Common Implementation Strategy for the Water Framework Directive.

Nickel RAR (2008) *EU risk assessment report on nickel, nickel sulphate, nickel carbonate, nickel chloride, nickel dinitrate.* Danish EPA on behalf of the EU. European Commission. [Online] Available from: http://echa.europa.eu/documents/10162/cefda8bc-2952-4c11-885f-342aacf769b3. [Accessed 8th August 2014].

OECD (2005) *Manual for investigation of HPV chemicals.* Chapter 4. [Online] Available from: http://www.oecd.org/chemicalsafety/risk-assessment/2483645.pdf. [Accessed 8th August 2014].

OECD (2006) Current approaches in the statistical analysis of ecotoxicity data: A guidance to application. OECD series on testing and assessment No. 54. ENV/JM/MONO (2006) 18.

OECD (2011) *Data compilation, selection and derivation of PNEC values for the aquatic compartment. Zinc example.* Metals Specificities in Environmental Hazard Assessment, Paris. [Online] Available from: www.oecd.org/chemicalsafety/risk-assessment/48720427.pdf. [Accessed 8th August 2014].

OECD (2012) *Manual for the Assessment of Chemicals.* OECD Cooperative Chemicals Assessment Programme (CoCAP). [Online] Available from: http://www.oecd.org/fr/env/ess/risques/manualfortheassessmentofchemicals.htm. [Accessed 8th August 2014].

OpenEpi 2.3 (2014) *Software.* [Online] Available from: http://www.openepi.com/v37/Menu/OE_Menu.htm. [Accessed 8th August 2014].

OpenStat (2014) *Free Statistics Programs and Materials.* Bill Miller. [Online] Available from: http://statpages.info/miller/OpenStatMain.htm. [Accessed 8th August 2014].

Origin (2014) *Graphing and analysis software.* [Online] Available from: http://www.originlab.com. [Accessed 28th July 2014].

PAST (2014) *Past of the Future. Free software for scientific data analysis.* [Online] Available from: http://folk.uio.no/ohammer/past. [Accessed 8th August 2014].

Péry, A.R.R., Flammarion, P., Vollat, B., Bedaux, J.J.M., Kooijman, S.A.L.M. & Garric, J. (2002) Using a biology-based model (DEBtox) to analyse bioassays in ecotoxicology: Opportunities & recommendations. *Environmental Toxicology & Chemistry* 21(2), 459–465.

PROAST (2014) *A software package for dose–response modeling.* RIVM. [Online] Available from: http://www.rivm.nl/en/Documents_and_publications/Scientific/Models/PROAST. [Accessed 28th May 2014].

R (2014) *R Project for Statistical Computing.* [Online] Available from: www.r-project.org. [Accessed 28th July 2014].

SalStat-2 (2014) *Statistical package.* [Online] Available from: https://code.google.com/p/salstat-statistics-package-2. [Accessed 8th August 2014].

SAS (2014) *Business Analytics System.* [Online] Available from: http://www.sas.com. [Accessed 28th May 2014].

SAS University (2014) *SAS University Edition.* [Online] Available from: http://www.sas.com/en_us/software/university-edition.html. [Accessed 28th May 2014].

SCHER (2009) *Risk assessment methodologies and approaches for genotoxic and carcinogenic substances.* The Scientific Committee on Health and Environmental Risks (SCHER), Scientific Committee on Consumer Products (SCCP) and Scientific Committee on Emerging and Newly Identified Health Risks (SCENIHR). [Online] Available from: http://ec.europa.eu/health/ph_risk/committees/04_scher/docs/scher_o_113.pdf. [Accessed 8th August 2014].

Schneider, K., Schwarz, M., Burkholder, I., Kopp-Schneider, A., Edler, L., Kinsner-Ovaskainen, A., Hartung, T. & Hoffmann. S. (2009) ToxRTool, a new tool to assess the reliability of toxicological data. *Toxicology Letters,* 189(2), 138–44. [Online] Available from: DOI: 10.1016/j.toxlet.2009.05.013. Epub 2009 May 27. [Accessed 8th August 2014].

SISA (2014) *Simple Interactive Statistical Analysis.* [Online] Available from: http://www.quantitativeskills.com/downloads. [Accessed 8th August 2014].

Slob, W. (2003) PROAST – a general software tool for dose-response modelling. RIVM, Bilthoven.

SOFA (2014) *Statistics Open For All.* [Online] Available from: http://www.sofastatistics.com/home.php. [Accessed 8th August 2014].

SPSS (2014) *IBM SPSS Statistics.* [Online] Available from: http://www-01.ibm.com/software/analytics/spss/products/statistics. [Accessed 28th May 2014].

STATA 13 (2014) *Data analysis and statistical software.* [Online] Available from: http://www.stata.com/products. [Accessed 28th July 2014].

StatCrunch (2014) *Software package.* [Online] Available from: www.statcrunch.com. [Accessed 28th July 2014].

Statext (2014) *Statistical software.* [Online] Available from: http://www.statext.com. [Accessed 8th August 2014].

STATGRAPHICS (2014) *Statgraphics online.* [Online] Available from: www.statgraphicsonline.com and http://www.statpoint.com/statgraphics%20online.pdf. [Accessed 8th August 2014].

Statist (2014) *Portable statistics program.* [Online] Available from: http://wald.intevation.org/projects/statist. [Accessed 8th August 2014].

STATISTICA (2014) *Analytics software products and solutions.* [Online] Available from: http://www.statsoft.com and http://www.statsoft.com/Products/STATISTICA/Product-Index. [Accessed 8th August 2014].

STATISTICA trial (2014) *Free trial of* STATISTICA *software package.* [Online] Available from: http://www.statsoft.com/support/free-statistica-10-trial. [Accessed 4th August 2014].

Statistical Software (2014) *Free downloads.* [Online] Available from: http://www.quantitativeskills.com/downloads. [Accessed 4th August 2014].

Statistical Solutions (2014) *Statistical Solution.* [Online] Available from: http://www.statsols.com. [Accessed 28th May 2014].

Statistics (2014) *The Institute for Statistics Education.* [Online] Available from: http://www.statistics.com. [Accessed 28th May 2014].

Statistics software information (2014) *Software information.* The Institute for Statistics Education. [Online] Available from: http://www.statistics.com/how-courses-work/software-information. [Accessed 28th July 2014].

StatPages (2014) *Web Pages that Perform Statistical Calculations.* [Online] Available from: http://statpages.org/index.html. [Accessed 28th July 2014].

StatPages Free (2014) *Information on free statistical software.* [Online] Available from: http://statpages.org/javasta2.html#Freebies ITT. [Accessed 4th August 2014].

Stats with cats (2014) *Blog.* [Online] Available from: http://statswithcats.wordpress.com/2010/08/22/the-five-pursuits-you-meet-in-statistics. [Accessed 28th May 2014].

StatSoft, Getting started (2014) [Online] Available from: https://www.youtube.com/watch?v= S9zG2GylAw4. [Accessed 4th August 2014].

StatTrek (2014) *Teach yourself statistics.* [Online] Available from: http://stattrek.com. [Accessed 28th May 2014].

STPLAN (2014) *Software download.* Division of Quantitative Sciences, Department of Biostatistics. https://biostatistics.mdanderson.org/SoftwareDownload/SingleSoftware.aspx?Software_Id=41. [Accessed 4th August 2014].

Swartjes, F.A. (1999) Risk-based assessment of soil and groundwater quality in the Netherlands: standards and remediation urgency. *Risk Analysis,* 19(6), 1235–1249.

*Tanagra* (2014) *A free data mining software.* [Online] Available from: http://chirouble.univ-lyon2.fr/~ricco/tanagra/en/tanagra.html. [Accessed 4th August 2014].

Terpstra, T.J. (1952) The asymptotic normality and consistency of Kendall's test against trend, when ties are present in one ranking. *Indagationes Mathematicae*, 14, 327–333. [Online] Available from: http://oai.cwi.nl/oai/asset/8258/8258A.pdf. [Accessed 28th May 2014].

TGD EQS (2011) *Technical Guidance for Deriving Environmental Quality Standards.* Common Implementation Strategy for the Water Framework Directive Guidance Document No 27. European Commission. DOI: 10.2779/43816 [Online] Available from: http://www.oekotoxzentrum.ch/expertenservice/qualitaetskriterien/doc/TGD-EQS_finaldraft.pdf. [Accessed 8th August 2014].

Torstensson, L. (ed.) (1994) Soil biological variables in environmental hazard assessment. Guideline MATS, Swedish EPA.

ToxRTool (2014) [Online] Available from: http://ihcp.jrc.ec.europa.eu/our_labs/eurl-ecvam/archive-publications/toxrtool. [Accessed 18th August 2014].

ToxRTool Instruction (2014) [Online] Available from: http://ihcp.jrc.ec.europa.eu/our_labs/eurl-ecvam/archive-publications/toxrtool/Instructions%20ToxRTool_FINAL.pdf. [Accessed 18th August 2014].

ToxRTool Download (2014) [Online] Available from: http://ihcp.jrc.ec.europa.eu/our_labs/eurl-ecvam/archive-publications/toxrtool/ToxRTool.xls. [Accessed 18th August 2014].

Toxstat (2014) *Statistics for toxicology.* [Online] Available from: http://www.scribd.com/doc/158074165/toxstat. [Accessed 28th July 2014].

UCLA (2014) *University of California Los Angeles.* [Online] Available from: http://www.ucla.edu and http://www.ats.ucla.edu/stat/mult_pkg/whatstat/default.htm. [Accessed 28th May 2014].

UDEL (2014) *Handbook of Biological Statistics.* University of Delaware. Online] Available from: http://udel.edu/~mcdonald/statintro.html. [Accessed 28th July 2014].

Unistat (2014) Statistics Software. [Online] Available from: http://www.unistat.com. [Accessed 28th May 2014].

Unistat – Bioassay (2014) *Unistat Statistics Software – Analyses of Bioassays.* [Online] Available from: http://www.unistat.com/guide/analysis-of-bioassays. [Accessed 28th May 2014].

Upton, G. & Cook, I. (2014) *The Oxford Dictionary of Statistical Terms.* 2nd revised edition. Oxford University Press.

Usablestat (2014) *Usable Statistics.* Stanford University. [Online] Available from: http://www.usablestats.com. [Accessed 28th May 2014].

US EPA (1995) *Supplemental Guidance for Assessing Susceptibility from Early-Life Exposure to Carcinogens.* EPA/630/R-03/003F. [Online] Available from: http://www.epa.gov/ttn/atw/childrens_supplement_final.pdf. [Accessed 8th August 2014].

US EPA (2002) *A Review of the Reference Dose and Reference Concentration Processes.* [Online] Available from: http://www.epa.gov/raf/publications/review-reference-dose.htm. [Accessed 8th August 2014].

US EPA (2005) *Guidelines for Carcinogen Risk Assessment.* EPA/630/P-03/001F; US EPA. [Online] Available from: http://www.epa.gov/raf/publications/pdfs/CANCER_GUIDELINES_FINAL_3-25-05.PDF. [Accessed 8th August 2014].

US EPA (2014) *Guidances and tools for risk assessment.* [Online] Available from: http://www.epa.gov/risk/guidance.htm. [Accessed 8th August 2014].

US EPA – Human Risk (2014) *Human health risk assessment.* [Online] Available from: http://www.epa.gov/reg3hwmd/risk/human/index.htm. [Accessed 8th August 2014].

US EPA – Eco Risk (2014) *Ecological risk assessment.*[Online] Available from: http://www.epa.gov/reg3hwmd/risk/eco/index.htm. [Accessed 8th August 2014].

Van den Berg, M., Birnbaum, L.S., Denison, M., De Vito, M., Farland, W., Feeley, F., Fiedler, H., Hakansson, H., Hanberg, A., Haws, L., Rose, M., Safe, S., Schrenk, D., Tohyama, C., Tritscher, A., Tuomisto, J., Tysklind, M., Walker, N. & Peterson, R.E. (2006) The 2005

World Health Organization reevaluation of human and mammalian toxic equivalency factors for dioxins and dioxin-like compounds, *Toxicology Science*, 93, 223–241.

ViSta (2014) *Visual Statistics System*. [Online] Available from: http://forrest.psych.unc.edu/research. [Accessed 4th August 2014].

Wikipedia (2014) *Statistics*. [Online] Available from: http://en.wikipedia.org/wiki/Statistics. [Accessed 28th May 2014].

WinIDAMS (2014) *Software package*. UNESCO. [Online] Available from: http://portal.unesco.org/ci/en/ev.php-URL_ID=2070&URL_DO=DO_TOPIC&URL_SECTION=201.html. [Accessed 4th August 2014].

Williams, D.A. (1972) The comparison of several dose levels with a zero dose control. *Biometrics*, 28(2), 519–531.

Williams, D.A. (1971) A Test for Differences between Treatment Means When Several Dose Levels are Compared with a Zero Dose Control. *Biometrics*, 27(1), 103–117.

XLSTAT (2014) *Analysis of Variance, ANOVA*. [Online] Available from: http://www.xlstat.com/en/ and http://www.xlstat.com/en/products-solutions/feature/anova-analysis-of-variance.html. [Accessed 28th July 2014].

XLSTAT Introduction (2014) [Online] Available from: http://www.youtube.com/watch?v= 4AZ8G_MqyiM&list=PLFAD0C5745D4A1F60. [Accessed 4th August 2014].

# Subject index

# Engineering Tools for Environmental Risk Management

*Editors: Katalin Gruiz, Tamás Meggyes & Éva Fenyvesi*

Engineering Tools for Environmental Risk Management: 1
Environmental Deterioration and Contamination – Problems and their Management
©2014
Editors: Katalin Gruiz, Tamás Meggyes & Éva Fenyvesi
ISBN: 9781138001541 (Hardback)
e-book ISBN: 9781315778785
Cat# K22815

Engineering Tools for Environmental Risk Management: 2
Environmental Toxicology
©2015
Editors: Katalin Gruiz, Tamás Meggyes & Éva Fenyvesi
ISBN: 9781138001558 (Hardback)
e-book ISBN: 9781315778778
Cat# K22816

**Forthcoming:**

Engineering Tools for Environmental Risk Management: 3
Site Assessment and Monitoring Tools
Editors: Katalin Gruiz, Tamás Meggyes & Éva Fenyvesi
ISBN: 9781138001565 (Hardback)
e-book ISBN: 9781315778761
Cat# K22817

Engineering Tools for Environmental Risk Management: 4
Risk Reduction Technologies
Editors: Katalin Gruiz, Tamás Meggyes & Éva Fenyvesi
ISBN: 9781138001572 (Hardback)
e-book ISBN: 9781315778754
Cat# K22818

Engineering Tools for Environmental Risk Management: 5
Integrated Environmental Risk Management – Case Studies
Editors: Katalin Gruiz, Tamás Meggyes & Éva Fenyvesi
ISBN: 9781138001589 (Hardback)
e-book ISBN: 9781315778747
Cat# K22819

Printed and bound by CPI Group (UK) Ltd, Croydon, CR0 4YY

25/10/2024

01779095-0001